Forensic Investigation
of Explosions

Forensic Investigation of Explosions

Edited by

ALEXANDER BEVERIDGE

UK   Taylor & Francis Ltd, 1 Gunpowder Square, London EC4A 3DE
USA   Taylor & Francis Inc., 1900 Frost Road, Suite 101, Bristol, PA 19007-1598

British Library Cataloguing-in-Publication Data

A catalogue record for this book is available from the British Library
ISBN 0-7484-0565-8 (cased)

Library of Congress Cataloging-Publication Data are available

Cover design by Youngs Design in Production

Typeset in Times 9/11pt by Santype International Ltd, Salisbury, UK

Printed by George H Buchanan, Bridgeport, NJ, USA

Contents

Contents

Contents

Preface

In deciding what topics to include in this book, I was influenced by my experiences on task forces set up to investigate major explosives incidents and by related civil and criminal proceedings. Explosion investigation can involve experts from many fields and information on the range and nature of such expertise is widely scattered. A primary objective, therefore, was to provide pertinent, up-to-date, multi-disciplinary information in one volume. The authors are all hands-on experts in their respective fields and even if information sought is not specifically discussed or referenced then the reader should at least know 'who to call'.

The chapters may be grouped under the following broad headings:

- explosives;
- explosions;
- detection of hidden explosives;
- processing scenes of explosions;
- forensic chemistry;
- aircraft sabotage investigation;
- forensic pathology;
- expert testimony.

The **first two chapters** present the historical development of explosive formulations and the basic physics underlying explosions. **Chapter 3** deals with prevention of explosions by the detection of hidden explosives. The authors discuss the current technology being applied in airline security to detect explosives in and on baggage. The technology ranges from canine detection through X-ray and CAT-scan instrumentation to the new generation of ion mobility spectrometer and chemiluminescence based 'sniffers' which can identify traces of explosives in seconds.

Chapters 4, 5, 6 and 7 focus on the essential 'dos and don'ts' in processing scenes of:

- criminal acts using explosives;
- aircrashes where explosive sabotage is suspected;
- gas explosions in buildings

In **Chapter 4**, scenes of an explosion are viewed through the eyes of a 'bomb squad' investigator with specialised training as a post-blast consultant. He focuses primarily on efficient and effective organisation and deployment of people, and on communications. In **Chapter 5**, a forensic chemist describes the very important consultative role which a scientist can play as a team member at an explosion crime scene. His primary perspectives are recognition and handling of exhibits which might bear residue, and how observations at the scene may facilitate subsequent laboratory examination. He also includes case examples and describes a team approach to forensic laboratory processing of the recovered evidence, including analytical protocols.

Chapter 6 deals with protocols at the scene of an air crash. An air accident investigator describes the international conventions governing air crash investigations and how these may influence the conduct of a criminal investigation when explosive sabotage is suspected. He also details how advances in modern technology can facilitate searching, recording and communications, especially when an aircraft has broken up at altitude and victims and wreckage are scattered over a wide area. This chapter is complemented by **Chapter 13** which illustrates the physical signatures which explosives impart to metals and fabrics and by **Chapter 14** on how sound spectrogram analysis of cockpit voice recorders can distinguish decompression due to structural failure from that caused by a bomb.

Gas phase explosions in buildings typically leave no significant chemical residues but do impart characteristic damage to the scene. In **Chapter 7**, a consultant engineer describes the key features of such explosions, details the investigative process and illustrates its practical application with two case examples.

Chapters 8, 9, 10, 11 and 12 and also **Chapter 5** deal with chemical analysis of explosives and explosive residues. The chemicals used in explosives cut across all of the classical boundaries, and consequently a combination of methods is required to analyse both unreacted explosives and their residues. There have been very significant developments in analytical methodology over the past three decades, but there is no agreement on one analytical protocol which can be applied in all residue examinations. I have asked the authors of **Chapters 5, 10, 11 and 12** to describe how they combine individual methods into routine analytical protocols in their laboratories.

Any methods used in such protocols must meet specific criteria for sensitivity, selectivity and reliability. Other influencing factors include:

- the type of explosives normally encountered in the service area;
- cost and frequency of use of instruments;
- familiarity, training and personal preference;
- the courts;
- challenges by opposing experts;
- personal contacts and professional visits;
- scientific literature, sales presentations and conference attendance.

In my view, the most important of these latter factors is 'the courts'. If a court of law does not accept the validity of a particular methodology/protocol, then there is no point in using it. If a court does accept a protocol, it can reasonably be argued that the protocol need not change until the *status quo* does – provided that quality control procedures are in place, detection levels are adequate and the protocol can be demonstrated to successfully identify test explosion residues of explosives used in the service area.

However, in many parts of the world, including North America and Japan, industrial changes have changed the *status quo* in recent years in that emulsion high explosives are rapidly replacing dynamites. Because emulsion explosives are formulated primarily from ammonium nitrate and hydrocarbons, the analytical protocols developed to detect nitrated organic explosives are ineffective at detecting emulsion explosive residues. The same issue applies to detecting residues from bulk ammonium nitrate-based blasting agents (e.g. ammonium nitrate/fuel oil; ANFO) which increasingly are being used in terrorist bombings. Consequently, forensic chemists are having to adjust and expand the methods in their general protocols. Emulsion explosives are discussed in **Chapter 1** and analysis of ammonium nitrate-based explosives is discussed and referenced in **Chapter 5**.

An unreported 'International seminar and workshop on explosive residue analysis protocols' held at the FBI Academy at Quantico, VA, USA in 1993 clarified which methods were most widely used for residue analysis in seven major forensic laboratories in Europe, North America and Japan (the authors who present protocols all did so at this conference). The most commonly used methods were:

(1) For residues from organic high explosives:

Gas chromatography/chemiluminescence (GC/TEA)   7/7
Gas chromatography/mass spectrometry (GC/MS)   5/7
Thin layer chromatography (TLC)   4/7
High performance liquid chromatography/chemiluminescence (HPLC/TEA)   3/7
High performance liquid chromatography/pendant mercury dropping electrode
  (HPLC/PMDE)   2/7

(2) For residues from low explosives:

Ion chromatography (IC)   6/7
Scanning electron microscope/energy dispersive X-ray analyser (SEM/EDX)   4/7
X-ray powder diffraction (XRPD)   4/7
Spot tests   4/7
Infrared spectroscopy (IR)   2/7
Capillary electrophoresis (CE)   2/7

Some other statistics were that the most commonly used methods for unknown particles and insoluble residues were IR, SEM/EDX and XRPD. The most commonly used solvents for extraction of debris were, for non-polar components: pentane and diethyl ether; for nitrated organic explosives: acetone, methylene chloride, methanol and methyl *tert*-butyl ether; and for inorganic components: water. The most commonly used clean-up procedures for organic extracts were column chromatography and thinlayer chromatography. For aqueous extracts, the best clean-up procedures were centrifuging and filtration through a membrane filter (not filter paper).

New methods discussed included ion mobility spectrometry (see **Chapter 3**) and super-critical fluid extraction.

Another key point raised at this conference is that there must be analytical confirmation of a positive screening test.

Some of the confirmation procedures were:

(1) Organic explosives
  To confirm TLC   IR; HPLC/ultraviolet detector (UV); GC/TEA; GC/MS; a com-
                       bination of HPLC/UV, HPLC/TEA and HPLC/PMDE
  To confirm GC/TEA   3 separate GC columns; HPLC/TEA; GC/MS; GC/MS/MS
(2) Inorganic components
  To confirm IC   CE; IR, SEM/EDX; XRPD

These methods are drawn primarily from chromatography and spectroscopy and so a chapter is devoted to each; **Chapter 8** (GC, HPLC, TLC, SFC, IC and CE) and **Chapter 9** (IR, GC/MS, HPLC/MS and GC/MS/MS) respectively. The applications of SEM/EDX, XRPD, spot tests and other methods used to analyse low explosives and insoluble residues are discussed in **Chapter 11**.

Chapters 10 and 12 deal with the critical areas of quality control and significance. Effective quality control procedures ensure that any traces of explosives found in debris originated from the scene and not from contamination or spurious results. Once traces of explosive are found at a scene, the significance of their presence must be ascertained.

As noted above, **Chapters 13 and 14** describe two very different approaches to answering the question 'was there an explosion?' in the context of an air crash.

In **Chapter 15**, a forensic pathologist describes and illustrates the valuable information which can be obtained by autopsy of victims of an explosion. In **Chapter 16**, an experienced trial lawyer details, in a case context of a terrorist bombing, how most effectively to prepare an expert to testify. The case illustrates the many other types of material analyses besides analysis of explosives which are applied in the investigation of an explosion.

In conclusion, it was a pleasure working with all of the authors and I thank them again for their dedication in producing their chapters in the face of the very heavy demands of their professions.

The Royal Canadian Mounted Police accept no responsibility for any use made of this material.

Sandy Beveridge
Vancouver, BC

Notes on Contributors

Maurice Baker

SM2 Division, Defence Research Agency, Fort Halstead, Sevenoaks, Kent, TN14 7BP, England

Maurice, a chartered engineer, retired in 1997 from the UK Defence and Evaluation Research Agency. His expertise includes failure analyses and studies of materials, pyrotechnics and explosives, but his particular specialisation is the study of the micro-effects of explosions on materials. He has now worked on twelve incidents of suspected aircraft sabotage. His corrosion expertise has been particularly relevant when aircraft have crashed into the ocean.

Ed Bender

Bureau of Alcohol, Tobacco and Firearms, 1401 Research Boulevard, Rockville, Maryland 20850, USA

Ed is a forensic chemist in the explosives section of the Bureau of Alcohol, Tobacco and Firearms (ATF). He is a member of the ATF national and international post-blast response teams and serves as an instructor at the Federal Law Enforcement Training Center in courses on Explosives Investigation and International Post-blast Investigations.

Sandy Beveridge

Chemistry Section, Royal Canadian Mounted Police, 5201 Heather Street, Vancouver, BC, Canada V5Z 3L7

Sandy, a Fellow of the Chemical Institute of Canada, joined the Royal Canadian Mounted Police Forensic Science Laboratories in 1967 and heads the Chemistry Section of the Vancouver Laboratory. His primary research interests lie in explosive residue analysis and he has testified, published and lectured extensively in this field.

Jim Connelly

Federal Aviation Administration, William J. Hughes Technical Center, Atlantic City International Airport, New Jersey, USA 08405

Jim joined the FAA's Aviation Security Research and Development Group in 1990. His current project involves the installation and evaluation of new luggage inspection systems at airports.

Bill Curby

Federal Aviation Administration, William J. Hughes Technical Center, Atlantic City International Airport, New Jersey, USA 08405

Bill, who has flown on the space shuttle 'Columbia', is a research chemist at the FAA Aviation Security Research and Development Group.

James (Rex) Ferris

Department of Forensic Pathology, Vancouver General Hospital, 855 W 12th Avenue, Vancouver, BC, Canada V5Z 1M9

Rex is Professor of Forensic Pathology at the University of British Columbia. Rex has been associated with a number of national and international associations including President of the International Association of Forensic Sciences. He is a Fellow of the Royal College of Pathologists and has given expert evidence in several countries and much of his current expertise relates to the interpretation of injuries.

Chris Foster

Dr J. H. Burgoyne & Partners, 39A Bartholomew Close, London EC1A 7JN, England

Chris joined Dr J. H. Burgoyne & Partners, consulting engineers and scientists, to specialise in fire and explosion investigation in 1972. He is currently still active in these fields and is well known internationally for his work in conducting investigations into the causes of explosions in buildings and ships. Chris is a regular contributor to technical courses on fire science and fire investigation.

Frank Fox

Federal Aviation Administration, William J. Hughes Technical Center, Atlantic City International Airport, New Jersey, USA 08405

Frank is a research chemist with the FAA, and has formerly held senior positions in chemistry and toxicology laboratories.

John Garstang

Special Engineering Services Section, Engineering Branch, Operations Directorate, Transportation Safety Board of Canada, Ottawa, Ontario, Canada K1A 1K8

John, a mechanical engineer, is a superintendent in the Transportation Safety Board of Canada (TSBC). His section is responsible for providing technical investigative services in diving, scene/vehicle reconstruction and analysis, human engineering, photogrammetry, image processing, documents, and fire and explosion.

Susan Hallowell

Federal Aviation Administration, William J. Hughes Technical Center, Atlantic City International Airport, New Jersey, USA 08405

Susan joined the Federal Aviation Administration in 1992 as a research chemist in the area of trace explosives detection. She supports the technical R&D needs of the FAA canine program manager and is also the technical lead on several international projects on counterterrorism.

Robin Hiley

Chemistry and Research, Forensics Explosives Laboratory, Defence Research Agency, Fort Halstead, Kent TN14 7BP, England

Robin works at the UK Defence Evaluation and Research Agency Forensic Explosives Laboratory, where he manages the chemistry and research area. He has concentrated recently on explosives trace analysis.

Bob Hopler

Powderman Consulting Inc., PO Box 86, Oxford, Maryland 21654-0086, USA

Bob recently retired as Corporate Manager of Technical Services with Dyno Nobel Inc., Salt Lake City, Utah, after almost 35 years in the explosives industry and has formed his own consulting company.

Jim Jardine

Surrey Provincial Court, 14340 57 Avenue Surrey, V3X 1B2

Jim received the professional honour of appointment as Queen's Counsel (QC) in 1984. His experience as a prosecutor has included several high-profile terrorist cases involving bombings. He was formerly in private practice in Vancouver and acted for both prosecution and defence. He was recently appointed a judge of the Provincial Court of British Columbia.

Bruce McCord

Forensic Science and Research Training Center, FBI Academy, Quantico, Virginia 22135, USA

Since 1989, Bruce has been a research chemist in the FBI Laboratory at Quantico, VA, where he has specialised in chromatography – in particular in the application of ion chromatography and capillary electrophoresis to analysis of low explosives. Other research interests include drug analysis and DNA typing using HPLC and capillary electrophoresis.

Bibhu Mohanty

ICI Explosives Canada, 701 Richlieu Boulevard, McMasterville, Quebec, Canada J3G 6H3

Bibhu is a senior science and technology associate at the Technical Centre of ICI Explosives Canada at McMasterville, Quebec. His primary interests are in the fields of detonation physics, explosion hazards and rock blasting. He also serves as an adjunct professor of mining engineering at McGill University, Montreal.

Gerry Murray

Explosives Section, Forensic Science Agency of Northern Ireland, 152 Belfast Road, Carrickfergus, Northern Ireland BT38 8PL

Gerry is a chartered chemist and Fellow of the Royal Society of Chemistry. Gerry was appointed an Officer of the Order of the British Empire (OBE) in 1994. He is Principal Scientific Officer in the Forensic Science Agency of Northern Ireland, has 23 years of experience in forensic explosives investigation, and has testified and lectured extensively in Northern Ireland and in other jurisdictions.

Notes on Contributors

Frank Slingerland

Vibacon Inc., Vibracoustic Engineering, 25 Eisenhower Crescent, Nepean, Ontario, Canada K2J 3Z8

Frank, a mechanical engineer, recently retired as Director of the Aeroacoustics Facility at the National Research Council of Canada to form his own company, Vibacon Inc., providing vibroacoustical engineering services to the aerospace industry in Canada and Europe.

Richard Strobel

Explosives Section, Forensic Science Laboratory, Bureau of Alcohol, Tobacco and Firearms, US Department of the Treasury, Rockville, Maryland 20850, USA

Rick attained his present position as Chief, Explosives Section, in 1989. His areas of expertise include forensic explosives, arson analysis and canine detection of accelerants and explosives.

Jean-Yves Vermette

Explosives Disposal and Technology Branch/Canadian Bomb Data Centre, Royal Canadian Mounted Police, Ottawa, Ontario, Canada K1A 0R2

Jean-Yves has been the post-blast consultant for the Explosives Disposal and Technology section of the Royal Canadian Mounted Police since 1989. He is very active in post-blast investigations in and outside Canada.

John Winn

SM2 Division, Defence Research Agency, Fort Halstead, Sevenoaks, Kent, TH14 7BP, England

John was Head of Metal Physics at the UK Defence Evaluation and Research Agency from 1986 until retirement in 1995. He spent his career with the Ministry of Defence and Defence Research Agency at Woolwich and Fort Halstead respectively. His interests included the research, development and application of X-ray and electron optical techniques to the study of a wide range of materials.

Shmuel Zitrin

Chemistry and Biology Section, Division of Identification and Forensic Science, National Israel Police Headquarters, Jerusalem, Israel

Since 1983, Shmuel has headed the Chemistry and Biology Section of the Division of Identification and Forensic Science of the Israeli Police. He is a Visiting Professor in the Hebrew University of Jerusalem, lecturing on forensic science at the Law Faculty. His areas of expertise include analysis of organic compounds, especially by mass spectrometry, but explosives have always been his major interest.

The History, Development, and Characteristics of Explosives and Propellants

ROBERT B. HOPLER

1.1 Introduction

The history of explosives and propellants, also known generally as 'energetic materials', begins with the material known as gunpowder or black powder, whether the intended use was for civil applications such as rock blasting, military uses in demolition, shell filling (bursting charges) and construction projects, or military and civilian propellant charges for shotguns, pistols, rifles, or artillery. The individual inventor of black powder will undoubtedly forever remain unknown, but numerous writers such as Drinker (1878), Munroe (1888), Marshall (1915), and Davis (1941, 1943), have described what is known about its development and evolution: therefore, there will be no such discussion here. Suffice to say that until the discovery of nitrated explosive compounds such as nitrocellulose by Schönbein and Böttger (independently of one another) and nitroglycerin by Sobrero (all occurring in 1846), the only explosive available for *any* purpose was black powder.

For purposes of this discussion, the subject will be divided into three categories:

- solid (particulate) propellants;
- military explosives;
- commercial explosives.

Also, in the attempt not to make this discussion overly broad, this chapter stresses US practice, explosive history, and developments.

1.2 Propellants

Propellants may be granular, solid, or liquid. The primary focus is on granular (particulate) material since they are the most commonly encountered by the forensic chemist.

Solid propellants are deflagrating materials designed to accelerate a projectile from its position of rest at the breech of a weapon to its full velocity as it exits the tube or barrel.

In the ideal (and designed-for case), the complete consumption of the propellant and the exit of the projectile will occur at the same instant. Propellant grains are thus chemically formulated and physically designed to achieve this end. The grains burn particle-to-particle at speeds below the speed of sound in the material: this defines the word 'deflagrating'. Historically such materials have been termed 'progressive' powders. In addition to burning particle-to-particle, each particle burns from its free surface inward or, in the case of perforated grains, also from the free surface outward. This characteristic enables the propellant designer to size and configure the grains or particles to be totally consumed at the optimum instant. Propellant grains may be found in a multitude of shapes and sizes, as might be expected given the varieties of weapons and desired pressures and projectile velocities.

1.2.1 Black Powder

Black powder is a mixture of three components, generally (and originally) charcoal, sulfur, and potassium nitrate. These are typically in the ratio of 15/10/75. Many variations to that ratio have been used: Cundill (1889) lists over 20 varieties, many with sub-varieties. Most of the differences, however, are insignificant. The one major development in the past 100 years is the use of sodium nitrate in some black powder grades.

Black powder has an inherent drawback as a military propellant due to the fact that it produces a solid reaction product. Because of this, a dense black cloud is produced upon firing a weapon. This has two adverse results: the position of a firing weapon is readily apparent, and after a number of rounds are fired the volume of battlefield smoke leads to confusion and general chaos. For this reason the development of a 'smokeless' propellant charge was the objective of every government's weapons laboratory. Upon the discovery of the nitration reaction this research intensified.

1.2.2 Smokeless Powder

The early history of the nitrated carbohydrates, which includes the 1833 discovery of nitro-starch (called *xyloidine* by its discoverer, Braconnot) and guncotton, called *pyroxyline* or *pyroxyle* by the chemist Pelouze, is thoroughly covered by Davis (1941, 1943).

Guncotton, nitrocellulose of high nitrogen content (13.35% to 13.45%), was the first nitrated material to be tried as a replacement for black powder, but it was too prone to accidents. However, its military use continued after it was found that the newly-invented mercury fulminate blasting cap would cause compressed guncotton to detonate, leading to its application as a demolition charge and shell filling. Its use was rather short-lived, however, due to the introduction of picric acid.

Research was continued on nitrocellulose of lower nitrogen content as a propellant material, and the first good smokeless rifle powder was produced by Vieille in 1886, for the French Government. This was nitrocellulose with ether-alcohol, kneaded in a bread-making type machine, rolled out into thin sheets, and then cut into small squares and dried (Military Explosives, 1924). This was a 'single base' smokeless powder (nitrocellulose only).

In 1888 Nobel invented a powder he called *Ballistite*, which was a low nitrated nitrocotton gelatinized with nitroglycerin: this became known as 'double base' powder. In the same year Cordite (given that name because it was extruded in the form of a cord or ribbon), a mixture of high nitrated guncotton, nitroglycerin, and vaseline, gelatinized by means of acetone, was developed by an English Committee (Marshall, 1915).

Later, 'triple base' smokeless powders were developed, containing nitroguanidine in addition to the nitrocotton and nitroglycerin of typical double base powders. Triple base powders were cooler-burning than the single- or double-base materials and use was mainly restricted to large caliber weapons.

Developments in smokeless powder since those early days have been primarily to improve stability, decrease the erosion of the barrel of the weapon, control pressures, decrease smoke output ('smokeless' powders are smokeless in comparison to black powder, but still produce

Table 1.1 Smokeless powder composition

	M6	M5	M15	Function†
Nitrocellulose (13.15% N)	87		20.00	e
Nitrocellulose (13.25% N)		81.95		e
Nitroguanidine			54.70	c, e
Nitroglycerine		15.00	19.00	e
Potassium Nitrate		2.15		f
Ethyl Centralite		0.60	6.00	d, g, k
Graphite		0.30		a, h
Diphenylamine	1*			k
Dinitrotoluene	10			e, g, i, j
Dibutyl phthalate	3			f, h
Sodium alum. fluoride			0.30	b
Total	100	100.00	100.00	

* Added to mix.

† a, antistatic; b, cooling rate modifier; c, coolant; d, deterrent; e, energizer; f, flash suppressant; g, gelatinizer; h, hygroscopicity reducer; i, inhibitor; j, plasticizer; k, stabilizer.

visible smoke), and to decrease the muzzle flash from a firing weapon. The geometry of powders may include flakes, tubes, cylinders, sticks, flattened balls, or spheres (see Chapter 11, Figure 11.21).

The reader desiring a detailed background in the field of smokeless powders and other propellant materials should refer to the 10 volume *Encyclopedia of Explosives and Related Items* (referred to below as *Encyclopedia*), and to the publication *Propellant Profiles* (1988).

Table 1.1 (*Encyclopedia*: Vol. 8, P407), illustrates typical formulations of the three general classes of smokeless powders. Some formulations may contain dye ingredients for the identification of particular brands, especially in commercial products.

1.3 Military Explosives

Just as black powder was the first propellant, so too was it the first military explosive. It was used for shell fillings, demolition, and military construction projects from the earliest times up until the invention of nitroglycerin.

Military explosives as discussed here are those used as the shell filling or 'bursting charge' in artillery rounds and those explosives used for demolition charges. Military construction projects typically use commercial-type explosives, except in field-expedient situations. The brief use of guncotton as a military explosive was noted above.

1.3.1. Picric Acid

By the early 1900s picric acid (2,4,6-trinitrophenol) had become the shell filling of choice of most of the world's military forces. The Russians were the first to work out methods of production and use as a shell filling, in 1894 (*Encyclopedia*: Vol. 8, P286). Picric acid went under many names; 'Lyddit' in England; 'Mélinite' in France; 'Sprengkorper' in Germany; and 'Shimoza' in Japan, to name but a few. Other picrates, especially ammonium picrate ('Explosive D', $C_6H_2(NO_2)_3 \cdot NH_4$) have been widely used in fillings for armor-piercing shells, due to their extreme insensitivity to shock. In World War I this material was used in large-caliber shells. In World War II a mixture of ammonium picrate and TNT was widely used in the press-loading of armor-piercing shells.

1.3.2   Trinitrotoluene (TNT)

During and after World War I the explosive trinitrotoluene (TNT, $C_7H_5N_3O_6$) became the dominant shell filling and demolition charge material. TNT has the advantage of being very easy to cast, since it has a wide spread between its melting and decomposition temperatures. One disadvantage is its extreme insensitivity in the cast form: where necessary, this is overcome in practice by adding a core of tetryl ($C_7H_5N_5O_8$). In order to conserve TNT for small caliber shells in World War I, a mixture of TNT and ammonium nitrate ('amatol') was developed. It was specified for use only in shells of 4.7-inch to 9.2-inch diameter (Crowell, 1919) but in actual practice it was used in all sizes.

For the same reason of conserving TNT, nitrostarch explosives were used very successfully in that war for hand grenades and trench mortar shells (Williams, 1920).

1.3.3   Tetryl

Tetryl (2,4,6-trinitrophenylmethylnitramine, *N*-2,4,6-tetra-nitro-*N*-methyl aniline, or picryl-methyl nitramine) was used in military boosters, but has generally been replaced by materials such as RDX and HMX.

The 'tetrytols' are mixtures of tetryl and TNT, which were utilized in boosters, demolition charges, shells, and shaped charges. The TNT generally ranged from 20 to 35 per cent of the mixture. An advantage of tetrytol is that it allows the casting of the explosive into munitions rather than requiring pressing. It is also more powerful than TNT, but not as sensitive as tetryl alone (*Encyclopedia*. Vol. 9, T165).

1.3.4   Pentaerythritol Tetranitrate (PETN)

PETN (pentaerythritol tetranitrate, $C(CH_2ONO_2)_4$), first prepared in 1891, became commercially available in the 1930s in detonating cord and blasting caps. It is a component in many military explosives, most notably pentolite, where it may comprise from 10 to 60 per cent of a mixture with TNT.

1.3.5   RDX and HMX

Between the World Wars a number of explosives were developed, and after the start of the second war a vast amount of explosives research took place.

One of the most important and useful military explosives is RDX (an acronym for 'Research Department Explosive'), which was discovered in 1899, but not used until World War II. It is also called cyclonite, hexagen, and cyclo-trimethylenetrinitramine.

HMX is another explosive used for military applications during and after World War II. The initials are said to stand for "High Melting Explosive', although other sources for the acronym are sometimes cited. It is also called cyclo-tetramethylenetetranitramine or octogen.

A vast number of explosives consisting of mixtures of various explosive compounds were developed by the combatants in World War II. Many of these combinations may include materials such as HMX, RDX, TNT, aluminum powder, wax, and plasticizers, with or without other ingredients for special properties. A few worth mentioning are Composition B (60% RDX/40% TNT/plus wax), Cyclotol (60–75% RDX/25–40% TNT), Torpex 2 (42% RDX/40% TNT/18% Al), and Composition C-4 (91% RDX/9% plasticizer).

1.3.6 Plastic Explosives

PBX (an acronym for 'Plastic-Bonded Explosive') explosives are a fairly recent development in the military area. These are characterized by high mechanical strength, high detonation velocity, excellent stability, and insensitivity to shock and high temperatures. They cover a wide range of formulations, usually containing some ingredients from the list of RDX, PETN, HMX, aluminum, binders such as nylon, polyurethane, rubber, or similar material, and a plasticizer. Two of the best known are C-4 (see above) and SEMTEX (RDX/PETN), an explosive manufactured in the Czech Republic, which has been widely used in terrorist bombings.

1.3.7 General

It is apparent that a great number of varieties of military explosives are available, each having unique features of detonation velocity, stability, brisance, and resistance to initiation by rifle bullet impact or shock. Some specialized explosives developed by the nuclear research laboratories even have the ability to withstand extremely high temperatures, such as those experienced upon reentry into the atmosphere by a ballistic missile.

Due to this great range of materials, this section is necessarily only a very brief overview of the field. The reader is referred to the references cited for further information. *Encyclopedia 1960–1983* and Williams (1920), both published by the US Army, should be referred to for details of military explosive materials. For general explosives chemistry, the 4-volume series by Urbanski (1964) is an excellent reference.

Table 1.2 shows the density and detonation velocity for some of the explosives mentioned

Table 1.2 Military explosives – density and velocity (*Research and development of material,* 1963)

Explosive	Density (g/cm^3)	Velocity $(m/sec, 1''\phi)$
Amatol (50/50)	1.55	6430
Composition C-4	1.59	8040
Composition B	1.68	7840
HMX	1.84	9124
Pentolite (50/50)	1.66	7465
PETN	1.70	8300
Picric acid		
(Pressed)	1.64	5270
(Cast)	1.71	7350
RDX	1.65	8180
Tetratol (20% TNT)	1.60	7385
Tetryl	1.71	7850
TNT	1.56	6825
Torpex	1.81	7495

above. These are all from one reference, but sources frequently will disagree on some of these characteristics due to physical form (e.g. grain size) or test methods. Thus, this table should only be used for a general idea of characteristics.

1.4 Commercial Explosives

As noted above, military explosives, propellants, and commercial explosives all have the same ancestor: black powder. It is interesting to note that although black powder was noted as being used in warfare in very early times, its use in mining was not recorded until 1627. After that date there is a multitude of references to its use, but not before. The reason? Probably that for its use in mining, someone had to devise a way to initiate the powder after the person was out of the area. In using the powder in a cannon, the gunner stands immediately beside the weapon in perfect safety – this is obviously not the situation with powder in a borehole! Once a 'delay' type material was developed (a powder-filled straw, for example), the breaking of rock became a great deal easier than it had been by hand or with fire-setting methods. A co-development was probably a metal drill sufficiently hard to bore a hole for the powder in the rock face.

1.4.1 Nitroglycerine

Alfred Nobel, noted previously for his invention of 'Ballistite,' took Sobrero's discovery of nitroglycerin (glyceryl trinitrate, NG) and made it a useful explosive by inventing the detonating blasting cap. This replacement for the spit from burning safety fuse was necessary because the **detonating** explosives need a shock wave for their reaction to start – the **deflagrating** explosives needed only a spark or flame (Hopler, 1992).

The first use of nitroglycerin in mining was in the liquid form, with it either poured into downward-looking boreholes or encased in cans and slid into horizontal or 'up' holes. In any case it was hazardous to ship and use. After tests with absorbent materials in an effort to make NG easier to handle, Nobel settled upon diatomaceous earth (also known as kieselguhr or diatomite) in 1866. He named his mixture (at first 75% NG/25% diatomaceous earth) 'dynamite.'

1.4.2 Dynamite

Over the years formulas for dynamite have been developed to fit every type of rock blasting. Varieties have been developed for severe water conditions, utilizing nitrocotton to gel the NG (the 'gelatins' are one branch of the dynamite family); for cohesiveness to enable loading into 'up' holes in mines; for safe usage in underground coal mines (the 'permissibles' having salts of various types to cool the explosive reaction as part of the formula) (Hopler, 1996); and for economics by making formulas with very high ammonium nitrate percentages (and thus no cohesiveness or water resistance), where severe field conditions do not exist. Literally thousands of different formulas might be found, but for years all had the basic commonality of having nitroglycerin as a component.

For all of its good explosive characteristics, nitroglycerin has the very undesirable property of having a high freezing point, in the area of $+50°F$ ($+10°C$). When dynamite is totally frozen it is relatively safe, unless one tries to break a cartridge. The danger (and the source of untold deaths in the first 50 years of dynamite) is when the miner tries to thaw the stick to use it in a borehole. Unless thawed slowly, with controlled temperatures, the material is very dangerous. Every explosives company catalog from the early days pictures 'thawing kettles' or other recommended devices for the thawing procedure. Many approaches were made to solving the problem. Nitrostarch was successfully introduced by duPont as the sensitizer for 'non-freezing' dynamites, and that material later served as the basis for all the dynamite products of the Trojan Powder Company. However, problems of low sensitivity and lack of water resistance prevented nitrostarch dynamites from ever becoming more than a 'niche' product. TNT and related compounds are also found in various companies' 'low freezing' or 'arctic' dynamites.

In the mid-1920s, the automobile industry was experiencing fast growth, and an anti-freeze superior to alcohol was needed. Ethylene glycol was found to be excellent for this application. That liquid had also been known for years as an excellent NG substitute when nitrated, and when co-nitrated with NG (at a ratio of 60% ethylene glycol/40% glycerin) the mixture had a freezing point of about −40°F (−40°C). The resultant mixture is ethylene glycol dinitrate/ nitroglycerin (EGDN/NG). With this development the freezing hazards of dynamite became a thing of the past in most of North America. Other proportions are also effective, and today economics (the relative prices of the two materials at any given time) generally dictate what the ratio will be. A proportion used by one US company for many years has been 83/17.

Other than the change to the EGDN/NG mixture as a sensitizer, dynamite formulations have changed relatively little since the 1880s when the gelatins were invented by Nobel. The notable advances have been in the area of underground coal mining explosives ('permissibles') wherein the addition of cooling salts was found to decrease the potential of igniting methane gas or coal dust, and in the introduction of a mixture of metriol trinitrate and diethylene glycol dinitrate (MTN/DEGDN) as a non-headache sensitizer substituting for EGDN/NG. Dynamites with this new material as a sensitizer have found a substantial market. Some typical formulations are shown in Table 1.3.

1.4.3 Liquid Oxygen Explosives

Dynamite realized some minor competition beginning in the 1920s when liquid oxygen explosives were introduced into the surface coal mines in the central US. This was a simple

Table 1.3 Dynamite formulations – generalized (Percent by Weight)

Material	Type*				
	1	2	3	4	5
Nitroglycerin	40.0	15.8	91.0	26.0	9.5
Nitrocotton	0.1	0.1	6.0	0.4	0.1
Ammonium nitrate	30.0	63.1		39.0	72.2
Sodium nitrate	18.9	11.9		27.5	
Wood pulp or nut meal	8.0	3.4	0.5	2.0	2.4
Balsa	2.0				
Starch or wheat flour		3.9	1.5	3.8	4.0
Guar gum		1.3			1.3
Microballoons (phenolic)				0.3	
Sodium chloride					10.0
Chalk	1.0	0.5	1.0	1.0	0.5
Total	100.0	100.0	100.0	100.0	100.0

* 1, NG dynamite ('ditching dynamite'); 2, 60% extra dynamite; 3, blasting gelatin; 4, 60% extra gelatin; 5, permissible dynamite.

material: a cloth cartridge containing carbon black was soaked in a vat of liquid oxygen until the material in the cartridge was totally saturated. The cartridges were immediately lowered into the borehole, stemmed, and shot. A drawback was that only a very limited number of holes could be loaded, since the evaporation of the oxygen was rapid. Even with this limitation, LOX (as the explosive was called) did a good job and was inexpensive. To enhance the economics, mine-site plants were built by oxygen companies. This explosive reached a maximum consumption in the US of about 25 million pounds in 1953, soon after which, due to developments discussed below, it disappeared from the scene.

1.4.4 Ammonium Nitrate (AN)

Ammonium nitrate had been an ingredient in explosives since the earliest days, primarily as an oxidizer in nitroglycerin dynamite mixtures. There had been patents issued wherein the AN was simply mixed with a fuel material, but no significant commercial products resulted.

Since production of the material began, whether for explosives, munitions manufacture, or fertilizer, the physical form of ammonium nitrate (AN) was granular: small crystalline particles produced in graining kettles or by other means to slowly dry a highly-concentrated liquor. In the mid-1940s, a new, much more economical production method began, called 'prilling'. This method uses the old shot-tower concept used for ages to produce lead shot. With this development, AN became available to the fertilizer trade as a small porous sphere which was free-flowing, absorbent, easily handled and stored, and economical. AN used as a dynamite ingredient continued (and continues to this day) to be of the granular variety for various reasons of formulation.

1.4.5 Ammonium Nitrate/Fuel Oil (ANFO)

In 1953 a large surface coal mine in Indiana began experimenting with prilled AN and carbon black or ground coal for use as a dynamite or LOX substitute (Hopler, 1993a) The efforts were extremely successful and were disclosed to the world by the company (Maumee Collieries) in May 1955. Other mining regions immediately picked up on the benefits of the material, and soon found that fuel oil worked better than the solid fuels. Thus ammonium nitrate/fuel oil (ANFO) was born. The result was that the entire explosives consuming and producing industries were converted almost overnight from dynamite-based materials to simple fuel/oxidizer mixtures, or 'Blasting Agents'.

A word should be mentioned here about the 1947 disaster at Texas City, Texas, since it is commonly stated that this occurrence brought about the 'ANFO Revolution'. The disaster involved the detonation of two ships loaded with bags of grained ammonium nitrate fertilizer, with their destination as Europe. (Kintz et al., 1948). The probable cause was a fire from smoking by stevedores loading the ship. The AN was rosin-coated for protection against moisture: thus it was a fueled material, not pure AN. The fact is that it was well-known that mixtures of AN and fuel were good explosives, and extensive research had been conducted in the 1930s by explosives companies on such mixtures. It had been found that these mixtures were relatively insensitive, were easily damaged by the slightest amount of moisture, and would only detonate in relatively large-diameter boreholes.

In 1947, at the time of Texas City, AN was not yet available in the proper form (prills) for its application as a simple oxidizer/fuel blasting agent. Basically, since it was already common knowledge that AN and fuel worked as an explosive, when the material became available in the prilled form in the 1950s the mining companies tried it. To state it simply, 'the rest is history'. In other words, the ANFO Revolution would have happened even if the Texas City disaster had never occurred.

ANFO dominates the commercial blasting industry today, but it is not the perfect explosive. It has no water resistance, has a relatively low density and detonation rate, and operates rather poorly in small-diameter boreholes unless emplaced in such a way that the particle (prill) size is decreased.

1.4.6   Slurry Explosives

These limitations were recognized as soon as ANFO was introduced, and the first products to evolve were the slurry explosives. These were invented in the late 1950s, and consisted of ammonium nitrate (usually combined with another oxidizer such as sodium nitrate), water, a gelling agent (usually guar gum), and a high explosive or propellant material for sensitization. Aluminum, in powdered, grained, or flaked forms, was often added for energy. All of the explosives companies were producing slurry prior to 1960. Each company had its favorite sensitizer, usually dictated by patent considerations, success in bidding on war surplus explosives, or the particular in-house expertise of each individual company. The *Encyclopedia* has an excellent compilation of patents in this field (*Encyclopedia*: Vol. 9, S139).

Although all water-based explosives products were called 'slurry' in the 1950s and early '60s, as time passed this product type was subdivided into two distinct classes:

- 'slurry', which was a thickened (usually by a guar gum, a polysaccharide) but not cross-linked product;
- 'water gel' which contained a cross-linking agent for the polysaccharide thickeners. The cross-linking agent forms a chemical bond. Early cross-linkers used borate ions, but other materials may be used.

As the surplus sensitizers became more and more difficult to obtain (and more expensive, if available at all), company scientists developed proprietary sensitizers. These included MMAN (monomethylamine nitrate), EGM (ethyleneglycol mononitrate), HMTAN (hexamethylenetetramine nitrate), and MEAN (monoethanol amine nitrate). With these developments, and the frequent use of paint-grade aluminum powders, water gels were able to be manufactured in small diameters to compete in the remaining dynamite markets. However, cost, and insensitivity when cold, were continuing drawbacks to the water gels.

Table 1.4 illustrates some very generalized water gel product formulations. It must be pointed out that there are hundreds of varieties of these products. Even one company's products will vary greatly depending upon the application, whether packaged (small and large formulas vary) or bulk.

1.4.7   Emulsion Explosives

Almost as soon as the water gels came into usage another development was made in the commercial explosives field. This new product was the emulsion explosive. Although first conceived in 1961, commercialization was not really successful until the early 1980s.

Whereas water gels/slurries are thickened aqueous solutions of oxidizers and/or fuels and sensitizers/energizers, emulsions are mixtures of two immiscible liquids with one liquid phase dispersed uniformly throughout the second phase. Explosive emulsions are generally 'water-in-oil' types, wherein microscopic droplets of the oxidizer phase are surrounded by a continuous fuel phase. This provides almost perfect water resistance to the product. Examples of this type of emulsion in the non-explosive world are butter, margarine, cold cream, and shoe polish.

Emulsion explosives are the ultimate result of the scientific refinement of composite explosives. Black powder, the first composite explosive, had particles that were relatively large. The reaction of the powder, although rapid, was a burning phenomenon on the surface of the

Table 1.4 Water gels – examples of formulations (Percent by Weight)

Material	Formulation					
	1	2	3	4	5	6
Ammonium nitrate	25.3	50.0	40.7	73.8	47.5	42.0
Water	22.5	17.7	13.6	8.9	18.0	15.0
Smokeless powder	25.0					
Sodium nitrate	15.0	15.1	15.8		15.0	
Calcium nitrate						20.0
Hexamine nitrate				14.2		
MEAN						5.0
Sodium perchlorate				2.0		
MMAN			25.9			
Ethylene glycol	1.5	7.2	0.6		1.2	
Diethylene glycol						2.0
Sugar					5.5	
Aluminum	10.0	8.0			12.0	12.0
Gum	0.7	2.0	1.1	1.1	0.8	1.0
Microballoons						1.0
Miscellaneous			2.3			2.0
Total	100.0	100.0	100.0	100.0	100.0	100.0

particles. Dynamite was the next major composite, with a liquid sensitizer, nitroglycerin, mixed with coarse particles of ammonium nitrate, sodium nitrate, and various carbonaceous materials, examples of which are wood flour, pulp, nut hulls, and sawdust. The chemical reaction of detonation was carried along by the NG, but also took place as a surface reaction on the solid materials, as did the burning reaction in black powder, but at shock wave ('detonation') velocity rather than at burning velocity. The detonation velocity was adjustable through the choice of particle size of the ingredients.

ANFO, the ammonium nitrate/fuel oil mixture consisting of 94% AN and 6% fuel oil, is probably the most used explosive material in the world today. ANFO is a prime example of a composite explosive, consisting of large particles of AN soaked with oil. Because of the particle size, all of the AN is not in intimate contact with oil, and thus all of the AN does not react within the detonation reaction zone. This results in ANFO being a less efficient explosive than it should be. An improvement in efficiency, and therefore detonation rate, can be made by reducing the size of the prills. Doing this, however, changes other very important characteristics, such as density, flowability, and certainly not least of all, economics.

In emulsion explosives, the oxidizer portion is drastically reduced in size, thus improving the efficiency of the detonation reaction. Rather than being dry prills, the emulsion oxidizer is a highly concentrated solution of ammonium nitrate and/or other salts. By the use of emulsifiers and precise processing methods, the 'particle size' of droplets of this solution is reduced to microscopic proportions, typically in the range of about 2 to 10 μm in diameter. Surrounding each droplet is a film of oil, wax, or other fuel material. The result is still a 'mixture' of fuel and oxidizer, similar to black powder, dynamite, and ANFO, but the particle size comes as close as possible to mimicking the intimacy of combination found in molecular explosives such as NG or TNT.

Emulsions depend upon 'hot spots' for their sensitivity. These may be air bubbles resulting from the emulsification process, but they more generally are artificially created by chemical

Table 1.5 Examples of emulsion formulas (Percent by Weight)

	1	2	3	4
Ammonium nitrate	78.0	70.7	58.0	47.0
Sodium nitrate		10.7	15.0	12.0
Water	13.5	7.3	17.0	8.5
Emulsifier	1.5	0.8	2.0	1.5
Oils/waxes	5.5	3.1	6.0	5.0
Aluminum		5.0		
Microballoons	1.5	2.4	2.0	6.0
MMAN				15.0
Sodium percholrate				5.0
	100.0	100.0	100.0	100.0

(1) and (2), Sudweeks (1985)
(3), Gehrig (1992)
(4), Ruhe and Wieland (1992)

gassing or by the addition of solid 'density control' materials such as perlite or microballoons. Microballoons may be of glass or phenolic in a wide range of crushing strengths. A combination of economics and the purpose for which the emulsion will be used dictates the density control type to be used. For example, a ditching application where the explosive may be subjected to high hydrostatic shocks from nearby-detonating holes, needs an emulsion with high-strength microballoons.

Emulsifier technology is the most important aspect of the competitive situation in the manufacturing industry. The choice of material and the processes used in the emulsion manufacture (temperatures of manufacture, cooling rates, etc.) are important factors in the final explosive, rheological, and shelf-life characteristics of the emulsion.

Table 1.5 shows some very generalized emulsion formulas.

Formulation (4) in Table 1.5 was one proposed by the US Bureau of Mines as a permissible explosive (for use in underground coal mines), and is shown only as an example of the diverse collection of materials which may be put into a formulation when economics is not a factor. Commercial companies presently manufacture a great number of emulsions for permissible applications, utilizing formulas closely resembling the others in the table.

The combining of emulsions and AN or ANFO is the latest development in the commercial explosives field. These combinations generally are called 'heavy ANFO' or 'blends'. The result of this has been to produce a 'waterproof ANFO' by having the emulsion material completely coat the dry AN prills or the ANFO. This combination allows loading into wet holes, and in addition enables the explosives engineer to tailor the detonation rate of the material to the rock in question. Emulsions, with their extremely small 'particle size' noted earlier, have an inherently high velocity, frequently not suitable for some blasting jobs. The ANFO/emulsion mixtures, however, may be entirely suitable for such situations.

The relative insensitivity of the ANFO and slurries, water gels, emulsions, and blends naturally led to their use as bulk-loaded explosives. Hopler (1993b) has written a thorough history of the development of bulk loading.

1.5 Summary

The development of emulsions and their applications with AN or ANFO brought about profound changes in the commercial explosives industry. Dynamite is still a viable product, but it is rapidly being displaced in most mining and construction applications. The number of dynamite manufacturing plants (called 'works' in the industry) in the US has gone from 34 in 1959 to 1 in 1996. In about the same period of time, bulk loading of explosives into the borehole (as opposed to the use of cartridges) has gone from nearly zero to possibly representing 95% of all commercial explosives usage.

This short summary of the vast subject of propellants and explosives may be built upon by the references cited.

References

CROWELL, B., 1919, *America's Munitions, 1917–1918*, p. 106, Washington, DC: Government Printing Office.

CUNDILL, J. P., 1889, *A Dictionary of Explosives*, Chatham, England: The Royal Engineers Institute.

DAVIS, TENNEY L., 1941, 1943, *The Chemistry and Technology of Explosives*, volume 1 and 2, New York: John Wiley.

DRINKER, H. S., 1878, *Tunneling, Explosive Compounds, and Rock Drills*, New York: John Wiley.

Encyclopedia of Explosives and Related items, 1960–1975, volumes 1–7, Dover, New Jersey: Picatinny Arsenal; 1978–1983, volumnes 8–10, Dover, New Jersey: US Army Research and Development Command.

GEHRIG, N., 1982, The future of slurry explosives, *Proceedings of the Eighth Conference on Explosives and Blasting Technique*, Society of Explosives Engineers, pp. 221–228.

HOPLER, R. B., 1992, Commercial detonators: A review of methods used, past and present, to compare their strengths, *Proceedings of the Eighteenth Annual Conference on Explosives and Blasting Technique*, Society of Explosives Engineers, pp. 191–203.

HOPLER, R. B., 1993a, Custom-designed explosives for surface and underground coal mining, *Mining Engineering*, **45**, 1248–1252.

HOPLER, R. B., 1993b, History of the development and use of bulk loaded explosives, from black powder to emulsions, *Proceedings of the Nineteenth Annual Conference on Explosives and Blasting Technique*, Society of Explosives Engineers, pp. 177–199.

HOPLER, R. B., 1996, The history and development of explosives for underground coal mining, *Proceedings of the Twenty-second Annual Conference on Explosives and Blasting Technique*, Society of Explosives Engineers, pp. 44–63.

KINTZ, G. M., JONES, G. W. and CARPENTER, C. P., 1948, *Explosions of ammonium nitrate fertilizer on board the S.S. Grandcamp and S.S. High Flyer at Texas City, Texas, 16, 17 April 1947*, R.I. 4245, Washington, DC: US Bureau of Mines.

MARSHALL, 1915, *Explosives*, p. 50, Philadelphia: P. Blakiston's Son & Co.

Military Explosives, 1924, Washington: Office of the Chief of Ordnance.

MUNROE, C. E., *Lectures on Chemistry and Explosives Delivered at the Summer Class of Officers of 1888 at the Torpedo Station*, Torpedo Station Print, 1888.

Propellant Profiles, 1988, 2nd edition, Prescott, Arizona: Wolfe Publishing.

Research and Development of Materiel – Engineering Design Handbook – Properties of Explosives of Military Interest – Section 1, 1963, Washington: Headquarters, US Army Materiel Command.

RUHE, T. C. and WIELAND, M. S., 1992, Rugged emulsion explosive formulation #37 – candidate permissible, *Proceedings of the Eighth Conference on Explosives and Blasting Research*, Society of Explosives Engineers, p. 70.

SUDWEEKS, W., 1985, Physical and chemical properties of industrial slurry explosives, *Industrial and Engineering Chemistry – Product Research and Development*, **24**, (3).

URBANSKI, T., 1964, *Chemistry and Technology of Explosives*, volume 1, New York: Mac-Millan; 1965, 1967 and 1984, volumes 2, 3 and 4, Oxford: Pergamon Press.

WATSON, S. C. and WINSTON, S. E., 1985, A new day for dynamite, *Proceedings of the Eleventh Conference on Explosives and Blasting Technique*, Society of Explosives Engineers, pp. 320–334.

WILLIAMS, W. B., 1920, *History of the Manufacture of Explosives for the World War, 1917–1918*, Washington, DC: US Ordnance Department.

2

Physics of Explosion Hazards

BIBHU MOHANTY

2.1 Introduction

The instantaneous release of energy from a relatively small volume of material can be viewed as an explosive event. This is achieved by changes in the chemical composition of the solid, liquid or gas, and the release of chemical energy. Depending on initiation conditions, charge geometry and chemical composition, this reaction can accelerate until a steady value (detonation) has been achieved, or decelerate (deflagration) and eventually die out. The distinction between true detonation and deflagration is not crucial at this stage, as both processes can lead to release of very large amounts of energy in a small fraction of a second. Most incidents involving dust or vapour cloud explosions (flour, sawdust, gasoline vapours, natural gas, etc.) involve only rapid combustion and not detonation. Most commercial explosives such as ammonium nitrate (AN)–fuel oil mixtures exhibit non-ideal behaviour, in that their sensitivity and severity of explosion falls off rapidly with decreasing diameter and lack of confinement. In this discussion, confined to condensed phase explosions, the words 'detonation' and 'explosion' are used synonymously and no distinction is made between commercial and military explosives.

The energy from an open-air explosion results in compression of the surrounding air, which gives rise to a rapidly propagating *shock wave* or pressure wave. Except in the immediate vicinity of the explosion, where the explosion fireball may pose a serious hazard, the major hazards resulting from an accidental explosion are due to the shock wave and high velocity fragments expanding from the explosion site. The main characteristic of the shock wave is an extremely sharp rise in pressure value in the front, followed by a slow decay. The front part of the wave is entirely compressive and the tail-end part of it is entirely tensile but of much lesser amplitude. The amplitude of the compressive shock in the immediate vicinity of this explosion could be in excess of one million pounds per square inch; it decreases very rapidly, however, as it travels away from the source of the explosion. Nevertheless, its amplitude and duration (both of which contribute to its damage potential) can be altered dramatically through multiple reflections caused by proximity of the explosion to ground surface or other rigid structures. This can result in significant increase in the amplitude of the 'reflected' shock compared to the 'incident' shock.

The additional hazards resulting from an explosion include the explosion fireball, secondary fragments, cratering, perforation and spalling. The effect of the shock wave on personnel and structures has been studied extensively in recent years, and guidelines established for its damage potential. There are currently several excellent monographs and treatises available on the subject of science and technology of explosives. There are also comprehensive design handbooks dealing with blast hazards and blast resistant structures. The purpose of this paper is not so much to summarize these available sources of information but to provide an up-to-date review of the inter-related topics and elucidate the scope and complexity of the subject.

2.2 Historical Developments

It is difficult to speak of modern explosives without referring to black powder. The discovery of black powder probably precedes its actual use in a systematic fashion by several centuries. Its essential ingredients (potassium nitrate, charcoal and sulphur) have been available since ancient times, and chance or deliberate ignition of a mixture of these ingredients may not have been so rare. However, the credit for its systematic use belongs to the Chinese, who packed these mixtures into bamboo tubes and used them as rockets for display and signalling purposes. It took several centuries before black powder became a standard military tool. Even then, the early 14th century cannons consisted simply of wooden tubes filled with black powder charge which expelled a stone projectile. The first use of the material in mining took place in Hungary in the early 17th century. Its use accelerated with the discovery of vast deposits of sodium nitrate in Chile in 1840.

Other related developments quickly followed. In 1846, reacting strong nitric acid with glycerol, a by-product of soap manufacture, resulted in an oily product called glyceryl tri-nitrate, which is more commonly known as nitroglycerin (NG). Practical use of NG was pioneered by the Nobel family in the years following 1859. Alfred Nobel also invented the blasting cap in 1863, which revolutionized the mining industry. But the behaviour of the NG-based explosives still remained highly unpredictable, resulting in numerous accidents and fatalities. After many years of work, Nobel finally discovered that kieselguhr, a diatomaceous earth, absorbed up to three times its own weight of NG to form a relatively dry, leak resistant paste, which came to be known as 'dynamite' (Meyers and Shanley, 1990). The word was derived from 'dynamos', the Greek word for 'power'.

Other momentous advances in the explosives technology include the development of safety fuse (essentially black powder core inside a tough yarn) by William Bickford in 1831, invention of the detonating cord (a sensitive high explosive core inside a thin plastic tube or textile yarn) in 1908 in France and further refined by Ensign–Bickford Corporation in USA, and the chance discovery of ammonium nitrate as being a very powerful explosive in 1947, when the ship *Grand Camp* carrying fertilizer grade ammonium nitrate (AN) blew up at its dock in Texas City following a fire! The place of AN in explosives industry has since been secure. The other significant developments in the explosive industry were the introduction of the slurry explosives in the late 1950s, and shock tube based detonators ('Nonel': a plastic tube with a wall coating of HMX and aluminum) in the early 1970s, and of water-in-oil emulsion explosives in the late 1970s.

2.3 Thermochemistry of Explosives

Explosive reactions can be slow or fast, the former characterized by low rates of reaction (a few centimetres to a few meters/second) and the latter by very high rates (up to several kilometres/second). The reactivity of a chemical depends on its chemical structure. All explosive chemicals such as nitrate ($-ONO_2$), nitro ($-NO_2$), chlorate (ClO_3^{-1}) and perchlorate (ClO_4^{-1}) are characterized by low thermodynamic stability.

The chemical compositions of typical molecular explosives are shown in Figure 2.1. The oxygen attached to these structures breaks away easily to combine with other elements such as carbon, hydrogen, sulphur, etc. to form more stable compounds. There are also some explosive compounds which are either highly oxygen deficient (e.g. trinitrotoluene (TNT)) or totally devoid of it (e.g. lead azide: PbN_3). But it can be said in general, that most explosives contain carbon, hydrogen, nitrogen and oxygen, and are categorized as 'C-H-N-O' explosives.

The major reactions in the explosion process are the following: C to carbon dioxide, H to steam and water, N to nitrogen gas, Al to aluminum oxide, S to gas or solid sulphates, ammonium nitrate to water, nitrogen and oxygen, and so on. On this basis let us examine the reaction process in three classical explosives: black powder, TNT and nitroglycerin.

$$8C + 3S + 10KNO_3 \rightarrow 3K_2SO_4 + 2K_2CO_3 + 6CO_2 + 5N_2$$
(Black Powder)

$$2[CH_3C_6H_2(NO_2)] \rightarrow 12CO_2 + 5H_2 + 3N_2 + 2C$$
(TNT)

$$4[C_3H_5(NO_3)_3] \rightarrow 12CO_2 + 10H_2O + 6N_2 + O_2$$
(Nitroglycerin)

In each case, the unstable union between nitrogen and oxygen is transformed into more stable compounds; nitrogen reuniting with itself, oxygen combining with carbon, hydrogen and sulphur. Actual reactions at the high temperatures and pressures prevailing in the reaction zone are, of course, more complex, and the relative gas fractions continually change as the temperature inside the explosion drops.

All explosive compounds can be considered to be composed of three components:

- fuel;
- oxidizer;
- sensitizer.

Carbon, hydrogen, sulphur, etc. provide the essential fuel for the oxygen in the oxidizer. Incorporation of a chemical or physical sensitizer enhances the ease with which the explosive can be made to react by means of an initiator. The molecular explosives such as nitroglycerin or TNT do not require a sensitizer, whereas others may require such components in order to

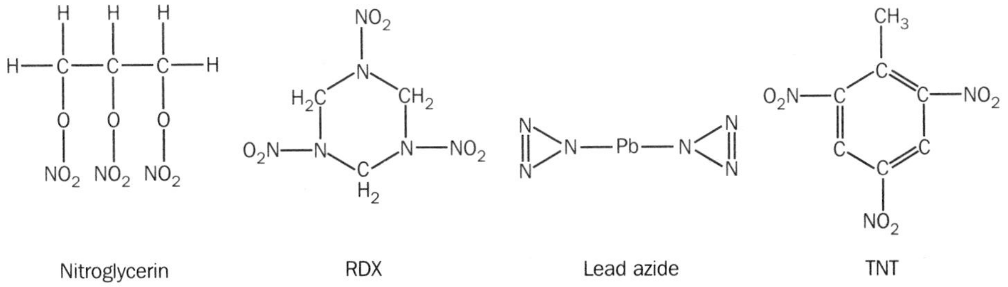

Nitroglycerin   RDX   Lead azide   TNT

Figure 2.1 Chemical composition of typical molecular explosives.

Table 2.1 Typical sensitizers in commercial explosives

	Chemical	Physical
Composite B	Perchlorates	
TNT	Monoethalamine Nitrate	Aluminium (fine)
RDX		Air bubbles
	Nitromethane	
Nitroglycerin		Glass or plastic microballoons
	Nitropropane	

attain a degree of sensitivity that is practical from a blasting point of view. A partial list of sensitizers is given in Table 2.1.

The bulk of the commercial explosives are not molecular explosives but are made of mixtures of these three essential components. As a result, they are considerably less sensitive (and therefore safer) than the molecular explosives (e.g. PETN (pentaerythritol tetranitrate), TNT, NG, etc.). Initiation of these explosives is therefore much more difficult, and requires adequate boosting. The common form of initiators and boosters are detonators (~ 1 g of PETN), detonating cords (4 g/m to 40 g/m of PETN), cast boosters (20 g to 1 kg of Pentolite (PETN/TNT: 50/50)), and cartridged explosives (200 g to 500 g of detonator-sensitive commercial explosives).

2.4   Types of Explosives

Explosion events could originate from several sources. These include dust explosions, ignition of flammable gases and initiation of condensed phase chemicals such as propellants and explosives, and finally, detonation of nuclear devices. Dust and vapour cloud explosions represent the most common form of accidental events and are discussed in Chapter 7. The basic mechanics of damage is similar to that due to chemical explosives, and so the present discussion will deal only with the latter.

The chemical explosives can be roughly grouped under two categories: military explosives or commercial explosives. There is however no sharp distinction except in their applications and their relative sensitivity to initiation. The military explosives, as well as the so-called primary explosives used in the manufacture of detonators, are normally composed of molecular explosives which require no additional ingredients to make them explode. Examples are: lead azide, lead styphnate, TNT, PETN, RDX (research department explosive), and various combinations of the latter three compositions such as 'Semtex' (RDX/PETN). In general, the molecular explosives have higher sensitivity and higher reaction rates than composite explosives. Despite their extensive military use and apparent familiarity, most of these explosives were developed only during the early half of the 20th century.

The development of commercial explosives preceded military explosives by several decades and continues to be an active area of research. Examples are: nitroglycerin (NG)-based explosives (dynamites), explosives with alternate sensitizers such as TNT, RDX and perchlorates, slurry explosives, emulsions and dry blasting agents such as ANFO (ammonium nitrate + fuel oil). All explosives other than NG-based ones may contain varying amount of aluminum for extra energy. Although NG-based and slurry explosives still have a significant market share, dry blasting agents such as ANFO and emulsion explosives and their variants have become the mainstay of most blasting operations.

The emulsion explosives have several outstanding advantages over other explosives including: its simplified composition (saturated AN liquor with fuel oil and an appropriate emulsifier), intimate mixing of fuel and oxidizer (droplet size ranging from 1 μm to 5 μm), its relative waterproofness, its ease of manufacture and high velocity of detonation. Emulsion explosives can be sensitized, for small diameter applications, with either chemical or physical sensitizers (air bubbles or glass or plastic microballoons), and can incorporate varying amounts of aluminum to give it additional strength. It can be manufactured in a wide range of densities – lower densities for small diameter application and higher densities for large diameter application. The emulsion explosive can be used by itself or mixed with ANFO or ammonium nitrate prills, or it can be used as a filler of intergranular spaces in ANFO.

The bulk of the commercial explosives in use today are either straight ANFO or its derivatives with emulsions.

2.5 Explosion Process

Heating a reactive material results in its exothermic decomposition. The resulting heat may further increase the rate of reaction and may eventually lead to a self-sustained reaction known as 'deflagration'. A rapidly travelling shock wave also can provide the initial source of heat in the material. Under certain conditions of initiation and confinement the deflagrating reaction can transit to a supersonic but steady rate of reaction, otherwise known as 'detonation'. In deflagration mode, the reacted materials flow away from the unreacted material, whereas, in the detonation mode, the detonation products flow with great velocity towards the undetonated explosive.

In the detonation process, a shock front or shock zone propagates at a characteristic velocity into the unreacted explosive at very high pressures and temperatures. Immediately behind the shock front is the chemical reaction zone where the original material is rapidly converted into reaction products. The width of the shock front and the reaction zone could be as low as a few millimetres depending on the nature of the explosive material and the boundary conditions. The chemical reaction zone is followed by a slower moving zone consisting of the detonation products. The mechanism is shown schematically in Figure 2.2. The pressure and temperature in the detonation zone could exceed several hundred thousand atmospheres and 3000°C.

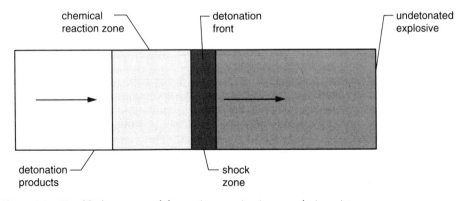

Figure 2.2 Simplified structure of detonation reaction in an explosive mixture.

2.5.1 Pressure of an Explosion

One can get some idea of the magnitude of the pressure associated with an explosion process by treating the gaseous explosion products as ideal gases. The ideal gas law is given by:

$$PV = nRT$$

where P is the pressure, V the volume, n the moles of gas, R the gas constant, and T the absolute temperature.

This pressure–volume–temperature relationship holds only at relatively low pressures where molecules of gas are relatively far apart. However, under explosion reaction conditions this is far from so, as seen in Table 2.2 for three representative gases. Real gases are seen to be much more compressible as the varying values of PV indicates.

To account for the behaviour of gases at temperatures and pressures characteristic of the explosion process, several 'equations of state' (EOS) have been proposed with varying degrees of success. Some typical examples of these are shown in Table 2.3.

2.5.2 Energy Release in Explosions

In simple terms, the amount of heat released can be taken to be the equivalent of the energy content in an explosive. The known heat of formation of compounds is used to predict the heat of explosion. The decomposition of nitroglycerin can be taken as an example.

Table 2.2 Pressure–volume relationship per mole of gas at 0°C (after Manon, 1976)

Pressure (atm)	Hydrogen		Oxygen		Carbon dioxide	
	V	PV	V	PV	V	PV
1	22.428	22.43	22.393	22.39	22.262	22.26
100	0.2386	23.86	0.2075	20.75	0.044 50	4.45
400	0.071 63	28.65	0.058 87	23.55	0.040 51	16.21
800	0.043 92	35.13	0.042 07	33.66	0.037 79	30.23
1000	0.038 37	38.37	0.038 86	38.86	0.036 87	36.87

Table 2.3 Some equations of state to relate pressure–volume–temperature relationship

Name	Relationship*
van der Waals	$(P + a/V^2)(V - b) = R$
Becker–Kistiakowsky–Wilson (BKW)	$PV = RT(1 + \chi e^{\beta \chi}); \chi = (\kappa \sum X_i k_i)/[V(T + \theta)^x]$
Beattie–Bridgman	$PV^2 = RT[V + B_0(1 - b/V)][1 - c/VT^3] - A_0(1 - a/V)$

* P, V, T, R, X, and k denote pressure, volume, temperature, gas constant, species mole fractions and individual species volumes. As expected only R is a constant in the above equations, and the fitting parameters a, b, c, A_0, B_0, C_0 and α, β, θ and κ must be determined from experimental data.

$$\Delta H_{p(explosion)} = H_{p(products)} - H_{p(explosive)}$$

$$4[C_3H_5(NO_3)_3] \rightarrow 12CO_2 + 10H_2O_{(g)} + 6N_2 + O_2$$

$$-\Delta H_p = 12(94) + 10(58) - 4(83)$$

$$= 367 \ kcal/g - mole \ at \ 0°C$$

$$or \ 1620 \ cal/g$$

The energy released in ANFO can be similarly calculated. A simple calculation goes as follows:

94.5% AN + 5.5% FO

$$3NH_4NO_3 + CH_2 \rightarrow CO_2 + 3N_2 + 7H_2O + 930 \ cal/g$$

For slightly different proportions of AN to fuel oil, the results would be as follows,

92% AN + 8% FO

$$2NH_4NO_3 + CH_2 \rightarrow CO + 2N_2 + 5H_2O + 810 \ cal/g$$

and

96.6% AN + 3.4% FO

$$5NH_4NO_3 + CH_2 \rightarrow CO_2 + 4N_2 + 2NO + 11H_2O + 600 \ cal/g$$

Clearly the maximum amount of energy is released with a 94.5/5.5 ratio of ammonium nitrate and fuel oil.

Oxygen Balance

The role of oxygen is crucial in optimizing the reaction products for maximum release of heat. As shown above, with few exceptions most molecular explosives are oxygen deficient. To achieve full oxidation it is necessary to add oxygen-rich ingredients in making explosive mixtures with these explosives. The ratio of oxygen present to the amount of oxygen required to achieve full oxidation is a measure of the oxygen balance. As the above equations show, ANFO with 8% fuel oil is oxygen-negative or has a negative oxygen balance as evidenced by the presence of CO in the reaction products. Similarly, lowering the fuel oil to 3.4% renders the ANFO oxygen-positive with extra oxygen being in the form of NO in the reaction products.

Most commercial explosive compositions are formulated ideally to have a zero oxygen balance.

2.5.3   Detonation Properties

Conservation of momentum across the shock front results in rapid acceleration of the reacted material in the direction of shock. This is termed as the particle velocity. The density, pressure, particle velocity and temperature in the reaction zone must all obey conservation laws of physics. The Chapman–Jouget (C–J) postulate states that the sound velocity in the reacted material plus the particle velocity be at least the same as the shock wave velocity, to ensure that the chemical reaction energy can be transported forward to sustain the shock front. The C–J plane defines the limit beyond which no further energy release can contribute to the propagation of the shock front. The corresponding pressure is known as the C–J pressure.

Table 2.4 Detonation parameters of some typical explosives

Explosive	Density (g/cm^3)	Detonation velocity (m/s)	Detonation Pressure (GPa)
ANFO	0.84	4600.00	4.50
Emulsion	1.20	5200.00	8.10
Dynamite	1.45	5400.00	10.6
TNT*	1.64	6930.00	21.0
PETN*	1.67	8000.00	30.0
RDX*	1.77	8700	33.80

* Dobratz and Crawford, 1985.

These characteristic pressures for some typical explosives are shown in Table 2.4, along with their density and velocity of detonation.

Besides density, the other critical parameter controlling detonation properties is the charge size or charge diameter. This applies particularly to the more insensitive commercial explosives, where the velocity of detonation is not only a strong function of charge diameter, but such explosives are not detonable below a certain critical diameter. For commercial explosives the critical diameter ranges from 20 mm to 200 mm depending on the density and composition. The other distinguishing features of commercial explosives are their generally lower density and lower velocity of detonation and pressure. However, despite their some-what reduced hazard potential in terms of damage to structures and injury to personnel, detonation of commercial explosives, deliberate or otherwise, still represents a major hazard.

2.6 Characteristics of Blast Waves

The original potential energy in the explosive upon detonation, is distributed among a number of distinct forms as a function of space and time. These consist of:

- wave energy;
- residual or waste energy;
- kinetic and potential energy of confining material or fragments;
- kinetic energy of source material;
- source potential energy;
- radiation.

The energy distribution is shown schematically in Figure 2.3. The energy in a propagating wave system is both potential and kinetic, but excludes that contained in the volume occupied by the explosion products or the quiescent medium between the blast wave and the products (Strehlow and Baker, 1976). Thus, at far field, the total wave energy – a character-istic of each explosive type – should remain constant with time. The residual or the waste energy represents the heat generation in the intervening air due to passage of the shock wave, which is irreversible.

The kinetic and the potential energy imparted to any confining medium or resulting frag-ments due to the initial expansion of the shock wave can be a significant fraction of the total energy. This is done through plastic deformation, heat transfer, etc. Similarly, some amount of energy is also transferred as kinetic energy to the explosive source material. All kinetic energy components must eventually reduce to zero as the fragments or the source materials come to rest. In a typical explosion process, a considerable amount of energy is still retained by the explosion gases which are still under relatively high pressures and temperatures. This is known as the potential energy of the product. Although this energy is eventually dissipated through expansion and cooling, the time scale in which this occurs is much larger than that associated with the propagation of the blast wave.

The final component of energy loss is represented by radiative heat losses. However, this is rather a small fraction of the total heat energy in chemical explosions compared to nuclear explosions. Figure 2.3 shows that only a fraction of the total available explosive energy appears as wave energy. A larger fraction of the total energy, however, may be transferred to the blast wave in the case of a slower rate of release of explosive energy.

The characteristic propagation and decay of blast wave in air is shown schematically in Figure 2.4(a). With greater time (or larger distance travelled), the blast wave undergoes systematic changes in amplitude, duration, and profile. In the far-field, the pressure signature assumes a pronounced negative phase. Figure 2.4(b) shows the features of an ideal blast pressure profile and its relevant features. This should be compared with an actual pressure profile recorded from a 14 lb near-surface ANFO charge at a distance of 7 feet, as illustrated in Figure 2.4(c).

The similarity between the ideal shock pressure profile and that from measurement is remarkable. The leading part of the blast wave is compressive with an extremely rapid rise in

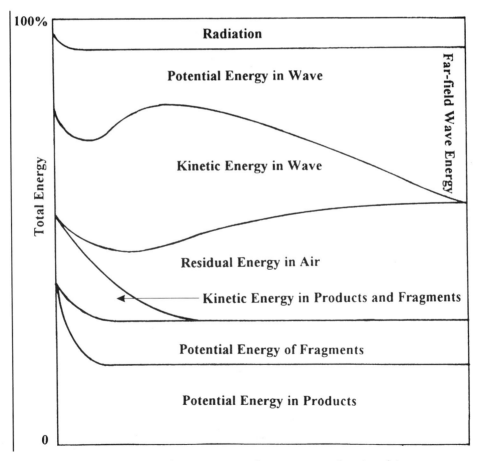

Figure 2.3   Energy distribution (schematic) in an explosion event as a function of time.

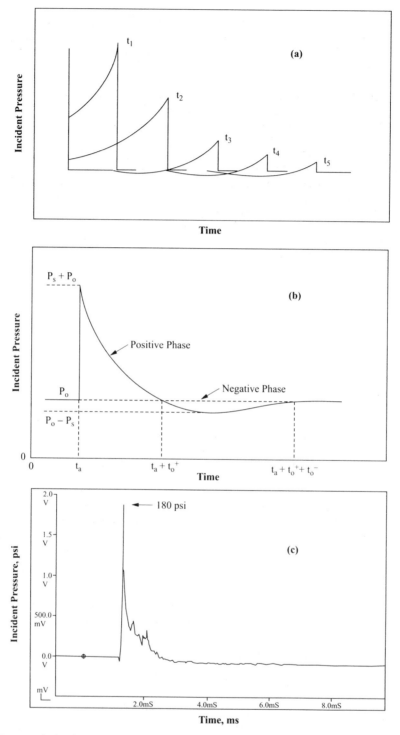

Figure 2.4 (a) Idealized propagation and decay of shock wave from an explosion at increasing distances; (b) typical shock pressure profile; (c) measured shock profile from a small near-surface ANFO charge (14 lb) at a distance of 7 ft from the source.

pressure level, which is followed by exponential decay of the pressure. Due to the inertial effect, the pressure eventually falls below the atmospheric level and then returns more or less to the ambient level after some time. The amplitude of the negative phase of the blast wave is only a fraction of the positive phase, although in the far-field conditions its duration is larger than the positive or the compressive phase. Initially, the velocity of the blast wave, due to the high pressures involved, could be several thousand meters per second, but with distance the shock decays to a regular sound wave in air with the characteristic velocity of ~ 330 m/s. From this it is evident that the shock wave from a larger or more intense explosion will arrive at a given site sooner than due to a weaker explosion.

A very important aspect of the shock wave in air is the high velocity wind associated with the shock front. The pressure resulting from the latter is called the dynamic pressure (Kinney, 1962). It is proportional to the square of the wind velocity and to the density of the air mass behind the shock front. It can be shown that the dynamic pressure becomes larger than the peak overpressure when the latter reaches or exceeds 70 psi. The calculated peak dynamic pressure and the corresponding wind velocity are shown in Table 2.5.

Since the dynamic pressure denoted by 'q' is proportional to the square of the wind velocity (u) and the density (ρ) of the shocked air, it can be written as,

$$q = 1/2(\rho u^2)$$

which is essentially the kinetic energy per unit volume of air immediately behind the shock front. It can be further shown that,

$$q = P_s^2/[2\gamma P_0 + (\gamma - 1)P_s]$$

which, for $\gamma = 1.4$ (the ratio of the specific heats at constant pressure and at constant volume for air) reduces to,

$$q = 5/2 P_s^2/(7P_0 + P_s)$$

where P_0 is the ambient pressure, and P_s the peak overpressure. The above expression will be used in calculating the reflected overpressure to be discussed in the next section.

The peak dynamic pressure and its duration play a critical role in the development of the drag force around a structure and the subsequent damage to it. The relationship between the shock pressure and the dynamic pressure is illustrated more graphically in Figure 2.5.

As expected, there is a small time lag, due to inertial effects, between when the shock and the dynamic pressure reach the atmospheric pressure. Also, unlike the shock, the dynamic pressure is never negative. Initially, the blast wind blows away from the explosion, but when the shock pressure goes below ambient, the wind direction reverses and it now flows, albeit at a much lower velocity towards the source of the explosion. Although the dynamic pressure

Table 2.5 Peak overpressure, dynamic pressure and maximum wind velocity in air at sea level for an ideal shock front (after Glasstone and Dolan, 1977)

Peak overpressure (psi)	Peak dynamic pressure (psi)	Maximum wind velocity (mph)
200	330	2078
100	123	1415
50	41	934
30	17	669
10	2.2	294
2	0.1	70

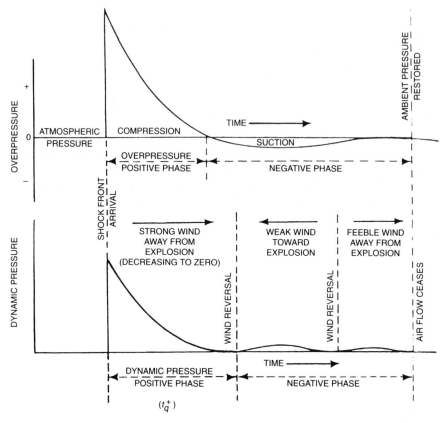

Figure 2.5 Shock pressure profile from a free-air explosion and the associated dynamic pressure at a fixed distance from the explosion (after Glasstone and Dolan, 1977).

has a longer duration than the shock pressure, it is of relatively low amplitude in the latter part and does not pose a significant hazard to a target. It can also be shown that it decays faster than the direct shock pressure.

2.6.1 Reflection of Blast Waves

When a shock wave impinges on a 'rigid' target, be it the ground surface or a building, it undergoes reflection and possibly diffraction. Assumption of a rigid target makes the analysis simpler at this stage, as it precludes any significant energy transfer through refraction or deformation of the target. Also for simplicity, only 'grazing' or normal incidence at right angles to the target is considered in this analysis. When a spherical explosive charge is detonated sufficiently far away from the ground surface or any other reflecting surface, the shock pressure expands spherically, and its characteristics such as, a peak, duration impulse, and arrival time, at a certain location, are known as *free-air* explosion parameters. This is in contrast to *surface* explosion parameters where the explosive charge is near or on the ground surface.

The explosion parameters are very different for these two conditions. In the former case, any point above the ground surface would experience two distinct shocks: one due to the incident or direct shock from the explosion and the other, a reflected shock from the ground surface (or any other reflecting surface). The latter would be delayed with respect to the direct shock because of the extra travel path involved. The explosion resting on the ground would produce only a single shock, but it will have significantly different characteristics than the *free-air* burst especially in its peak amplitude.

A free-air spherical explosion upon reflection from ground surface also gives rise to an important modification to the shock profile. As the reflected shock has to travel in pre-shocked air (caused by the direct shock) it travels at higher velocities than its own amplitude would demand. This eventually leads the reflected shock to merge with the direct shock, forming what is known as a triple point (Figure 2.6). The region between ground surface and the triple point is called the Mach region, and the corresponding shock front is known as the 'Mach stem'. Both the height and the peak value of the Mach stem have critical bearing on blast loading of structures. For the surface burst, there will no region of reflection, but the amplitude of the shock would be considerably higher than the free-air case. In fact, it would be the same as in the Mach stem, with the blast wind blowing more or less horizontally close to the ground surface.

Irrespective of whether the explosion is a free-air or a surface burst event, when the shock wave impinges on a rigid target it undergoes reflection and diffraction, the extent of the latter depending very much on the shape and size of the target. The pressure in the shock wave impinging on a target is known as the 'incident' or 'side-on' pressure (P_s), which upon reflection from a target is known as the 'reflected' pressure (P_r). There are additional complications for oblique impacts, so only the case of normal incidence for a plane shock (i.e. shock front parallel to a flat-faced target) is considered here. It can be shown that for such a case, the instantaneous reflected pressure, P_r, is given by:

$$P_r = 2P_s + (\gamma + 1)q$$

Upon using the penultimate equation and assuming $\gamma = 1.4$ for air, the above reduces to

$$P_r = 2P_s(7P_0 + 4P_s)/(7P_0 + P_s)$$

This shows that for very strong shocks (i.e. $P_s > P_0$), the instantaneous 'reflected' pressure can be eight times that of the 'incident' shock pressure at normal incidence. Conversely, for weak shocks (i.e. $P_s \ll P_0$), the reflected pressure is only twice that of the incident pressure, which is the simple acoustic case. Baker (1973), however, has suggested that the relationship $P_r = 8 P_s$ for strong shocks is an oversimplification since it is based on the assumption that the air behaves like an ideal gas at such high temperatures and pressures. The maximum

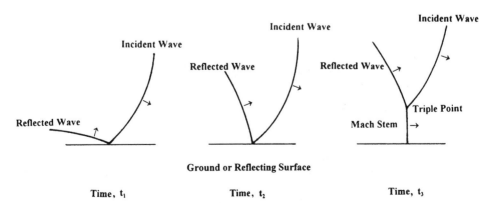

Figure 2.6   Formation of the triple point and Mach stem upon reflection of shock wave.

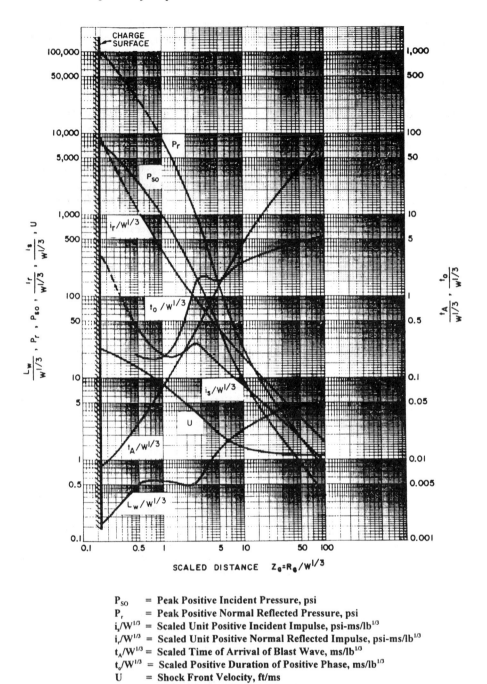

P_{SO} = Peak Positive Incident Pressure, psi
P_r = Peak Positive Normal Reflected Pressure, psi
$i_s/W^{1/3}$ = Scaled Unit Positive Incident Impulse, psi-ms/lb$^{1/3}$
$i_r/W^{1/3}$ = Scaled Unit Positive Normal Reflected Impulse, psi-ms/lb$^{1/3}$
$t_A/W^{1/3}$ = Scaled Time of Arrival of Blast Wave, ms/lb$^{1/3}$
$t_o/W^{1/3}$ = Scaled Positive Duration of Positive Phase, ms/lb$^{1/3}$
U = Shock Front Velocity, ft/ms
W = Charge Weight, lbs
$L_w/W^{1/3}$ = Scaled Wave Length of Positive Phase, ft/lb$^{1/3}$

Figure 2.7 Shock wave parameters (positive phase only) for a free-air spherical TNT explosion at sea level (after Joint Departments of the Army, the Navy and the Airforce, 1990).

28

value of the reflected shock in reality could be much higher than this. It is evident however from the above equation, that the reason for the reflected pressure being greater than $2 P_i$ is due to the presence of the dynamic pressure associated with the blast wind discussed earlier.

The free-air measurement of the blast wave parameters form much of the basis for estimation of explosion hazards. A very large number of tests have been performed over decades involving mostly spherical TNT and Pentolite explosive charges. By means of the principles of scaling and equivalency, to be discussed in the next section, the blast parameters for other explosives can be calculated for various quantities and at various distances.

Typical shock wave data for a free-air spherical TNT explosion are shown in Figure 2.7 as a function of 'scaled distance'. The latter is defined as the ratio of distance over the cubed root of the weight of explosive $(x/w^{1/3})$. The parameters shown are for the positive phase only. These include the peak incident and reflected pressure (P_s and P_r), the corresponding impulses (i_s and i_r), the arrival times (t_a), the duration of the positive phase (t_0), and the wave length (L_w). Similar curves can be generated also for the negative phase of the blast wave. The common practice is to represent them in scaled terms (i.e. scaled with respect to the cube-root of the charge weight), so that shock wave parameters can be obtained for any combination of charge weight and distance. It is clearly evident that the reflected pressure is many times higher than the incident pressure even at relatively low incident pressures (~ 10 psi). One has to be almost at a scaled distance of 50 away (i.e. 50 ft from a 1 lb charge or 500 ft from a 1000 lb charge) before the reflected pressure becomes sufficiently small and the blast wave can be considered an ordinary acoustic wave.

The shock parameters shown in Figure 2.7 change significantly for a surface or near-surface burst. The same applies when the charge shape deviates from spherical symmetry. Three such examples for hemispherical, spherical and cylindrical charge of composition B (RDX/TNT/wax: 56/40/4) are shown in Figure 2.8 for surface explosions. The variation of both incident pressure and impulse with shape is clearly evident at identical scaled distances. Similar variations occur with more common shapes such as hoppers, bags or flat beds of explosives, but values for these can normally be extrapolated from the above three cases.

2.6.2 Scaling and TNT Equivalency

An accidental explosion can involve a variety of conditions such as specific types of explosives, charge geometry, height of burst and atmospheric pressure conditions. Since the source conditions in most accidents are not known very accurately, estimation of the size of the explosion or the incident blast overpressure for assessing damage (or designing against damage) would be a formidable task. To overcome this, the techniques of 'scaling' and 'TNT equivalency' are widely used to estimate the size of an accidental explosion or design blast-resistant structures. This allows calculation of blast parameters from different explosive–distance combinations from standard tables of such parameters available for TNT and Pentolite (see Figure 2.7).

Scaling

The self-similarity of air shock from detonation of a variety of explosive types has been mentioned earlier. The most common practice of explosion scaling is based on 'cube root' scaling, otherwise known as Hopkinson's law (Baker, 1973). It states that for any pressure generated at a distance R_1 from a reference explosion of W_1, in weight, the same pressure would be registered at a distance R_2 from a different explosive of W_2 in weight, provided

$$R_2/R_1 = (W_2/W_1)^{1/3}$$

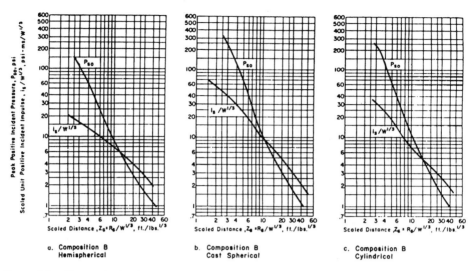

Figure 2.8 Changes in shock wave parameters due to change in shape of the explosive source for Comp. B (after Joint Departments of the Army, the Navy and the Air Force, 1990).

Similar relationships apply to time (t) and impulse (I),

$$t_2/t_1 = R_2/R_1 = (W_2/W_1)^{1/3}$$

$$I_2/I_1 = R_2/R_1 = (W_2/W_1)^{1/3}$$

The above means that two different weights of the same explosive would give the same pressure at the same 'scaled distance' (i.e. $z =$ distance/(charge weight)$^{1/3}$). Similarly, an observer stationed at a distance 'λR' from the centre of an explosive charge with characteristic dimension 'λd' would experience a similar blast wave with amplitude P, duration λt and impulse λI, to a reference explosive having the characteristic dimension 'd' and the same pressure, P. Hopkinson's law has been shown to hold true for a variety of explosives and energy yields.

The conditions described above apply to similar types of explosives, spherical charge geometry, and identical atmospheric conditions during detonation. Except for the latter, these can be easily controlled or duplicated. Another scaling law, known as Sachs scaling law is applied to correct for changes in the atmospheric conditions. The corresponding multipliers to pressure, distance, time and impulse (S_p, S_d, S_t and S_i respectively) based on Sachs law are given below.

$$S_p = (P_a/P_0)$$

$$S_d = (W_2)^{1/3}/S_p^{1/3}$$

$$S_t = (W_2)^{1/3}/S_p^{1/3} \cdot (288/T_a + 273)^{1/2}$$

$$S_i = (W_2)^{1/3}/S_p^{2/3} \cdot (288/T_a + 273)^{1/2}$$

where, P_a is the ambient pressure, P_0 the standard sea level atmospheric pressure (14.7 psi or 101.3 kPa), T_a the ambient temperature in °C.

The equations and the conditions described above apply strictly to spherical or nearly spherical charges, whereas most real explosions involve non-spherical charge geometries. With an asymmetrical explosive source, the azimuthal variation of shock wave parameters becomes very significant. A considerable body of experimental and empirical data exists to correct for the effect of asymmetry and generate equivalent spherical charges, which can then be used with the standard TNT blast parameter charts (Baker et al., 1980). However, caution must be exercised in any extrapolation of available data to a subject explosion whose geometry and orientation are known only approximately.

TNT equivalency

Since extensive experimental data exist for TNT explosions, it is customary to find a 'TNT equivalency' for a subject explosive, and then use only the TNT blast parameter to assess or predict damage, design appropriate blast-resistant structures or station personnel. The two approaches for estimating TNT equivalence are: (a) by calculation, and (b) by experiment.

The former is based on the calculated energy equivalence between TNT and the subject explosive. This is based on the calculated heats of explosion and assumes that the detonation reaction is an ideal one and all the energy calculated is actually released. The alternative method is to arrive at the equivalency from specific structural damage studies and relate it to known damage levels from specified overpressures. This approach can however lead to large errors since the degree of damage such as collapse of walls or breakage of windows cannot be accurately correlated with the levels of overpressure. The best approach is through actual measurement of shock pressure parameters (pressure and impulse) followed by comparison to standard charts of the equivalent weight of TNT which would yield the same pressure or impulse at the same distance. The equivalency can be expressed as

$$N_p = (W_{TNT}/W) = (Z/Z_{TNT})^3 P_s = \text{constant}$$

$$N_i = (W_{TNT}/W) = (Z/Z_{TNT})^3 i_s = \text{constant}$$

where W is the weight of the subject explosive, and z is the scaled distance (distance/weight$^{1/3}$), and N_p and N_i represent equivalency based on incident pressure and incident impulse respectively.

Unfortunately, the calculation method and the two experimental methods (pressure and impulse) do not always yield the same TNT equivalency (Esparza, 1986). A comparison among the three approaches is shown in Table 2.6.

As the data show there is considerable discrepancy among the three methods. On the experimental side, there is usually much greater uncertainty in the measurement of impulse than peak pressure. Equivalency based on the peak incident pressure is therefore the preferred method, but structures designed to resist impulse load would require the use of TNT equivalency based on impulse.

Table 2.6 Comparison of TNT equivalency by different methods for selected explosives* (after Esparza, 1986)

Explosive	TNT equivalency		
	Pressure	Impulse	Calculated energy
TNT	1.0	1.0	1.00
Composition B	1.2	1.3	1.09
Pentolite	1.5	1.0	1.09
PBX-9404	1.7	2.0	1.11

* For incident pressure range of 100–1000 psig, corresponding to scaled distance range of 3.5 to 0.74 ft/lb$^{1/3}$ with spherical charges.

Calculation of equivalency based on standard TNT curves yields reliable results for high explosives exhibiting ideal or near-ideal detonation behaviour. It is much more difficult to arrive at similar equivalency figures for the non-ideal detonations characteristic of some of the commercial explosives. These may exhibit relatively long run-up distances before they achieve steady detonation velocity, or there may be significant reaction behind the C–J plane due to granularity or inhomogeneity of the explosive matrix. The same applies to other types of explosions such as pressurized bursting vessels, dust and vapour cloud explosions. The calculated energies from the latter types is normally much higher than actually released because the calculations must assume optimum dispersion of oxidizer and fuel elements in the dust or vapour cloud – a situation which rarely exists in practice. These types of explosions usually have lower peak pressures of much longer duration, and their damage potential is significantly different than those due to molecular explosives.

2.7 Types of Hazards

The severity of hazards from a deliberate or accidental explosion depends on a number of factors. Different considerations apply if one is seeking protection of structures rather than people. Even here, one has to define the degree of protection sought, e.g. minor structural damage may be acceptable but not injury to people. The hazard could result from any number of factors such as air shock, ground cratering, seismic vibrations, collapse of buildings, fragments and missiles, ground ejecta, and explosion fireball and thermal radiation. Of these, cratering and fireball do not pose a serious hazard as they are confined to the immediate vicinity of the explosion. Thermal radiation is not a major factor in chemical explosions. Seismic vibrations caused by ground shock can result in some damage to structures located at a considerable distance from the site of the explosion, but at these distances the damage is minor and does not pose a threat to people.

It is nevertheless possible to estimate the crater dimensions from a knowledge of the ground conditions and the fireball dimensions from a knowledge of the explosive composition and its detonation characteristics. Thus in terms of severity, air shock and fragment and missile hazards pose the most serious threat to both structures and personnel in an explosion. The interaction of shock waves with structures and the nature of biological hazards will be described in greater detail in later sections. However, the pressure resistance values shown in Table 2.7 for various structural elements, which were first published almost three decades ago, provide a good starting point.

As will be shown later, the response of a structure depends usually on a combination of peak pressure and impulse (i.e. integral of pressure–time profile), and it is quite likely that different structures at the same site would have to be designed to withstand an impulsive load rather than peak pressure load and vice versa. The figures in Table 2.7 do show the easy fragility of even solid structures when exposed to relatively low level blast overpressures.

The fragment and missile hazards represent a different and very difficult problem in terms of their quantification and prediction. These also include ground ejecta from surface bursts, as well as 'spalling' of material due to an intense shock impacting on the opposite face of a structural member or machinery. A large amount of experimental data is available which tries to relate fragment characteristics (i.e. size, distribution and velocity) with the amount of explosive involved in the detonation. These studies have been carried out, of necessity, with simple geometries with both contact and decoupled charges. The basis for all predictions is tied to the 'Gurney Energy Constant', which is defined as a characteristic fragment velocity specific to each explosive. It has been experimentally determined that, at least for simple geometrical configurations in mild steel, the initial velocity (V_0) of a metal fragment can be related to the weight of the explosive and the metal casing by the following,

$$V_0 = (2E')^{1/2} f(W, W_c)$$

Table 2.7 Conditions of failure of peak overpressure-sensitive elements (after Brasie and Simpson, 1968)

Structural element	Failure	Approximate incident blast overpressure (lb/inch^2)
Glass windows, large and small	Shattering usually; occasional frame failure	0.5–1
Corrugated asbestos siding	Shattering	1–2
Corrugated steel or aluminum panelling	Connection failure followed by bucking	1–2
Wood siding panels, standard house construction	Usually failure occurs at the main connections allowing a whole panel to be blown in	1–2
Concrete or cinder block wall panels 8 m or 12 m thick (not reinforced)	Shattering of the wall	2–3
Self-framing steel panel building	Collapse	3–4
Oil storage tanks	Rupture	3–4
Wooden utility poles	Snapping failure	5
Loaded rail cars	Overturning	7
Brick wall panel 8 inch or 12 inch thick (not reinforced)	Shearing and flexure failures	7–8

where V_0 is the initial velocity, $(2E')^{1/2}$ is the Gurney energy constant, W is the weight of explosive and W_c is the weight of steel casing. Examples of the Gurney energy constant (in ft/s) for several explosives are shown in Table 2.8.

For a cylindrical container weighing W_c (lb) and filled with explosive weighing W, the last equation takes the form:

$$V_0 = (2E')^{1/2}[(W/W_c)/(1 + W/(2W_c))]^{1/2}$$

For a thin-walled cylinder and an explosive mass much larger than that of casing, the maximum initial velocity becomes:

$$v_{max} = 1.41(2E')^{1/2}$$

The Gurney Energy relations for other simple geometrics are summarized in Figure 2.9. The velocity of a fragment will depend on its mass, its shape and the distance travelled from the explosive source. Initial velocities can be calculated for different combinations of container geometry and explosive weight and type, for specific fragment mass. Assumption of a certain fragment shape and use of the appropriate drag coefficient can also yield a velocity profile over distance for specific fragment mass. Considerable analytical and experimental work has been carried out to estimate fragment size as well as mass distribution from exploding steel casings (Joint Departments of the Army, the Navy and the Air Force, 1990), which

Table 2.8 Gurney energy constant $(2E')^{1/2}$ for selected explosives

Explosive type	density (g/cm³)	$(2E')^{1/2}$ (ft/s)
Composition B	1.72	9100
PBX-9404	1.84	9500
PETN	1.76	9600
RDX	1.77	9600
TNT	1.63	8000
Nitromethane	1.14	7900
Tritonal	1.72	7600

can be combined with velocity information to arrive at critical impact parameters which a specific target must be designed to withstand.

Despite the wealth of information available, calculation of fragment size and velocities is still extremely difficult except for the simplest geometrics and mild steel casings. Although the more complex but common types of containers or parts of machinery can be broken down to individual cylindrical or spherical components, the calculated values must be used with a great deal of caution. The primary fragments issuing from an exploding vessel or container can also give rise to secondary fragments upon impact against other structural elements or machinery parts. Prediction of the characteristics of these secondary fragments or missiles is generally much more uncertain. However, once the missile characteristics have been defined, the response of structural or biological targets to their impact is more easily handled. The response of human targets will be discussed in a subsequent section. They are more vulnerable to both primary and secondary fragments.

Type	Cross-sectional shape	Initial fragment velocity v_o	Maximum v_o
Cylinder	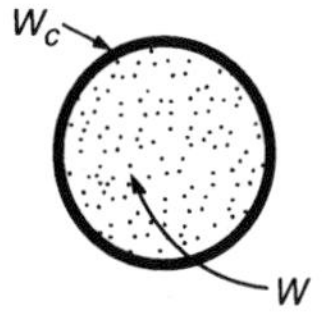 W_c ⬤ W	$\sqrt{2E'}\left[\dfrac{\dfrac{W}{W_c}}{1+\dfrac{W}{2W_c}}\right]^{1/2}$	$\sqrt{2E'}\,\sqrt{2}$
Sphere	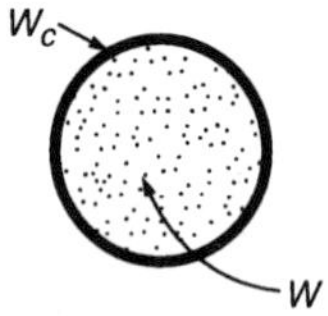 W_c ⬤ W	$\sqrt{2E'}\left[\dfrac{\dfrac{W}{W_c}}{1+\dfrac{3W}{5W_c}}\right]^{1/2}$	$\sqrt{2E'}\,\sqrt{\tfrac{5}{3}}$
Plate	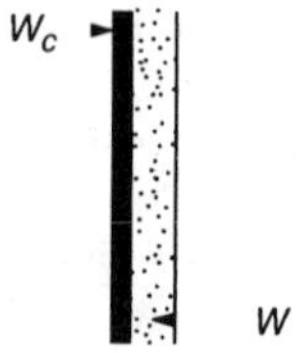 W_c ▮ W	$\sqrt{2E'}\left[\dfrac{\dfrac{3W}{5W_c}}{1+\dfrac{W}{5W_c}+\dfrac{4W_c}{5W}}\right]^{1/2}$	$\sqrt{2E'}\,\sqrt{3}$

Figure 2.9 Gurney energy relations for calculation of fragment velocity for simple geometries.

2.7.1 Missile Impact on Concrete

A steel fragment impacting on a concrete target can simply bounce back with a reduced velocity, or, with sufficient initial velocity, can cause perforation and spalling of the target. The rate of penetration of the missile depends on its mass and striking velocity. Initially the only effect on the concrete target surface is the formation of a crater due to dislodgement of material at the point of contact. As the velocity increases to 300 m/s (1000 ft/s) or more, the fragment usually penetrates beyond the bottom of the crater. This may be accompanied by 'spalling' of concrete on the rear side of the target. The velocity of spalled fragments can be high enough to cause injury or lead to sympathetic detonation of explosive material stored beyond the concrete wall. With sufficient striking velocity the fragment can penetrate the target and may eventually lead to perforation of the concrete wall. Empirical design equations have been obtained from experimental programs on concrete perforation, and these can be used with reasonable accuracy, provided the velocity and mass of the striking fragment are known. Based on these relations, it is predicted, for example, that a 4 in (10 cm) diameter fragment travelling at 3000 ft/s (900 m/s) will easily perforate a 12 in (30 cm) thick concrete wall.

2.8 Interaction of Blast Wave with Structures

The pressure exerted by the shock front on a target is known as the shock load, where all the phenomena discussed earlier (reflection, diffraction, dynamic pressure, etc.) may become operative. Except for explosions at great heights, all surface or near-surface events have pronounced or exclusively horizontal forces. The three types of loading that occur in this interaction with a target are: compression loading, diffraction loading and drag loading. The relative significance of these various loading conditions depends on the amplitude and duration of the shock wave, and also on the type of construction, the geometry of the structure, and its orientation with respect to the shock front. When the explosion occurs at considerable height above a relatively low-rise structure, the shock load initially is all compressive and almost simultaneous on all parts of the structure. In contrast, the loading characteristics from a surface or near-surface explosion are more complex.

When a steep shock front impacts on the face of a structure, it undergoes reflection with the resulting pressure build-up on the face being at least twice that of the incident pressure. However, for a structure of finite dimensions, the incident pressure wave continues on in the original direction of propagation and eventually engulfs the entire structure. This is known as 'diffraction loading'. For a relatively small structure with little or no openings, diffraction loading results in compression of the entire structure by about the same pressure, as contained in the incident overpressure. However, for a larger structure and a near-surface explosion, the diffraction effect would lead to differential loading of the front section of a structural member (e.g. roof or side-wall) compared to its rear section. The resulting relative displacement of a structural member during diffraction of the shock wave can have the same or greater damage potential than when the wave has finally engulfed the entire structure. Diffraction loading will continue until the positive phase of the shock has finally traversed the length of the structure and the pressure has fallen to ambient level. Presence of openings such as doors and windows or collapse of any structural member during the diffraction process would lead to lower pressures and rapid equalization of pressure within and outside the structure. The resulting pressure rise in the interior of the structure is known as the 'leakage pressure'.

When the shock wave impacts on a structure there is also an additional load due to the dynamic pressure caused by the strong winds behind the shock front. This is known as 'drag loading'; which has a longer duration than the positive phase of the incident shock. For an explosion source with a long positive duration phase, the structure could be subjected to drag

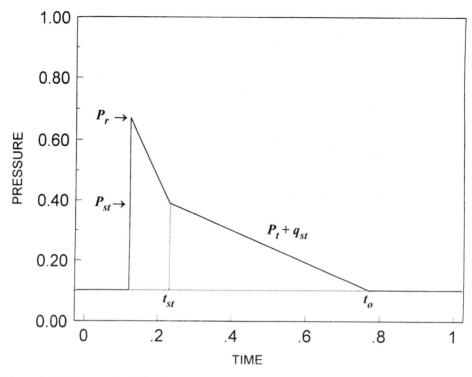

Figure 2.10 An example of blast load calculation on the front wall of a box-like structure (pressure and time in arbitrary units).

loading for a significantly longer duration than due to diffraction. Large buildings in general will respond mainly to the latter, whereas columnar structures such as smokestacks, electric transmission towers and truss bridges respond mainly to drag forces. On the other hand, large buildings with substantial openings or with weak sides or roofs which collapse soon after the arrival of the shock, become drag-type structures. The drag loading of a structure depends on shock wave parameters (amplitude and duration) as well as on the shape of the structure. The influence of this shape factor on the drag coefficient is less for streamlined objects (grain silos, smokestacks) than for irregular or box-like structures. The actual pressure generated is a product of the drag coefficient and the calculated value of dynamic pressure from the blast winds. Also, since the duration of the shock is proportional to the energy yield or size of an explosion for the same threshold overpressure, a high-yield explosion would cause more extensive damage at the same scaled distance than a low-yield one.

2.8.1   Calculation of Blast Load

The response of a structure would vary considerably, depending on whether the explosion is external or internal. In the case of the former, the structure would be designed to mostly resist 'pressure'; whereas, for the latter it would be designed to mostly resist 'impulse'. When an external blast pressure impacts on the outside of structure (e.g. the front wall of a box-like structure with little or no opening, and a shock front moving perpendicular to the plane of the wall), the pressure is immediately amplified to assume the value of the reflected pressure P_r. However, this decays rapidly to what is known as the 'stagnation pressure' P_{st}, which is the sum of the incident overpressure P_s and the dynamic pressure times the appropriate drag coefficient for the structural element. For the simple case considered (normal impact with $P_s < 50$ psi), the drag coefficient is unity, and the time, t_s, it takes the pressure to reduce to P_{st} is 3 S/U, where S is equal to the height of the wall or its half-width, whichever is less, and U is the shock velocity. The stagnation pressure is thus

$$P_{st} = P(t_s) + C_p q(t_s)$$

where q is the dynamic pressure and C_p the drag coefficient. Beyond t_s, the pressure decays to the ambient pressure, the total duration being equal to that of the positive phase of the incident overpressure. The blast loading profile for this example is shown in Figure 2.10. Similar profiles can be worked out for the roof, the side and the rear walls.

For an internal explosion, the pressures are normally much higher and the duration shorter. In this case, the 'impulse' loads on the walls and the roof would have to be calculated. Based on extensive experimental studies, a scheme has been worked out to calculate the impulse loads for specific geometries of a cubicle and explosive locations (Joint Departments of the Army, the Navy and the Air Force, 1990). An example of this scheme is shown in Figure 2.11. This also takes into account the additional loading on the walls due to multiple reflection of shock within the structure. The blast loading curves of course become more complex for other types of structures and shock impact angles.

2.8.2 Blast Resistant Structures

The design of blast resistant structures depends on the amplitude (P) and duration (t_0) of the blast load, and the reaction time (t_m) to maximum deflection of the structure. The various

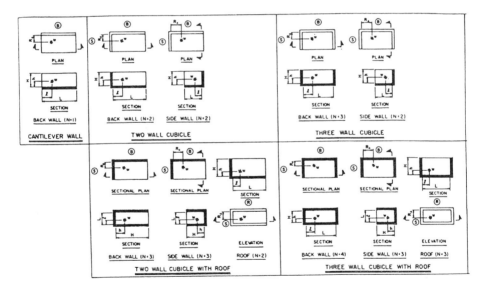

1. B denotes Back Wall, S denotes Side Wall and R denotes Roof.
2. Numbers in parentheses indicate number, N, of reflecting surfaces adjacent to surface in question.
3. h is always measured to the nearest reflecting surface.
4. ℓ is always measured to the nearest reflecting surface except for the cantilever wall where it is measured to the nearest free edge.

Figure 2.11 Cubicle configuration and design parameters for calculation of impulse load for an internal explosion.

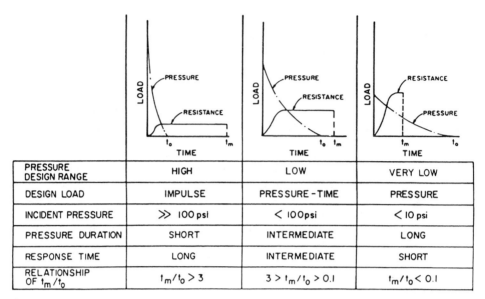

	HIGH	LOW	VERY LOW
PRESSURE DESIGN RANGE	HIGH	LOW	VERY LOW
DESIGN LOAD	IMPULSE	PRESSURE - TIME	PRESSURE
INCIDENT PRESSURE	$\gg$ 100 psi	$<$ 100 psi	$<$ 10 psi
PRESSURE DURATION	SHORT	INTERMEDIATE	LONG
RESPONSE TIME	LONG	INTERMEDIATE	SHORT
RELATIONSHIP OF t_m/t_o	$t_m/t_o > 3$	$3 > t_m/t_o > 0.1$	$t_m/t_o < 0.1$

Figure 2.12 Parameters defining pressure design ranges (after Joint Departments of the Army, the Navy and the Air Force, 1990).

pressure design ranges employed are shown in Figure 2.12. In close ranges, the blast pressure is very high, and its duration of the positive phase relatively short compared to the response time of the structure. If $t_m/t_0 > 3$, the structure is designed for impulse load. At the intermediate pressure ranges, with the response time in about the same range as the duration of the positive phase, the structure responds to both amplitude and duration of the blast load. Thus, when $3 > t_m/t_0 > 0.1$, the structure is designed to respond to the combined effect of both pressure and impulse. In the low-pressure regime, the blast load is of low amplitude (<10 psi), but its duration very long when compared to the response time of the structure. Thus, for $t_m/t_0 < 0.1$ the structure responds primarily to the peak pressure.

For a typical structural element, the duration of load t_o described above, can be calculated from the blast parameters (Figure 2.7) by the following empirical expression:

$$t_0 = (t_a)_f - (t_a)_a + 1.5(t_0)_f$$

where t_0 = duration of load in (ms), $(t_a)_f$ = time of arrival of the blast wave at the point in the element furthest from the explosion, $(t_a)_a$ = time of arrival of blast wave at a point on the element nearest to the explosion, $(t_0)_f$ = duration of the blast pressure at a point in the element furthest from the explosive. The increased value of $(t_0)_f$ in the above equation is intended to account for multiple reflections within the structure.

A blast-resistant structure in the high-pressure range may still have to be designed against peak overpressure rather than impulse, if the duration of load is substantially larger than the response time of the structures. In these cases, the blast load may be approximated by a fictitious triangular pressure profile with a peak value of pressure P_f from duration t_0 and average impulse i_b from the following empirical equation,

$$P_f = [2i_b/t_0]$$

The resistance of a structure to blast pressure depends on the massiveness of the structure and the type of reinforcement employed. A load-bearing masonry wall has little resistance to lateral forces of the blast load. Although most buildings are designed to withstand moderately strong winds, these loads start to build up gradually and eventually maintain a relatively steady state. The loads applied to a structure from an explosion are very transient in nature and therefore, inertia of the structure assumes, unlike in the case of static load, a very significant role.

The other key element in the design of a blast resistant structure is ductility, which is the material's ability to absorb energy inelastically without failure. Initial deformation of a load-bearing material occurs in the elastic range. However, with increasing pressure, the deformation becomes inelastic and the material fails when the stress exceeds its ultimate strength. Most common masonry materials without any reinforcement behave in a brittle fashion, i.e. they cannot accommodate any plastic deformation. So long as the pressure is low and within the elastic range, these materials can deform elastically, and recover fully once the load is removed. A ductile material, in addition to its ability to absorb elastic energy, can sustain large deformation in the inelastic range, and in the process absorb a large amount of energy before failure. Thus, a structure built with ductile elements can resist large blast loads before collapse, compared to its elastic mode of deformation, although it does not recover its original shape. The ductility in a concrete structural element is provided by steel reinforcement. An example of steel reinforcement employed in blast resistant structures is shown in Figure 2.13. The extent of reinforcement would depend on the blast load, and the degree of protection sought. Extensive design guidelines have been developed to construct blast-resistant wall and roof elements to withstand specified pressure and impulse loads (Joint Departments of the Army, the Navy and the Air Force, 1990).

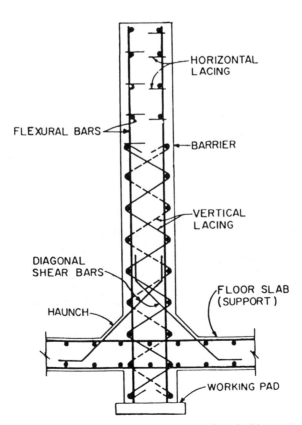

Figure 2.13 Typical construction of reinforced concrete members for blast resistant structures.

Earth or sand-filled barricades or a combination of reinforced concrete and sand panels can also be used as blast-resistant structures. These aim at utilizing the inertia effect rather than ductility in resisting blast loads. All these approaches apply to non-contact explosive charges. For explosive in contact with a structural element, it is almost impossible to prevent cratering damage or even perforation of the element. In this case, spalling of material from the other side of the same element becomes an important design consideration.

2.8.3 Hazards to Personnel

Injury to people can result from both direct and indirect blast effects. The former are caused either by direct air shock on the subject or by collapse of structural members such as the roof and the walls of a building. The indirect blast effects can be further subdivided into secondary effects and tertiary effects. High velocity fragments issuing either from the exploding vessel or the building containing the explosive represent the major secondary effect. The tertiary effects include the leakage pressure build-up within the structure, the secondary fragments due to spalling and window glass breakage, and whole-body displacement and subsequent decelerative impact against a rigid structure.

Human tolerance to blast pressure against serious injury is surprisingly high. In a high-shock environment, tests have shown that release of air from disrupted alveoli of the lungs into the vascular system is the most likely cause of death. The response depends on the body weight and on both amplitude and duration of the shock; a longer duration shock being more hazardous than a shorter duration one for the same level of shock. The probability of lung damage against blast pressure as a function of both peak incident pressure P_s and incident impulse i is shown in Figure 2.14 as a function of body weight w. This is based on extensive experiments on animals of different body weights, and considered fully applicable to determining the probability of lung damage in people (Richmond et al., 1968). The figure

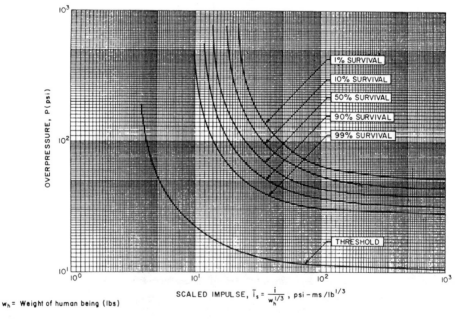

w_h = Weight of human being (lbs)

SCALED IMPULSE, $\bar{i}_s = \dfrac{i}{w_h^{1/3}}$, psi – ms /lb$^{1/3}$

Figure 2.14 Pressure and impulse for lung damage due to a blast wave (after Joint Departments of the Army, the Navy and the Air Force, 1990).

shows that for very long duration shock, any pressure in excess of 70 psi would lead to certain fatality in a person of average weight (160 lb) whereas, for a very short duration shock it would require about 200 psi to achieve the same effect. Similarly, the limiting pressure for threshold lung damage for a long duration shock would be only about 10 psi, whereas, for the short duration shock it may exceed 30 psi. In calculating the peak overpressure, one should also take into account the dynamic pressure associated with the shock front and any reflections in case the subject is standing next to a wall. The response would change depending on whether the subject was standing in open air, or lying prone on the ground or standing next to a rigid wall facing the blast.

In the low pressure regime, temporary hearing loss or eardrum rupture would be the other kind of injury from direct shock. It has been established that temporary hearing loss could result with an overpressure level lower than 1 psi, and threshold eardrum rupture could occur at an overpressure of 5 psi; at 15 psi overpressure the rupture rate would rise to 50%. Depending on the orientation of the person with respect to the explosion and the presence of any adjacent reflecting surface, the actual incident pressure could be lower than these values.

Although air shock can cause severe injury by itself and lead to numerous fatalities due to destruction of plants and buildings, its effect is confined to the immediate vicinity of the explosion, as the intensity of shock wave decays very rapidly with distance.

The secondary effects represented by high velocity fragments constitute the other major hazard from explosions, both in the immediate vicinity and at large distances. Human tolerance to such fragments is very low. The primary fragments such as metal pieces from a vessel are characterized by high velocity but usually small size. Available published literature shows that the efficiency of penetration of human skin by a fragment depends on a fragment's area to weight ratio, and its velocity. Increased area requires higher velocity. For small fragments weighing ~15 g the corresponding velocity for 50% probability of skin penetration is 30 m/s (100 ft/s). Heavier fragments require much higher velocities. However, fragments weighing a kilogram or more ($\geqslant$ 2 lb) cause serious injuries at velocities as low as 3 m/s (10 ft/s). This would include all secondary fragments from equipment and building.

The tertiary effects of an explosion can also lead to serious injuries or even death. A structure may be designed to withstand high-shock loading, but any opening in the structure will give rise to leakage pressures within, which although much lower than the exterior loads, may still be sufficient to cause large scale destruction of the interior walls, ceilings and other types of supports, which in turn can result in missile impact on human subjects. The leakage pressure can also be sufficient to cause eardrum rupture, and whole-body displacement of workers caused either by overpressure or shaking of the structure. An unrestrained person can be thrown off balance when subjected to a horizontal acceleration of 0.45 g or more. This can also result in impact against hard surfaces resulting in serious injuries. The injury caused by window glass breakage is also a serious concern, as the overpressure required to break window glass can be less than 0.5 psi. Under certain atmospheric conditions, the air shock from a blast can be highly asymmetrical over large distance with its intensity being many times greater in one direction than the other at the same radial distance.

2.8.4 Blast Protection

The best protection against the effects of an accidental explosion is increased distance between the explosion site and the potential targets. This is the basis of the Table of Distances, which are protocols enforced by most countries to isolate potential explosion sites from road sites, public places, harbours, populated areas, etc. These 'Tables' have been generated after decades of analysis of planned experiments and accidental explosions, including damage data from the two world wars. They provide an extremely high degree of protection to property and people, according to specified categories and weight of explosive involved. The categories are defined by the degree of protection sought (e.g. schools on one hand, and plants within an explosive manufacturing operation on the other, would require very different separation distances from the same potential explosion site). There are differences among the different 'national' Tables of Distances but they pertain only to details, the respective distances are more or less the same for similar situations.

As expected, the above distances are very large, and cannot be accommodated in routine manufacturing or storage requirements. This mandates additional protective measures around sensitive areas. The various protective measures are designed to provide one of the following:

(1)   complete containment of internal explosions;

(2)   complete protection of explosives sites, people and machinery from an external explosion; and

(3)   reduction of blast effects.

Both steel and reinforced concrete structures can be used to provide complete containment. However, for steel chambers it is practical for containing the explosion effects of only a small amount of explosives (typically less than 2 lb). In steel chambers backed with regularly reinforced concrete the containment limit can be increased to around 30 lb. For specially reinforced concrete the practical limit is set by the design limit of $W/V < 0.15$ lb/ft^3, where W is the weight of explosive and V the internal volume of the containment structure. The above design limit shows that it is not feasible to design a containment structure which can contain the explosion effects of very large amounts of explosives, even with specially reinforced concrete structures. An example of the latter with laced steel reinforcement designed to withstand high impulse load is shown in Figure 2.14. Beyond a certain explosive limit these become 'suppressive' rather than 'containment' structures.

There are several types of suppressive structure designs. These range from specially designed 'suppressive shields' to simple barriers and barricades; the exact choice depends on the degree of suppression required. As the name suggests, these designs do not provide complete containment of a blast, but serve to reduce the effect of an internal explosion on the outside of the structure. In the suppressive shield design, used to attenuate internal explosions, the wall members are composed of sandwich panels of perforated metal sheets around a framework of properly spaced structural angle iron members. The role of the former is to attenuate the air shock, whereas the juxtaposed angle iron members help arrest the fragments. These can be prefabricated in appropriate sizes and assembled on site.

Barriers and barricades provide a high degree of safety against the primary and secondary blast effects from donor explosion sites. They also prevent sympathetic initiation among various explosives sites. Sand-filled wooden barricades, earth fills and barricades, and appropriately designed reinforced concrete walls belong to this category. At some distance from the explosion site a properly barricaded building can provide complete protection against both missiles and air shock from an external explosion. The region immediately behind a vertical barricade or reinforced concrete wall also can provide shock attenuation by a factor of five or more compared to the direct shock.

The openings in a structure add further complication to the development of both external and internal blast loads. For an internal explosion, the openings reduce the blast load on the walls, but also in this process focus the overpressures in the direction of openings. This has to be taken into account in locating structures and personnel exterior to the building. With reduced blast loads on the walls, the building can be designed more economically. However, the openings have to be substantially large for the pressure to vent. Often lightweight walls serve as potential vent 'blow-out' panels, but there are strict guidelines on their minimum size and maximum weight for them to retain their 'frangibility'. These panels must yield within the time-frame of the explosion to provide effective venting and reduction of blast loads. The calculation of blast loads described in section 2.8.1 must take into account the effects of venting.

Provision of venting or creation of new openings due to partial collapse of roofs, walls or doorways, due to an external explosion, also gives rise to a pressure build-up within the receptor structure. These leakage pressures can have significant hazard potential, and the design of the receptor structure and its interior layout and stationing of personnel must be considered with care. It has been shown that window glass breakage can be extensive at a much greater distance than anticipated as a result of an explosion. The usual practice in such cases is to replace the glass by non-brittle plastic panels or provide additional reinforcement to the glass panes.

2.9   Conclusion

Although the energy content in the shock wave constitutes only a fraction of the total energy released in an explosion, its interaction with biological and non-biological targets represents the major hazard. When the explosive is contained in a metal casing or container or the explosion occurs within a structure, the resulting fragments also pose a serious hazard to people and property. The magnitude of shock and its temporal characteristics are controlled by the detonation behaviour of the explosive, and are further modified by geometry, height of burst, and both terrain and atmospheric conditions. Calculation of blast loading on structures requires accurate characterization of the magnitude and decay properties of the shock. However, the conditions of self-similarity of shock properties is not always satisfied, especially for non-ideal explosives and for explosives involving inhomogeneous mixtures of reactants as in dust and vapour cloud explosives. In such cases the principles of scaling and TNT equivalency cannot be applied with the same rigour. The effect of post-detonation reaction and after-burning behind the C–J plane and of the extended fireball from such non-ideal detonations is poorly understood at present, therefore it is difficult to assess their contribution to the damage potential of an explosion event. More work is necessary to validate the principles of scaling and equivalency for such commercial explosives through measurement of shock parameters in the near-field, and especially with non-spherical geometries.

The body of data currently available on fragment characteristics (size, velocity and distribution) is largely statistical in nature and relates only to very simple geometries (spheres and cylinders) and military explosives. No comparable data is available for commercial explosives and for realistic geometries of containers and equipment associated with explosives manufacture, transport and storage. The existing database would also have to be extended to metals other than mild steel. However, current data, along with Gurney Energy values, does provide an excellent base for quantification of damage potential from primary fragments. Generation of secondary fragments and their characteristics and attempts to relate them to the explosive source are understandably more difficult areas of study and would require a great deal more effort than the primary fragment field. In contrast, the biological response to both shock and fragments is well understood, and the available information comprehensive.

The same applies to tertiary hazards such as response to vibrations and whole-body displacements. Adequate protective measures against such hazards can be implemented easily. The hazard from flying glass from broken window panes is more difficult to quantify partly due to wide variation in strength properties, which is characteristic of glass, and the unpredictable nature of shock wave intensity at large distances from the explosion site.

The design of protective structures to withstand blast loads is well-advanced, and once the characteristics of the blast load have been determined, appropriate blast-resistant structures can be designed with a high degree of confidence. There is a wide choice of available designs varying from containment and suppressive types to simple barricades. The choice will depend on cost and the degree of protection sought. Design of containment or suppressive structures against internal explosives still remains somewhat problematic as most practical structures must incorporate openings and venting. The characteristics of the vent panels required to be truly effective in an explosion event with very fast rising shocks, rarely match those of commonly available materials normally employed in vent panels. The effect of such deviations on the internal pressure build-up on the walls and the roof of a cubicle or other type of structure is not very well known. The same applies to leakage pressure build-up both inside and outside cubicles, barricades and other types of structure. Despite these limitations and the need for much additional research, the field of explosive containment must be considered a mature technology, and it is possible to provide a high degree of safety against the effects of accidental explosions in most cases. In all this, the pre-eminent role of employing 'preventive' measures as opposed to 'protective' measures cannot be over-emphasized.

References

BAKER, W. E., 1973, *Explosions in Air*, pp. 1–268, Austin: U. of Texas Press.

BAKER, W. E., KULESZ, J. J., WESTINE, P. S., COX, P. A. and WILBECK, J. S., 1980, A manual for prediction of blast and fragment loadings on structures, *final report DOE/TIC-11268*, Amarillo, Texas: US Department of Energy.

BRASIE, W. C. and SIMPSON, D. W., 1968, Guidelines for estimating damage from chemical explosions, *Proceedings of the Symposium on Loss Prevention in the Process Industries, 63rd National Meeting AIChE, St Louis, MO*, pp. 91–102, New York: Association of Industrial Chemical Engineers.

DOBRATZ, B. M. and CRAWFORD, P. C., 1985, *LLNL Explosives Handbook: Properties of Chemical Explosives and Explosive Simulants*, Livermore, CA: Lawrence Livermore National Laboratory, University of California.

ESPARZA, E. D., 1986, Airblast measurements and equivalency for spherical charges at small scaled distances, *Proceeding of the 22nd Department Defense Safety Seminar, 26–28 August 1986*, pp. 2029–2057, Alexandria, VA: Department of Defence, Explosives Safety Board.

GLASSTONE, S. and DOLAN, P. J., 1977, *The Effects of Nuclear Weapons*, 3rd edition, Washington, DC: US Department of Defense.

JOINT DEPARTMENTS OF THE ARMY, THE NAVY AND THE AIR FORCE, 1990, *Structures to Resist the Effects of Accidental Explosions, TM 5-1300/NAVFAC P-397/AFR 88-22*, Washington, DC: US Government Printing Office.

KINNEY, G. F., 1962, *Explosive Shocks in Air*, New York: McMillan.

MANON, J. J., 1976, The chemistry and physics of explosives – Part II, *Engineering and Mining Journal*, 60–68.

MEYERS, S. and SHANLEY, S., 1990, Industrial explosives – a brief history of their development and use, *Journal of Hazardous Materials*, **23**, 183–201.

RICHMOND, D. R., DAMON, E. G., FLETCHER, E. R., BOWEN, I. G. and WHITE, C. S., 1968, The relationship between selected blast-wave parameters and the response of mammals exposed to air blast; prevention of and protection against accidental explosion of munitions, fuels and other hazardous mixtures, *Annals New York Academy of Science*, **152**, 103–121.

STREHLOW, R. A. and BAKER, W. E., 1976, The characterization and evaluation of accidental explosions, *Progress in Energy and Combustion Science*, **2**, 27–60.

3

Detection of Hidden Explosives

JAMES M. CONNELLY, WILLIAM A. CURBY, FRANK T. FOX AND
SUSAN F. HALLOWELL

3.1   Introduction

The detection of hidden explosives has become a subject of major interest in recent years. Incidents involving the explosive sabotage of airplanes, terrorist attacks on buildings, and suicide bombers attacking crowds of people or busses, have been making headlines with far more regularity than anyone but the terrorists would like. In the United States, the Federal Aviation Administration (FAA) has the lead responsibility in developing technologies, procedures, and training to detect explosives. There are agencies with similar responsibilities in most countries. Naturally, the primary concern is the detection of explosives being secreted on to aircraft. Many of the technologies being developed for this purpose, however, also have application to other areas where one needs to protect against explosive attacks. X-ray systems are used to protect court room entrances and trace based systems are used for detecting persons smuggling explosives across borders or other control points.

There are two broad categories of technological development: bulk and trace. Bulk technologies attempt to detect the main mass or 'bulk' explosive inside the luggage or other concealing container. Trace detection technologies attempt to detect minute explosive quantities present on the outside of the container due either to contamination or vapors emanating from the explosive. The following section discusses some of the current technologies and issues in each of these two areas. The focus is on the US.

3.2   Bulk Detection of Explosives

Bulk detection deals with the detection of the main charge (or bulk explosives) of the improvised explosive device (IED). The general technique is to direct a form of radiation at the object under inspection, detect the resulting radiation emanating from the object and determine whether that received signal carries the signature of an explosive. The most common example of this technique is the basic X-ray inspection system seen at most airport security check points.

The first wide scale deployment of bulk detection technologies occurred in the early 1970s and consisted of X-ray systems and metal detection systems. These systems were installed in an attempt to prevent the major security threat of that time, armed hijacking. The X-ray systems were used to inspect carry-on luggage for concealed weapons. Because of the high attenuation of weapons and the low clutter of the carry-on bags, the systems were quite effective in exposing weapons and reducing the incidence of hijacking. Searching for explosives presented a more difficult challenge.

Explosives attacks on commercial aircraft date back to 1949, when a woman paid two ex-convicts to kill her husband by placing a bomb on a Philippine Airlines flight. Since 1970, over 58 terrorist bombing attacks have been carried out against commercial aircraft, and since 1985 three wide body aircraft were completely destroyed. Of particular note, was the destruction of Pan Am Flight 103 by a small bomb concealed inside a radio in a checked suitcase. All 259 persons aboard the Boeing 747 aircraft were killed as well as 11 residents of the town of Lockerbie, Scotland. This incident brought attention to the fact that the basic X-ray system was inadequate for detecting explosives. The X-ray attenuation of explosives is similar to that of other items in the baggage. Furthermore, the system was now applied to the larger and more densely packed checked baggage. This low signal strength combined with the higher level of clutter made the job of detecting explosives very difficult. New systems needed to be developed.

In the US, the FAA significantly elevated the priority of security research in 1990. They established the Aviation Security Research and Development Service and significantly increased the funding of technologies with potential for explosives detection. Since the mid 1970s, the FAA had been sponsoring development of a Thermal Neutron Analysis (TNA) based detection system (Buchsbaum et al., 1991). Unfortunately, original design specification were inadequate to deal with a bomb the size of that estimated for the Pan Am 103 disaster. The FAA considered a redesign, but eventually abandoned TNA in favor of other technologies.

The efforts by the FAA, other governments and some independent sources brought about a number of explosives detection devices which are currently being installed at airports around the world. The following sections address these technologies, as well as some currently under development, discussing their basic function and their relative strengths.

3.2.1 Basic X-ray System

The original systems installed at the airports in the 1970s were basic transmission X-ray systems. The system consisted of a standard X-ray tube that emitted X-rays in a range of energies based on the voltage applied across the tube. The X-rays are collimated into a fan beam before passing through the luggage. On the opposite side of the luggage, a linear set of detectors reads the X-rays that made it through the bag. The bag is moved perpendicular to the direction of the line of detectors by a conveyor belt and a 2-D image is collected one line at a time.

The intensity detected at each detector is determined by the attenuation of the objects lying in the ray path subtended by that detector. The number of X-ray photons exiting a substance, I_{out}, is related to the number of photons rays entering the substance, I_{in},

$$I \frac{I_{out}}{I_\epsilon} = e^{t\mu}$$

where t is the thickness of the substance and μ is the linear attenuation coefficient (Knoll, 1989). When N substances overlap the path, the attenuation is a multiplication of the individual attenuations such that

$$I \frac{I_{out}}{I_\epsilon} = \prod_{i0}^{N1} e^{t_i \mu_i}$$

where t_i and μ_i are the thickness and linear attenuation coefficient of the ith substance respectively. Because overlapping objects result in the product of the attenuation of both objects rather than one object simply obscuring the other, the image differs significantly from an optical image of the suitcase. Everything in the bag is represented in the image.

This all-inclusive nature of a transmission X-ray is both an advantage and a disadvantage. On the one hand, the signal for each item in the bag is captured. On the other hand, there is so much clutter from the innocuous items in the bag, that the presence of an explosive threat is difficult to detect. This is especially true as one progresses from thin carry-on items such as a briefcase to very thick, densely packed suitcases and boxes that represent checked luggage as can be seen in Figure 3.1. The dark mass in the upper left is a TNT simulant concealed in a boot. The dark mass in the lower right is a large bottle of Pepto-Bismol.

3.2.2 Energy X-ray

The first evolution of the X-ray system occurred in the late 1980s with the introduction of the dual energy X-ray system. Dual energy systems exploit the fact that μ is a function of X-ray energy. By looking at the ratio of an object's attenuation at two different energies, one can estimate the μ which in turn indicates the object's effective atomic number or 'Z'. Low 'Z' materials absorb low energy X-rays well, but not high energy X-rays. High 'Z' materials absorb both high and low energy X-rays (Kolla, 1995). Where a thick stack of paper might show up the same as a thin sheet of steel on the basic system, a dual energy system could tell them apart.

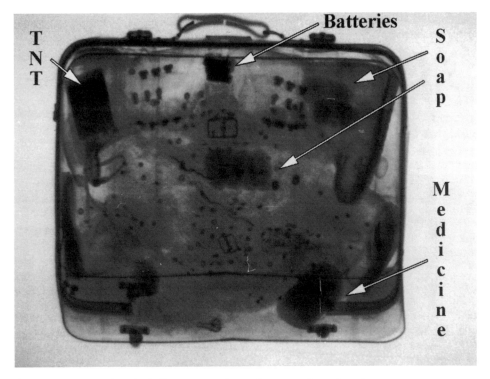

Figure 3.1   Output from basic X-ray system (normally gray scale as shown).

The earliest dual energy systems accomplished this by measuring the transmitted X-rays with two sets of detectors. One array receives all the X-rays. The other array is filtered such that only the higher energy X-rays are detected. The high energy value is subtracted from the unfiltered value to yield a low energy value. The high energy to low energy ratio is compared against a look-up table to classify. As a result, the corresponding pixel is classified as metallic (high Z), organic (low Z), or transitional. The pixel is then assigned a color based on its class and an intensity based on the intensity of the X-rays received. Consequently, that thick stack of paper might show up as a dark orange while the sheet of steel might show up as a dark blue.

While this was a major improvement in the detection of weapons, it was not much help for explosives. Since most baggage contents are organic, the metallic weapons stand out even better. The organic explosives, however, are still hidden by a sea of organic background clutter. Figure 3.2 demonstrates this with the dual energy image of the same bag from Figure 3.1. The majority of this image would show up as a light orange with the TNT, the soaps and the Pepto-Bismol showing up a darker orange. The batteries show up as green, as do some of the TNT pixels, indicating that the lower energy X-rays did not penetrate well. The metal fixtures on the bag (the zipper, wheels, etc.) show up blue.

3.2.3 Automated X-ray System

Automated X-rays systems exploit the concept of dual energy X-rays and the fact that there is far more information received at the detectors than can be displayed on an image monitor

Figure 3.2 Output from dual energy X-ray system (normally three colour).

and appreciated by the human eye. These automated systems process this information, deciding without human intervention whether the bag is safe. If the system cannot declare the bag safe on its own, it alerts its operator, and provides the operator specific information about why the bag is a concern. The operator then examines the information and the X-ray image to decide whether further investigation is necessary. Automated X-ray systems are being installed in airports throughout Europe. The Vivid Rapid Detection System is an example of an automated X-ray system (Eilbert and Krug, 1992).

One of the major shortcomings of the dual energy technique is that it only divides the image pixels into three classes: organic, inorganic (or metallic), and transitional. This means that most of the pixels in the image cannot be discriminated from explosives. This shortcoming is partially attributable to the uncontrolled nature of the X-ray source. X-ray output varies as the tube temperature or applied voltage changes. These variations result in errors in the measurements that reduce the precision of the 'Z' estimate. Vivid Technologies, Inc. (Manufacturer of the Vivid Rapid Detection System) utilized a highly controlled source to correct this problem. Additionally, rather than use two sets of detectors to achieve measurement at two energies, the source voltage is varied from a high to a low setting as each line is collected. This in conjunction with a series of shuttles that move reference materials into and out of the beam as a real time check of beam stability allows their Rapid Detection System (RDS) to make much more precise estimates of effective 'Z' of the materials along the path subtended by each detector.

In addition to these improvements in system architecture, Vivid developed very effective algorithms to identify objects in the image and determine their 'Z' by estimating their contribution to the reading at each detector. As discussed above, an X-ray transmission attenuation image is the multiplication of the attenuation of each object in the image. The result is as if the contents of the bag were compressed down into the plane of the belt. Consequently, it is difficult to identify where objects stop and start. The RDS employs algorithms that look at a pixel and its neighbors and attempt to determine whether the measurements at the pixel are due to the attenuation of one material or of several overlapping materials. In the process, the system estimates a 'Z', density, and mass for each object. The objects' characteristics are compared against those of explosives and close matches are flagged as potential threats. The system also looks for indication that extremely dense materials that could shield an explosive from detection are present. The image presented to the operator is primarily gray level based on the attenuation for each pixel. Pixels that exhibit characteristics consistent with explosives are painted a brownish orange. Pixels that contain shielding materials are painted yellow. Objects judged suspicious are painted red and an alarm is sounded. If significant shielding is present the system alarms as well, identifying the object in yellow. The operator then examines the images, concentrating on the alarmed objects, and determines whether further inspection is necessary.

Figure 3.3 shows the same bag from Figures 3.1 and 3.2 when run through a Vivid RDS. For this figure, white is substituted for red, and grey with a black outline is used in place of the brownish orange. The system alarms and indicates that the TNT is the source. It also finds pixels of interest in several of the bars of soap and the Pepto Bismol, but does not consider them a threat.

The major advantage of the automated X-ray system is that the operator is no longer expected to examine all bags. Bags that do not alarm the system are forwarded without the operator's intervention. Bags that do alarm require action by the operator, but the operator can concentrate on specific, troublesome areas of the bag. This changes the emphasis of the job from pure vigilance to problem solving. The second advantage of the system is speed. The Vivid system advertises a through put of 1200 bags per hour. A limitation of the automated X-ray systems is the loss of spatial information along the beam path. The nature of transmission X-ray collapses three dimensions into two. As bags become thicker with more objects, the collected image becomes more complicated and precise measurement of the attenuation of a particular object is restricted.

Figure 3.3 Output of a Vivid DS automated X-ray system (normally gray scale with red and orange highlights).

3.2.4 Computed Tomography

Computed tomography (CT), widely used in the medical field, has been adapted to explosives detection. In computed tomography, X-ray transmission information is collected at multiple angles around the item being inspected within a particular plane (usually perpendicular to the plane of the conveyor belt and the direction of the belt motion). This information is then used to create a 'slice plane' image. The image represents the X-ray attenuation of the objects in the collection plane and appears as if the luggage had been sliced by a guillotine. The value of each image pixel is a function of both the density and the 'Z' of the object just as in the basic X-ray system; however, in this case the value is unaffected by neighboring objects. (Ideally, the voxel is not affected by its neighbors. In practice, very dense, high 'Z' materials can cause artifacts that affect other voxels.) Therefore, a more precise measurement of a given object's attenuation can be obtained. The InVision ®CTX 5000 SP uses this technique in detecting explosives.

The CTX 5000 SP is actually a combination of two major systems. The first system is a dual energy transmission X-ray. The image collected by this system is processed to determine what regions of the bag could contain a significant mass of explosives. (Regions with only light clothing could not conceal an explosive; however, a region containing a denser object such as a bar of soap would have to be checked further.) These regions are designated for slice collection. Further, certain explosives can be detonated in extremely thin, or sheet configurations. These configurations might not show up on the transmission X-ray image, so additional slices are requested to prevent any significant area from being overlooked. The bag is then moved into the second system, the CT unit. The CT unit has a rotating gantry with an X-ray source mounted on one side and a detector array mounted on the other. The bag is positioned so the plane of interest is aligned with the system's slice plane. As the X-ray and

detector array rotate around the bag, information is collected from adequate angles to reconstruct the plane. The bag is then moved to the next plane of interest. Once reconstructed, the
voxel values of objects in the slice plane are compared against a range of explosives. Additionally, the system looks for potential shielding and informs the operator. Suspicious objects
are flagged and the operator is alerted.

The system provides its information to the operator using two screens. The first screen is
used to display the transmission image. Any objects flagged as alarms are highlighted using a
red or yellow box. This screen also indicates where the system collected slice information. The
second display shows the slice images. The operator is able to look at the series of slices and
correlate them to their location in the transmission image. On this screen, alarmed objects are
painted red. The operator is also provided information about the suspected class of explosives
and the mass. Figures 3.4 and 3.5 show the example bag (inverted in the scan image) when
run through a CTX 5000 SP. In Figure 3.4, the vertical lines indicate where CT slices were
taken along the bag's axis and the shortened set of lines indicates the current slice. White
lines (red on the display) indicate that the slice intersects a suspicious object. The TNT is
highlighted in a white box (red on the display) indicating that it is the first suspect substance
found. The Pepto Bismol is also highlighted by a gray box (yellow on the display) indicating
that it is also suspect.

Figure 3.5 shows the slices. The large slice displayed in the middle is the current slice. The
small slices along the bottom are the neighboring slices to the left of the current slice, while
the slices along the right side of the screen are the slices to the right of the current slice.
Notice the oblong shape in the large slice image. This is the 'sliced' image of a boot top
(barely noticeable in the upper left portion of Figure 3.4). The corresponding boot sole may
be seen in the neighboring slice to the left. The TNT is both boxed and highlighted in white

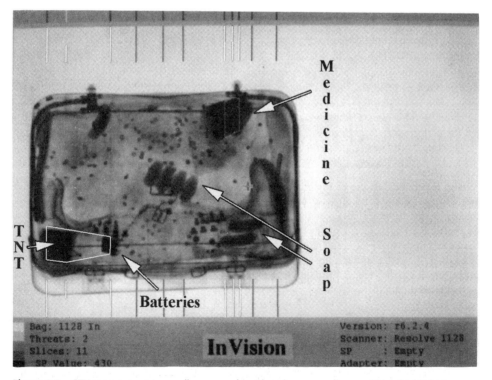

Figure 3.4 CTX scan output (normally gray scale with red, green, yellow and blue highlights).

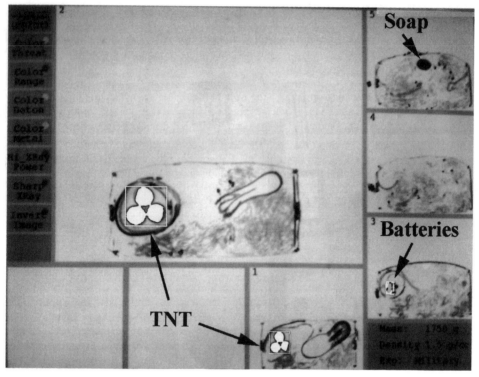

Figure 3.5 CTX slice images (normally gray scale with red, green, yellow and blue highlights).

(red on the display). The batteries are boxed in white, but highlighted in grey (yellow on the display) indicating that they are very dense and are blocking most of the system's X-rays.

The major strength of the CT based system is its ability to discriminate between innocuous items and real explosives. The CTX 5000, predecessor of the CTX 5000 SP, is the only system FAA certified to date. To achieve this the system had to exceed a minimum detection rate while not exceeding a maximum false alarm rate for a wide variety of explosives as specified in the FAA's 'Criteria for certification of explosives detection systems' (1993). Based on the performance of its predecessor, the CTX 5000 SP has been installed at several airports in Europe, the United States, and the Middle East. Additional installations are expected in Asia. As of this writing, the company was successful in submitting the CTX 5000 SP for certification.

The major limitations of the current CT system are its complexity and its throughput. While the CTX 5000 and 5000 SP have been remarkably reliable to date, the large amount of moving parts require significant maintenance measures not needed for other technologies. The advertised throughput of the system is 300 bags per hour. This is considerably slower than the automated X-rays and consequently would require more machines for the same load of bags. Both these issues are the subject of current development efforts. These efforts hope to produce CT based systems with simpler manufacture and maintenance requirements as well as higher throughput.

3.3 Trace Explosives Detection

The preparation of the type of bomb which involves the secretion of explosives into the cavities of a portable item such as a suitcase, laptop computer, radio 'boom-box' or telephone very frequently leads to inadvertent contamination of the outer surfaces. This contamination can result from the activity of the original bomb maker, or intermediaries in the process of delivery to the final destination. Consideration given by aviation security has been focused to a large degree on detection of hidden explosive devices in portable items, as these are most easily placed onto the aircraft or other location. In addition to surface contamination, explosives vapors can be detected. Many explosives have sufficient volatility inherent to the primary explosive or a component of the mixture comprising the explosive to allow detection. This process is enhanced by the use of taggants (Danylewych-May and Cumming, 1993). (Taggants will be discussed in section 3.3.4.)

The earliest trace explosives detection system is the dog. Dogs have been used in bomb detection since the early 1970s. More recently, a variety of instruments based on laboratory analysis techniques have been developed. The next sections examine the trace detection problem and discuss the dog and the instrumental approaches as well as some current issues in the testing and training for each.

3.3.1 The Trace Detection Problem

Explosive chemicals can arrive at the sampling section of the detection system *via* one of two routes:

- captured directly from the air;
- dislodged from a surface.

Whether airborne or on a surface, the explosive chemicals will be in one of five distinct forms:

(1) vapor, as separate explosives molecules in free space form;
(2) an aerosol containing explosives molecules dissolved in, or forming small fluid droplets;
(3) microparticles containing molecules of the explosives chemicals attached to small inert particles in the air;
(4) microfragments which are crystal fragments and crystals of the actual explosives chemicals;
(5) clusters of microfragments and/or microparticles.

Capture and release of explosives chemicals

The properties of explosive chemicals determine how best to search for and to capture them. Regardless of the technique used, it must be designed to capture as close to all of the available explosive chemicals as possible. Also, the technique must allow for the release into the chemical processing section of the detection system, upon demand, as close to all of the captured explosive chemicals as possible.

Military plastic explosives tend to have very low vapor pressures, e.g. at $300°K$: RDX (cyclonite) 6.4×10^{-9} torr, PETN (pentaerythritol trinitrate) 2.0×10^{-5} torr, TNT (2,4,6, trinitrotoluene) 9.6×10^{-6} torr. Almost all of the molecules of these explosives do not have enough energy to move beyond the first boundary layer of the explosive's surface before they fall back. A very few have enough energy to move away and enter free space above the surface, thus becoming a vapor entity. Air in the land environments in which we live contains in the order of 50 000 inert microparticles (10^{-7} m or smaller in diameter) per cubic centimeter of air. As this air passes into the first boundary layer of these explosives, molecules of the explosives attach to the inert particles rather than falling back into the masses of the explosives. If these microparticles containing the explosives molecules are captured and heated, the molecules of the explosives will be driven off and can be treated as a true vapor by the chemical processing section of the explosives detection system. All particles of less than 10^{-6} m in diameter will not settle out of the air rapidly, but rather, will be carried in the air until they strike a surface. Microfragments and clusters of explosives are always released into the air when plastic explosives are manipulated. This is illustrated in Figure 3.6 for the plastic explosive Semtex H.

Figure 3.6 is a photomicrograph of a cluster of explosives contained in a sample of Semtex H which has been frozen and fractured. The crystals of explosive which are normally coated by polymer in the commercial product, are clearly seen. The rectangular crystal at the lower left of the cluster and the one behind it are crystals of PETN. The rest of the crystals are forms of RDX. The crystals are resting on the polymer (cellulose nitrate) that enclosed them to make up the final form of the high explosive. Note that the crystals are held, but have apparently lessened adhesion to the binding polymer. If a piece of commercial Semtex H is roughly manipulated, crystals of PETN and RDX can be similarly freed or loosened. In other words the polymer coat is partially removed, although perhaps not to the degree caused by freeze fracture. In the preparation of the 'controlled deposition' standards this effect is minimized by the conditions used.

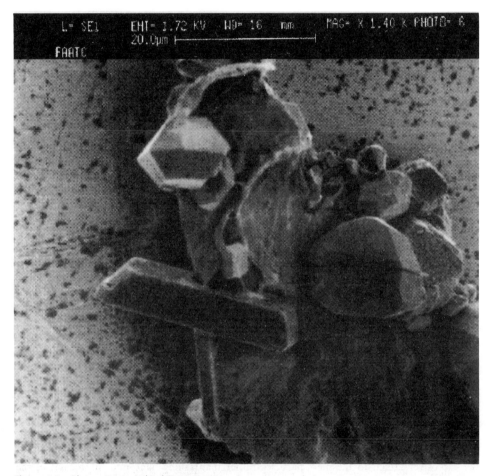

Figure 3.6   Photomicrograph of Semtex H.

Released fragments are caught on the skin and clothing and items surrounding the explosives manipulation site. Anyone making an IED will have these microforms and clusters on them and around the IED. It is almost impossible to package the IED without having airborne forms of the explosive remain on the outer surfaces of the package. Anyone coming close to the bomb maker will also become contaminated with the explosive which has been manipulated. The low vapor pressure explosives remain around for extended periods, and are hard to wash away. For this reason sampling techniques which dislodge aggressively and then capture one or more of the five forms of the explosive will have the best chance of processing, detecting and sending the alarm that the explosive has been found in the sample. This may happen only if the explosive can be desorbed completely into the chemical processing section of the system. Traces of the explosive may be found in any of the aforesaid forms. We apply the title of 'Trace Detection' to the whole subject of detecting the existence of IEDs by chemical analysis methods.

Capture devices, passive and active

Samples of explosive chemicals must be accumulated and held on some form of collection apparatus which mates with the chemical processing section of the Trace Detection system. Passive collection devices must stop and hold vapors, particles and aerosols. The most recent high efficiency passive collection devices have been filter/selective adsorber combinations. These complex desorbable filter units are designed to release explosives vapors into the chemical analyzer section of the Trace Detector by increasing (ramping up) the heat of the filter device in a pre-established range, and at a controlled heating rate. The maximum heat must be high enough to vaporize all of the explosive, but lower than the heat necessary fragment the molecules of the explosive.

Samples of explosive can be accumulated actively when the forms of the explosive are actually airborne. In this case it is appropriate to use electrostatic precipitators, or other capture devices that harvest the explosives from large volumes of air and deposit all of the harvest into a small volume of liquid or dry aggregate. These concentrators are active, in that they need an input of extra applied energy to operate efficiently. Preconcentrators must not cause the explosives to be decomposed, or altered as a result of the preconcentration process. Preconcentrators are necessary when the sensitivity of the Trace Detector system is lower than that needed for aviation security and explosives detection work.

Marking chemicals

If manufactures of explosives throughout the world will all put an easily detected marker chemical into all of the explosives they produce, and if the marker chemical is one that is inexpensive, has a long enough detection half life, and does not lower the effectiveness or detonation or deflagration property of the explosive, it will be much easier to detect terrorist IEDs which contain commercially produced explosives. This was the thinking of several nations faced with countering terrorist threats. Several countries have made this a reality. Of the four marker chemicals considered, the United States, for one, has elected to use DMNB (2,3 dimethyl 2,3 dinitrobutane). This chemical is presently being produced in quantity to fill the needs of US manufacturers of explosives and ammonium nitrate fertilizer. How effective this strategy for countering terrorist bombings will be remains to be learned.

Interferents and obscurants (masking agents)

Trace Detectors employ different analytical technologies to determine the presence of explosive molecules. The ideal detector section of a Trace Detection system should never make false negative decisions when an explosive is masked by the presence of another chemical in the same sample. Likewise, the systems should not be fooled into a false positive situation by identifying a nonexplosive chemical as an explosive. Trace Detection systems are evaluated by the FAA to help manufactures produce more effective detector models.

Other materials which may have to be analyzed

(1) *Liquid and gas explosives* Stabilized forms of some liquid explosives have made their appearance in letter bombs and in containers sent through the mail or carried in hand luggage. These liquids have a fairly high vapor pressure. The containers carrying these explosives can be sealed and washed so that no trace of the explosive remains outside the container. The mechanism to set off the explosive can be placed alongside the main liquid charge after the bomb has passed through the security screen. The FAA has developed Bulk Detection systems to counter these threats. Devices that release explosive gases have configurations which alert the security screeners that they are present. These threats are countered by applying Discovery System techniques rather than Detector System techniques

(2) *Incendiaries and toxic gases harassment agents* Smokes, flames, tear gases and toxic chemicals produced or released into in an aircraft full of passengers when it is flying at 35 000 ft above the ground can be devastating. Lives may, or may not be lost. The aircraft may or may not be lost. However, terrorists have made a statement that will be immediately newsworthy throughout the world. Some of these threats can be easily detected with presently available Trace Detection systems, some with Bulk Detection systems, and some require the use of Discovery systems.

3.3.2 Dogs in Explosives Detection

The efficacy of using dogs' olfactory capabilities for detection of bombs and explosives and narcotics was demonstrated by the US Air Force in 1971 (Lemish, 1996). Since then, dogs have become an important component (and in many cases, the only component) of explosives detection systems used for both military and civilian agencies. Despite rapid technological advances in the area of explosives detection, the dog and its handler remain the most widely used detection system (*Technology against Terrorism: the Federal Effort*, 1991). The FAA has had a K-9 Explosives Detection Team Program for 24 years. Dogs seem to be the popular choice for explosives detection by the US military and multiple federal and local law enforcement agencies, throughout the US.

Dogs play a prominent role in the security of numerous foreign countries, and are extensively used in Canada, the UK and the Republic of Germany, to mention only a few. The popularity of dogs seems to arise from the fact that they have been a primary part of explosives detection for over a quarter of a century, and were in use long before the development of the current generation of trace explosives detection equipment. Additionally, the dogs can serve two security functions by being dually trained as both guard and explosives detection dogs.

Canines are equal or superior to machine technology in the areas of sensitivity and selectivity, collection and transfer efficiency, mobility, user friendliness, processing time, and going to source operations. These detection characteristics of canines have not been systematically assessed, and little has been done in the way of scientific studies directly comparing trace detection technology to K-9 detection for explosives. This lack of a scientific body of knowledge, along with the stigma of using a 'low' technology has resulted in some hesitance to expand the use of dogs to different operational scenarios. Despite this sense of conflict between the two communities of K-9 users and non-users, both instruments and dogs will play important parts in present and future explosives detection efforts, and will continue to be used both alone and in combination.

Dogs v. instruments

There are a number of important differences between these two detection methodologies that will dictate when, where, and how they are utilized. There are clearly some operational advantages in using dogs. During an alert situation, such as a telephonic threat a dog team can be readily deployed to a site. Once at the site, the dog can perform a search pattern that allows him to search a large area, and, if a scent of an explosive substance is noted, the canine can follow the concentration gradient, and go to the source of the scent. The dogs are taught to give a passive response, and will sit when they detect explosives odorant.

Machines come in various sizes, some transportable, some not. High quality of data is typically produced by machines that are not highly portable. Many of the detection devices force the operator to make a choice between sampling for volatile explosives or nonvolatile explosives, which is an artificial construct that the dog is not forced to make in order to analyze for a substance. Trace detection equipment is typically programmed *a priori* for the explosives of interest. Most equipment is designed to indicate which explosives have been detected, so the operator will know, for instance, that he has a hit for a plastic explosive, rather than for nitroglycerine. A dog cannot indicate which odorant he has hit on. K-9s are typically trained on up to a dozen different odors from military or commercial high or low explosives. So if a dog hits on an area, the threat could be smokeless powder, military explosives, or any other number of things that the dog has been trained to detect. Arguably, the claim can be made that canines do have a distinct advantage over machines since they can generalize odors.

The use of canines requires a serious programmatic commitment, well beyond that required to purchase an explosives detector. The dog and handler must undergo rigorous training. The FAA uses the 341st Training Squadron at Lackland AFB, Texas to acquire and train dogs (Cormier et al., 1995). The dogs are bought on consignment and trained to smell one of two spice fragrances. Dogs are chosen based upon stamina and disposition, along with other attributes, and are typically either German Shepherd or Belgium Malinois, although the breed is not specified for this program. The dog's training lasts an average of 40 days whereas the handler's training lasts for 12 weeks. After this initial training, the dog must be trained on explosives at least once a week. Dog handlers insist on training their dogs with typically encountered quantities of explosives. This means that any facility using dogs must have a bunker to store different bulk quantities of explosives, nominally about half pound quantities of each explosive required by the trainer.

This is in contrast to ownership of a commercial trace explosives detection device. Although the machine must be calibrated prior to use, it can sit dormant for months at a time without degradation of performance. Additionally, highly specialized handling skills are not required to operate it, and calibration can be readily performed with a dilute solutions of explosive materials rather than hazardous, regulated bulk quantities of authentic explosives.

Finally, dogs have a discrete work cycle, and performance will degrade if the dog loses interest in the search. Some of the differences between using canines and instrumental detectors for detection of explosives have been summarized in Table 3.1.

Dog training issues

There are also many apparently subtle nuances to conducting an effective canine program. For instance, the claim is repeatedly made that dogs must be trained on everything from 'trace' amounts of explosives to 'gross' amounts; i.e. if trained on one concentration, the dog will 'miss' other concentrations. Cross contamination issues are also critically important.

Table 3.1 Subjective comparisons – dogs *vs.* instruments: factors in explosives detection

Factor	Canines	Instruments
Sensitivity	Unknown	Characterized
Training	Specialized	Basic
Transportable	Self portable	Some portable, some not
Selectivity	Classes of compounds	Chemicals selected for analysis
Processing time	Instantaneous (ms)	1 s to 2 min: typical 10 s
Sampling	Particle or vapor Extremely efficient	Usually one or the other Inefficient
Calibration	No field methods	Well established kits
Duty cycle	Short: one or two hour shifts	24 h shift
Search	Dogs can go to source	Most acquire remote sample
Maintainability	High maintenance	Low maintenance

Type I cross contamination occurs when explosives training aides that are used with canines become contaminated with some other odorant not associated with the explosive, but associated with the aide. Commonly, if the aide is repeatedly handled by the trainer, it becomes scented by the trainer, and this becomes the primary cue for the dog. This effect can be eliminated by handling the aide with gloves, but this must be done with caution. The use of surgeon's gloves will soon imprint the dog to search for odor of latex. However, gloves used by food handlers appear to be a good choice. Additionally, most programs attempt to obtain a fresh explosive stock at least once a year on which to train their dogs in order to minimize this problem.

Type II cross contamination occurs when the explosives themselves become cross contaminated. This usually occurs when explosives are stored together in a bunker: the most volatile explosives permeates the other explosives. A typical problem is the pervasiveness of EGDN, which is a primary ingredient in many dynamites. EGDN is readily absorbed by other explosives, which can serve as a secondary emitter. It is not yet clear how significant this problem is or how much it effects the efficacy of a dog program, but clearly, appropriate segregated, sealed storage of each explosive type in a controlled fashion will minimize this problem.

Testing of dogs

From the scientific standpoint, the performance of the canine remains the least understood of available sensor systems. Few scientific studies have been conducted on canine performance. Little is known about the effect of disease, aging, nutrition, or exposure to toxic materials on olfactory abilities. Although many organizations have preferences to particular training methods or deployment procedures, this has not been independently validated. There are questions about the suitability of using canines for lengthy, continuous screen operations due to a relatively short duty cycle. Specific questions about the capabilities of the canine as a sampling and detection instrument has yet to be determined. In particular, the following questions need to be addressed:

- What substances can the dog detect?
- How sensitive are dogs to specific substances?

- How well can dogs distinguish between different concentrations?
- How well can dogs distinguish between different substances?
- Does the presence of one odor affect detection of others?
- What are the constituents of the target odors that dogs detect?
- How important is the composition of training substances with regard to its vapor, composition, including impurities, contaminants, and degradation by-products?

At the heart of any scientific study designed to evaluate canine olfaction is the ability to dose an animal with a known quantity of scent, and measure the animals sensory capabilities. Of these two tasks, the latter has proven to be much more straightforward, and is a well established scientific field (Stebbin, 1970; Gescheider, 1985; Passe and Walker, 1985). Two basic approaches have generated substantial research literature. One is based on respondent (also called 'classical' or 'Pavlovian') conditioning, and the other is based on operant conditioning.

In a joint services study funded by Office of Special Technology, the US Secret Service, and the FAA, we have chosen operant conditioning to study olfaction thresholds to explosives. This study is being conducted at the Institute for Biological Detection Systems at Auburn University (IBDS) (Williams et al., 1997a, 1997b). This research program focuses on collecting fundamental psychophysical information on canine detection capabilities using characterized explosive vapor generation techniques similar to those used to evaluate explosive detection equipment. In addition to determining olfactory functions and thresholds for various substances, one of the objectives of this research is to determine the components of explosives that canines cue on as a result of odor discrimination training.

In this paradigm, an environment is arranged in which a trained behavior will only be followed by positively reinforcing consequences when it occurs in the presence of that class of stimuli. In this study, dogs were individually trained and tested in experimental chambers large enough to permit the dog to stand and move around slightly (Figure 3.7). One end of the chamber contained an interface panel locating two levers and an aperture allowing insertion of a dog's muzzle into a 1 l scent chamber protruding on the back of the panel. A dish for delivery of food reinforcers was located at the bottom of the panel. Dogs were trained for olfactory testing using a modified backward chaining procedure. Over a period of weeks, the dogs were trained to press the left lever if they detect no odorant at the scent chamber, and the right lever if they detected an odorant. The final performance obtained by this training procedure was as follows. In the presence of a broken tone, the dog inserted its muzzle into the scent chamber; withdrew its muzzle when a continuous tone sounded; pressed the left or right lever depending on whether it detected unscented or scented air, respectively; consumed food reinforcers, if present; and waited during an inter-trial interval for the next trial to begin. Incorrect responses were followed by a 15 s blackout. Daily test sessions were usually 1 h long and were composed about 80 trials evenly divided between odor and blank trials presented in quasi-random order.

Odor stimuli were delivered by a vapor generator instrument or olfactometer which provided a quantitative vapor delivery to the scent chamber. Although these delivery systems have been used to study olfaction for over a century (Tucker, 1963; Takagi, 1989; Passe and Walker, 1985; Prah et al., 1995) significant engineering control upgrades, and rigorous analytical methodology was necessary to validate the performance of the olfactometer (Hartell et al., 1995a). The olfactometer used here contained a multistage dilution system with temperature and humidity control, the precise control of odorant concentrations via the use of mass flow controllers, the use of odorless components, the use of pure stimulus compounds, and the inclusion of design parameters to create an environment which is relatively inert. This system has been proven to provide a consistent and quantitative delivery of vapor (Hartell et al., 1995a, b). Additionally, an effective cleaning protocol was developed for minimization and/or removal of chemical adsorption.

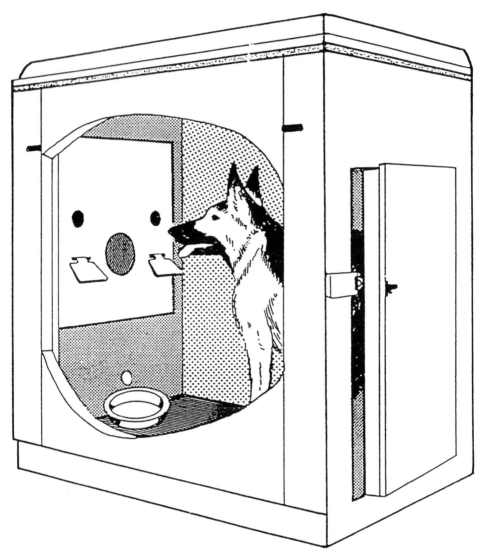

Figure 3.7 Canine testing apparatus.

This device delivered to the scent chamber for each odor trial samples of vapor from odor source material diluted by clean air supplied by a purge gas generator. On non-odor or blank trials, the device delivered clean air only. Vapor analysis were performed by gas chromatography/mass spectrometry (GC/MS) on a Hewlett-Packard GCS-1800A system. A unique feature of the Auburn olfactometer is the ability to use complex odorants. The primary objective of this described study was to provide the critical reference basis in the form of detection threshold on one of a series of commonly encountered explosives, nitroglycerin (NG) double-based smokeless powder. Using six canines, the testing protocol results in over 4700 total test trials per dog, half of which were odor trials, conducted in about 60 sessions over a period of more than 2 months.

Figure 3.8 shows the detection accuracy for smokeless powder for each dog across concentrations measured in parts per trillion. All dogs show the same general function. Using these data, the absolute threshold of detection was calculated, a statistical concept equivalent to the

Detection of NG Smokeless Powder

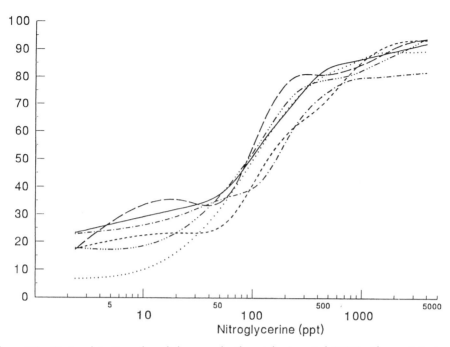

Figure 3.8 Canine detection of smokeless powder (from Johnston et al. (1995) with permission).

concentration at which each dog is accurate on 50% of odor trials. These values are indicated for each dog below the horizontal axis and range from 89 ppt to 161 ppt. It should be noted that this particular experiment does not indicate which substance(s) that dogs are actually cueing on as the recognition odorant, but presents dog response as a function of NG concentration present in the smokeless powder. However, this data does reveal characteristics that may be important in the detection of NG smokeless powder under field conditions. The between-dog variability seemed to be relatively low. All dogs detected the odor at good levels of accuracy across a range of concentrations, and conversely, false positive performance was consistently low for all dogs. It should be noted that these results describe the performance of trained dogs tested under laboratory conditions in which odor stimuli were carefully controlled and measured. This study is significant in that it has produced detection data collected with the same controls and statistics employed to evaluate explosives detectors, and to our knowledge, is the first of its kind. Work is currently being conducted on other explosive compounds to determine both detection limits and constituents of recognition in odorants.

Dogs will continue to play an active role in explosives detection programs. Likely, this role will expand to different venues. One promising approach utilizes the collection of a sample remotely using a sample tube and a pump, followed by presenting the sample tube to the dog for analysis. This mimics how many of the trace explosive detectors are used in that the sample is acquired remotely and brought to the detection device, which in this case is the canine. This approach, called MEDDS in other countries and renamed Checkmate in the US, was developed and used in South Africa. The advantage of this approach is that large containers can be screened, and analysis can be 'batched' such that several sample tubes from different locations can be cleared at once. The dog can be handled in a controlled environment, removing distracters that can occur on typical searches. This technique has not been validated yet with respect to sampling efficiency to various explosives.

As more research is done validating dog explosives detection performance, we will better appreciate the remarkable nature of canine olfaction. Although not as characterized as instrumental systems, it remains a model system for efficiency and selectivity for explosives detection. It is not by accident that vendors of explosive detectors try to liken their systems to electronic noses. Arguably, however, no system will approach the capabilities of the canine nose for decades, as we are just now beginning to understand the underlying science behind the remarkable sense of olfaction, that benefits from millions of years of evolution.

3.3.3 Instrument Based Trace Detection

Over the past approximately 10 years, a variety of commercial instruments have become commercially available for the detection of trace amounts of explosives. These devices have a spectrum of capabilities, features and prices. A comparison chart shown below in Table 3.2 lists those most commonly available. Sampling time of 5–15 s is a feature which is common to virtually all of the instruments. Total analysis time is approximately 5–20 s for most of the instruments. This speed is crucial to keep up with passenger flow in a busy airport concourse.

The detectors in most of the instruments consist of enhanced versions of classical techniques. They are robust and highly sensitive to extremely low levels of explosive. One of the reasons for the high sensitivity is the low efficiency of wipe or suction sampling to remove plastic and other particulate explosives from surfaces. Eventually a dramatic improvement in this area may lead to less sensitive detector requirements, and subsequent lower instrument cost.

All of the instruments are designed to be portable, by use of built-in wheels for the heavier instruments. Nearly all operate from a 110 VAC outlet.

In general, all explosives detection systems that identify the presence of specific chemicals contain the following four stages:

(1) sampling;
(2) concentration;
(3) analysis;
(4) discrimination.

The critical stage in a chemical identification detection system is sampling. A knowledge of the ways that the explosive chemical of interest may present itself must be known so that full advantage can be taken in designing a capture device that will present the greatest quantity of the chemical of interest to the chemical processing section. If too little of the explosive chemical is available to the analysis section, the alarm logic section will not alert the presence of the explosive chemical, and the IED will not be found. This is a false negative encounter. If the chemical processing section identifies a chemical as some explosive, and the chemical is not explosive, the alarm section will sound the alert. This is a false positive encounter. False positive encounters are disruptive to the orderly and timely processing of air travelers. False negative encounters can be fatal.

These stages are discussed in the following sections as well as a method for producing consistent trace detection testing standards.

Sampling

The first step in the detection process is the transfer to the preconcentrator of the trace analyte from the suspect object presumably containing a bomb. The analyte may be particu-

Table 3.2 Trace explosives detection devices comparison

Commercial instruments	Analysis time (s)	Operational time (s)	Specificity	Sampling method	Weight (lb)	Cost $K	Explosives detected	Taggants	Technology	Comments
Thermedics — EGIS	18	23–28	yes	wipe/vacuum		400	170 — Vap TNT, DNT, NG; Part. RDX, PETN	EGDN	Chemiluminescence	GSA, Desorb unit to replace current configuration
Barringer — Ionscan 350/400	5	10–15	yes	wipe/vacuum		59	50–80 — Vap TNT, NG, DNT; Part. PETN, RDX; TOVEX, HMX, TETRYL, NITRATES	EGDN, DNB	IMS	Dual mode drugs/explosive GSA
Scintrex — EVD-1DC	90	105	yes	pump		35	35 — Vap. EGMN, MMAN, NG, DNT	DMNB, MNT, EGDN	GC–ECD	Dual column, vapor only; Separate vapor sampler
Scintrex — EVD-3000	10	15–20	no	pump/swab		6	23 — Vap. NG, HMX, EGMN, MMAN, DNT; Part. PETN, RDX, TNT	DMNB, EGDN, MNT	Chemical electrolytic	No special gas or radioactivity vapor and particle; Integrated
Scintrex — EVD-8000	60	75	yes	pump/swab		48	58 — Vap. NG, NB, EGMN, MMAN, DNT; Part. PETN, RDX, TNT	DMNB, EGDN, MNT	GC–ECD	Separate vapor sampler
IonTrack — Itemizer	5	10–15	yes	wipe/vacuum		80	65 — RDX, PETN, NG, TNT, DNT		Ion Trap MS	Explosives and drugs; GSA, needs gas
IonTrack — Vixen Viper 97	10	20	yes	sniff		10	24 — RDX, PETN, TNT		GC/ECD	
IonTrack — Exfinder	<1	<3	no	sniff/wipe		2	4 — NG	EGDN	ECD	
CPAD — EDSS Tracker	5	10–15	yes	vacuum/wipe		>100	120 — RDX, PETN, NG, TNT, DNT		IMS/IMS	Explosives only
EDSS-GPS II	10	15–20	yes	vacuum/wipe		>100	150 — RDX, PETN, NG, TNT, DNT	EGDN, MNT, DMNB	IMS/IMS	Explosives and ICAO taggants (increases analysis time)
EDSS with metal detector							2		magnetometer	optional metal detector feature
Graseby — Plastec	3	8–13	yes	Wipe or vacuum		40	45 — RDX, PETN, TNT, NG	No	IMS	
Graseby — GVD4				sniff		10	6		Corona Discharge	
Graseby — GVD6			no	sniff		22	17 — RDX, PETN, TNT, NG		IMS	
Graseby / Lockheed Martin — PD5 / MVS-200	5–10	10–30	Yes	sniff/wipe		35	45 — RDX, PETN, TNT, dynamite, DG	some	GC/SAW	No data
Prototype Units										
Amerasia — SAW			yes	vacuum			RDX, PETN, TNT		Surface acoustic wave	Project finished, needs Front collector, priming problem
SAIC — Phase II			yes	vacuum			RDX, PETN, TNT		IMS	Project Completed
NRL — Biosensor			yes	high volume particulate collector			RDX, PETN, TNT		biosensor	Under Development

Compiled by Paul Jankowski; Trace Detection Program, Federal Aviation Administration, William J. Hughes Technical Center, Atlantic City, NJ late matter from a wipe or vapor emanating from the object. Development of efficient sampling methods has been elusive. The sampling techniques shown in Table 3.2 typically consist of the operator simply wiping the test object with a cotton glove, followed by vacuuming the glove to collect particles from it onto a collector or preconcentrator. This is often a metal or fiber filter which is subsequently heated to desorb the explosives into the analyzer. Sampling in this way typically has removal efficiencies of from less than 1% up to as much as 10% of plastic explosives contaminating the surface.

Research efforts to develop an improved sampling method have previously been limited to deposits made from solutions of explosives dissolved in organic solvents or particles of siliceous material which have been coated with explosive crystals by evaporation of a solution (Neudorfl et al., 1993; Rodacy, 1993). Unfortunately, dissolution and recrystallization of plastic explosives yields a material without the particle size or adhesion character of the original plastic. Use of actual fingerprints as a test standard also presents difficulties because of the inconsistent nature of the deposit due to differences in finger size, pressure used, nature of the bulk explosive touched, and many other factors (Cheng et al., 1995; Langford, 1995).

Concentration

Sampling may result in the collection of particulates, gases, or liquids depending on the methods used. Subsequent to the sampling step it may be necessary to concentrate the explosives for analysis.

The usual wiping of the surface to obtain a sample results in the collection of dust, skin flakes and lint. These particulates can carry attached explosives, and can be treated simply by placing into a heat desorber to drive off the volatile explosives either into a cold trap for further concentration or directly into an analyzer.

Detection of explosives in a gas or air stream is typically done either by vacuuming a surface with collection of the particulates on a filter, or by dislodging particulates with an air stream as would occur in a walk-through portal. The air stream in the portal case can have the particulate collected by a filter or by using plates or coils which have an electric charge or polymer coating. In either case the ultimate sample is particulate. The low volatility of most high brisance explosives has reduced the need for direct analysis of an air stream for vapors of explosives.

Collection by sampling into a liquid is a newly developed technique called 'Space Charged Atomizing Electrostatic Precipitation'. In this procedure, air is sampled into a glass collection tube countercurrent to a fine liquid spray. The spray is produced by a two-fluid nozzle in which high pressure air is directed at small drops of liquid exiting a stainless steel needle. The needle is connected to a DC power supply producing a high positive charge on the spray. The walls of the collection tube are grounded which attracts the charged spray and collected aerosol to the tube walls. The liquid flows by gravity to a well at the bottom of the tube and exits at a single collection point. Capture efficiencies range from 65% to 100% for micron to submicron particles and for gases.

Analysis

The overriding characteristics for trace explosives detection instrument analyzers are speed and sensitivity. During the early development of these instruments the detectors were confined largely to gas chromatography (GC) combined with electron capture (ECD) or to mass spectrometry (MS). The use of triple quadrupole MS eliminated the need for GC but resulted in large and expensive instruments which required large and noisy vacuum pumps. In the end these were not practical and today most commercial instruments use a chemiluminescence (TEA) detector combined with GC, or an ion mobility spectrometry (IMS) detector with or without GC. One instrument uses a surface acoustic wave detector combined with a GC.

At present the major manufacturers are: Thermedics, Barringer, CPAD, IonTrack, Graseby, and Scintrex. The US Aviation Security Act of 1996 (H.R.3539) specifies the deployment of commercially available explosives detection instruments in US airports. The products of the instruments manufacturers discussed here may soon be in widespread use.

Discrimination

The premier analytical technique for discrimination of unknown compounds is mass spectrometry. There are several research groups and companies actively developing small ion trap or time-of-flight mass spectrometers for the detection and identification of explosives. These devices may be available commercially in the near future. At present, discrimination of the explosives detected is limited to the retention time in a gas chromatograph as determined by a non-discriminating detector or the flight time of fragments in the plasmagram in an IMS.

These techniques may be embellished by the use of cold traps, catalytic converters and multiple GCs as in the Thermedics EGIS instrument, or by a technique called 'ion trap IMS' as in the IonTrack instrument, or by combining a GC with IMS as in the CPAD instrument. Discrimination is important in bomb detection because it suggests what further to look for either by direct inspection, suspect questioning or by use of another analytical approach such as penetrating radiation. Simple instrument response that an 'explosive' is or is not present is less useful. The demands on the analytical repertoire of modern instruments have expanded in response to the changing methodology of terrorism. Ideal instruments should be capable of detecting not only military and commercial explosives but also flammable liquids and chemical warfare agents, plus they should have the additional flexibility of detecting other substances as they appear. And, ideally, all of these substances should be detected rapidly and at extremely low levels of analyte contained in samples that may have complex matrices. Very few analytical techniques are sufficiently robust and versatile to do this with a rapid analysis time of a few seconds.

Development of consistent trace detection testing methods

Accurate and repeatable trace amounts of explosives must be placed onto test surfaces in order to accurately assess the sampling efficiency and sensitivity of a trace detection instrument to low levels of explosives The amount must resemble the type of sample that would normally be presented to the test instrument operator in the real world. That is: what amounts of plastic explosives will occur on the outer surfaces of luggage due to inadvertent contamination by a bomb maker?

Experiments have been performed by building bombs in suitcases or other portable items under controlled conditions such as: extremely careful to avoid contamination, less careful, and not careful at all. By measuring the amount of explosive on the outer surface, these tests yielded data as to the amount to be expected and the probable location.

The advantage of the development of a repeatable accurate method of depositing explosives onto surfaces is that a standard can then be created at specified levels or amounts of explosive. These 'controlled deposits' provide a consistent means to test and compare the operational characteristics of commercial instruments.

Many attempts have been made to achieve a consistent deposit of traces of explosives for testing purposes which retains the characteristics of the original material. This is especially the case with plastic explosives because the RDX and PETN crystals are coated with a thin layer of polymer in the manufacturing process. This coating causes a degree of adhesion to surfaces which is different from the crystals alone. Instruments which sample for trace amounts of explosives by wiping can register different levels of sampling efficiency depending on whether the test explosive does or does not have the polymer coating. For this reason preparing deposits of plastic explosives by simple dissolution in an organic solvent is inadequate because the result is a deposit of crystals without the polymer coating.

Experience has shown that depositing less than 1 μg of plastic explosive, for example 100 ng, allows the operator to perform only one test of the instrument using a wipe sampling method or sampling in a walk-through portal. The small amount of material is apparently simply rubbed off. It becomes crucial, therefore, to be able to prepare many sets of samples, each with the same amount of explosive deposited. In this way the precision of the instrument tested, its sampling efficiency, the lower limits of detection and the minimum alarm level can all be accurately assessed. In addition the 'controlled deposit' method provides a means to have quality control and test material available for instrument development at the manufacturer site. Test samples prepared by the deposition process are useful for trace explosives detection instrument testing and development in the laboratory, at the manufacturer site and at field operational sites.

The test for the consistency of the amount of explosive deposited by multiple fingerprints on a surface is to extract and quantitate alternate prints. These data permit the assumption that an 'average' unassayed alternate print in the imprint series prepared in the same experiment has a known amount deposited. Unfortunately, the error in the amount of explosive analyzed from print to print is greater than the amount of explosive removed by the wipe method of sampling. This disparity is due to the very small amount of explosive removed using standard, and inefficient, wipe sampling techniques. An 'average' amount of explosive per fingerprint calculated in this way has too much variation to allow determination of sampling efficiency or to use this method to prepare explosive standards. In addition to fingerprints not being reproducible, another drawback is the inability to reproducibly vary the amount deposited (Fox et al., 1995b).

We have now developed a method for the consistent, accurate production of deposits of plastic explosives which have the same sampling characteristics, namely 'stickiness', as the bulk explosive. The method works well with the major commercial and military plastics. The technique takes advantage of the fact that plastic explosives are manufactured as crystals of explosive which are stabilized and protected by coating with plastic, such as polyisobutylene or styrene-butadiene. The result is encapsulated crystals which are made into the final product by pressing the particles together into bricks or sheets, along with a little oil or other agent to aid flexing. Normal production of a fingerprint occurs by touching the bulk explosive and removing some of the coated particles mechanically from the bulk because they stick to the skin. The particles are then transferred to the next object or surface touched.

Our method takes advantage of the encapsulation resulting from the manufacturing process. The adhesion forces between the coated particle in the compressed bulk are overcome mechanically. A small piece of plastic explosive is agitated with water causing particles of the coated explosive to be shaken loose from the bulk and suspended in the water. By analogy, touching the bulk causes mechanical removal of particles due to pressure and adhesion. The explosive is not dissolved by the water agitation process, but particles from the bulk are merely suspended by gentle mechanical means. Variations on the agitation process include stirring, sonication and shaking. Once the explosive is suspended in its normal form of coated crystals, the actual amount is determined by usual methods including HPLC, GC and MS. Dilution is then made to an appropriate level.

In order to prepare a deposit consisting of the desired amount of explosive, an aliquot of the diluted and quantified suspension is pipetted onto the surface of a test object and the water evaporated. The dried residue consists of plastic coated explosive crystals in the original state as would occur with a natural fingerprint or smear. The amount of dried residue deposited can easily be varied by controlling the degree of dilution of the original suspension of explosive particles. We have prepared reproducible deposits ranging from high micrograms to low nanograms. The concentration of the deposit is verified by back-extraction of a test surface and quantitation of the extract. Particle size of each preparation is verified by scanning electron microscopy.

Our goal was to achieve a consistent particle size range in a standard deposit. Comparability of the particle size range of the deposit to actual fingerprints was also considered important. If large particles appeared, then a large and unusable variation in the amount deposited occurred. Consistent analysis with RSDs of less than 10% were obtained on back extraction and analysis of the deposits. In addition, a close comparison of particle size between a fingerprint and a deposit of the suspension of C-4 was observed as shown in Figures 3.9, 3.10 and 3.11 (Fox, 1995; Fox et al., 1995a, 1995b).

The controlled deposition of several plastic explosives on some typical carry-on electronic items was used on commercial instruments at the 1995 ICAO meeting of the 'Ad Hoc Group of Specialists on the Detection of Explosives', held at the FAA Technical Center in Atlantic City, NJ. Six commercial instruments were tested over several days using the standard sampling technique proposed by the trace detection instrument manufacturer. Participants in the exercise included manufacturer representatives and an international group of persons experienced in trace explosives detection (Fox and Jankowski, 1995).

The older technique of simply dissolving the plastic explosive bulk material in an organic solvent for deposit by recrystallization on evaporation, removes the plastic coating and inevitably changes the sampling characteristics observed. Coating inert material with a solution of plastic explosive, and evaporating the solvent has the same drawback (Elias, 1995).

The deposition technique provides a means to produce an accurate standard – so far limited to plastic explosives – which can be used for testing and evaluating trace detection instrument systems, including the efficiency of the sampler and the effect of interferents. In addition, deposits onto fabric may provide a standard for testing walk-through portals designed to detect traces of explosives on passenger clothing. Several portals are in the

Figure 3.9 C-4 fingerprint particles, SEM image, 600 × .

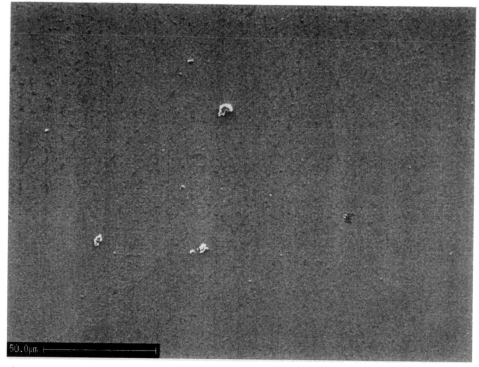

Figure 3.10   C4 deposit particles, SEM image, 600 × .

developmental stage. Explosives-removal techniques in these portals are dependent on air flow onto a collector or intimate sampling by means of suction of the surface of passenger clothing as they pass along tubes with suction holes along their length. In both types of collection the sample ultimately ends up on a prefilter, is desorbed, and analyzed by means similar or identical to those shown in Table 3.2. Advances in portal development were presented at an FAA-sponsored portal conference in January 1995 (Jankowski, 1995).

Testing checked and carry-on luggage, as well as passenger garments and hands is obvious immediate application of trace-based, hidden explosives detection, but there are other applications. Trace detection based technologies are under development at several centers for cargo inspection. Ultimately, the application may be adapted to postal needs and to other transportation modalities. Trace detection is, of course not limited to explosives hidden in bombs, but may be used to detect a wide variety of other chemicals used in hidden devices. Such hidden devices have been used recently in Japan for toxic gases. There is a long list of flammable and toxic substances which may be hidden and yet detected by trace methods.

3.3.4 Taggants

The process of explosives detection and identification of explosives can be enhanced by the use of taggants, which are substances which are more easily identified than the explosive itself. A full range of explosives must be considered. The exact material to be used either admixed with, or present as an integral part of explosives, has yet to be defined. Under a new US law, the 'Antiterrorism Effective Death Penalty Act' of 1996, the US. Treasury Depart-

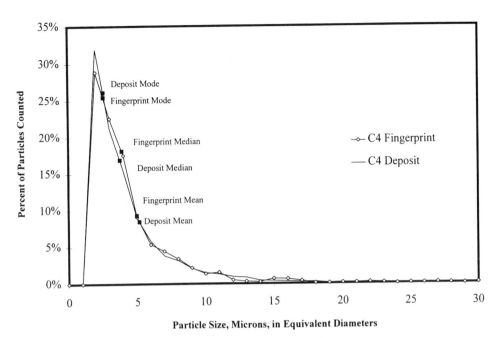

Figure 3.11 Comparison of C4 fingerprint and deposit particle sizes.

ment can suggest regulations to tag explosives. The goal is tracers which substantially assist law enforcement, do not pose a risk to human life or safety, do not substantially impair the quality of the explosive for their intended lawful use and do not have substantially adverse effects on the environment. Moreover, the cost of adding tags should not outweigh their benefits. A full range of explosives and their uses must be considered, including at least 500 types of explosives that are on the market, from blasting caps to triggering devices on air bags. Consideration must also be given to potentially explosive materials such as ammonium nitrate fertilizer. The US National Research Council (NRC) is organizing the 'Committee on Marketing, Rendering Inert and Licensing of Explosives Materials' that will carry out the study.

In addition to microscopic particles called taggants, chemical markers based on isotope labeling are now available for post-blast identification. Other approaches are being proposed. The Institute of Makers of Explosives (IME) based in Washington, DC, has indicated that adding taggants could easily double the cost of commercial explosives. Given the stakes, the NRC study will clarify which explosives should be tagged for maximum benefit to law enforcement.

Tagging options

Chemical & Engineering News, 9 September 1996, p. 10, listed the following taggant options:

(1) Multicolored, multilayered plastic particles, such as 'Microtaggant', produced by Micro-trace Inc., Minneapolis MN, USA, and 'HF 6', produced by Swiss Blasting A.G., Bulach, Switzerland. Code is based on color sequence of particles.

(2) Rare-earth elements embedded in synthetic granules blended with fluorescent pigments, such as 'Expotracer', produced by Plast Labor, Bulle, Switzerland. Code is based on melting point of the matrix, and composition of embedded elements.

(3) Isotopically labeled trace components of material being tagged; such as markers produced by 'Isotag LLC', Houston, TX, USA. Code is based on extent of substitution and proportions of variously substituted compounds.

(4) Inert chemicals identifiable by specific antibodies and detected by immunoassay techniques, such as those developed by Bar-code Inc., Cambridge, Mass., USA. Code is based on the identity of the chemical.

(5) Polymeric microbeads with different diameters and colors. Code is based on the population of various sizes and colors.

(6) Slow release microcapsules containing perfluorodimethylcyclohexane or perfluoromethylcyclohexane to enhance detectability.

3.3.5 *Trace Detection and Forensics*

Aviation security trace detector systems are designed to process passengers at the rate which will not cause delays in aircraft departure. These devices can measure the amount of explosives in a sample, identify the explosive or explosives, and record the findings in both analog and digital form in less than 1 min. These capabilities make trace explosives detectors extremely valuable in the analysis of post-blast debris to detect traces of unreacted explosives in bulk debris.

Since aviation security trace detector systems are available worldwide, they can be put to the alternative use of aiding searches of post-blast debris.

When an IED is set off, traces amounts of unburned explosive may be flash-deposited on surfaces near to the blast location. Careful searching for traces of the chemicals involved in the blast can potentially allow authorities to gain knowledge about the IED itself, and may even provide clues as to who caused the blast. As is noted in Chapters 4 and 5, aviation security trace detection systems have become a useful tool in the forensic analysis of bomb episodes. Typically a large number, from dozens up to hundreds of debris pieces are examined for traces of unburned explosive using commercially available trace detection instruments such as the Thermedics EGIS and the Barringer 400 (see Table 3.2). This process can greatly facilitate chemical analysis for explosives traces by indicating to the chemists where to concentrate their initial efforts – and indeed, what explosives to look for. The FBI have reported positive results from tests using an IMS instrument to detect traces of explosives on hands and in post-blast debris. Detection levels of 200 pg were achieved for common explosives (Fetterolf and Clark, 1992). As a result of this work, an IMS instrument has been incorporated into routine laboratory protocols where it is used, along with MS methods, to confirm the presence of nitrated organic explosives indicated by GC/TEA analysis (S. G. Burmeister, personal communication).

The sampling method typically is the standard manufacturer-recommended method of wiping the surfaces or vacuuming. Explosives detected include high explosives such as RDX, PETN, NG, TNT, EGDN and DNT. The sampling methods used at a crash site may need to be modified depending on the circumstances. Recent evidence seems to indicate that trace amounts of plastic explosives may be washed away in seawater over a period of time (Fox and Sisk, 1996). However, Higgs and Hayes (1982) have reported that NG in seawater has a half-life of 92 h – half of that in air.

Earlier reports of the study of debris from aircraft crashes have relied largely on metallurgical evidence (Tardif and Sterling, 1967, 1969; Clancey, 1968; Higgs et al., 1978). But as the authors of Chapters 6, 13 and 14 ably describe, aircraft accident investigation is all about multidisciplinary team-work, and the extension of the application of trace detection instruments from aviation security to the crime scene and the forensic laboratory is a new facet of exactly that.

3.4 Summary

Governments must remain proactive in their efforts to stop all terrorist plans to harm the flying public and aircraft. Applications of new technologies, development of better equipment to find threat items, and the constant improvement of techniques to utilize these assets continues daily. The use of explosives by terrorist groups is by far the most common and frequent action resorted to. Thus, detection of all types of explosives remains the most critical work of the aviation security program.

References

BUCHSBAUM, S., KNIZE, D., FEINSTEIN, L., BENDAHAM, J., and SHEA, P., 1991, An Approach to Improving TNA EDS, in Khan, S. M. (Ed.) *Proceedings of the First International Symposium on Explosive Detection Technology*, pp. 70–81.

CHENG, C., KIRKBRIDE, T. E., BATCHELDER, D. N., LACEY, R. J., and SHELDON, T. G., 1995, In situ detection and identification of trace explosives by raman spectroscopy, *Journal of Forensic Sciences*, **40**, 31–37.

CLANCEY, V., 1968, Explosive evidence in an airplane accident, *Canadian Aeronautics and Space Journal* (Nov), 337.

CORMIER, S., FOBES, J. L., HALLOWELL, S. F., and BARRIENTOS, J. M., 1995, Systems Analysis of the Federal Aviation Administration's K-9 Program, *DOT/FAA/AR-95/123*, Atlantic City: Federal Aviation Administration.

CRITERIA FOR CERTIFICATION OF EXPLOSIVES DETECTION SYSTEMS AS DEFINED UNDER 14 CFR 108.20, 1993, *Federal Register*, **58(174)**, Docket 27026, 47804–47814.

DANYLEWYCH-MAY, L., and CUMMING, C., 1993, Explosives and taggant detection with ionscan, in Yinon, J. (Ed.), *Advances in Analysis and Detection of Explosives*, pp. 385–401, Dordrecht: Kluwer Academic Publishers.

EILBERT, R. F., and KRUG, K. D., 1992, Aspects of image recognition in Vivid Technology's dual-energy x-ray system for explosives detection, *Proceedings of the SPIE Conference on Application of Signal and Image Processing in Explosives Detection Systems*, SPIE, **1824**, 127–143.

ELIAS, L., 1995, NRC procedures for laboratory evaluation of trace explosive detectors. *Proceedings of the Third Workshop of the International Civil Aviation Organization (ICAO) Ad Hoc Group of Specialists on the Detection of Explosives*, Atlantic City: Federal Aviation Administration.

FETTEROLF, D. D., and CLARK, T. D., 1992, Detection of trace explosive evidence by ion mobility spectrometry, *Journal of Forensic Sciences*, **38**, 28–39.

FOX, F. T., and SISK, S., 1996, Loss of trace amounts of plastic explosives in seawater, *Proceedings, The Second Explosives Detection Technology Symposium and Aviation Security Conference*, Atlantic City, NJ: Federal Aviation Administration.

FOX, F. T., and JANKOWSKI, P., 1995, Personal communication, data reduction, *Proceedings of the Third Workshop of the International Civil Aviation Organization (ICAO) Ad Hoc Group of Specialists on the Detection of Explosives*, Atlantic City: Federal Aviation Administration.

FOX, F. T., 1995, Evaluation protocol for trace explosives detection instruments, *Proceedings of the Fifth International Symposium on Analysis and Detection of Explosives*, in press.

FOX, F. T., GREEN, D., SISK, S., DIBARTOLO, R., and MILLER, J., 1995a, Particle characterization of trace explosives solid samples, *Proceedings of the Fifth International Symposium on Analysis and Detection of Explosives*, in press (preparation: Bureau of Alcohol, Tobacco and Firearms (ATF), National Laboratory Center, 1401 Research Blvd., Rockville MD, 20850).

FOX, F. T., SISK, S., DIBARTOLO, R., GREEN, D., and MILLER, J., 1995b, Preparation and characterization of plastic explosives suspensions as analogs of fingerprint derived contaminants for use in certification of trace explosives detection systems, *Proceedings of the Third Workshop of the International Civil Aviation Organization (ICAO) Ad Hoc Group of Specialists on the Detection of Explosives*, Atlantic City: Federal Aviation Administration.

GESCHEIDER, G. A., 1985, *Psychophysics: Method, Theory, and Application*, Hillsdale, NJ: Lawrence Erlabaum Associates.

JOHNSTON, J. M., WAGGONER, L. P., WILLIAMS, M., JACKSON, J. M., JONES, M. L., MYERS, L. J., HALLOWELL, S. F., PETROUSKY, J. A., and DESO, W., 1995, Canine olfactory capability for detecting NG smokeless powder, *Proceedings of the Fifth International Symposium on Analysis and Detection of Explosives*, in press (preparation: Bureau of Alcohol, Tobacco and Firearms (ATF), National Laboratory Center, 1401 Research Blvd., Rockville MD, 20850).

HARTELL, M. G., MYERS, L. J., WAGGONER, P., KUHLMAN, M., HALLOWELL, S. F., and PETROUSKY, J. A., 1995a, Design and testing of a quantitative vapor delivery system, *Proceedings of the Fifth International Symposium on Analysis and Detection of Explosives*, in press (preparation: Bureau of Alcohol, Tobacco and Firearms (ATF), National Laboratory Center, 1401 Research Blvd., Rockville MD, 20850).

HARTELL, M. G., PIERCE, M. Q., MYERS, L. J., HALLOWELL, S. F., and PETROUSKY, J. A., 1995b, Comparative analysis of smokeless powder vapor signatures derived under static versus dynamic conditions, *Proceedings of the Fifth International Symposium on Analysis and Detection of Explosives*, in press (preparation: Bureau of Alcohol, Tobacco and Firearms (ATF), National Laboratory Center, 1401 Research Blvd., Rockville MD, 20850).

HIGGS, D., and HAYES, T. S., 1982, Post-detonation traces of nitroglycerine on polymeric materials: recovery and persistence, *Journal of the Forensic Science Society*, **22**, 343–352.

HIGGS, D. JONES, P., and MARKHAM, J., 1978, A review of explosives sabotage and its investigation in civil aircraft, *Journal of the Forensic Science Society*, **18**, 137–160.

JANKOWSKI, P., (Ed.), 1995, Trace portal detection portal workshop technical note, *DOT/ FAA/CT-TN95/05 Revision A*, Atlantic City: Federal Aviation Administration.

KNOLL, G. F., 1989, *Radiation Detection and Measurement*, 2nd ed. New York: John Wiley.

KOLLA, P., 1995, Detecting hidden explosives, *Analytical Chemistry*, **67**, 184–189.

LANGFORD, M., 1995, Development of a print deposition series model, *Proceedings of the Third Workshop of the International Civil Aviation Organization (ICAO) Ad Hoc Group of Specialists on the Detection of Explosives*, Atlantic City: Federal Aviation Administration.

LEMISH, M. G., 1996, *War Dogs: Canines in Combat*, Washington DC: Brassey's Inc.

NEUDORFL, P., McCOOEYE, M., and ELIAS, L., 1993, Testing Protocol for Surface Sampling Techniques, in Yinon, J. (Ed.), *Advances in Analysis and Detection of Explosives*, pp. 373–384, Dodrecht: Kluwer Academic Publishers.

PASSE, D. H., and WALKER, J. C., 1985, Odor psychophysics in vertebrate, *Neuroscience and Biological Reviews*, **9**, 431–467.

PRAH, J. D., SEARS, S. B., and WALKER, J. C., 1995, Modern approaches to air dilution olfactometry, in Doty, R. L. (Ed.), *Handbook of Olfaction and Gustation*; chapter 9, New York: Marcel Decker.

RODACY, P., 1993, The minimum detection limits of RDX and TNT deposited on various surfaces as determined by ion mobility spectroscopy, *Report 1993, SAND-92-0229, NITS Order No. DE 93018521*.

STEBBIN, W. C., (Ed.), 1970, *Animal Pyschophysics*, Englewood Cliff, NJ: Prentice-Hall.

TAKAGI, S. F., 1989, Standardised olfactometry in Japan – a review over 10 years. *Chemical Senses*, **14**, 25.

TARDIF, H., and STERLING, T., 1967, Explosively produced fractures and fragments in forensic investigations, 1967, *Journal of Forensic Science*, **12**, 247–272.

TARDIF, H., and STERLING, T., 1969, Detection of explosive sabotage in aircraft crashes, *Canadian Aeronautics and Space Journal*, **12**, 19.

Technology Against Terrorism: The Federal Effort. 1991, OTA-ISC-481, Washington, DC: US Government Printing Office.

TUCKER, J., 1963, Physical variables in the olfactory stimulation process, *Journal of General Physiology*, **46**, 453.

WILLIAMS, M., JOHNSTON, J. M., WAGGONER, P., JACKSON, J., JONES, M., BOUSSOM, T., and HALLOWELL, S. F., 1997a, Determination of the canine detection odor signature for NG smokeless powder. *Proceedings of the Second Explosives Detection Technology Symposium and Aviation Security Technology Conference*, Atlantic City, NJ: Federal Aviation Administration.

WILLIAMS, M., JOHNSTON, J. M., WAGGONER, P., JACKSON, J., JONES, M., BOUSSOM, T., HALLOWELL, S. F., and PETROUSKY, J. A., 1997b, Determination of the canine odor detection signature for a selected nitroglycerin-based smokeless powder. *Proceedings of the 13th Annual Security Technology Symposium and Exhibition Government.* Industry Exchange, Virginia Beach, VA: American Defense Preparedness Association.

VALENTIN, G., 1848, *Lehrbuch der Physiologie des Menschen*, Braunschweig: Friedrich Bieweg und Sohn.

4

General Protocols at the Scene of an Explosion

JEAN-YVES VERMETTE

4.1 Introduction

The immediate aftermath of an explosion is usually chaos. Explosions release large volumes of gas into the environment at high velocity, frequently resulting in major damage, injury and death. A rapid, well planned and systematic response by emergency and investigative personnel is essential in order to aid victims and to determine what happened.

This chapter describes effective protocols for conducting a criminal investigation at the scene of a bombing incident.

4.2 Investigative Objectives

The criminal investigation of a bombing incident has the same objective as any other onvestigation: to identify the person(s) responsible and bring them before the courts. In this sense, a prosecutor will normally require the investigator to provide evidence that the person(s) charged had the motive, opportunity and means to commit the crime (Figure 4.1).

In bombing incidents, the investigation at the scene should quickly provide two of three essential pieces of information which pertain to motive, opportunity and means:

(1) the target(s);
(2) the victim(s);

but the third piece of information usually takes much longer:

(3) the design of the explosive device(s).

The most common motives underlying the use of an explosive device are: experimentation, murder, suicide, intimidation, vandalism and destruction of evidence. Information on motive is derived by basic police investigative procedures.

Thorough processing of the crime scene may allow the investigator to determine what caused the device to go off, and that knowledge will help to develop leads which might establish that the suspect had the opportunity to commit the offense. Determination of how an improvised device was constructed and initiated provides vital information towards achieving this objective.

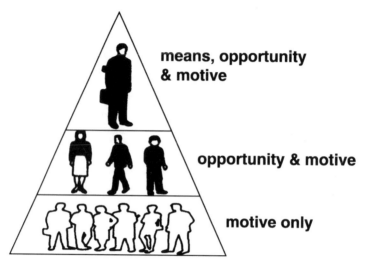

Figure 4.1 Motive, means, and opportunity.

To show that a suspect had the means to commit a crime, the investigator must link the suspect to the device, e.g. by establishing that the suspect acquired the explosive and/or components used in its fabrication. This objective is pursued by police investigation directed in considerable measure by results from examination of materials recovered from the blast scene.

4.2.1   Guidelines for Investigators

The following basic questions which should be an integral part of all investigations into the illegitimate use of explosives (Vermette, 1996).

- What
    - materials were used to fabricate the device?
    - skill or expertise did the bomber need?
    - object was damaged by the device?
    - object was the intended target of the device?
    - happened: an accident or a criminal act?

- Who
    - was the actual victim?
    - was the intended victim?
    - made the device?
    - placed the device?

- Why
    - was a device used?
    - was the device placed where it was?
    - was the device constructed as it was?

- When
    - was the device made?
    - was the device placed?

— was the device initiated?
— did the device explode?

- Where

  — was the device made?
  — was the device placed?
  — were the device components obtained?

- How

  — was the device placed?
  — was the device made?
  — was the device initiated?

It is important that investigators avoid:

- doing a superficial investigation in the belief that the situation could never be adequately understood;
- arbitrarily eliminating all but one line of inquiry.

4.3 Improvised Explosive Devices (IEDs)

Knowing the design of the explosive device can assist the investigator get into the mind of the bomber. The initiation system of the device indicate what the bomber was trying to do. For example, both timed and victim operated devices require knowledge of the intended victim's movements or behaviour patterns.

A typical IED consists of:

- an explosive;
- an initiating system;
- a container.

4.3.1 Explosive

As discussed in Chapter 1 and elsewhere, there are many explosives available for illegitimate use. The purpose of the act and the availability of explosives or their precursors in a particular geographic area determine what is used. Each type of explosive, however, leaves characteristic damage and chemical signatures which can be interpreted by skilled scene examination and by forensic science.

4.3.2 Initiating System

Bombing investigators must have a basic understanding of typical IED circuitry and of the three methods of initiating a device:

(1) Time delay
    The device incorporates a mechanical, electrical or chemical timing mechanism. Once the device has been set up, it will function after the designated time has elapsed.

(2) Victim operated
    The device (often called a 'booby trap') is designed to function when the victim does something which triggers the initiation system.

(3) Controlled by the bomber
     The device is initiated by the bomber, usually by a remote method.

Most commonly, the initiation system is electrical and damaged fragments almost always survive the explosion, for example:

- timer parts;
- batteries;
- circuit boards;
- electronic parts;
- wire;
- detonator fragments.

Typical non-electrical components are:

- chemical residues;
- burning fuse remains;
- adhesive tapes;
- monofilament line.

Because initiation systems are close to the explosive, they are often good sources of explosive residues – and indeed of fingerprints which survive an explosion. Such items should be treated accordingly.

4.3.3   Container

Containers range from pipes through stereos to parcels and suitcases. Some, as with low explosives, provide the containment necessary for an explosion to occur. Other containers provide means of transportation, disguise and fragmentation. All provide valuable evidence if identified and possible sources determined.

For example, a suitcase might help to identify an individual who was seen taking it to the scene and might also provide insight into intent. A container might also provide names, addresses, fingerprints, DNA and postmarks.

4.3.4   General

The technical expertise of an explosives disposal technician is important in determining a device's initiating system, especially when the device did function and was recovered in pieces.

The materials used in the construction of a device and the quality of the workmanship are also important evidence. Making a device does not necessarily require previous experience or specialized training. A wealth of printed material is readily available which provides details concerning the construction of explosive devices. Access to manufactured explosives is also not a prerequisite. Many explosives can be improvised, and commercial explosives can be stolen. Fortunately, many bombers have ill-conceived ideas about making explosive devices. Often devices fail to function and are recovered intact. Quite a few function prematurely and make the would-be bomber the victim.

Idiosyncrasies in construction can also be meaningful. A peculiarity found in the construction of one device could turn up in a device in another incident. Although this is not positive proof that the same person is involved, it is a viable investigative lead. Some serial bombers actually leave an intentional 'signature' to frustrate investigators.

4.3.5 Recovery of Intact Devices

Fortunately, some devices do not explode. A number of devices are recovered intact because they were:

- not meant to function;
- improperly designed or fabricated;
- discovered and deactivated by an explosives disposal technician or a lucky member of the public.

When the device is recovered intact, incident analysis proceeds quickly. Details concerning the fabrication of the device, the materials used and other evidence of investigative interest are more easily discernible. The investigator can readily determine how the bomber gained access to the area. Processing of this type of crime scene concentrates on collection of evidence like forced entry tool marks, footprints, and fingerprints. Such evidence is often more difficult to determine post-blast.

Device deactivation

When an explosives disposal technician must neutralize a device, the prime consideration is safety. However, to the extent possible, render-safe procedures should: avoid unnecessary destruction or contamination of forensic evidence.

The most useful physical evidence is obtained from a deactivated device which can be taken apart. However, if safety issues necessitate disruption of a device, then every effort should be made to avoid using methods which could inhibit subsequent laboratory examination. Ideally, the bomb technician, a senior investigator and a scientist should meet to discuss how to maximise collection of pertinent physical evidence and minimise the risk of contamination. At the very least, photographs, and if possible X-rays of the intact device, should be taken, the scientist must be made aware of any explosives used, the technician should wear gloves and disruption should not take place on a contaminated explosives range.

4.4 The Scene of a Bomb Incident

Explosive device investigations frequently are challenging and time-consuming. However, a well planned approach and disciplined effort can yield positive results.

Determining what actions are to be taken at the scene of an explosion depends on many variables: e.g. the type of target, the amount of damage, the location, the weather and people problems.

4.4.1 Immediate Aftermath

The aftermath of an explosion raises three special problems:

- threats to human safety;
- obstacles to crime scene protection;
- people problems.

Threats to human safety

Practically everyone at the scene of an explosion wants to go to the seat of the explosion. However, unless there is a compelling reason to enter the immediate area, such as assisting victims, officers attending the scene should not enter and should prevent others without specific and immediately required expertise from doing so. The site of an explosion is an extremely unsafe area, not only for those who must work there, like firefighters and rescue personnel, but also for anyone in the vicinity such as neighbours, sightseers or unwitting passers-by. Hazards may include unexploded devices, chemicals, biological fluids, downed power lines, fractured utility mains and structural damage.

Obstacles to crime scene protection

Protecting the scene of an explosion can be complicated by the number and variety of agencies which customarily respond. Without firm and immediate control by the first responders, it would be easy to have firemen, rescue squads, utility personnel, ambulance drivers, sightseers and property owners all milling about at the scene to the detriment of physical evidence.

There are non-human complications as well – one is fire. During firefighting operations, evidentiary considerations take a back seat. Some evidence may be hopelessly contaminated or destroyed by the fire or the firefighting activities. If possible, however, firefighters should have been trained, and if necessary redirected, to use minimally destructive fire suppression procedures; for example, misting instead of dousing, and removing smouldering items for extinguishing elsewhere.

If possible, dry chemical fire extinguishers should not be used since the chemicals can interfere with laboratory analysis for explosives residues. But if a dry chemical extinguisher has been used, then a small sample, like a swab from the nozzle, should always be taken and submitted as a control sample to the laboratory.

A different type of problem occurs when an explosion disrupts transportation – for example, an explosion in a vehicle at a busy intersection may result in a conflict between clearing the intersection and protecting the crime scene.

Each incident presents its own peculiar crime scene protection problem. Many scenes cover a huge area; the explosion of an aeroplane in flight, for example, can result in debris miles apart, and a major explosion in an urban area might strew parts of the building for blocks. Sometimes the complete securing of an explosion crime scene is not feasible; however, all possible protection must be put in place at the earliest possible moment.

People Problems

Securing the blast scene is often complicated by sightseers and the wishes of individuals with a vested interest, e.g. relatives of victims, property owners, lawyers and insurance adjusters. Diplomacy and firmness are the orders of the day.

4.5 Protocol for First Responders

The following outlines the main points to be considered by the first responders to the scene of an explosion.

(1) The job of the first responders is to tend to the injured and protect the scene.
    Advance education and training of police, fire and ambulance personnel in explosive scene protocols pay large dividends. Examples are:
    • minimal fire suppression;

- no unauthorised or non-essential personnel allowed on-site;
- demarcated lines of travel for emergency personnel who must enter the scene.

(2) Jurisdiction must be established immediately – one person must assume control of the scene and the criminal investigation.

(3) Boundaries are established: normally the apparent outer limit of scattered debris plus an additional distance of 20%.

(4) The scene must be declared safe by a bomb technician before the physical investigation can commence (Hall, 1991).

(5) Structural integrity and safety hazards such as gas leaks must be assessed and dealt with by qualified personnel.

(6) Witnesses must be identified and questioned before they disappear.

Several of these points are expanded upon in Section 4.7.

4.6   Objectives at the Blast Scene

The objectives of the crime scene search are to determine:

- what happened;
- how and why it happened;
- responsibility.

In order to accomplish these objectives following a major incident, efforts must be co-ordinated, directed and controlled. Three mutually supportive operational phases of investigation are:

- co-ordination and control;
- 'normal' police investigation: 'outside investigation';
- specialist investigation of the scene: 'inside investigation'.

The procedure is shown diagramatically in Figure 4.2.

Ideally, a post-blast response team will be on call and able to attend rapidly to assist local authorities. This topic is dealt with in detail in Chapter 5.

4.6.1   Co-ordination and Control

Without proper co-ordination and control, investigative effort may be aimless, definitive objectives may never be reached, or leads developed through investigation may never be followed up. The co-ordination and control function guides and directs all investigative activities. If several agencies are involved in the investigation, it may not be possible to maintain a single, united command of all activities. The functions of the various agencies should, however, be co-ordinated, and the criminal investigation should be paramount.

Once the crime scene has been secured and sufficient officers have arrived to interview witnesses, the second phase of the investigation should begin: compilation of available information from those who have already reported to the scene.

Information exchange

A protocol must be created for the mutual exchange of information to minimise unnecessary duplication of effort and organizational conflicts. The importance of communication between the 'inside' and 'outside' investigation through the medium of the control centre is vital.

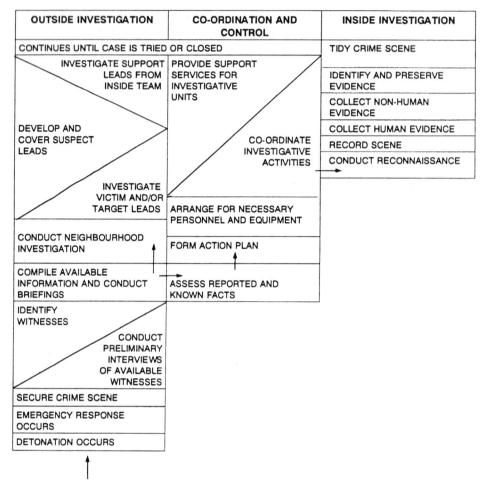

OUTSIDE INVESTIGATION	CO-ORDINATION AND CONTROL	INSIDE INVESTIGATION
CONTINUES UNTIL CASE IS TRIED OR CLOSED		TIDY CRIME SCENE
INVESTIGATE SUPPORT LEADS FROM INSIDE TEAM	PROVIDE SUPPORT SERVICES FOR INVESTIGATIVE UNITS	IDENTIFY AND PRESERVE EVIDENCE
		COLLECT NON-HUMAN EVIDENCE
DEVELOP AND COVER SUSPECT LEADS	CO-ORDINATE INVESTIGATIVE ACTIVITIES	COLLECT HUMAN EVIDENCE
		RECORD SCENE
		CONDUCT RECONNAISSANCE
INVESTIGATE VICTIM AND/OR TARGET LEADS	ARRANGE FOR NECESSARY PERSONNEL AND EQUIPMENT	
CONDUCT NEIGHBOURHOOD INVESTIGATION	FORM ACTION PLAN	
COMPILE AVAILABLE INFORMATION AND CONDUCT BRIEFINGS	ASSESS REPORTED AND KNOWN FACTS	
IDENTIFY WITNESSES		
CONDUCT PRELIMINARY INTERVIEWS OF AVAILABLE WITNESSES		
SECURE CRIME SCENE		
EMERGENCY RESPONSE OCCURS		
DETONATION OCCURS		

Figure 4.2 Chronology of investigative activities.

All details, regardless of how inconsequential they may appear, are reported, evaluated and recorded. Frequent conferences, formal and informal, facilitate the exchange of information and also relieve boredom and boost morale.

Personnel deployment

The rule of thumb is to assign personnel to those areas where their specific skills can best serve the objective of the investigation.

In searching for evidence at the scene it is essential that the searchers know from experience exactly what they are looking for (see below).

The investigation co-ordinators must ensure that the morale of investigators working at the scene of the bombing remains high and that they do not lose their concentration during the search. The mental attitude of the inside workers is very important. In a major blast scene where there has been substantial physical damage a great deal of boring physical labour has to be done in hazardous conditions and perhaps in presence of the remains of victims. Inside searchers must be constantly on the alert for evidence, much of which may be minute and not readily recognizable. It is easy to become bored and to become merely a labourer, moving rubble from one pile to another – or to suffer post-traumatic stress.

The leaders of scene investigations should not be hurried as they may have to give the searchers frequent rest breaks. Briefings to ensure that these inside investigators are receiving privileged information are essential to the maintenance of high morale and the necessary level of attentiveness.

Consideration may be given to alternating personnel between inside and outside duties. However, such a policy should not conflict with the overall continuity of the investigative activity – and it is re-emphasized that effective searchers know from experience exactly what they are looking for.

Experience by conducting test explosions: Necessary experience is gained from practical tests on explosives ranges and by attending actual scenes of explosion first as an apprentice and then as a journeyman. By way of illustration, Figure 4.3 shows the explosion of dynamite inside a baggage container. Figure 4.4 shows the instant fireball produced by a plastic explosive. The scale can be judged from the side of a baggage container on the left of the photograph (this charge was in front of, not inside, the container).

This type of test gives experience in searching under optimal conditions for material known to have been exposed to the effects of the explosion of a known explosive.

Figures 4.5 and 4.6 show the before, during and after effects of a natural gas explosion in a house. The nature and distribution of the debris from a gas explosion is significantly different from that of high explosives – and there is no solid residue left over. Similar tests conducted with various explosives in buildings, cars and anywhere else subject to bombing attacks bring home the salient features to investigators and scientists.

Figure 4.3 Detonation of dynamite in a baggage container.

Figure 4.4 A fireball produced by a charge of plastic explosive.

Figure 4.5 A natural gas explosion.

Figure 4.6   The aftermath of a natural gas explosion – the basement.

By knowing what the original device and explosive consisted of, the efficiency and effectiveness of search procedures, team-work, communications and chemical analytical protocols can all be checked and enhanced.

Equipment

Planning for the procurement of necessary specialized equipment should be an early consideration of the control centre. However, all the equipment which may be required cannot be anticipated and consequently many equipment requirements have to be established as events unfold and needs are assessed and up-dated.

The following is a list of some of the tools which might prove useful:

(1) Hand tools: shovels, rakes, brooms, bolt cutters, wire cutters, wire gauges, hammers, sledgehammers, screwdrivers, wrenches, saws, hacksaws, chisels, magnets, flashlights, knives, measuring tapes, pliers, calipers.

(2) Light equipment: wheelbarrows, metal trash cans, power saws, cutting torches, ladders, portable lighting equipment, metal detectors, plastic sheets, Kraft paper, buckets, photographic and video equipment, sketching materials, ropes, hooks, pulleys, sifting screens.

(3) Heavy equipment: truck, front-end loader, bulldozer, crane, shoring materials, backhoe.

(4) Personal equipment: hard hats, safety goggles, work gloves, coveralls, work shoes, foul weather clothing.

(5) Other equipment: computer, survey equipment, helicopter for aerial photography and photogrammetry.

Creature comforts

Like many major police incidents, blast scenes may require intensive work for an extended period of time. If this is the case, consideration must be given to the physical comfort of the workers – to things like water, food, toilets and other items of personal convenience. Maximizing the workers' comfort will help maintain high morale and expedite the investigation. When personnel have to travel a great distance just to obtain a drink of water, a cup of coffee or a meal, much valuable time is lost and the workers are conscious of being inconvenienced.

Control centre personnel and facilities

The efficiency of the co-ordination and control centre is vital to the success of the investigation. The personnel involved are chosen for their skills in co-ordinating investigative efforts. Investigative skill and experience are essential. If there are a number of different agencies involved in the investigation, at least one representative from each should be available for liaison purposes.

The centre itself should be physically located in an area which is accessible to the crime scene without being so close that its presence would interfere with the crime scene investigation.

One experienced person should be designated to deal with the news media.

4.7   Outside Investigation

The first steps are to secure the scene and to identify and interview witnesses.

4.7.1   Secure the Scene

If the blast took place in an open space, the area in which fragments are found and a surrounding buffer zone should be secured (Figure 4.7). Debris produced by a blast can be hurled to considerable distances. Officers assigned to secure a blast scene must ensure that there are no fragments lying outside the secured zone. Access must be restricted solely to those who must be there as noted above.

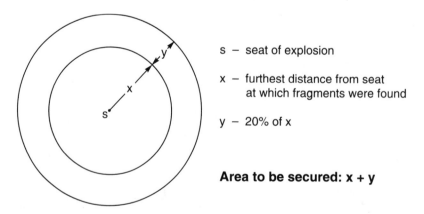

s – seat of explosion

x – furthest distance from seat at which fragments were found

y – 20% of x

Area to be secured: x + y

Figure 4.7   Explosion scene perimeter.

4.7.2 Witnesses

After the scene has been secured, the officers must concentrate on identifying and interviewing witnesses. Many witnesses at the scene who could provide valuable information are often passers by and once they have left the scene, they and their information might be lost forever. If it is impossible to conduct detailed interviews of witnesses, it is important to establish their identities and enough information to facilitate later contact.

This information should include: (1) each witness' name, address, telephone number(s); and (2) some other type of identification such as a social security number or driver's license.

In some instances, the witness list may include the name of someone who will be identified through subsequent investigation as a suspect.

If it is at all possible, the witnesses should be asked for information which could assist the investigation. Here are some appropriate questions (Vermette, 1996):

- Were you here when the explosion occurred?
- How long were you in the area prior to the blast?
- Where were you when the blast occurred?
- Did you see what happened?
- What was the victim doing?
- What colour was the flash or smoke?
- Was there more than one explosion?
- Did you notice any particular smell?
- What did the explosion sound like?
- Who went in or out after the blast?
- Did you notice anything unusual in the area prior to the blast?

The outside investigation unit should also try to identify vehicles in the immediate area insofar as this is possible, at least recording all the license plate numbers.

4.7.3 Building Plans

When structures are involved an investigator should be assigned to locate floor plans of the bombed facility to provide to the inside investigative team. Sketches, photographs and drawings would also help the on-site investigation and later the reconstruction of the crime scene.

4.7.4 Victims

A member of the outside investigative team should be dispatched to the hospital where the injured were taken, to seize and preserve all pertinent items such as clothing and foreign items removed from the victims' bodies.

If any of the victims died as a result of the blast, and if local practice permits, the investigator should attend the autopsy of each one and recover all evidential items from the body.

Forensic pathology of the victims of an explosion is dealt with in Chapter 15.

4.7.5 Other Assignments

As the investigation develops, normal assignments for the outside investigative unit would include some of the following:

GROUND FLOOR

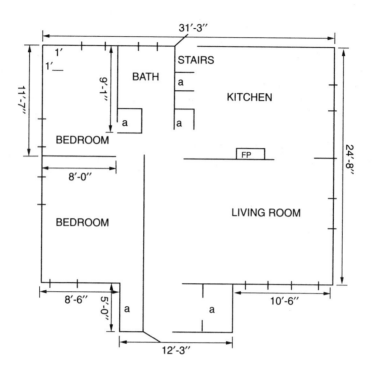

SIDE VIEW

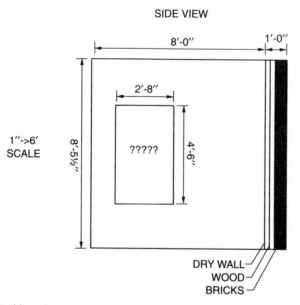

Figure 4.8 Building plans can greatly assist investigators at the scene to get their bearings.

- conduct follow-up interviews of witnesses;
- secure photographs of the external blast scene and the crowd from news media or other logical sources and review them for suspect and witness leads;
- obtain pre-blast photographs of the victim(s) and/or the damaged structures;
- if not already done, secure building blueprints or similar documents to assist the on-site investigators (Figure 4.8);
- if necessary, review utility record;
- commence background investigation of victims;
- investigate possible motives, means and opportunity;
- pursue leads as information on the device and explosive are obtained by the on-site investigation.

4.7.6 *Search of a Suspect*

Searches may involve: (1) the suspect; and (2) his or her vehicle, home and place of work.

If the suspect is quickly identified, there could be traces of the explosive on the suspect's hands, so swabs and fingernail scrapings should be taken (Figure 4.9). Isopropanol is a suitable solvent (see Chapter 10, and Lloyd, 1986). Nitroglycerine degrades on the skin (Lloyd, 1986) whereas other explosives such as RDX are absorbed by the skin and might be detectable a week afterwards (Lloyd and King, 1989). Clothing, too, can yield more evidence of explosives. All the skin swabs, fingernail scrapings and seized clothing should be preserved in clean, airtight containers.

The suspect's vehicle could also provide valuable evidence, particularly if it had been used to haul bulk explosives shortly before the search. However, the suspect's home and work area usually provide the most valuable evidence, particularly if tools used in fabricating the explosive device are recovered. If tool-marks on explosive device parts recovered by the inside investigators may later be positively matched to a tool of the suspect, the case against that individual is greatly enhanced.

4.7.7 *Search of a Suspect's Premises*

Investigators searching premises should be aware of all pertinent information recovered from the scene of the bombing: IED components such as containers, identified explosives and chemicals, timers, batteries, wires, and tapes.

A team approach involving an explosives disposal technician and a laboratory forensic specialist is desirable (see Chapter 5). Indeed, if it is suspected that there is a bomb-making facility, then significant precautions must be taken. The following outlines a useful protocol.

Protocol for search of a suspected clandestine laboratory

The following protocol (A. D. Beveridge and W. K. Jeffrey, personal communication) has worked well in practice.

Basic team composition:

- police;
- forensic scientists;
- explosive disposal technicians.

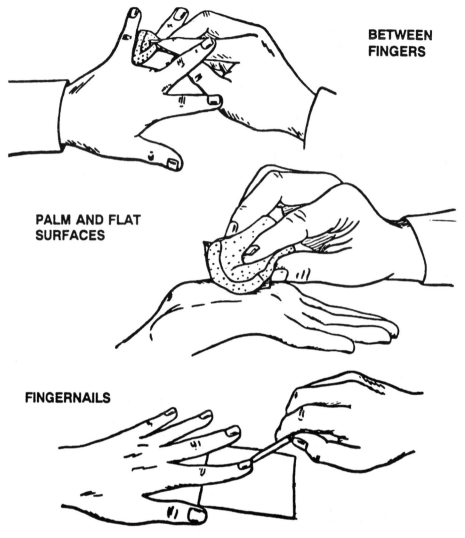

BETWEEN FINGERS

PALM AND FLAT SURFACES

FINGERNAILS

Figure 4.9   Trace quantities of explosives may be recovered from a suspect's hands by swabbing them with swabs moistened with alcohol and by taking fingernail scrapings.

The team may also include, depending on intelligence information and observations, specialist personnel from:

- occupational health;
- fire department;
- utilities;
- animal humane society;
- ambulance service.

Protocol:

- team meeting;
- principals removed, area secured;

- suspected chemist returned to identify hazards;
- explosives disposal technicians check for booby traps;
- cupboards and drawers opened by suspect or technicians;
- photographs taken, scene recorded;
- chemicals, apparatus, notebooks, literature examined by forensic scientists;
- bomb-making material examined by technicians;
- initial opinions by scientists and technicians given to investigators;
- one police officer designated to seize exhibits;
- chemical exhibits seized under direction of forensic scientists and transferred to laboratory.

4.8   On-site ('Inside') Investigation

The crime scene investigation should not be started until a plan has been formulated and the area cleared of safety hazards. There are four basic activities:

- leading;
- recording;
- searching and collecting;
- identifying and preserving.

4.8.1   Leading

The leader's most important job is to ensure that the specialists have everything that they need to do their job and can work unimpeded and uninterrupted.

The leader should also arrange for the assistance of individuals and things capable of helping the investigators to reconstruct the pre-blast scene. If the target of the explosive device was a vehicle, an identical vehicle or one very similar should be obtained for comparison. If the blast occurred inside a building, someone who is familiar with the building and its contents should be made available to assist in reconstructing the scene, the logic being that the pieces of the device are items which cannot be accounted for otherwise. Since the investigator does not know what does and does not belong at the scene, someone who is familiar with the area can render significant assistance.

Moving the scene

In certain circumstances the leader may decide that it is necessary to move the entire scene – for example, when the device has detonated in a public area which has to be rapidly cleared, or in an outside area during adverse weather conditions.

Moving the crime 'scene' can be done without presenting serious evidentiary problems. The leader would inform the control centre of the decision to move and together they would arrange to have the entire 'scene' collected and transferred to a suitable area such as a secure warehouse or aircraft hangar.

Ideally, everything should be picked up and moved together, i.e. the bombed object plus whatever bombing debris is lying in the general vicinity. To expedite this removal, a grid can be superimposed over the blast area, and the evidence and accompanying rubble placed in clean, unused collecting bins, square by square, and transported to the area where the detailed examination will be done. This collection procedure preserves at least some semblance of the spatial relationships of evidentiary items (see Figure 4.10).

4.8.2   Recording

The crime scene should always be recorded through the use of photographs, videotapes, sketches and notes. Photographs do not replace sketches and sketches do not replace photographs. Photographs should be taken of the scene in order to give the entire perspective. Often this is more easily done by videotaping or, if applicable, by aerial photography. Photographs should also be taken of the immediate area and close-up photographs should then be taken of key items of evidence. Specific details of current technology which can be used for position location at crime scenes covering a large area are discussed and illustrated in Chapter 6.

The recording of the crime scene is done concurrently with the investigative search of the scene. Photographs should show perspective and detail like specific damage and personal injury. Photographs and sketches should complement each other. When possible, the photographer(s) should work closely with the crime scene artist(s). It is desirable to have at least two individuals making sketches.

Figure 4.11 illustrates a view the scene of a suitcase explosion in the baggage handling area of Narita airport in 1985 which killed two baggage handlers. Figure 4.12 illustrates how the scene was divided into grids using existing lines in the concrete floor. Figure 4.13 illustrates how individual fragments were recorded in one rectangular grid of the scene. Further details on this case may be found in Chapter 16.

4.8.3   Searching and Collecting

Collecting evidence involves the search for and recovery of evidentiary items which fall into two categories: (1) bomb related items; and (2) other evidentiary items.

The search should be continued until the entire area has been covered thoroughly. The search methods depend on the number of searchers, the degree of destruction caused by the device, and the physical layout of the area. Figure 4.14 shows familiar search patterns (Figures 4.10, 4.12 and 4.15 illustrate aspects of grid patterns).

When relevant material is observed, it is photographed *in situ* (see Figure 4.13) prior to collection.

When there are large quantities of debris, as when a building has collapsed, then sifting of debris is essential. This may be done on-site or at a remote search facility. Sifters are vital for a thorough debris search at a crime scene. They may be commercially manufactured or home-made. All of the debris in a particular area should be sifted, the final mesh being as fine as window screens. Sifting should be done by teams of two to help preclude monotony and error. If possible, the area searched should be covered at least twice by different searchers. Varying the workers' job assignments and double-checking helps to promote the most productive searches.

Larger, bulky objects, such as cushions or doors, which may contain items of evidence, should be X-rayed if they appear to have been penetrated. Subsequently they should be broken apart and searched.

While conducting a search, the searchers should be advised to guard against three particular pitfalls:

• taking a blasé approach to their duties;

Figure 4.10   A grid system surrounding a bombed car.

Figure 4.11   Narita airport, Japan: bomb scene 1985.

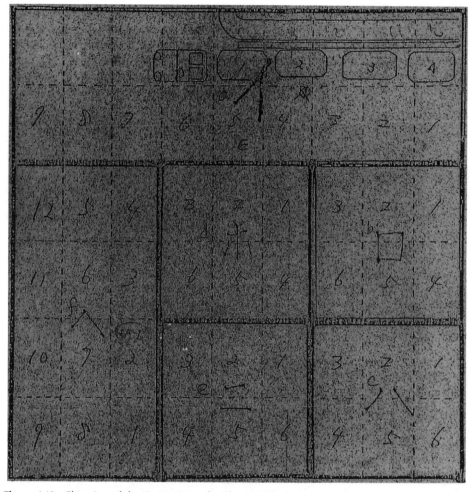

Figure 4.12 Plan view of the Narita airport bomb scene.

- jumping to hasty conclusions which may later prove to be incorrect (i.e. do not stop searching after the recovery of just a few items of evidence);
- concentrating only on explosive device-related evidence.

Contamination

As discussed in detail in Chapters 10 and 12, it is essential to be able to satisfy a court that no question can be raised about possible contamination – especially when dealing with traces of high explosives.

The responsibility lies on everybody, but particularly the team leader. Thus:

- explosive disposal technicians and military personnel should not handle evidence which may be examined for traces of high explosives;
- evidence should not be handled with gloves, tools etc. which have been contained in a 'bomb truck' or any facility or vehicle ever exposed to explosives;

Figure 4.13 Individual items, Narita airport bomb scene.

- material should not be contained in bags, cans, etc. which have been stored or transported in a 'bomb truck' or any vehicle or facility ever exposed to explosives;
- military equipment and transportation should not be used in explosives investigations if there is any possibility that they have previously been used to handle or transport explosives or have been involved in discharge of munitions;
- anything used to sweep, vacuum, contain, contact, swab, touch or move trace evidence must be clean and unused or proven by analysis to be uncontaminated.

Figure 4.15 illustrates the simple action of measuring a bomb crater. However, just this simple act depicted could contaminate the crater with traces of explosive if the gloves, hands, boots or clothing of the two people involved were to have been in recent contact with explosives. In the ideal world, the measurers would be wearing new latex gloves and a clean oversuit, and be using a new tape etc. Figure 4.10 should be re-examined with the same critical eye (e.g. were the shovel and rake against the truck clean and unused, or certifiably free from explosives traces).

To minimize exhibit continuity problems and for efficient exhibit control, one police officer only should be responsible for pickup and retention.

4.8.4 Identifying and Preserving

Ideally, the inside search should be conducted by explosives disposal technicians and forensic scientists who know what to look for in post-blast debris. However, many police forces may sometimes have to handle post-blast searches without expert assistance and must therefore have a basic understanding of typical IED construction and of the sorts of evidence to look for at a post-blast scene.

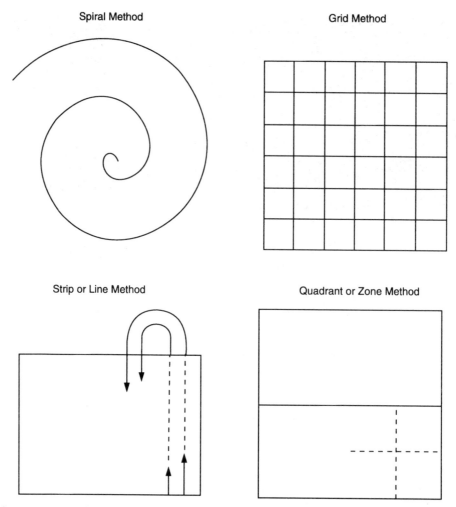

Figure 4.14 Search patterns.

The best place to seek residue from an explosive is at the seat of the explosion or in/on objects which were very near the seat at the time of the detonation. Occasionally the explosive filler used can be identified by packaging materials which survived the blast. These may include dynamite wrappers, paper backing from military C-4 explosive and plastic and metal clips from slurry and emulsion explosives.

When material like soil, plastic etc. is seized for residue examination, standard samples of the same material should also be taken for comparison purposes. This is illustrated in Figure 4.16.

While much can be learned by assessment of damage at a bomb scene, final determinations should not be made until results of the laboratory examination are available.

4.9 Forensic Evidence

Forensic evidence may:

- determine the explosive used;

Figure 4.15   Measurement of a bomb crater.

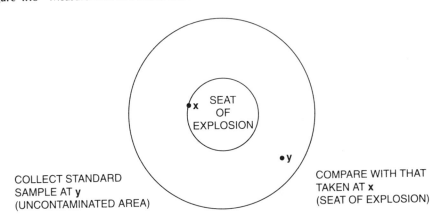

Figure 4.16   Taking standard samples.

- determine sources of IED components;
- compare physical evidence from the crime scene to the suspect;
- corroborate statements;
- link criminal cases.

4.9.1 *Physical Evidence of Potential Forensic Significance*

The following are examples of forensic evidence commonly found whole or in pieces at the site of an explosion:

- power sources: these can vary in size from car batteries through lantern batteries and flashlight batteries to Polaroid film batteries;
- circuitry: printed circuit boards, wires;
- timers: electronic, mechanical and chemical;
- switches: these are used for arming or firing an IED, and can be commercially manufactured or improvised; there are a variety of designs besides actual mechanical switches (e.g. pressure mats, clothes pegs, clocks, timers, electronic relays);
- detonators and igniters: (such as flash bulbs) which are rarely completely destroyed even if the IED has functioned as intended; parts of detonator wires, the detonator itself, safety fuse and 'hot wire' assemblies often are recovered;
- explosive charges: explosives can be commercially manufactured or home-made and even after detonation, can often be identified from wrapper fragments or residues found on other IED components or adjacent items;
- adhesive tapes: adhesive tapes are used in the construction of many bombs; they usually survive the explosion;
- IED containers: fragments of containers used to transport or hide the explosive, or provide the containment necessary to produce an explosion, are likely to be found at the blast site and often provide useful information. Examples are suitcases, stereos, clothing, automobiles/trucks, pipe fragments, cardboard boxes, and written or printed material, e.g. a stereo manual could indicate recent purchase;
- bomb-making equipment: items such as rubber gloves or rolls of tape, left in the explosive device by the bomber to be destroyed by the explosion, which will often give away the type and style of bomb, especially booby traps.

Such materials and others limited by the imagination of the bomber can be recognized, reconstructed, sourced and related back to the bomber. The Narita airport bombing illustrated in Section 4.8 above, is a good example (for details on prosecution of the case see Chapter 16). Nine items from the explosion scene in Japan were successfully related back to the bomb maker in Canada: adhesive tapes, stereo tuner (narrowed to 4 of 4000 by writing on cardboard fragments), ether can, smokeless powder, dynamite residue (recovered after 5 years from a victim's clothing), wires, timer, detonator, battery (Beveridge, 1993).

It is usually useful to sketch a map of the area in which evidence is lying, indicating where each item was found. One copy of this marked map can then be forwarded with the forensic evidence and another can be part of the incident report.

4.9.2 *Preparing Forensic Evidence*

As noted above forensic evidence should be identified and photographed *in situ*. Collection should be made by one person only with full regard to quality assurance, i.e. there must demonstrably be no opportunity for contamination to occur. Exhibits should be contained separately with the exception of sweepings or as directed by the attending forensic scientist, and handled only with disposable gloves and/or tweezers.

Large items must be dealt with in the most suitable manner; wrapping in sturdy plastic sheeting and sealing with tape will take care of most objects. If something must be dismantled, this should be done in such a way that the scientist in the laboratory can readily reconstruct it.

Explosive samples should be handled and transported in accordance with the government regulations. All forensic evidence specimens must be properly labelled in accordance with standard operating procedures.

4.9.3 Submitting Forensic Evidence

Forensic evidence must be handled by the minimum number of people. Even if it was necessary to deploy a search party to find evidentiary items, one individual should log the evidence as it is found and keep possession of it.

There must be a signed record of where and in whose possession the evidence has been from its initial collection until its appearance in court as an exhibit.

Forensic evidence must never be left unsecured or unattended.

4.9.4 Disposal of Bulk Explosives

If possible, the destruction of the bulk explosive evidence should be recorded by witnesses, still photographs and videotapes for use in future court proceedings to demonstrate the destructive force of the material.

Consideration could also be given to recording the type of effect the device would have had on an identical target. For example, the quantity of explosive being destroyed could be initiated in the same location inside a duplicate vehicle. Such a technique, of course, is not always applicable, and any deviation from the actual situation may be challenged in court. The conditions of the offense should be duplicated as exactly as possible on the test range, and the entire procedure should be accurately recorded and witnessed.

4.9.5 Forensic Analysis

See this publication, Chapters 5, and 8 through 12.

The Royal Canadian Mounted Police accept no responsibility for any use made of this material.

Acknowledgement

Figures 4.1–4.6 and 4.14–4.16 are reproduced with permission of the Royal Canadian Mounted Police.
© Royal Canadian Mounted Police, 1997. Published by Taylor and Francis with permission.

References

BEVERIDGE, A. D., 1993, Investigation of explosions, in Freckelton, I. and Selby, H. (Eds), *Expert Evidence*, chapter 84, Sydney: Law Book Company.

HALL, R. A., 1991, Booby traps associated with violent crime investigations, *Journal of the Forensic Science Society*, **31**, 255–257.

LLOYD, J. B. F., 1986, Glyceryl dinitrates in the detection of skin contact with explosives and related materials of forensic science interest, *Journal of the Forensic Science Society*, **26**, 341–348.

LLOYD, J. B. F., and KING, P. M., 1989, 1991, *Detection and persistence of traces of Semtex and some other explosives on skin surfaces*, in *Proceedings Third Symposium on Analysis and Detection of Explosives*, pp 9.1–9.14, Berghausen, Germany: Fraunhofer-Institute für Chemische und Technologie (ICT).

VERMETTE, J.-Y., 1996, *Explosion Investigation*, internal publication, Explosives Disposal and Technology Branch, Royal Canadian Mounted Police, Ottawa.

Further Reading

Required reading for anyone seeking information on systematic explosion scene investigation:

NFPA 921 Guide for Fire and Explosion Investigations, 1995, by the US National Fire Protection Association, 1 Batterymarch Park, PO Box 9101, Quincy, MA 02269-9101, chapter 13.

Useful background, but out-dated forensic science:

Yallop H. J., 1980, *Explosion Investigation*, Forensic Science Society, Harrogate, England, and Scottish Academic Press, Edinburgh.

The scene of an air crash:

Garstang, J. R., this publication, Chapter 6.

The scene through the eyes of a forensic scientist:

Strobel, R. A., this publication, Chapter 5.

The scene of a gas explosion:

Harris, R. J., 1983, *The Investigation and Control of Gas Explosions in Buildings and Heating Plant*, London: Spon.

Foster, C. D., this publication, Chapter 7.

5

Recovery of Material from the Scene of an Explosion and its Subsequent Forensic Laboratory Examination – a Team Approach

RICHARD A. STROBEL

> 'I see no more than you, but I have trained myself to notice what I see.'
> Sherlock Holmes
> *The Adventure of the Blanched Soldier*

5.1 Introduction

An explosion, whether accidental or intentional, typically results in serious damage to property and harm to people. Rather than the well rounded investigative expertise of a Sherlock Holmes, explosion investigations require personnel highly experienced in post blast investigation representing a variety of highly specialized disciplines. As the investigation moves from the scene to the laboratory, additional experienced forensic expertise will be required. This chapter discusses the team approach to the investigation of the scene of an explosion from the perspective of a forensic scientist, and then looks further at the application of a team approach to the laboratory aspects of an investigation.

5.2 The Team Approach to Investigation of the Scene of a Bombing

The forensic scientist is usually thought of as being tied to the laboratory examining evidence using various instruments, microscopes and chemicals. In the investigation of an explosion the observations of the scientist at the scene and assistance in the collection of explosive residues on site can aid both the overall investigation and subsequent laboratory examination. The sheer variety of chemical compounds for which the chemist must search and the unlimited and complex matrix of materials from which the chemist must recover residues make for a formidable task.

5.2.1 Incident Response Planning

Whether the incident is a simple 'test bomb' made by children and detonated in an open field, or a major terrorist attack involving multiple jurisdictions and crossing national borders, response planning is essential. The individual steps involved in the post-blast bombing investigation remain the same for any incident but the resources required vary widely. The crime scene may cover an area varying from a school washroom to several square blocks as in the bombing of the World Trade Center, New York, in 1993, or multiple counties as in the Pan Am 103 bombing over Lockerbie, Scotland, in 1988. Personnel needs can vary from as little as one investigator to as large as deployment of the nation's armed services to search large areas. Once the event has occurred, the investigating agency which has done no planning can do little more than play 'catch up' as the investigation unfolds. Preplanning on the part of investigative agencies largely dictates whether the investigation runs the investigators or vice-versa.

5.2.2   Response Teams

An available and experienced response team capable of instant mobilization is the most effective means of providing the necessary resources. Such teams are composed of trained and experienced personnel in all appropriate disciplines. Additional resources and specialized disciplines can be added as circumstances dictate. The following illustrate the salient features of some North American teams:

- Bureau of Alcohol, Tobacco and Firearms (ATF): ATF, of the US Treasury Department, has primary federal jurisdiction in criminal bombings in the United States. The National Response Teams, established in 1978, are constructed of 10 Special Agents, a forensic chemist, an explosives technician and a team supervisor. Four teams service the country such that the closest can be on-site and working within 24 hours of activation. Within the body of the team lie the varied expertise necessary for proper examination and documentation of the crime scene. Agents specializing in photography, crime scene sketching, interviews, evidence collection and preservation are trained together to provide for primary and back-up duties in those areas. This flexibility permits a shift of resources in response to rapidly changing situations. Additionally ATF has applied this same concept to its International Response Teams which are available to aid foreign governments upon request. This 'team concept' is illustrated by the duties of the various members of the team as listed in the 'Explosives Investigation Guide' in Appendix 1 (Garner et al., 1986).
- Federal Bureau of Investigation (FBI): The FBI, which has responsibility for terrorist bombings in the US, established specialist Evidence Response Teams in 1993 to undertake evidence collection at major crime scenes. A forensic chemist is available to attend scenes of major explosive incidents falling within FBI jurisdiction.
- The Royal Canadian Mounted Police (RCMP): The RCMP has a National Post-blast Response Team built around a post-blast consultant and a forensic chemist. Other disciplines are added as required. Local resources are utilized for crime scene recording, interviews etc.

Such teams aid local authorities by supplying the necessary specialized expertise to quickly provide the maximum amount of information. This allows a fast-track approach to the investigation. The team's forensic scientist(s) provides expert advice and guidance on recognition, collection and preservation of physical evidence.

5.2.3   Field-portable Explosives Detectors

Most laboratory instrumentation is not readily portable and given that material can be shipped almost anywhere overnight, a temporary on-site laboratory is rarely used in most bombing investigations.

However, for major bombing scenes, sensitive and selective field-portable explosives detectors developed for aviation security can be used with great effectiveness on-site (Chapter 3, Section 3.3.5). Post-blast investigators and/or forensic scientists can quickly screen large quantities of bombing debris in order to prioritize evidence for subsequent laboratory examination and provide highly reliable indications of the identity of the explosive used.

For example, the EGIS explosives detector is a 'field portable' instrument which uses the same principles of detection and specificity as laboratory instrumentation now in standard use in many forensic explosives laboratories. The EGIS is capable of detecting traces of all common high explosives of interest: EDGN, NG, TNT, PETN, RDX and others (Fine, 1984; Hnatnicky, 1994).

If the number of laboratory personnel available are limited, their primary focus and talents should be utilized in evidence collection and preservation. Generally, laboratory analysis can be performed more efficiently and effectively at a later time. If the most valuable evidence is not collected, no amount of effort in the laboratory can overcome the deficiency.

5.2.4   Initial Evaluation of the Incident

Initial reports received through normal telecommunication channels may often prove inaccurate. They can, however, provide the foundation for clarifying questions to assess personnel and equipment needs, for example, special equipment for protection from hazardous materials. Current television news reports are a good source of initial scene information and can assist the expert in the initial scene assessment.

It may be useful initially to dispatch one member of the forensic team to assess personnel and equipment needs and to time the supply of additional resources. It may be days or weeks before access can be made to portions of the scene. For example, the scene at the World Trade Center bombing was in total darkness and structurally unsound for four days before lighting was installed and several weeks before the structure was stabilized. Such scenes may have to be worked in stages such that only a few team members can be deployed at a given time. However, personnel should always be deployed on the basis of 'more rather than less'.

Experience has proven that many bombing cases break very quickly and the investigation may run in many directions at once. Multiple search warrants, multiple explosion scenes and the collection of massive amounts of evidence make the need for adequate personnel critical.

Questions before leaving for the scene

Prior to traveling to the scene the responding expert may need to initiate actions necessary to ensure evidence preservation. A pre-planned check list might ask the following questions:

- How are the victims being handled?
  Are they being documented, and considered as sources of physical evidence?
- Will rescue efforts be coordinated with the investigation?
- Are precautions underway to preserve explosive residue evidence against environmental loss – e.g. cover the crater, prevent accidental removal of materials?
- Are render safe procedures being employed against unexploded or misfired explosive devices?
- Have suspect(s) been apprehended and are appropriate measures being taken to preserve explosive residue evidence? Have clothes been retained and properly packaged, vehicles and possessions seized and preserved?
- Is execution of a search warrant imminent?

Questions about possible accidental cause

It is important during initial scene evaluation to be open to the possibility of any accidental causes for the explosion. If the initial investigation proceeds without a clear indication as to whether the explosion is accidental or intentional, the process will soon become bogged down with no clear direction. The determination of the exact cause of the explosion is the primary priority for the investigative team – dictating all future activities. Some clarifying questions can narrow the possibilities:

- Was there natural gas service to the location?
- Was any construction work in progress in the building or its environs?
- Could the explosions have been caused by another crime – e.g. a fuel/air explosion caused by the accidental ignition of fumes from an intentionally poured accelerant or a clandestine drug lab?
- Was there accidental structural collapse?

Questions on-site

Following the deployment of the team, the forensic scientist will need to work with the response team leader to assess the scene by answering the following:

- What physical hazards will team members face when working the scene?
- What chemical and biological hazards may be present?
- What actions must be taken to preserve the scene for explosive residue evidence?
- What actions did the fire department or other emergency responders take?
- Is the team following appropriate procedures to ensure that the crime scene is not contaminated?

The crater may be the most fruitful area for the collection of residues and device components. In the Oklahoma City bombing the crater was the geographical focus for much of the rescue effort. It was preserved by building a wooden cover spanning its diameter.

5.2.5   Hazards

The forensic chemist contributes to the safety and welfare of the team by identification of hazardous conditions and reporting to the team leader.

Chemical

Scenes may produce a wide array of chemical hazards beyond the obvious physical hazards caused by the explosion itself. Everything from PCB-contaminated transformer oil to airborne asbestos may confront the team. Identification, evaluation and mitigation of those hazards may fall to the chemist if no industrial health personnel are available. The greatest threat, however, is posed by unreacted improvised explosives and by the clandestine explosives laboratory. The latter will be discussed in Section 5.6.

Biological

Biohazards may not be immediately obvious. Consequently, the cause and effect relationship between exposure to a biohazard and resultant illness may not be evident. HIV, hepatitis, and a host of virus strains could infect a scientist exposed to bomb evidence contaminated with biological fluids. The simplest safeguard is to assume that contaminated material is present and to take appropriate prophylactic measures. Airborne biohazards may also present a problem in scenes of massive bombings where victims' remains cannot be readily recovered.

Physical

Building collapse is a serious hazard in massive bombings. Directed demolition of the structure may be necessary. Structural engineers may be employed to monitor the remaining building structure as was done in the Oklahoma City bombing in 1995, where numerous such engineers monitored the structural integrity of the Federal Building. A system of loud air horns warned rescue personnel and investigators of any changes in structural stability.

Secondary devices

Any bombing scene may have a secondary device targeted at rescuers and investigators (Hall, 1991). The scientist should not enter the scene until bomb disposal experts have determined that no such devices are present. This procedure is often referred to as the 'primary sweep'.

5.3   Types of Response

The nature of the incident determines the nature of the response.

5.3.1   *Live Explosive Devices – Implications of Render Safe Procedures*

Rendering safe a live device often involves the use of explosives. The objective is to preserve as much evidence as possible without impairing safety. Pre-planning and communication with the forensic scientist is essential. The explosive used in the render safe procedure must be chosen so that the forensic scientist may readily exclude it from the explosive used in the device. For example, a pipe bomb usually contains a low explosive such as black or smokeless powder and thus could effectively be disrupted with detonating cord, which contains the explosive PETN. Unless the residue can be attributed, without doubt, to the original contents of the bomb, no conclusion can be reached by the forensic scientist.

Environmental contamination from the site of disposal – e.g. an explosive range or a bomb containment vessel, must also be considered. The scientist and the bomb technician must communicate closely to ensure that there is no doubt that recovered residue can be attributed to the device.

5.3.2   *The Post-blast Explosion Scene*

The successful conclusion of a post-blast investigation is directly correlated to the quality and quantity of information collected.

Initial observations

The observation process begins before the forensic scientist has hung up the phone from the initial call. All future information will continuously undergo a process of re-evaluation and verification. Initial hypotheses may change as they are tested by the collection and evaluation of further facts. Once on-site, systematic observation and evaluation of the entire crime scene will take place. Means and methods vary depending upon circumstances of the scene but generally follow a pattern where the scientist evaluates from the least damaged areas to the most with the seat of the explosion being the final goal. This approach is often useful because any singular observation may yield information leading the investigator in the wrong direction. Conclusions must only be made when all the information is available.

Typically, the scientist, team leader and post-blast experts 'walk through' the scene, following the primary sweep, to get an overall impression, and then commence evaluation at the periphery and work towards the seat of the explosion. Observations are directed towards answering the following questions:

- Does the explosive indicate a diffuse blast seat or a point source?

- Is the blast damage indicative of a high or low explosives?

- Are obvious components of the explosive device present (e.g. pieces of fragmented pipe) and where relative to the seat of the blast are they located?

- What type of damage and discoloration are on witness materials (indigenous to the scene and in a position to directly receive blast damage)?

- What is the nature, size and geographic distribution of the blast debris?

- At what distance from the blast seat are the furthest items of debris located? If the furthest items are beyond the established crime scene perimeter, immediate action will have to be taken to extend the perimeter.

- What odors are present? A sulfurous odor indicating a sulfur-containing explosive such as black powder? Are numerous people in the vicinity of the blast complaining of headaches? Often a dynamite-based explosive will leave airborne traces of nitroglycerine, a known vasodilator, capable of producing headaches.

Identification of the explosive through damage assessment

One of the most important tasks requiring the set of skills that only the forensic scientist possesses, is the preliminary identification of the explosive used in an incident. This determination can provide important information early on for the investigator and may play a major role in the direction the investigation will take. Another value in the identification of the explosive is to provide investigative leads for the police team.

Explosives are usually only available through certain restricted and regulated commercial routes and outlets. For instance, the identification of the chlorate explosive mixture and subsequent findings of match heads in the remains of a pipe bomb might suggest juvenile involvement in a bombing incident. The investigator could use this information coupled with such information as to the target of the bombing to conclude that the incident is prank-like in nature. Alternatively if RDX, the explosive ingredient in the military plastic explosive C4, is identified in the remains of the pipe bomb, a very different circumstance is at hand. The availability and occurrence of C4 in explosives incidents in the United States is very low and would require a more sophisticated firing train and knowledge of explosives to be employed in the incident. The investigator would therefore take on a very different investigative track asking: What is the source of the C4? Is a military base nearby? Where was the necessary detonator obtained? This information coupled with information concerning the bombing target will result in a very different investigative course. This reduces and focuses the canvassing activities for the investigators.

Further value in the identification of the explosive lies in the information's use in providing a *corpus delicti* for the crime itself. Opposing legal counsel may imply that the explosion was accidental and suggest that there was no crime at all.

A final value in identification of the explosive is in its ability to link suspect and crime. If the same explosive is found on the suspect's possessions, premises, tools, clothing, vehicle or in his possession an extremely strong association with the crime is established.

A series of diagnostic observations will aid the team in a tentative determination of the explosive used. Determining the explosive used begins with a series of diagnostic questions through which the team must proceed. These observations focus on three primary effects of an explosive upon its surroundings:

- pressure;
- heat;
- fragmentation.

These blast effects are discussed in detail under the Section 5.5.1. Both high and low explosives exhibit these effects but in very different manners.

Low explosive damage   Low explosives function by deflagration (very rapid burning) rather than detonation (shock wave) (Bender, Chapter 11). Deflagration occurs at speeds where the reaction front proceeds through the unexploded material at velocities less than the speed of sound. Typically these explosives need some form of confinement; low explosive pipe bombs comprise the majority of the explosive devices encountered in the United States.

Finding the remains of a container can verify the use of a low explosive, as can searching the immediate seat of the blast. If a low explosive was used, the degree of damage directly attributable to blast pressure should be minimal and highly localized. The vast majority of the damage produced is fragmentation damage from flying pieces of the container. Most witness material in close proximity to a bomb would exhibit evidence of impact from these flying sections of container, and only those in the most immediate vicinity of the blast would exhibit damage from the pressure effects of the deflagration of the explosive itself.

The quantity of low explosive affects the area over which characteristic damage is visible but not the characteristics of the damage.

Primary fragmentation should be distinguished from secondary fragmentation. Primary fragmentation originates from the explosive device, e.g. pipe fragments, nails and battery casing. Secondary fragmentation is material from the environment which is picked up and propelled by the force of the blast. Damage produced by secondary fragmentation is minimal in cases where low explosives are used.

As discussed in Chapter 11, some low explosives, notably double base smokeless powder, can detonate given the proper conditions and exhibit post-blast effects more characteristic of a high explosive.

High explosive damage   Use of high explosives can be recognized by characteristic blast effects which exhibit a great deal more blast pressure damage than low explosives. This power results from the reaction front traveling through the unreacted explosive in excess of the speed of sound. The characteristic manner in which that pressure is applied on the environment is termed its *brisance* or shattering effect.

The pressure from a high explosive is applied in massive amounts and in a very short span of time. A crater is typically produced at the seat of the blast and damage is caused by shattering the environment by pressure and shock. This effect is evident regardless of whether or not the explosive was confined. This brisance produces terminal blast effects unique to high explosives and some effects which are unique to specific explosives within this group. High explosive detonation velocities range from 3000 feet per second to over 25 000 feet per second. The degree of brisance evidenced on the surrounding witness material correlates with increasing velocity. A careful and studious analysis of these observations at the bombing scene can allow the experienced investigator to estimate the type of high explosive used.

5.4   Post-blast Residue Collection

The forensic scientist's ultimate goal is to find traces of the original explosive material not consumed in the blast. The type of residue, the manner of its physical form (solid, liquid or gas) and the means and manner of collection, depend on which explosive was used. Because of these variables, an assessment of the type of explosive used based on observations at the scene, as discussed above, is very important. That determination provides the starting point from which flow all subsequent steps in the residue collection process. By consulting and/or participating in collection and preservation of suspected residue-bearing material, the forensic scientist can ensure that the best evidence gets to the laboratory.

Residual traces of explosives can remain for long periods of time. The explosive oils nitroglycerine and ethylene glycol dinitrate have been recovered as long as five years after initial deposit (Beveridge, 1991). Free-flowing particulate explosives, like black and smokeless powders, remain indefinitely if not physically removed. Longevity of the explosive traces is also related to the material on which it is absorbed or deposited – for example, nitrate esters have a great affinity for protein-based or like materials such as wool, or for plastics such as acrylics, polyester or nylon carpeting (Higgs et al., 1978; Higgs and Hayes, 1982).

5.4.1   Low Explosive Residues

The mechanism and chemistry of the deflagration of most low explosives is relatively simple and the principles of collection and preservation are easily conveyed to crime scene investigators. With the proper collection of these residues, properly trained and equipped laboratory personnel can achieve a very high rate of success in reaching a conclusion as to the explosive used (Bender, Chapter 11). The key to success lies primarily with the collection of residues at the scene of the explosion.

Low explosives require confinement, typically a pipe, and initiation either with a fuse or 'hot wire' capable of reaching the initiation temperature of the explosive. The reaction then proceeds with a particle to particle transmission of the deflagrating front. As the pressure and temperature rise, the rate of the reaction also increases until the pressure exceeds the bursting strength of the container. At this point, pressure is rapidly released, the reaction rate rapidly decreases and unconsumed particles are ejected from the container. Such recovered particles permit the forensic scientist to provide the investigator with the probable manufacturer and brand of explosive which may narrow the search of possible retail sources. If a suspect is subsequently identified, recovered particles can be compared to powder in the suspect's possession.

Collection

The most efficient manner in which to collect low explosive particles is physical sweeping (clean, unused broom and dustpan) or mechanical vacuuming (clean, unused) of the blast seat. Subsequent microscopic examination in the laboratory frequently reveals unconsumed low explosives.

5.4.2   High Explosive Residues

Where high explosives have been used, collection and preservation of residues are more complex and require the full technical expertise of the forensic scientist. The first obstacle is the sheer variety of explosive formulations and physical forms which are available today: pure compounds, emulsions, slurries, mechanical mixtures, gels and various combinations of liquids and solids. Each has its own effect on collection of post-blast residues or residues on a suspect or his possessions, and on issues relating to contamination between items of evidence.

Because the reaction rate proceeds through the unreacted explosive at a velocity exceeding the speed of sound, the unreacted explosive remaining after a blast is typically in trace quantities and invisible. The exception is when the explosive does not function as intended resulting in visible amounts of remaining explosives. This condition is termed a 'low order' detonation and can result from the improper initiation of the explosive, the condition of the explosive or improper design of the firing train. Sometimes, the mixture of the explosive may be faulty through inadequate mixing or an individual component may be missing; this most frequently occurs with improvised explosives.

Collection

Identification of the post-blast residues from high explosives is necessarily a laboratory process. The primary goal of the field work is collecting witness materials which have a high probability of yielding explosive residue in laboratory examination. This becomes a formidable task considering the wide array of variables affecting the survival of these residues. A massive bomb results in a vast crime scene with corresponding probable 'dilution' of residues. Such an investigation would be impaired without the on-site assistance of a consultant forensic chemist.

Recognizing materials likely to retain residues which are at the same time close to the original explosive charge but not subjected to the residue-destroying thermal effects of the explosion requires experience. Such experience may be gained as an apprentice at actual scenes, and as an active participant in test explosions (Beveridge, 1986).

Vapor at the blast scene and on items of evidence may be sampled for explosive vapors by passing air through a suitable adsorbent material such as Tenax® or charcoal using a portable pump (Chrostowski et al., 1976; Wardleworth and Ancient, 1983; Deak et al., 1989). This technique is particularly useful when dynamite has been used since it contains explosive oils which have sufficient vapor pressure to be collected in this manner. However, the decline in the use of dynamite is diminishing the utility of this type of vapor analysis. On the other hand, developments in the application of detectors developed for aviation security are proving increasingly useful for detection of a wider range of explosives as described in Section 5.2.3. However, the wider availability and industrial use of water gel, emulsion and high strength blasting agents which do not contain nitrated organic explosives are pushing the post-blast analysis techniques used in criminal bombing investigations in other directions (Section 5.11).

Once the analytical challenges have been met and chemical residues identified, the next challenge is to assign scientific significance to those findings. For instance, to determine the relevance of finding chemicals such as ammonium nitrate and fuel oil, which are found in some explosive formulations, but also have widespread 'ordinary' uses, the laboratory finding must always be placed within the context of where the residue was found. Thus, finding ammonium nitrate (a fertilizer) may not be significant in a soil sample but could be very significant if found on the surface of a road sign directly facing the site of explosion of a large bomb. Likewise, finding fuel oil in the blast crater may be of little value if the bomb detonated in a diesel-fuelled truck.

Therefore investigation and documentation of the scene cannot be divorced from the analytical findings of the laboratory if the laboratory findings are to be scientifically sound and useful. This topic is discussed in detail by Murray in Chapter 12.

As described below, all material showing explosive damage, including device parts and material from the environment, should be submitted to the laboratory – or if a surface suspected of bearing residue cannot be moved, it should be swabbed and the swabs submitted.

Some effective swabbing techniques are described by Hiley in Chapter 10. Many factors must be taken into account prior to selection of a particular method, e.g. the explosive sought, the substrate on which it is located and the matrix in which the sample resides. This is an area in which consultation with experienced colleagues and careful review of the scientific literature can significantly assist reaching an optimal decision.

5.5   Device Components

Debris must be searched for the remains of components of the explosive device. Protocols are described by Vermette in Chapter 4. The targets of search protocols are typically damaged and altered in appearance by the explosion. Recognition of such fragments can be assisted by assessment of such factors as blast damage and fragmentation.

5.5.1   *Blast Diagnostics*

'Blast diagnostics' is a term applied to assessment of the blast effects of various explosives upon witness material surrounding the explosive charge. In some jurisdictions the scientist undertakes this work; in others a post-blast expert or team does it.

Blast effects for particular explosives apply their characteristic signatures only a very short distance from the location of the explosive charge. However, since components of the explosive device are likely to be in close proximity to the explosive charge, they are highly susceptible to evidencing the post-blast signature effects. Thus, if searchers recover explosion-damaged material, there is a high probability that it was a device component even if its precise nature is not immediately apparent.

Blast diagnostics seeks evidence of the three primary effects of an explosive on the component parts of a device:

- thermal: the temperature effect of the explosive upon its surroundings at the instance of explosion;
- fragmentation: the effects of the fragments of the explosive device upon its surroundings;
- pressure: the amount of pressure exerted by the explosion of the explosive upon its surroundings (often viewed as the peak amount of that pressure applied over a given time – *brisance*).

In order to perform blast diagnostics in a given bombing case, the scene must be assessed for indications as to whether the explosive was high or low, and the general class of explosive within the group.

The crime scene investigator must make a thorough and complete examination of the entire blast scene before forming preliminary conclusions. This includes comparing damage from the furthest reaches of the scene to the blast seat. High explosives in contact with a surface essentially excavate the blast seat whereas low explosives do not.

The scene investigator may have to make allowances for a charge which is detonated some distance from a surface, for example, a bomb in a vehicle. This distance is referred to as 'standoff'. In such instances, the investigator will have to reconstruct the seat of the blast. For example, if a high explosive charge was placed in the bed of a truck, the investigator will need to gather and then reconstruct the fragments of truck bed to characterize the nature of the blast signature and subsequently the explosive type.

After an accurate and complete assessment of the scene and an estimate of the explosive type used, the investigator may begin to examine the blast debris to identify the remaining component parts of the explosive device itself. Few among even the most experienced investigators can recognize every possible material which can be used in the construction of an explosive device, especially as small, fragmented pieces. Indeed, often only a portion of the individual components such as timers, detonators and power sources are ever recovered. The conclusions which an investigator reaches as to whether a particular small fragment constitutes a portion of the explosive device will rely heavily upon the blast damage being consistent with that on more obvious device components.

In the case of low explosives there should be a distinct lack of thermal effect on the immediate blast seat. The exception to this will be instances where fire follows the initial blast. Residue and odors characteristic of the common low explosives such as black or smokeless powder will be evident. Black powder produces a sulphurous odor plus grey to black caked-on residue on the interior surfaces of the container and possibly on nearby materials. On the other hand, burning of smokeless powder produces almost no odor and surrounding materials exhibit no visible residue.

High explosives exhibit little visible residue with the exception of high explosives composed primarily of organic compounds such as TNT, PETN and RDX which have a negative oxygen balance. In those cases significant quantities of carbon, seen as blackening, will be evident on materials close to the explosive. In contrast, lower velocity high explosives such as ditching dynamites, ANFO, emulsions and water gels exhibit almost no visible residue. This observation can be further supported by the absence of thermal damage to the material surrounding the blast seat. Significant thermal damage, such as melting of plastic materials and other synthetics, will be evident in materials close to the explosive.

5.5.2 Pipe Bombs: Metal Fragmentation Diagnostics

The pipe bomb is the most common and least complex of all explosive devices. The pipe, end caps, fuse and powder are all readily available without restriction in many countries including the United States. Because pipe bomb construction varies little and the fillers are easily recovered, much is known about their terminal blast effects. The 'fragmentation diagnostics' of the explosive filler can be observed in the field by the examination of the fractured pipe and end caps. Pipe is an engineered product meeting certified test standards for performance including pressure handling capabilities. Consequently, the pipe provides for the perfect vehicle for studying the pressure effects of the explosive filler since its pressure containment properties remain fairly consistent.

Major factors affecting the terminal blast effects observed on a pipe bomb are:

- the container;
- the explosive filler;
- the quantity of filler;
- the means of initiation.

The size of the pipe plays a role only when the pipe reaches the larger lengths and diameters. The filler determines the velocity of detonation or deflagration. Some fillers may undergo transition from deflagration to detonation velocities either as a function of the means of initiation or the nature of the explosive itself. As noted, double base smokeless powder initiated with a detonator will produce damage to the pipe resembling that of dynamite. However, many double base powders possess a critical diameter above which they are capable of reaching detonation velocity when initiated with ordinary burning fuse. The most significant factor is the size of the pipe. The critical dimensions are diameters of ca. 2 inches and/or lengths of 12 inches or more. Figures for critical diameters for various powder brands exist with the powder manufacturers.

The amount of explosive filler may also influence the blast effects. If a pipe is only partially filled, damage to the pipe fragments may vary over very short distances along its length. Thus a 12 inch pipe which is only filled $\frac{1}{4}$ of its length with dynamite will produce pipe fragments with both the characteristic fragmentation for dynamite and fragmentation characteristic of a lower velocity explosive. Therefore the scene investigators must collect as many fragments as possible before reaching any firm conclusions.

111

Next, the investigator must observe the number of fragments and their size distribution, taking care to identify which fragment is attributable to each type of pipe fitting or component. End caps, for example, are manufactured by casting and break apart without evidence of stretching. On the other hand, pipes are manufactured by extrusion and stretch to produce thinned metal. Similarly, a piece of pipe coupling, which is manufactured with threads the entire length of its cast body, will produce an effect based upon that construction: the threads act as serrations enhancing the fracturing of the coupling during the fragmentation.

When confined in a pipe, the highest velocity explosives such as C-4, TNT and PETN, typically produce fragments which are small and approximately equal in size. Other fillers produce fewer fragments with a wide variation in size as illustrated in Chapter 11.

After examining the size and distribution of fragments, individual fragments should be examined. Fractures exhibiting a 'squared off' appearance at right angles to the inner and outer surfaces of the pipe indicate a low velocity explosive such as black or single base smokeless powder initiated by flame. As deflagration velocities increase, the fractures begin to form at 45° angles to the plane defined by the parallel inner and outer surfaces of the pipe. This is caused by the stretching of the metal prior to its failure in the form of a fracture. The pipe fragments also begin to display a thinning from the original dimensions. At detonation velocities, the pipe fractures begin to form an alternating pattern of the 45° fractures. The 45° fracture will first begin on the pipe's inner surface and then alternate with the 45° angle beginning at the outer surface and proceeding inward. This switching pattern repeats itself along the length of the fracture in a stepwise fashion. This fracture pattern is termed an 'alternating stepwise fracture' pattern.

As velocity of detonation increases, the surface of the fragment itself should display characteristic erosion effects. This effect is caused by the thermal erosion of the metal as it evaporates and is a function of the thermal output of the detonation of the explosive. This erosion creates a ripple effect on the pipe's surface as the flowing blast wave passes over its surface evaporating the metal. This occurs only at the detonation velocities attained by high explosives such as C-4, Semtex, and TNT. Confined in a pipe, dynamites will also exhibit the alternating stepwise fracturing and thinning of the metal. The surface erosion however, will be absent for all but the highest velocity dynamites.

Another effect evident with high detonation velocity explosives is the 'blueing' of the metal. This is caused by the oxidation of the metal's surface in the presence of intense heat. It is often seen on the edges of fragments which have passed through the hot gases generated by the thermal effect of the explosive. Because only the highest velocity explosives produce thermal effects of duration sufficient to produce blueing, its presence is a good indicator of a high velocity explosive such as PETN, RDX, TNT, or HMX. This thermal effect will also produce characteristic effects on other witness materials (see Baker and Winn, Chapter 13).

5.5.3 Thermal Effect on Other Witness Material

Witness materials such as plastics and synthetic fabrics are likely to exhibit thermal effects. Because the thermal output of any explosive is very short in duration and only dispensed in close proximity to the initial charge, the effect is very useful for identification of materials which were adjacent to the original charge. This information greatly assists reconstruction of the device.

5.5.4 Victim Activities

The activities of the victim will often provide clues to the type of explosive device involved. The acts of turning on the ignition, stepping on the brakes or opening a door provide clues to the investigator as to the possible means of initiation. In vehicle bombings, information regarding the proximity of the victim to the explosive charge is of vital importance to the reconstruction of the original location of the device within the vehicle. This can be of great importance in reconstructing events leading up to the blast and even the activities of victims of the blast.

This was illustrated in the bombing of a government witness in a drug case. The victim was seated in the front passenger seat of the target vehicle which exploded as it cruised along an interstate highway. The victim sustained severe trauma to the lower extremities, especially the buttocks. This along with the damage to the vehicle clearly indicated that the explosive charge was placed under the passenger seat within the interior of the vehicle. The driver of the car, the girlfriend of the victim, escaped with only minor injuries.

Forensic examination of the vehicle and scene yielded few components which could be associated with any part of the explosive device. The only device component found was the remains of the detonator leg wire. But this wire yielded all the information necessary to solve the mystery. Examination of the thermal effect on the wire showed that virtually the entire length except for the last several feet of plastic insulation on both wires exhibited identical thermal damage. By matching the degree of thermal damage to areas of the vehicle with similar damage to plastic material, the original location of the wire was disclosed – it had been a coil within inches of the charge.

Further examination of the less damaged ends of the wire showed that equal lengths of both detonator wires were stripped of insulation. Comparison of the questioned wire ends with known samples of the same brand of leg wire showed the same type of toolmarks present for the stripping of the insulation and cutting of the wire ends. The lack of any evidence of material attached to the wire such as solder or signs of physical attachment such as twisting of the wire indicated that it had not been attached to any circuit or power source. Thermal damage to the wire insulation indicated that it was some distance from the explosive and not in the same location as the bulk coil of wire. Measurement of this barely damaged end coincided with it having been pulled up from under the seat and placed between the knees of the victim. This was the only location which could have afforded the necessary protection from the thermal effects of the explosive. Corroborating evidence was developed later in the form of a damaged 9 volt battery recovered from the roadway where the explosion occurred.

Later during interrogation, when confronted with the evidence developed from the forensic examination, the vehicle's occupants confessed, verifying the observations made in the laboratory. The whole incident had been a 'lovers' quarrel' with the male victim attempting suicide and murder by touching the end of a detonator wire to a 9 volt battery detonating a US Military M112 block of C4 (1.25 lb.) under his seat. This entire investigative journey was set into motion by the initial observations and interpretation of the blast effects at the crime scene.

5.5.5   Investigative Evidence

Criminal investigation is an exercise in gathering facts and information. Forensic science provides, a cold, sober, means to collect and assess those facts. The forensic scientist must perform in an unbiased manner because his or her first and ultimate duty is to assist the court in its function, the administration of justice.

The scene of a bombing always contains physical evidence originating from the device. The challenge for the investigator is to recognize and collect it and interpret its secrets. Every component of the device originated from somewhere; determining the origin could lead to the bomber.

5.5.6 Explosives

Only the forensic scientist can identify the explosive. This determination provides a wide range of important information for the investigator and may play a major role in the direction of the investigation. For example, some types of explosives are available through restricted and regulated commercial (and military) outlets. Further, if the same explosive is found on the suspect's possessions, premises, tools, clothing, vehicle or in his possession a strong link to the crime may be established.

5.5.7 'Signatures'

The forensic scientist also performs a significant role in recognition, recovery and identification of 'trace evidence' during microscopic examination of debris.

One of the most valuable leads is the discovery of a bomber's signature – a definitive design feature or choice of materials which point to one person. It could be an improvised explosive, the design of the firing circuit, a combination of components, a target type or an obvious identifier. For example the 'UNABOMER' in the United States included the initials, 'FC' on an internal component of his devices. The initials were placed in such a manner that they would survive the blast and be discovered during the examination of the device remains. It is interesting to note that the inclusion of these initials became superfluous since the UNABOMER's device designs were so unique that investigators could conclude that the bomb was his handiwork after only a cursory examination of the evidence.

The determination that bombings are a serial event allows for a more exhaustive view of the bomber to be drawn by use of advanced psychological profiling techniques. These techniques can be of great assistance in modeling the personality of the bomber, predicting future activity and suggesting investigative techniques which could be brought to bear on the case.

5.5.8 Commercial Products

As the forensic scientist and the investigator examine the fragments of the explosive device, decisions to trace some of the materials must be made. Retail establishments maintain large databases to assess and determine their marketing strategies. Those databases can also provide information on where specific items can be purchased. Many companies employ serial or production codes and data on products to assess what happened when the product is defective or requires a widespread consumer recall. Some data indicate product shelf life.

Another investigative lead can be found in product and date shift codes on commercially available materials. Additionally, many retail establishments maintain large 'point-of-sale' databases to track supply and demand. These databases can likewise provide information on the availability of a particular item in stores in a particular area of interest. Armed with such information, the investigator can prioritize personnel if a decision is made to canvass retail outlets for device components. The successful resolution of many major bombing cases have been aided by the cooperation of the retail industry.

5.5.9 Associative Evidence

Explosive devices are designed to enable the bomber to be far away at the time of explosion. Without direct eyewitnesses or evidence such as fingerprints or DNA, the court case becomes circumstantial – that is, the evidence is associative. However, a case based on strong circumstantial evidence has the potential to be just as persuasive to a jury as an eyewitness. To bring a circumstantial case to the courtroom requires all the skills of the prosecutor, investigator and forensic scientist. More will be discussed about the specific skills of various forensic disciplines in Section 5.6.1.

5.6   Clandestine Explosive Manufacturing – 'Bomb Factories'

An illicit explosives manufacturing facility is almost always a hazardous scene to process. The explosives which are easiest to manufacture are also the most sensitive to initiation by spark, heat or friction. Improvised explosives can be manufactured with as little effort as combining two chemical solutions and filtering out the explosive crystals. Often the makers have little education in the chemistry of explosives and take little care to ensure the quality of their product. This increases the hazard to investigators. Indeed, the hazards of a given situation cannot be assessed without a thorough knowledge of the improvised explosives, the literature for the manufacture of those explosives, the methods of manufacture and the raw materials involved.

5.6.1   Protocol for 'Taking Down' a Clandestine Laboratory

See Vermette, Chapter 4, Section 4.7.7.

5.6.2   Clues at the Scene

Small clues coupled with knowledge of synthetic chemistry yield valuable information in cases involving improvised explosives.

- yellow fingertips give away the maker of picric acid;
- white crystals in a jar in the refrigerator may be acetone peroxide;
- urea and nitric acid may indicate the manufacture of urea nitrate high explosive;
- hexamine, sulfuric acid and hydrogen peroxide may indicate the manufacture of the explosive HMTD (hexamethylene triperoxide diamine);
- sulfuric acid, nitric acid, glycerine and baking soda indicate manufacture of nitroglycerine: consequently, any container of clear liquid should be considered to be nitroglycerine until proven otherwise.

Interpretation of such observations and the response to them falls to the chemist. Clandestine laboratories also frequently contain unmarked containers. The contents must be analyzed and possible uses for the chemicals as precursor reagents or explosive or incendiary materials must be recognized by the chemist.

A primary manufacturing process for explosives involves nitration, usually by a nitric/sulfuric acid mixture. Alternatively, various oxidizers such as chlorates, nitrates, perchlorates, etc. may be mechanically combined with diverse fuels like charcoal, sulfur, sucrose, and aluminum powder. Such mixtures must be handled with the greatest of care.

5.6.3   Literature

Unfortunately, the ability to manufacture even the most 'sophisticated' of the improvised explosives is within the grasp of almost anyone regardless of age or experience as a chemist. Step-by-step instructions are available in specialized books available via mail order, and on the Internet. Instructions usually involve readily obtainable chemicals or common household substitutes. Many manuals feature instructions which are intended for use 'in the field' providing for methods of manufacturing under very primitive conditions. However, the methods and products are far from fool-proof and can present as hazardous a challenge to the manufacturer as to the investigation team.

Therefore, just as a chemist must know what to look for in post-blast debris by experimenting on an explosive range, so too must she know what to look for in a clandestine laboratory by maintaining a thorough knowledge of the literature.

The key to bringing a hazardous situation to a safe and successful conclusion is advance intelligence, a thorough knowledge of chemistry and rigid application of a protocol as noted above.

5.7 The Role of the Forensic Scientist in the Laboratory – the Team Approach

Much of the information available in the debris of a bombing must wait to be unravelled in the laboratory. The forensic scientist may use systematic observations in the field to generate theories about what explosive was used, but only a systematic analysis will determine the actual composition.

5.7.1 Application of Forensic Laboratory Disciplines to Explosion Investigations

One of the key resources to the successful outcome of a bombing investigation is the proper utilization of all the tools of forensic science.

Chemistry

A primary role of the forensic chemist in a bombing investigation is identification of the explosive used. Analytical techniques vary with the circumstances, but the key is a systematic analytical protocol which will capture any explosive. Details are given below. Intact explosive samples must also be analyzed to verify their identity.

Analysis of related materials from the explosive device are also conducted within this discipline. Materials such as wire insulation, lubricants, batteries, glue, adhesive tapes, electronic parts, plastic and metal fragments are among those which are typically identified, compared and sourced by the chemist. These materials may also be referred to experts in other disciplines; for example, tape will be referred to latent fingerprint specialists and writing to questioned document examiners.

Toolmarks

Toolmarks are valuable direct evidence in a bombing investigation. These are present on bombs in most cases and are capable of surviving the blast depending on its degree of destructiveness. Almost without exception, the construction of an explosive device requires the use of tools and their use inevitably results in the deposition of toolmarks. If a component of the bomb can survive the blast, so too can the toolmark. Therefore, components such as screws, cut ends of wire, wrench marks from the tightening of pipe caps and marks deposited by the action of wire strippers on insulation are all useful for later association with the tools of the bomber.

Examiners are capable of characterizing the toolmarks observed on a device component and rendering an opinion as to what type or class of tools made the mark. This provides an investigative lead which will allow the investigator to narrow the scope of his search in subsequent search warrants and also provide for the basis for search warrant applications.

Trace evidence

Trace evidence covers a wide variety of microscopic evidence which can link the environment of the bomb construction with evidence at the bombing scene. Materials as diverse as carpet fibers located on tape which held batteries to the bomb, to DNA deposited by the bomber on the stamps placed on a letter bomb are often recovered and subject to analysis.

Metallurgy

Many components of an explosive device consist of metal. Metals not only retain toolmarks but their elemental composition may link them to materials in the bomber's possession. Metal shavings from a work bench can be linked through elemental composition to the pipe in a pipe bomb, or shavings on a drill bit linked to an access hole for fuse or wire. The techniques of metal analysis are fully described in Chapter 13.

Document examination

Documents in the form of threatening letters or demand notes or even a mailing address may accompany an explosive device. Components in an explosive device may consist of items from which a document examiner may obtain investigative information.

In a series of bombings in the Midwestern region of the United States in 1988, threat notes accompanied booby-trapped devices. The notes were hand-written with a 'magic' marker felt pen. Some of the marker bled through the top page onto subsequent pages of the note pad on which the threats were written. The bombings comprised 22 incidents over a five state area over 2 months. The devices, owing to their random placement and random time of discovery, did not immediately display a pattern to investigators.

A document examiner was able to determine the sequence in which the devices were placed by examination of the bleed-through patterns on the notes allowing the investigators to establish the trail the bomber took throughout the Midwest and alert authorities to missing bombs. After capture of the bomber, the notes were directly linked to the pad of paper in the bomber's possession and to the handwriting of the bomber himself.

In the Narita airport bombing in 1985, a letter 'M' from a handwritten stencil on the cardboard box which contained the bomb inside a metal-cased tuner survived the explosion and narrowed the box to one of six from the original starting point of one in four thousand (Beveridge, 1991).

Latent fingerprints

Many of the components of an explosive device survive the blast although their physical form is usually quite altered. Latent fingerprints, too, can survive a blast. The thermal effect of the explosion is most detrimental to survival of a fingerprint – especially those deposited on nonporous surfaces. The likelihood of a print surviving on the surface of a pipe is diminished by not only the nature of the surface but because of the thermal 'evaporation' of the latent print's components.

Fingerprints are more likely to survive on porous materials such as tape, paper and cardboard. If these materials survive the explosion and its thermal effects, the recovery of latent fingerprints should always be considered as a possibility. Among the most fruitful areas to examine for fingerprints are the adhesive side of tapes which often are used as a component in explosive device construction.

5.8 Explosives Technician – Device Reconstruction

In many law enforcement organizations the expertise relative to the explosive device design, construction and effects may be handled by personnel with explosive ordnance disposal backgrounds or those skilled in the dismantling of military ordnance and improvised explosive devices (bomb squad).

As noted in Chapter 4, it is essential that the chemist be consulted prior to rendering safe a live explosives device to minimize contamination problems.

Although rarely the case, if the explosive used cannot be identified, it then becomes the responsibility of the explosives technician to determine that the device indeed contained an explosive. He can reach that conclusion using available evidence such as an X-ray of the device prior to render safe procedures or the post-blast signature indicating an energy output of the bomb greater than that of the explosive used in the render safe procedure. As an expert in improvised explosive device construction, the explosives technician can lend expertise to the recognition of device components both intact or post blast.

Another significant role is construction of a simulated device for demonstration purposes in court.

5.9 Cross Contamination of Evidence – the Issue and the Precaution

When a laboratory scientist takes an explosives case, he or she must not only document the chain of custody but also evaluate the integrity of the evidence by questioning those who seized, packaged and transported the evidence. If the integrity of the evidence cannot be assured, then the hours of work in the laboratory will be for naught.

For many explosives, contamination only occurs when evidence comes in physical contact with contaminated surfaces. The simplest manner in which to eliminate contamination in these cases is to use disposable material for everything that comes in contact with the evidence. Disposable lab bench covers, disposable lab-ware, containers for sampling and sample preparation and other apparatus all ensure against contamination. The best insurance for avoiding contamination between debris from the blast and material seized from suspects is isolation. It is best to maintain physical separation by keeping debris and associative evidence physically separated in different parts of the laboratory, different rooms or different buildings.

The explosive oils, particularly ethylene glycol dinitrate (EGDN) and nitroglycerine (NG), primary components of dynamite, possess high vapor pressures – particularly EGDN. This allows them to diffuse through the air and also through any packaging material not impervious to the vapors. Plastic wrapping of the type typically used for evidence packaging is not impervious to these vapors. It must be assumed that EGDN and NG from a bombing where dynamite was used could easily migrate through plastic containers and into, for example, the clothing seized from a suspect if it also were to be packaged in plastic. The volatile components of dynamite have a very strong affinity for a variety of materials to which they bond in an almost irreversible manner. There the material will remain regardless of environmental factors until removed by chemical means.

Therefore, any container or even an environment such as a room or building which is not vapor tight, must be considered as having the potential to allow for cross contamination. The issue of quality control is dealt with in depth in Chapter 10.

5.10   Initial Evaluation of the Evidence – Formation of the Forensic Team

5.10.1   Establishing Priorities

When a bombing case enters the laboratory, priorities must be assigned to maximize the usefulness of the forensic laboratory's capabilities.

Questions to be asked are:

- Are fatalities involved?

- Is this a part of a serial bombing case?
  If the case is part of a serial bombing, immediate action needs to be taken in order to put maximum effort into the apprehension of the bomber before the next incident.

- Have investigators already identified a suspect?
  In this instance lab priorities may be shifted toward corroborating evidence to rule the suspect in or out. The lab may also be required to provide information to form probable cause for the issuance of a search warrant. The scope of the warrant may also be determined by laboratory findings . . . is the explosive one which would leave residues capable of recovery in a specified amount of time? . . . would the warrant include a search for tools which were identified by the laboratory as having been used in the construction of the device?

- Is the device typical of those used by a juvenile?

- Is the device something that has been seen of late on computer bulletin boards?

- Is it a bomb commonly used by organized crime groups?

- Are any of the components uniquely traceable?

If there are no suspects, laboratory examination may provide investigative leads to one.

5.10.2   Investigative Evidence

A variety of investigative information can be obtained from the laboratory analysis of explosive devices.

These include.

- the sophistication level of the bomb maker;

- the source for the design concepts of the device: is this design copied from those available in the literature? Does it indicate military knowledge or training?

- similarity to other devices: ATF maintains a database, EXIS (Explosive Incidents System) containing pertinent information on all domestic bombings since 1975. The database can retrieve information based on a host of categories such as target, device components, motive, etc. In excess of 52 780 incidents are currently logged in the system.

- sources for the individual components of the device: often the laboratory has contacts in the industries which produce commonly used parts of an improvised bomb including explosives manufacturers, electronics stores such as Radio Shack®, model rocketry suppliers, pipe and fittings suppliers, manufacturers of low explosives such as black and smokeless powder, adhesive tapes, batteries, etc.

- Is the explosive commercial or improvised? If improvised, what is the knowledge source? From where could the required chemicals be obtained? What level of training/knowledge is required for the manufacture of the explosive?

5.10.3 Associative Evidence

Once a suspect is identified the search for physical evidence switches from investigative to associative – that evidence which will directly or indirectly tie the suspect to the crime. The evidence can range from direct, such as a suspect's latent print on a fragment of tape, to the more circumstantial such as the purchase of the components used in the device.

Circumstantial evidence, though not conclusive in its individual instances, can become very persuasive if it is present in large quantity. Associative evidence is often developed by laboratory examination in the various forensic disciplines mentioned previously.

5.11 Information Management – Integrating the Laboratory Information with the Field Investigation

Communication is essential. In Chapter 4, Vermette outlined a protocol for crime scene management whereby a coordination and control center ensured that information passed readily between the crime scene investigators and the 'outside' investigators. A similar protocol is required to ensure that the results of laboratory analyses and conclusions are transferred immediately to the field investigators.

5.11.1 The Laboratory Task Force

Designation of a person to be responsible for information flow both within the laboratory and to investigators in the field may be required – particularly in the early stages of a bombing investigation where information gathering moves fastest.

Most bombings have their best chance of being solved if information is plentiful and timely. Efficient information flow is the crucial ingredient. An example is the 1993 bombing of the World Trade Center in New York City. Two days after the bombing a team of New York City Bomb Squad, ATF Explosives Enforcement Officers and ATF Forensic Chemists on assignment to retrieve explosive residue samples for analysis encountered a fragment of frame rail from a heavily damaged vehicle. From the blast signature on the metal frame it was concluded that this was a portion of the vehicle which held the explosive device. Following documentation, the frame piece was immediately transferred to the New York City Police Department crime laboratory where a team of technical experts were able to recover a portion of a vehicle identification number. This number was immediately traced to the rental agency where a suspect was promptly identified and arrested. The information, which originated in the blast crater had rapidly passed to personnel who could recognize, decipher and immediately put to use the investigative lead.

5.11.2 Modes of Communication

Information flow is critical in all areas of the investigation: the bombing scene, command centers and the forensic laboratory.

Just as all laboratory findings may pass through a single coordinating member of the laboratory it is critical that this also occur at the bomb scene.

ATF National Response Team guidelines are similar to those described in Chapter 4. The Team Leader receives information from the lab, interview teams, each on-site post-blast processing team, headquarters personnel and intelligence. It is also the Team Leader's responsibility to make sure each member of the team gets all the information needed. Daily briefings at the beginning and end of the day are the most effective means of accurately and efficiently transmitting information.

Depending on the incident, many of the investigators and other personnel may be working in widely separated geographic areas. For this reason other forms of communication become important. Telephone conference calls, fax transmissions, teleconferences and even recorded videos may all be necessary to keep everyone informed. Video briefings from the laboratory as results become available have been particularly effective in getting information to field investigators. In many cases as the investigation proceeds, investigators will be out of touch with command centers and may miss some of the daily briefings. The video briefing allows all investigators access to laboratory information in a timely manner regardless of whether they miss other briefings. They also allow the laboratory personnel to conduct in-depth visual briefings showing important pieces of physical evidence to each investigator. If search warrants are under way, the visual briefing with descriptions of items of evidence to collect can be invaluable.

5.11.3   Search Warrant Planning – the Right People, Information and Equipment

Increasingly the judicial system looks to scientifically analyzed physical evidence at trial. Juries expect unbiased testimony from forensic scientists for information in reaching a verdict. One of the best means for tying suspect and crime is through physical evidence. In bombing cases, much of the physical evidence will be circumstantial.

Most of the associative physical evidence collected in a bombing case result from warrants served on the suspected bomber seeking materials used in the bomb, materials scrapped during device construction, tools used in construction or residues deposited during construction. Logically, the people within an organization with the greatest expertise in these areas, forensic specialists, will have the breadth of experience to maximize the identification and collection of these items during execution of a search warrant.

At this crucial stage of investigation, it is critical to assemble the most experienced personnel. All team members must be carefully briefed, made aware of the expertise of others on the team, have clearly defined responsibilities and the means to provide for the proper flow of information while the search is on. This is crucial when multiple simultaneous searches are taking place in different geographic locations.

5.11.4   The Evidence Advisory – a Listing of All Associative, Circumstantial and Direct Physical Evidence

The 'Evidence Advisory' is a list of all associative, circumstantial and direct physical evidence items that should be sought during the search. The list is prepared by laboratory personnel in each of the forensic disciplines which have examined physical evidence. Evidentiary weight for subsequent judicial proceedings is not an issue.

Items run the range from explosive residues to receipts for component parts of the device to original packaging for a component.

An item such as black electrical tape, common to everyone's home and common to many explosive devices, may be of little evidentiary weight in court. But if the tape roll has a unique torn end matching tape from a bomb, the evidence becomes compelling. Tape could also yield trace evidence which could be matched to the bomb such as hair, carpet fibers, material on the workbench which adhered to the edge of the adhesive layer. Much of this evidence could be overlooked if not noted on the Evidence Advisory checklist.

5.12   Systematic Analysis of Explosive Residues

The flow diagram illustrates a generalized analysis scheme for post blast identification.

The types of explosives and explosive residues received by the laboratory for analyses can be either simple or complex. Most are complex. Explosives generally fall into two classes, low explosives and high explosives. These can be further classified as commercial, military or improvised, and can consist of inorganic, organic or organic/inorganic mixtures. Typical examples are:

- inorganic: black powders, pyrotechnics, improvised explosives, flash powders.
- organic: plastic explosives (military), TNT, cast boosters, smokeless powders, NG, PETN, RDX, HMX, detonating cords, improvised explosives;
- organic/inorganic: dynamites, water gels, emulsions, blasting agents, binary and improvised explosives.

The amount of residue generated in an explosion is generally related to the amount and type of explosive used; however, there are several factors that determine the quantity of residue received in the laboratory for analyses. These include whether the explosive functioned as designed, the type of containment, the location, weather conditions, whether or not a fire occurred, the method of fire suppression and the experience of the investigator in collection, preservation and packaging of samples.

It would be highly unlikely that the laboratory would ever receive two post-blast explosive cases in which exhibits were identical in all aspects. Therefore, it would be impractical to try to set up a rigid standard analytical procedure to analyze non-standard exhibits. Thus, while the following general analytical scheme in the flow diagram can be applied to most cases, there will be times when it must be modified to meet the circumstances and facts of a particular case, the experience of the examiner, and/or results of preliminary testing. The scheme and narrative address both bulk and post-blast explosives residue, but are equally applicable to the identification of trace amounts of an explosive such as may remain in an emptied container or which may transfer as a result of a chance contact.

The flow diagram was developed by the ATF Laboratory Panel for the Quality Assurance Program for the analyses of explosives. The scheme is intended to serve as a general outline of the sequence of the examination. Not all of the tests listed in the scheme will be required for every examination and other unlisted tests may be applicable as required by the circumstances of individual cases..

A pertinent example is analysis of emulsion explosives (see Chapter 1). Protocols and useful background information on composition and analysis of water gels and emulsions have been published by Midkiff and Walters (1993) and Bender et al. (1993). Analysis of residues from these explosives is the subject of on-going research and testing and is requiring significant adjustments to protocols for analysis of residues from high explosives. For example, the primary screening method for high explosive traces, GC/TEA, does not detect the components of emulsions because they do not contain nitrated organic molecules, whereas GC/FID does detect hydrocarbon oil components. However, since these are widely distributed products, the issue of significance then arises. The other target compounds in residue are inorganic salts, primarily ammonium nitrate, and insoluble components. Such compounds also occur widely in the environment.

Flow chart for the examination of explosives and residues

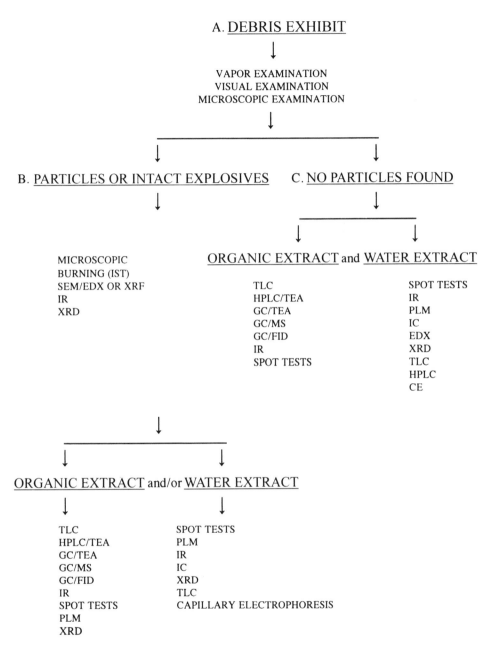

A. DEBRIS EXHIBIT

↓

VAPOR EXAMINATION
VISUAL EXAMINATION
MICROSCOPIC EXAMINATION

↓

B. PARTICLES OR INTACT EXPLOSIVES   C. NO PARTICLES FOUND

↓   ↓

MICROSCOPIC   ORGANIC EXTRACT and WATER EXTRACT
BURNING (IST)
SEM/EDX OR XRF   TLC   SPOT TESTS
IR   HPLC/TEA   IR
XRD   GC/TEA   PLM
                     GC/MS   IC
                     GC/FID   EDX
                     IR   XRD
                     SPOT TESTS   TLC
                                            HPLC
                                            CE

↓

ORGANIC EXTRACT and/or WATER EXTRACT

↓   ↓

TLC   SPOT TESTS
HPLC/TEA   PLM
GC/TEA   IR
GC/MS   IC
GC/FID   XRD
IR   TLC
SPOT TESTS   CAPILLARY ELECTROPHORESIS
PLM
XRD

Key to abbreviations: CE, capillary electrophoresis; EDX, energy dispersive X-ray; FID, flame ionization detector; GC, gas/liquid chromatography; HPLC, high performance liquid chromatography; IC, ion chromatography; IR, infrared spectroscopy, primarily Fourier transform infrared (FTIR) technology; IST, ignition susceptibility test; MS, mass spectrometry; PLM, polarized light microscopy; SEM, scanning electron microscope; Spot Test, wet chemical tests (often performed in a depression or 'spot' plate); TEA, thermal energy analyzer; TLC, thin layer chromatography; XRF, X-ray fluoresence; XRD, X-ray diffraction.

5.12.1   General Procedures to be Used for Post-blast Explosives Residue Identification

Debris

Before attempting the analysis of evidence, the analyst should study the transmittal letter to become familiar with the situation involving the explosion or call the investigator if more details are needed. Exhibits should be inventoried and correlated with the evidence log.

   If he or she has not attended the scene, then by becoming familiar with the situation surrounding an explosion and post-blast investigation, the analyst may develop an idea as to the type of explosive residues that may be present.

(1) Vapor examination   In cases where high explosive is suspected and commercial dynamite may have been used, selected samples can be subjected to vapor collection for volatile organic explosives such as NG, EGDN, metriol trinitrate (MTN) and diethylene glycol dinitrate (DEGDN).

   The collected sample can then be identified by TLC, GC, HPLC, IR, etc.

(2) Visual examination   A visual examination can be very beneficial in identifying device components such as container parts, batteries, timers, wires, switches, igniters, fuses, detonator components, remote control receiver components, tape, and other associative trace evidence.

(3) Microscopic examination   A thorough and diligent search of the debris with the aid of the microscope will more often than not yield some type of particle(s) associated with the explosive used such as black powder, smokeless powders, and components of improvised explosive mixtures.

   Also, on occasion, bits of unconsumed high explosive materials may be found, particularly where a low order explosion has occurred.

Particles

Unconsumed explosive particles can be readily analyzed by one or more of the methods described in Appendix 2, *Recommended analytical protocols in the ATF Explosives Section; part I: General procedures to be used for intact explosive*, depending on the type and amount of particles found (IST, SEM/EDX, XRF, IR, TLC, Spot Tests, HPLC, GC, IC, XRD, etc.).

No particles found

If no particles are found, the examiner should prepare the debris samples for organic and/or aqueous extractions.

(1) Organic extract   At this point, if either NG, EGDN, MTN or DEGDN has been identified by the sorption methods described under vapor examination, then the examiner could proceed directly to extracting the debris with water and examine the extract for traces of ammonium and/or sodium nitrate or other ingredients. If volatile components have not been recovered, then the debris sample should first be extracted with one or more organic solvents (see Appendix 2 for discussion on solvent selection) and then with water.

   The organic extracts can then be analyzed by one of the following methods: TLC, GC, HPLC, IR. The results are compared to standard explosive compounds analyzed under the same conditions. If the result is positive an independent chromatography method or a spectroscopic technique should be used for confirmation. Details may be found in Appendix 2, the Preface and in Chapters 8 through 12.

(2) *Water extract* The portion of the debris selected for water extraction should be rinsed with water and filtered through a membrane filter – not ordinary filter paper which can actually add ions of interest to the extract. These extracts can be slowly dried (in air or under a stream of nitrogen) and the residues analyzed by methods drawn from chemical spot tests, PLM, IC, IR, XRD, SEM/EDX, XRF, TLC and/or HPLC. Some typical protocols are discussed in Chapter 11, in Section 11.3.3.

5.13   Summary

Optimal results from forensic investigation of major crimes involving explosives are best achieved by well trained and coordinated teams of specialists.

References

BENDER, E. C., CRUMP, J., and MIDKIFF, C. R., 1993, The instrumental analysis of intact and post blast water gel and emulsion explosives, in Yinon, J. (Ed.), *Advances in Analysis and Detection of Explosives*, pp. 179–188, Dordrecht: Kluwer Academic Publishers.

BEVERIDGE, A., 1986, Explosives residue analysis in the mid-1980's – An expanding and challenging role for the forensic scientist, *Journal of Energetic Materials*, 4, 29–76.

BEVERIDGE, 1991, Analysis of explosives, *Proceedings of the International Symposium on Forensic Aspects of Trace Evidence*, pp. 177–190, Washington, DC: US Government Printing Office.

CHROSTOWSKI, J., HOLMES, R., and REHN, B., 1976, The collection and determination of ethylene glycol dinitrate, nitroglycerine and trinitrotoluene explosive vapors, *Journal of Forensic Sciences*, 21, 611–615.

DEAK, J. S., CLARK, H., DAGENAIS, C., JONES, S., McCLURE, D., and RICHARDSON, B. W., 1989, Post-blast residue analysis in the RCMP Laboratories: some practical observations, *Proceedings Third Symposium on Analysis and detection of Explosives*, pp. 18-1–18-19, Berghausen: Fraunhofer-Institut für Chemische Technologie (ICT).

FINE, D., 1983, Description of a nitro/nitroso specific detector for the trace analysis of explosives, *Proceedings of the International Symposium on the Analysis and Detection of Explosives*, pp. 159–168, Washington, DC: US Government Printing Office.

GARNER, D., FULTZ, M., and BYALL, E., 1986, The ATF approach to post-blast explosives detection and identification, *Journal of Energetic Materials*, 4, 133–148.

HALL, R. A., 1991, Booby traps associated with violent crime investigations, *Journal of the Forensic Science Society*, 31, 255–257.

HIGGS, D., and HAYES, T., 1982, Post-detonation traces of nitroglycerine on polymeric materials: recovery and persistence, *Journal of the Forensic Science Society*, 22, 343–352.

HIGGS, D., JONES, P., MARKHAM, J., and NEWTON, E., 1978, A review of explosives sabotage and its investigation in civil aircraft, *Journal of the Forensic Science Society*, 18, 137–160.

HNATNICKY, S., 1994, Selection and use of explosives detection devices to check hand-held luggage, *Journal of Testing and Evaluation*, 22, 282–285.

MIDKIFF, C. R., and WALTERS, A. N., 1993, Slurry and emulsion explosives: new tools for terrorists, new challeneges for detection and identification, in Yinon, J. (Ed.), *Advances in Analysis and Detection of Explosives*, pp. 77–90, Dordrecht: Kluwer Academic Publishers.

TARDIF, H., and STERLING, T., 1969, Detection of explosive sabotage in aircraft crashes, *Canadian Aeronautics and Space Journal*, **15**, 19–27.
WARDLEWORTH, D. F., and ANCIENT, S. A., 1983, A novel method for the recovery of volatile explosive traces, *Proceedings of the International Symposium on the Analysis and Detection of Explosives*, pp. 405–408, Washington, DC: US Government Printing Office.

Further Reading

YALLOP, H., 1980, *Explosion Investigation*, Harrogate, England: The Forensic Science Society.
YINON, J., and ZITRIN, S., 1993, *Modern Methods and Applications in Analysis of Explosives*, Chichester: John Wiley.

Appendix 1: ATF Explosion Investigation Guide

Team Leader

(1) Select and assemble personnel.

(2) Select and assemble equipment.

(3) Utilize arrival techniques.

(4) Check hazards at scene and determine if security squad is needed.

(5) Coordinate the ATF team with other persons at scene.

(6) Set up command center with proper equipment.

(7) Assign the immediate and general search units.

(8) Assign the immediate and general investigative units.

(9) Coordinate activities with other agencies.

(10) Conduct conferences, evaluate evidence collected, and assemble court exhibits.

Photographer

(1) Select and assemble equipment.

(2) Take methodical and logical sequences of photographs.

(3) Photograph the blast seat and blast damage showing measurements.

(4) Photograph evidence as found and in composite inventory.

(5) Photograph extended general area, including aerial photos if possible.

(6) Photograph a series of reconstruction.

(7) Photograph the crowd and vehicles for investigative purposes.

(8) Photograph blueprints, previous photographs of structures, and maps of the area.

(9) Coordinate with other special investigators.

(10) Make photographs of suspects.

Evidence Technician

(1) Select evidence collecting kit and place evidence blanket at the control center.

(2) Prepare evidence log and coordinate it with the schematic artist's evidence control sketch.

(3) Record the receipt of all properly marked and packaged evidence from search units, and insure that evidence collected is properly identified by the finder.

(4) Sort evidence when possible by using the categories on the evidence blanket.

(5) Maintain custody and control of collected evidence at the scene.

(6) Prepare the composite and collected evidence for shipment to the laboratory.

(7) Prepare detailed requests for laboratory analysis.

(8) Properly transmit evidence to laboratory.

(9) Receive and review the laboratory reports.

(10) Coordinate with the team leader and other investigators.

Schematic Artist

(1) Select and assemble equipment.

(2) Diagram immediate blast area.

(3) Diagram general area.

(4) Identify evidence found by assigning numbers and indicating the numbers on the evidence control sketch showing location found.

(5) Show necessary measurements of heights, lengths, and widths.

(6) Make artist's conception of scene prior to blast with the help of witnesses showing where furniture was arranged or how the structure was before the explosion.

(7) Prepare a legend on the diagrams.

(8) Inventory collected evidence with the evidence technician and ensure that all evidence is noted on the control sketch.

(9) Properly mark and identify the evidence control sketch and other diagrams for proper court presentation.

(10) Coordinate with the team leader and other investigators.

Immediate Area Investigative Unit

(1) Select and assemble equipment, including camera, fingerprint kit and routine processing aids, such as hand swabs and plastic bags for processing clothing of defendants and/or suspects.

(2) Interview local officers, firemen and all possible witnesses at the scene.

(3) Determine the owner of the property, the victim of the explosion, and if any persons were injured in the blast.

(4) Obtain names of any persons that are normally on the premises, such as employees, watchmen, or janitors.

(5) Provide the names and location of all persons or groups who should be interviewed by the general investigative unit. This list could include the injured persons who were taken to a hospital or rescue workers who have departed from the scene.

(6) Attempt to identify all persons at the explosion scene or coordinate with the photographer to record as much as possible by photographs or movie film.

(7) Record descriptions and time of sounds, color of smoke, and any odors noticed by the witnesses.

(8) Question the witnesses and record facts pertaining to the general activity of the scene prior to the explosion.

(9) Question the witness and record facts pertaining to anything unusual about the activity or any facts concerning unidentified packages, items, persons, or vehicles.

(10) Reconstruct the immediate area activity, if possible, and coordinate with the team leader and other investigators.

Immediate Area Search Unit

(1) Select and assemble equipment and properly place sifter screen.

(2) Stay alert for structure hazards or a second bomb before and after entering blast area.

(3) Locate seat of explosion.

(4) Coordinate with schematic artist and photographer before disturbing the crater or immediate blast area.

(5) Measure and record the size, depth and shape of the crater or damage.

(6) Collect samples from the blast seat and retain necessary control samples.

(7) Search and sift the seat of the explosion.

(8) Divide the immediate area into a search pattern and make a methodical search. Search from the seat of the explosions to an expanded area that overlaps with the general area search unit.

(9) Individually record and package evidence found and follow the routine procedure with the photographer, schematic artist, and the evidence technician.

(10) Reconstruct the immediate area scene, if possible, and coordinate with the team leader and other investigators.

General Area Search Unit

(1) Select and assemble equipment and select a search pattern.

(2) Stay alert for hazards and remain cautious about second bombs.

(3) Check all surrounding buildings, vehicles and objects for damage by missiles from the explosion. Tag these locations for the photographer and the schematic artist.

(4) Search area for evidence from the explosion.

(5) Search area of ingress and egress for associative evidence, such as footprints, tire tracks, torn clothing, blood, hair, fingerprints or other evidence that may relate to a violator.

(6) Search rooftops and trees or other high places that may have caught debris from the explosion. (Document wind blast effects and glass breakage in surrounding area.)

(7) Determine the extent of the outer perimeter of thrown missiles and evidence and indicate this finding to the schematic artist and the photographer.

(8) Adjust the outer perimeter of the search pattern as necessary.

(9) Individually record and package the evidence found and follow the routine procedure with the photographer, schematic artist, and the evidence technician.

(10) Reconstruct the general area scene, if possible, and coordinate with the team leader and other investigators.

General Area Investigative Unit

(1) Select and assemble equipment, including camera, fingerprint kit, and routine processing aids, such as hand swabs and plastic bags for processing clothing of defendants and/or suspects.

(2) Review maps and evaluate ingress and egress and select a methodical pattern for canvassing the area.

(3) Determine possibility of deliverymen being in the area, such as mailmen, milkmen, and paper boys, and make a list of their names and addresses.

(4) Canvass neighborhood for witnesses.

(5) Canvass business premises that may be related to ingress and egress, such as all-night service stations, cafes, taverns, toll bridges, or others.

(6) Follow team leader assignments as developed from the scene and make over-all evaluation from conferences.

(7) Check sources of raw materials that relate to the explosion.

(8) Record descriptions of suspects, suspect vehicles, and suspect premises for future use.

(9) Prepare a suspect list with necessary facts relating to the investigation. Follow routine criminal investigative procedures if the explosion is suspected of being a criminal act.

(10) Maintain communications with the team leader and coordinate with other investigators.

Appendix 2: Recommended Analytical Protocols in the ATF Explosives Section

Part I – General Procedures to be Used for Intact Explosive Identification

Low Explosives

A. Smokeless powder

Smokeless powders are generally categorized as single base (nitrocellulose) or double base (nitrocellulose and nitroglycerine).

Morphology: Visual and microscopic examination of the particles to describe the shape and characteristic features. An effort should be made to identify the brand(s) using LC/UV to identify the chemical additives and/or the smokeless powder library or database.

IST (Ignition susceptibility test): A particle is ignited to verify the characteristic burning properties of smokeless powder. (In some cases, a single powder particle precludes IST and complete morphology description.)

Chemical analysis: The presence of nitroglycerine or nitrocellulose can be confirmed by IR, TLC, HPLC or GC to establish if the powder is single/double base.

B. Black powder

Black powder is composed of potassium nitrate, sulfur and charcoal. Commercial products are generally glazed and produced in specific granulation size ranges. Improvised black powders are not normally glazed and particle size variations are greater. Pyrodex, a modified black powder, is distinguished by morphology and chemical composition.

Morphology: Visual and microscopic examination of the particles to describe the shape and characteristic features. An effort should be made to identify the granulation size using laboratory standard powders.

IST (Ignition susceptibility test): A particle is ignited to verify the characteristic burning properties of black powder or Pyrodex. (In some cases, a single powder particle precludes IST and complete morphology description.)

Chemical analysis: The presence of potassium nitrate and sulfur should be identified by appropriate examination methods (spot tests, IR, XRD, EDX, IC or HPLC). Potassium perchlorate or cyanoguanidine should be confirmed by similar techniques if Pyrodex is identified.

C. Flash powder

Flash powders are mixtures of inorganic oxidizing agents and fuels. Commonly encountered oxidizing agents are perchlorate, chlorate, and nitrate salts. Common fuels are finely divided metal powders, with sulfur powder and carbonaceous filler materials frequently added.

Microscopic: Describe color, homogeneity, appearance of metal, crystalline inclusions and other filler materials.

IST (Ignition susceptibility test): A small amount of the powder is ignited, either unconfined or rolled in a piece of tissue paper, and the burning properties are described.

Chemical analysis: The identity of the inorganic oxidizing agent(s) is confirmed by appropriate examination methods (spot tests, IR, XRD, EDX or IC). The metallic components and sulfur should be identified using appropriate examination methods (spot tests, EDX or XRD).

D. Improvised low explosives

Improvised explosives are generally mixtures of inorganic oxidizing agents and fuels.

Microscopic: Describe color, homogeneity and general physical appearance.

IST (Ignition susceptibility test): A small amount of the mixture is ignited, either unconfined or rolled in a piece of tissue paper, and the burning properties are described.

Chemical analysis: The identity of the inorganic oxidizing agent(s) is confirmed by appropriate examination methods (spot tests, IR, XRD, EDX or IC). The fuel component is identified by appropriate examination methods (spot tests, GC, IR, XRD, EDX or IC).

High Explosives

A. Organic high explosives

Explosives in this category are commonly nitroaromatics (TNT), nitramines (RDX), or nitrate esters (PETN). They can be found as free flowing crystalline powders (PETN in detonating cord), solid cast material (TNT booster), homogeneous mixtures (RDX and TNT in 'military dynamite') or in an organic matrix (RDX in C-4).

> *Visual and microscopic*: Describe color, consistency and general physical appearance.
> *Chemical analysis*: The sample is examined by IR or XRD or chromatographic techniques such as TLC, HPLC, or GC.
> *Matrix or binder*: When appropriate, the polymeric matrix and/or plasticizer should be identified using IR, GC or HPLC techniques.

B. Dynamite

Only those explosive formulations containing organic nitrate esters (NG, EGDN, MTN, DEGDN, etc.) will be considered as dynamite. In addition to these ingredients, common commercial dynamites contain carbonaceous filler material, inorganic oxidizing salts such as ammonium and sodium nitrates, and occasionally sulfur, nitrocellulose, salts, or microballoons.

> *Visual and microscopic*: Describe color, consistency, presence of prills, sulfur particles, and fillers.
> *Chemical analysis*: Organic nitrate esters are confirmed by an organic solvent extract analyzed by TLC, IR, GC or HPLC. The identity of the inorganic salts and sulfur (if present) are confirmed by appropriate examination methods (spot tests, IR, XRD, EDX or IC).

C. Blasting agents, slurries and emulsions

Explosives in this category include ANFO, binary explosives, watergels, and a number of other explosives. The majority contain ammonium nitrate as the oxidizing agent combined with a sensitizer or fuel.

> *Visual and microscopic*: Describe color, consistency and general physical appearance.
> *Chemical analysis*: (i) *Oxidizing agent* – Appropriate examination methods (IR, TLC, XRD, EDX, spot tests, IC); (ii) *Sensitizer* – A sensitizer, if present, is typically an amine salt (such as monomethylamine nitrate), aluminum powder, or microballoons. The identity of the amine salt is established using a chromatographic or appropriate instrumental analysis (IR, TLC, GC, HPLC, IC or XRD). The metal powder is identified by EDX or spot test. Microballoons, either polymeric or glass, are isolated from the explosive matrix and identified microscopically.
> *Fuel*: A fuel, if present, is typically a petroleum distillate, for example in ANFO; nitromethane, for example in binary explosives, or a high molecular weight oil or wax. The identity is confirmed by IR, GC, or other appropriate instrumental analysis.

D. Primary (initiating) high explosives

These shock, flame and friction sensitive explosives are typically styphnates, azides or fulminates, organic diazo compounds (diazodinitrophenol), or organic peroxides.

> *Visual and microscopic*: Describe color, and general physical appearance.
> *IST (Ignition susceptibility test)*: When appropriate, a *very* small particle is ignited and the burning characteristics are described.
> *Chemical analysis*: IR or XRD is sufficient to identify most primary explosives. In some situations a chromatographic or second appropriate instrumental analysis (TLC or EDX) may be required. In cases involving unusual or novel explosives where reference materials are not available, additional examinations may be necessary to fully characterize the substance.

6

Aircraft Explosive Sabotage Investigation

JOHN H. GARSTANG

6.1 Introduction

The in-flight bombing of a large commercial civil airliner is particularly difficult to investigate. Significant additional complexities and problems are often associated with these types of investigations. Most modern jetliners cruise at high altitudes (i.e. near six statute miles) and at high speeds (i.e. near Mach 0.85 or approximately 500 knots/575 miles per hour). The destruction of an aircraft of this type in cruise flight can potentially scatter 200 tons of structure, 100 tons of fuel, and 50 tons of cargo and passengers over a wide area. Further, an explosion may be intentionally timed to occur over inhospitable terrain such as over an ocean, desert, or remote forested mountainous region.

This chapter provides background information on a few subjects and techniques that may be of assistance when preparing for, or investigating, cases of explosive sabotage in aircraft. The material presented is far from being a complete technical treatise. Short briefs are given on different topics in order to highlight some key points. The subjects covered are not new but are nevertheless not common knowledge in the investigation community where they can be most effectively applied. Hopefully, the discussions and references will stimulate interest in acquiring additional knowledge in the subjects referred to, for possible incorporation into current or future investigations, and perhaps into other fields of endeavour.

Fortunately, the number of explosive sabotage cases world wide represents only a small fraction of the total number of commercial civil airliner losses. International statistics indicate that from 1961 to 1994 there were approximately 1553 fatal airline accidents which resulted in 41 730 deaths. This is an average of 47 fatal accidents and 1265 deaths annually. These figures do not include occurrences and deaths associated with sabotage, hijacks, or military action against civil targets. During the same time frame, there were approximately 58 cases and 1812 deaths where aircraft were damaged or destroyed by the detonation of an explosive device within the aircraft. In other words, approximately one to two aircraft bombings have occurred each year on average over these 33 years. However, during the last few years the number of aircraft bombings has fallen below average.

The cited accident and explosive sabotage statistics are based on provisional data compiled largely from information submitted by different States to agencies such as the International Civil Aviation Organization (ICAO). Not all States are members and some States have only recently joined this organization. Consequently, some data are lacking and the statistics represent conservative estimates (Figure 6.1).

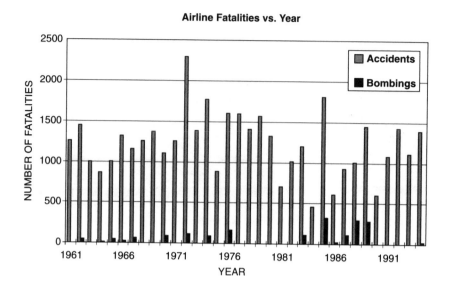

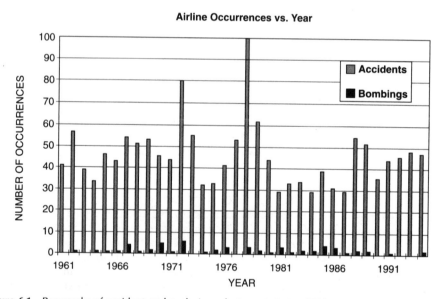

Figure 6.1 Bar graphs of accident and explosive sabotage statistics, 1961 to 1994. Notes: (1) Airline accident information based on provisional data from "Flight International" Airline Safety Reviews. (2) Airline bomb information based on provisional data from the US FAA Office of Civil Aviation Security and Newton 1985.

These statistics might seem to suggest that the number of fatal airline accidents and aircraft explosive sabotage cases is not too alarming. For instance, in 1994 the number of fatal accidents (47) happens to equal the cited annual average accident rate. The number of fatalities was only slightly higher than the average (1385 deaths as opposed to 1265). These numbers should not, however, be taken to imply a status quo. Quite apart from the tragic loss of life, 1994 was the worst year on record for monetary losses to the aviation insurance industry, approximately $1.9 billion. From 1984 to 1994, losses exceeded $10 billion (O'Toole, 1996), in part due to transport aircraft becoming larger and more expensive.

The loss of a single modern transport aircraft can potentially cost hundreds of millions of dollars if all associated expenses are taken into account (e.g. investigation cost, lost revenues, insurance settlements, etc.). More importantly, a single criminal act of explosive sabotage in a large transport aircraft can instantly kill 200 to 400 innocent people. These figures could significantly increase with the future introduction of 'super-jumbo' transports. Maximum takeoff weights could exceed 1 million pounds and passenger capacities could be over 500 (Proctor, 1996). According to an Airbus Industry market forecast, more than 1300 500-seat-plus transports valued at $300 billion are expected to be delivered to major air carriers in the next 20 years (Fiorino, 1996).

6.2 Conventions and Jurisdictions

The loss of a commercial civil airliner triggers an immediate response by a host of organizations pursuing different responsibilities and mandates. Initially, all converge into a chaotic scene of mass destruction under exigent circumstances. First responders typically might include firefighting, ambulance, police, military, and government transportation personnel. These are often quickly followed by experts and lawyers representing the airline, its employees and passengers, by airframe, engine, and systems manufacturers, by insurance underwriters and by investigators etc., along with the media, interested onlookers and relatives of the victims.

During the early stages of the investigation, it is usually not known whether the event was an accident or a criminal act, and jurisdictions of various parties inevitably overlap. It naturally follows that unless a single, integrated, command authority is immediately established (staffed with the appropriate specialists and representatives), control of the scene and the proper protection and processing of evidence, as well as the focus of the investigation itself, may be compromised.

The importance of advance planning, practice exercises, and other review processes to validate emergency readiness plans, cannot be over emphasized. Key investigative agencies must understand each other's mandate and wherever possible integrate functions and responsibilities. The answers to questions such as 'who is in charge?' and 'how will the evidence be handled?' should be clearly known prior to arriving at the scene of a disaster.

Fortunately, the international community has agreed to certain protocols for air crash investigations.

6.2.1 Aircraft Accident Investigation Organizations

Aircraft accident investigation organizations typically inherit the lead investigative role in aircraft occurrences. They have the requisite specialized skills, knowledge, and equipment to identify and reconstruct the technically complex systems used in large modern civil airliners. These agencies routinely liaise with and solicit expert assistance from many sources within the industry such as operators, manufacturers, and specialists in air traffic control, meteorology, aerodynamics, aerospace medicine, and aviation pathology.

Most air crashes are attributed to accidental rather than criminal causes. Thus, if there is no immediate reason to suspect a criminal act, there is a natural tendency for criminal authorities to let the accident investigation agency carry out most of the specialized investigative tasks, particularly technical ones. However, when a criminal act is suspected or there is loss of life, criminal authorities normally establish an immediate physical presence at the scene and set up close working liaisons with the accident investigation agency. As described by Baker and Winn in Chapter 13, trained forensic teams may be dispatched, but more commonly the expertise of the accident investigators is entrusted and relied upon to find, identify, and notify the appropriate authority should anything suspicious or of a criminal nature be found. This approach is based upon the sound principle that aircraft accident investigation expertise is needed. However, it also has its limitations or weaknesses, which will be discussed later.

6.2.2 *International Civil Aviation Organization (ICAO)*

In most countries or States of the world, aircraft accident investigation work should be conducted in a manner that meets well defined standards. The International Civil Aviation Organization (ICAO) is the specialized agency the United Nations has designated to be responsible for establishing international standards, recommended practices and procedures covering the technical, economic, and legal fields of international civil aviation operations (including security). ICAO was founded with the signing of the 'Convention on International Civil Aviation' on 7 December 1944, in Chicago, USA and is headquartered in Montreal, Canada. The organization has seven regional offices located in Mexico, Lima, Dakar, Paris, Cairo, Nairobi, and Bangkok. To date, 184 States are members (source: ICAO).

ICAO Annex 13

Article 26 of the Convention on International Civil Aviation obliges States to institute an Aircraft Accident Inquiry in accordance with ICAO procedures. 'Standards and Recommended Practices for Aircraft Accident Inquiries' were first adopted pursuant to Article 37 of the Convention and were designated as Annex 13.

The following four ICAO documents provide background information of relevance to criminal investigations:

- International Standards and Recommended Practices, Aircraft Accident and Incident Investigation, Annex 13 To the Convention on International Civil Aviation (ICAO, 1994);
- Manual of Aircraft Accident Investigation (ICAO, 1970);
- Manual of Civil Aviation Medicine (ICAO, 1985);
- Security Manual for Safeguarding Civil Aviation against Acts of Unlawful Interference (ICAO, 1996).

ICAO Annex 13 defines obligations and actions required by contracting States which include specific clauses relating to the State of design, manufacture, occurrence, operator, and registry. Information regarding key subjects such as notification procedures, protection of evidence, custody, removal of aircraft, autopsy examinations, and report requirements are contained within the document. In general, Annex 13 imposes obligations on States to investigate an aircraft accident or serious incident when the aircraft involved is of a maximum mass of over 2250 kg (4960 lb). The document defines who may participate or attend an investigation and in what capacity (i.e. investigator-in-charge, accredited representative, advisor). A copy of ICAO Annex 13 definitions is attached as Appendix.

The *Manual of Aircraft Accident Investigation* complements Annex 13 with technical information, examples of current investigative techniques, and recommended procedures. Guidance material for organizing, conducting, and controlling an investigation is also outlined. It provides an internationally recognized comprehensive reference to which all parties are expected to adhere.

The *Manual of Civil Aviation Medicine* (ICAO, 1985) provides detailed guidance on toxicological testing as well as more explicit reference material with regard to the conduct of autopsies.

The *Security Manual for Safeguarding Civil Aviation against Acts of Unlawful Interference* (ICAO, 1996) provides guidance material on ICAO protection standards and recommended practices. This document, together with the more detailed civil aviation security (AVSEC) plans (which it advocates be developed by each State and for each airport), provides criminal investigators with important security information. Investigators should also be aware of the specific policies and actions taken by the airline operators involved in the occurrence.

Criminal investigators may obtain analyses of past occurrences and relevant statistical data from the ICAO 'Committee on Unlawful Interference'. The Committee publishes reports on acts of unlawful interference which are available only to authorized personnel. Each member State typically has a designated representative on the Committee (usually from an office related to Civil Aviation Security within a Department of Transportation). This representative, or the appropriate Civil Aviation Security Office, can access the security information and specialist assistance an investigator might require. Similarly, criminal authorities can also remotely monitor (to a very limited extent) the progress of various national and international investigations to determine if aircraft accident investigation personnel working at the scene suspect an unlawful act.

Chapter 5, paragraph 5.11 of ICAO Annex 13 requires:

If, in the course of an investigation it becomes known, or it is suspected, that an act of unlawful interference was involved, the investigator-in-charge shall immediately initiate action to ensure that the aviation security authorities of the State(s) concerned are so informed.

Criminal investigators should be aware that a preliminary report on an act of unlawful interference is required to be completed by each State and forwarded to ICAO within 30 days of the occurrence. A final report must be completed and forwarded to ICAO within 60 days of the occurrence. The content and particulars of these reports are outlined in the ICAO *Security Manual for Safeguarding Civil Aviation against Acts of Unlawful Interference* (ICAO 1996). The ICAO legislation requiring these reports is outlined under Annex 17, Article 11 of The Hague Convention or Article 13 of the Montreal Convention.

(1) ICAO Annex 13 and aircraft accident investigation agencies Civilian aircraft accident investigation organizations typically are part of a State government Department of Transportation or an independent agency. The mandate of these agencies often reflects ICAO policy in pertinent areas. It is significant to note that the objective of an aircraft accident/incident investigation as defined in chapter 3, paragraph 3.1, of ICAO Annex 13 is:

The sole objective of the investigation of an accident or incident shall be the prevention of accidents and incidents. It is not the purpose of this activity to apportion blame or liability.

The Annex then recommends in chapter 5, paragraph 5.4.1:

Any judicial or administrative proceedings to apportion blame or liability should be separate from any investigation conducted under the provisions of this Annex.

For example in Canada (which is an ICAO signatory State), an independent federal government agency, the Transportation Safety Board of Canada (TSBC), was created to carry out accident investigation and related duties for the marine, rail, commodity pipeline, and air modes of transportation. The mandate of the TSBC (Canadian Transportation Accident Investigation Safety Board Act, 1985) is to advance transportation safety by:

- conducting independent investigations, and if necessary, public inquiries into transportation occurrences in order to make findings as to their causes and contributing factors;
- reporting publicly on its investigations and inquiries and on the related findings;
- identifying safety deficiencies as evidenced by transportation occurrences;
- making recommendations designed to eliminate or reduce any such safety deficiencies;
- conducting special investigations and studies on transportation safety matters.

The Board does not assign fault or determine civil or criminal liability and its findings are not binding on the parties to any legal, disciplinary or other proceedings. The Board's sole objective is to advance transportation safety. As in some other States, the Board is not part of, and is independent of, the government department responsible for regulating the transportation industry. The Department of Transport is the agency responsible for aviation regulation development and enforcement in Canada. The above-noted Act establishing the TSBC provides this independence and permits the Board to grant observer status to persons having a direct interest in the subject matter of an investigation. Examples include representatives of the transportation company, the operating crew, the manufacturer(s), and the Department of Transport. However, the Board may remove from a TSBC investigation any observer who contravenes any imposed conditions or who, in the Board's opinion, has a conflict of interest.

The Board can choose which occurrences it elects to investigate, as long as it does not contravene ICAO requirements as set out in documents such as ICAO Annex 13. In essence, the policy of the TSBC is to investigate only the occurrences which have a reasonable potential to result in safety action or which generate a high degree of public concern for transportation safety. How the TSBC elects to conduct its investigations is in part set by ICAO (being a signatory State), by the State's legislation (in this case the Act), and in part by the agency's management who set procedures based on their interpretation of the guidelines provided to them.

(2) Air accident investigation mandates   As previously noted, air accident investigators are often relied upon to find, identify, and notify the appropriate authority should anything suspicious or of a criminal nature be found. This approach is based upon the sound principle that their specific expertise is needed in an air crash investigation. However, this approach also has its limitations and weaknesses. It is important that restrictions imposed by the mandate of the aircraft accident investigation organizations for the State(s) involved in an aircraft occurrence are clearly understood. As stated in ICAO Annex 13, it is *not* the purpose of this type of an investigation to apportion blame or liability. Further specific reference may be made that mandates do not include determination of 'criminal' liability (e.g. TSBC).

If an act of unlawful interference is known from the onset to have caused an aircraft occurrence, or evidence is uncovered to indicate this during the course of an investigation, the services and expertise of the aircraft accident investigation organization may be completely withdrawn, depending on the State. Similarly, even if no evidence of an act of unlawful interference has come to light, the accident investigation organization may be hesitant to, or may elect not to, openly share or release all records and information uncovered during the investigation to criminal authorities due to a perceived mandate conflict. This approach can be justified under chapter 5, paragraph 5.12 of ICAO Annex 13. It states:

> The State conducting the investigation of an accident or incident, wherever it occurred, shall not make the following records available for purposes other than accident or incident investigation, unless the appropriate authority for the administration of justice in that State determines that their disclosure outweighs the adverse domestic and international impact such action may have on that or any future investigations:
> (a) all statements taken from persons by the investigation authorities in the course of their investigation;
> (b) all communications between persons having been involved in the operation of the aircraft;
> (c) medical or private information regarding persons involved in the accident or incident;
> (d) cockpit voice recordings and transcripts from such recordings; and
> (e) opinions expressed in the analysis of information, including flight recorder information.

These records shall be included in the final report or its appendices only when pertinent to the analysis of the accident or incident. Parts of the records not relevant to the analysis shall not be disclosed.

Note. – Information contained in the records listed above, which includes information given voluntarily by persons interviewed during the investigation of an accident or incident, could be utilized inappropriately for subsequent disciplinary, civil, administrative and criminal proceedings. If such information is distributed, it may, in the future, no longer be openly disclosed to investigators. Lack of access to such information would impede the investigative process and seriously affect flight safety.

The national laws of the States involved dictate jurisdictional procedures. In some States, policies have been drafted to outline what steps are to be taken to obtain the release of aircraft accident investigation material to law enforcement agencies. Information such as witness statements, cockpit voice recorder and air traffic service recordings might only be released in response to a warrant.

Although aircraft accident investigators may be experts in their specialized fields of work, they may not necessarily possess the knowledge or expertise to recognize or identify evidence of in-flight explosive sabotage. As highlighted elsewhere in this book, material evidence and explosive signatures from a blast often possess subtle characteristics and may require special evidence handling, collection, and interpretation skills. Chapter 11 of the ICAO *Manual of Aircraft Accident Investigation* does provide a short review of some techniques that can be applied to the investigation of explosive sabotage. As emphasized in ICAO chapter 11 and in this chapter, technical investigation by specialists is of the utmost importance. However, depending on the States involved, little, if any, emphasis may be placed by the accident investigation organizations on this subject since it may be classified by some as a 'criminal' matter, falling outside of their mandate or responsibility. Yet some States still entrust and rely on the capabilities of the accident investigation organization to act as the lead agency (sometimes working alone) to investigate an aircraft occurrence of unknown cause. This approach is often based on the rationale that if an act of unlawful interference has taken place, such as explosive sabotage, the evidence of this event would be obvious or readily detected by accident investigators, irrespective of their lack of specialist training. Although this may work in some instances, success is far from being assured. The evidence may not necessarily be obvious nor may it be readily detected by an investigator without proper specialist training and experience. Proper education and a close on-going working relationship between the key parties involved are crucial to a successful investigation. A team approach is ideal.

(3) Investigation procedures and practices   As highlighted previously, aircraft accident investigation agencies typically operate with the mandate not to apportion blame or liability. In contrast, criminal judicial authorities operate with the mandate to prosecute offences. Depending on the States involved, aircraft accident investigation procedures and practices may be much more lax in some regards and may not meet the standards required by the criminal courts.

Consider the handling of evidence (i.e. concerns with regard to contamination, alteration, interference: the very bases of continuity of evidence). The common police practice of establishing a specific exhibit custodian to coordinate and track the movement of evidence to satisfy a tribunal that each piece has been accounted for, as prescribed by law, at each stage of the proceedings, may not be fulfilled. Similarly, the way in which aircraft accident investigators take notes and photographs, conduct interviews, record measurements, mark exhibits, and prepare samples for laboratory analysis may be unacceptable for court purposes but may fully meet the administrative requirements of the air accident investigation organization. Criminal investigators should, whenever possible and in advance, review involved agency procedures and practices to identify deficiencies or areas of concern. Some State agencies produce their own manuals in addition to the ICAO documents referenced. For example, the

TSBC issues a *Manual of Investigation Operations* to its staff outlining standards and practices to be followed during the conduct of an investigation.

ICAO does compel accident investigation agencies to co-ordinate activities with judicial authorities as outlined in chapter 5, paragraph 5.10 of ICAO Annex 13 which states:

> The State conducting the investigation shall recognize the need for co-ordination between the investigator-in-charge and the judicial authorities. Particular attention shall be given to evidence which requires prompt recording and analysis for the investigation to be successful, such as the examination and identification of victims and readouts of flight recorder recordings. Note 1: The responsibility of the State of Occurrence for such co-ordination is set out in 5.1. Note 2: Possible conflicts between investigating and judicial authorities regarding the custody of flight recorders and their recordings may be resolved by an official of the judicial authority carrying the recordings to the place of readout, thus maintaining custody.

Ideally, actual examination of wreckage for residues and explosive damage should be carried out by the criminal authorities jointly with the accident investigation agency(s). If this cannot be arranged, work of this nature should be carried out by criminal authorities before any other experts examine the wreckage.

With good will, full explanation of the specific requirements of a criminal investigation, and communication between all parties, this can readily be achieved. In international situations, police liaison officers play a crucial role in such negotiations. A small well-trained law enforcement team who know exactly what they are looking for and how to document, photograph and process evidence, can assess a hanger full of wreckage by starting their work earlier than other experts and allowing them to follow on behind. This way, important forensic evidence can be identified, protected and if necessary isolated and seized without significantly impairing examinations by others. A typical team with which the author has worked consisted of a forensic chemist (chemical residues), a post-blast expert (explosives damage), an engineering failure analyst (metallurgical assessment of damage) and police officer (liaison and exhibit control) (Figure 6.2).

In summary, in order to ensure a well coordinated response and to minimize potential jurisdictional conflicts and problems, it is important that key parties make all reasonable efforts to review each others procedures to ensure compatibility, and enter into agreements to clarify any points of contention or issues of concern. Ideally, readiness plans and investigation teams should be formulated jointly by all parties in advance to ensure that a proper mix and balance of specialized expertise and equipment is present on investigation teams.

6.3 The Scene

There are few scenes that are as horrific and as catastrophic as that created by the in-flight bombing of a large civilian airliner. Investigators realize that a scenario of this type does require significant additional resources and expertise. However, the enormity and complexity of the situation may not be fully appreciated. The exemplary investigation of the Pan American World Airways Boeing 747-121 aircraft, registration N739PA, clearly demonstrates the magnitude of the problems faced and the resources required to successfully investigate an occurrence of this type.

On 21 December 1988 the aircraft was being operated as a commercial civil airliner on a flight designated as PA103 from London Heathrow to New York. The British Air Accidents Investigation Branch Report Synopsis for this occurrence (Charles, 1990) concludes that:

> the detonation of an improvised explosive device led directly to the destruction of the aircraft with the loss of all 259 persons on board and 11 of the residents of the town of Lockerbie.

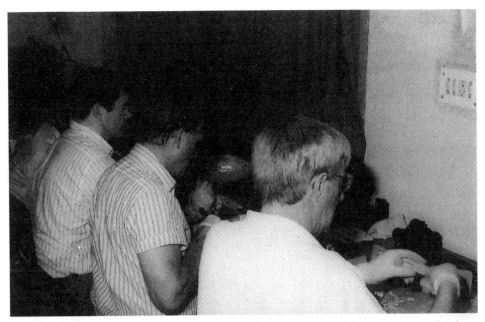

Figure 6.2 On-site team examination of small wreckage fragments which are being passed, right to left, from a forensic chemist (A. Beveridge, the editor), to a post-blast expert (Vermette, author, Chapter 4) and an engineering failure analyst (Garstang, author, this chapter).

The explosive device detonated while the aircraft was in cruise flight at level 310 (31 000 feet). Large portions of wreckage fell on the town of Lockerbie, Dumfriesshire, Scotland, while other parts landed in the countryside east of the town along two trails (Figure 6.3). The longest trail extended approximately 130 km to the east coast of England. Bodies and wreckage were recovered from an area estimated to be 845 square miles (The Lord Advocate, 1989).

The first responders to the scene included 29 Fire Brigade vehicles staffed by 188 personnel, who were deployed in fire fighting and body search operations. A contingent of 40 ambulances, with 115 personnel, were immediately dispatched to the same area. Over 5000 police officers from 13 different police forces were involved during the first week of operations, with the largest number involved on any one day being 1043. Liaison was established with many foreign agencies. Police vehicles alone logged some 2 million miles.

The British Air Accidents Investigation Branch (AAIB) responded with a team that ultimately involved 20 AAIB inspectors (not all of which were sent out to the scene), supplemented with participants from 22 other organizations. The military were also tasked. During the first three weeks of the investigation, 804 army personnel were engaged in search operations. In addition, 319 personnel from the Royal Air Force (RAF), operating 52 vehicles, carried out aircraft wreckage recovery operations. Support was provided by 16 RAF/Navy aircraft which logged 396 flight hours.

Dogs from the Search and Rescue Dogs Association (SARDA) were also used to search for bodies. During peak operations, 34 dogs and handlers together with an additional 10 personnel were used. Over 100 bodies were recovered by SARDA and police dog teams. Supplementary radio communications were provided by Raynet, a voluntary emergency measures organization of licensed radio amateurs, who provided an average of 80 operators per day for 10 days, commencing operations shortly after the occurrence took place. Numerous other voluntary organizations also assisted such as the Women's Royal Voluntary Service (WRVS). Hundreds of WRVS personnel supplied essential food services during the operation.

141

Figure 6.3 Forward fuselage including the flight deck from Pan Am 103 found approximately 4 km east of Lockerbie, Scotland.

Police officers travelled thousands of air miles to destinations across the world in connection with this investigation. Enquires were conducted in 70 countries worldwide. Over 700 documents such as statements, forensic reports, legal documents, and newspaper reports were translated (Dumfries and Galloway Constabulary, 1989). In February, 1996, the Dumfries and Galloway Constabulary estimated that over 14 000 statements had been taken, over 20 000 names had been recorded, over 18 000 items of property had been processed and over 35 000 photographs had been taken. Incomplete estimates of the cost of the aftermath work and police enquiry were set at approximately £17 million.

The insurance hull value for the aircraft was approximately US$32 million (source: Airclaimes Ltd, London). Insurance hull values for aircraft of this type can vary considerably depending on age, equipment installed, etc. For example, the approximate insurance hull value for the Air India Boeing 747-237B aircraft, registration VT-EFO (that was allegedly destroyed by a bomb over the North Atlantic on 23 June 1985, killing all 329 occupants onboard) was approximately US$95 million (source: Airclaimes Ltd, London). Hull values for new generation aircraft such as the Boeing 747-400 and the Boeing 777 could be in the order of US$150 million.

6.4   Specialized Investigative Tools and Techniques

The referenced Pan Am scenario shows that investigators could easily become overwhelmed by the diversity and magnitude of the work. Vast quantities of information need to be processed and analyzed under tight time constraints. The possibility that key pieces of data and critical interrelationships could pass undetected or not be fully comprehended, greatly increases in these types of major case investigations. Loss of control of the expanding volumes of material developed by the investigation could even lead to the loss of some forms of evidence (i.e. by not taking appropriate action at the appropriate time). In these situations it becomes even more imperative that each piece of evidence be properly tagged, documented and stored in a manner that will facilitate follow-up tracking, retrieval and analysis work. The objective is the output of sorted, coherent, meaningful data to substantiate and/or corroborate findings, deductions, and/or conclusions essential to the investigation process. Clearly some efficient, powerful, investigative tools and/or techniques that support these activities become invaluable assets in occurrences of this type.

The consolidated use of the following tools may be of particular value to explosive sabotage investigations. They may also be of help in other major case scenarios, as well as in emergency dispatch and damage control activities or search and rescue operations. The tools that will be discussed are:

- electronic office;
- project management software;
- geographic information systems;
- remote sensing;
- photogrammetry;
- global positioning systems.

Some topics that warrant specific mention and will also be discussed are:

- cockpit voice recorder and flight data recorder analysis;
- trajectory analysis of wreckage;
- reconstruction of wreckage;
- post-blast structural analysis;
- documents.

6.4.1 Electronic Office

The increase in computing power brought about by the development of new hardware and software has now made it possible to perform tasks on personal computers that previously would have required the use of expensive mainframes. Most of the tools and techniques that will be discussed involve the use of the electronic office, that is, operations that are based on the use of personal computers (i.e. electronic data handling). With the advent of powerful laptop computers, the electronic office has become truly portable.

The use of personal computers can greatly assist investigators. All of the benefits and power of working with electronic information can be taken advantage of. For example, large data processing tasks can often be more efficiently handled through computer-based automation than by paper-based manual processes. The latter is not only labour intensive but is also hampered by the sheer volume of the paper products involved. In an electronic office environment, massive data files can be conveniently copied, stored, and backed up on portable media such as CD-ROMs, optical disks, removable hard drives, or digital tape systems. Security can be implemented through computer encryption techniques or restricted access provisions. Electronic information can be rapidly retrieved, edited, and shared. Data can be quickly distributed via electronic mail through local or wide area networks, or globally through the Internet. Electronic information can also be sent or received by radio modems or through mobile phone connections (i.e. cellular or satellite communications). Transmissions of this nature can link remote field units to operational centres. These electronic capabilities can be exploited to great effect. They can be used to assist in managing work as well as to aid in analyzing data.

6.4.2   *Project Management Software*

In the Lockerbie disaster, the Home Office Large Major Enquiry System (HOLMES) was brought on-line within a few hours of the occurrence. This custom designed computer-based system (manufactured by McDonnell Douglas) provided a means to integrate case management functions with multi-tasking operations. The system also facilitated the registering of different types of evidence such as statements, messages, and property listings, and provided some analysis capability.

As expected, since the time of this occurrence (late 1980s), computer based technologies have continued to evolve at an exponential rate. There are now several off-the-shelf 'project management' software packages commercially available. These could be used by the whole investigation team to coordinate and track all operations, no matter how diverse. Several are designed to operate both on and off computer area networks. Most can be adapted to support operations at remote sites using portable laptop computers. Items such as resource lists, tasks, workloads, costs, and time lines can all be effectively linked together.

The application of graphic user interfaces in these software packages has greatly simplified their use. Most allow the users the capability to literally pick up data in the form of graphical objects to move, stretch, and adjust them until all the criteria is met to develop or alter a plan. Through computer area networks or via portable media, proposals can be distributed, the status of items can be updated and monitored, and/or reports can be quickly issued. This proficient software tool can assist the investigation team in organizing, estimating, and scheduling all work to develop one overall integrated plan, which can be continually modified to suit changing circumstances.

6.4.3   *Geographic Information System (GIS)*

Incidents of in-flight explosive sabotage can occur at any time, in any season, any where. Because the wreckage will likely be strewn over many miles and over a variety of different types of terrain, most field operations will be related to, dependent on, or linked in some manner to a form of geographic information. For example, it is essential that the geographic location (i.e. map coordinates) of scattered items be determined in order to dispatch investigators or salvage teams to them.

From the very onset of operations, one type of computer based system that is well suited to handle spatially referenced data is a Geographic Information System (GIS). A GIS is defined as a computer-based system that provides the following four sets of capabilities to handle georeferenced data (Aronoff, 1991):

- input;
- data management (data storage and retrieval);
- manipulation and analysis;
- output.

The United States National Science Foundation defines a GIS as 'a computerized database management system used for the capture, storage, retrieval, analysis, and display of spatial (e.g. locationally defined) data'.

A good GIS can integrate many different types of data. Information that may originally reside on paper (i.e. charts, documents, etc.), or in imagery (i.e. pictures, video, sonar records, etc.), or on other forms of media (i.e. magnetic tape audio recordings, event data, etc.) can be input into a GIS. Analogue data (information represented in a continuous form such as in a photograph) must be converted into digital data (information represented by discrete digits or numbers) in order to be input into a GIS. Images are commonly converted into a raster format (a computer graphics coding technique that portrays a picture as a series of digitally coded elements), whereas items such as maps are commonly converted into a vector format (a computer graphics coding technique that portrays data as a series of digitally coded lines). Various analogue to digital conversion devices exist such as digital cameras, digitizing tablets, flatbed and drum scanners.

GIS files may sometimes be found at agencies involved in activities such as map production, exploration, natural resource and urban/rural land use monitoring and management. GIS is not limited to civilian use. It is also used by the military for national security applications that might include intelligence, defence mapping, battlefield operations and facilities management. Through advanced contingency planning, and/or possibly under the exigent circumstances created by a mass disaster, criminal investigators may be able to acquire specialized GIS services from the military or other sources in Government departments. Military intelligence and mapping units are particularly well set up to provide services in crisis situations. Some commercial companies are capable of offering similar services but often at a much smaller scale and scope.

Digital chart of the world (DCW)

The 'Digital Chart of the World' (DCW) is a good illustration of a mapping product currently available for potential GIS use. It highlights what can be achieved using basic computer equipment. It provides a reference point from which investigators can envision the dramatic expansion of capabilities which the new generation of computer technology and applications now offer.

DCW is a 1:1 million scale comprehensive spatial vector database of the world. It was created through a cooperative project involving the military mapping agencies of Australia, Canada, the United Kingdom and the United States. DCW represents one of the largest spatial databases specifically designed for use on a rudimentary personal computer equipped with a CD-ROM drive (IBM PC/AT or compatible: minimum 80286 CPU with math co-processor). The vector data can be seamlessly accessed by simple viewing software. It consists of nearly two gigabytes of topologically layered information which is divided into 17 thematic and topographic layers with 31 feature classes. Six continental regions of the world are contained on four CD-ROM disks. Some of the layers include populated places, utilities, roads, railroads, aeronautical data, transportation structures, ocean features, land cover, vegetation, drainage, topographic relief, etc.

DCW contains its own software for graphic display and basic query functions, along with the ability to down-load user selected data sets for subsequent processing in GIS or Computer Aided Design (CAD) systems. An unlimited number of spatial overlays can be created (Natural Resources Canada, 1996a). It provides an excellent source of base information for study and analysis in any number of different investigative endeavours. In explosive sabotage cases, this type of data (at a larger scale for the scene in question) amalgamated with other information could be extremely valuable.

GIS and field evidence collection

During the field evidence collection phase of the investigation, the importance of getting the right information to the right people, at the right time, is crucial. Due to concerns such as transient evidence, weather, resource and cost limitations, success hinges on the ability of the investigation team to not only properly process the scene, but to do so quickly and efficiently.

A vital first step is to determine the geographic boundaries of the scene by systematic reconnaissance over a wide area. Thousands of scattered pieces of evidence must be rapidly found, positions must be accurately determined, recorded, and plotted. Teams comprised of the appropriate number and type of specialists need to be mobilized and efficiently dispatched to these sites to secure, document, and process the evidence. Special arrangements have to be made for collection and preservation of human remains, chemical residues, metallurgical signatures, etc. Unique equipment and services may be required to recover objects such as large, jagged pieces of wreckage or heavy items such as engines. Failure to properly tackle these issues could result in evidence being lost, contaminated, or reduced to a questionable value.

Some typical questions requiring immediate answers include:

- What communications and transportation networks exist at the scene?
- What types of terrain and geographic features will be encountered (i.e. mountains, ravines, forests, swamps, rivers, lakes, oceans, deserts, snow fields, cities, towns)?
- What sites are best suited for bases of operations or staging areas?
- How will weather effect operations?
- How can the scene be most efficiently divided up and processed?
- What will the lines of supply be?
- What specialized equipment is needed by field teams at what locations?

The integrated use of GIS with the other technologies discussed can provide investigators with answers in a timely and efficient manner.

Once spatially referenced data (i.e. evidence tied to geographic coordinates) has been input into a GIS, it can be very effectively linked or merged with other information gathered by the team. This information could be in the form of databases, spreadsheets, reports, computer aided design files, map and hydrographic chart libraries, graphics, etc. GIS offers investigators the extraordinary capability to clearly display, query, organize, and compile tremendous amounts of diverse information in a variety of ways to suit specific investigative needs. It also offers investigators the ability to rapidly output and distribute results, *en masse*, to all of the investigation team via electronic office networks. Outputs can be further tailored to meet individual requirements in a variety of custom formats whether they be in hard or soft copy form (i.e. text data can be tied to symbols and colour coded imagery or maps). This may be displayed or printed in whatever scale or colour scheme is deemed most appropriate.

To illustrate, look at what a GIS can do with telephone listings (both white and yellow page sections) that contain ZIP codes in the addresses, and with listings of ZIP codes cross-referenced to map coordinates. In some instances, both of these types of data sets are commercially available for entire countries and continents.

Once the geographic boundaries of the scene have been determined and the location of targets pinpointed (refer to remote sensing, photogrammetry and GPS), these listings could be analyzed to assess the distribution of much needed resources and assets. The information could also be used to efficiently mobilize, dispatch and manage them throughout the course of the investigation.

This could be done by a GIS by first linking millions of individual address records from telephone books to map coordinates, by matching ZIP codes. These records can then be queried in several different ways such as by country, province, city, area code, phone number, business type, name, and of course by map coordinates and ZIP codes. A host of goods and services such as accommodations, restaurants, supply and storage facilities, hospitals, airports, railway/bus/automobile service stations, harbours/docks, diving contractors, and heavy equipment operators, can be quickly compiled. Results can be incorporated with products such as airborne images, DCW, city and street atlases, and aircraft/ship/vehicle registries. Text information such as names, addresses, phone/facsimile numbers, and details on goods and services can then be rapidly related in a GIS to maps and images of the scene. Details can be plotted right down to street detail if desired. Routing by the quickest, shortest, and most efficient means (via various stops or layovers) can be determined. Tasking decisions can then be intelligently made and managed. These tools empower investigators with the proper knowledge to take command of the situation and respond accordingly.

6.4.4 Remote Sensing

Even if no GIS files for the scene in question exist, they can be quickly compiled. A wealth of GIS information can be rapidly gathered from remote sensing equipment (Hord, 1986). As defined in the *Manual of Remote Sensing* (Reeves, 1975), remote sensing is:

> in the broadest sense, the measurement or acquisition of information of some property of an object or phenomenon, by a recording device that is not in physical or intimate contact with the object or phenomenon under study; e.g. the utilization at a distance (as from aircraft, spacecraft, or ship) of any device and its attendant display for gathering information pertinent to the environment, such as measurements of force fields, electromagnetic radiation, or acoustic energy. The technique employs such devices as the camera, lasers, and radio frequency receivers, radar systems, sonar, seismographs, gravimeters, magnetometers, and scintillation counters.

Most current remote sensing is directed at the surface of the earth to facilitate the managing of the earth's resources (Ryerson, 1996a). Since GIS is also used extensively to support such activities, the two technologies are commonly used together and as a consequence, applications are often intertwined (Maclean, 1994).

A few examples of remote sensing that could be pertinent to explosive sabotage investigations involving land and/or water operations are:

- photography;
- synthetic aperture radar (SAR);
- sonar;
- remote operated vehicle (ROV);
- underwater laser imaging.

Photography

Photography deserves special mention since it was initially used to pioneer remote sensing and because it still provides users with significant capabilities today. Remote sensing records electromagnetic energy that emanates from features. A specific feature will typically radiate energy at numerous wavelengths, which are largely dependent on the compositional properties of the feature itself. Each feature (i.e. species of tree, type of man made object) will therefore tend to present a unique spectral 'signature'. If this 'signature' is known, it can in principle be used to identify or find similar features on remote sensing imagery (such as photographs). Multiband imagery (imagery taken in such a way that several filtered narrow ranges of wavelengths are captured simultaneously) can be used to accurately analyze the composition of energy radiated or absorbed by a target. This in turn can be used as a tool for detection and/or identification.

Investigators can capitalize on this technology at bomb scenes. Whether it be panchromatic, colour, false colour, or multiband etc., the use of aerial photography can greatly assist in locating and identifying scattered pieces of evidence (Figure 6.4). It also provides a comprehensive source of information on the scene itself (Philipson, 1996, Graham et al., 1996). The value of this type of imagery has been demonstrated in aircraft mass disaster investigations such as the Arrow Air Douglas DC-8-63 N950JW crash in Newfoundland, Canada where 256 personnel died (Del Vecchio et al., 1989), and the Pan Am Boeing 747-121 N739PA crash at Lockerbie (Charles, 1989). New digital imaging technology can supplement more traditional photographic data collection processes, particularly in instances where speed takes precedent over high resolution requirements (Miller and Richardson, 1995).

An often overlooked resource is remote sensing imagery archives. It is common practice for countries and companies to maintain depositories for airborne imagery (i.e. national air photo libraries). This imagery frequently is the primary source from which most topographic maps are compiled. Although the imagery may not have been taken recently, it provides investigators with an instant reference to build upon during the early stages of an investigation. Records showing seasonal and climatic changes can sometimes be reviewed. This may assist in evaluating the impact of weather in areas of operation.

Imagery archives may also contain commercial remote sensing satellite data. In general, these satellite systems provide superior large area coverage but at lower resolutions, and without the same degree of mission flexibility as aircraft. Some higher resolution systems that may offer imagery suitable for investigation use might include Commonwealth of Independent States payloads employing the KVR-1000 and TK-350 cameras. These satellites offer panchromatic photography with an optimum ground resolution of 2 m and 5 m, respectively (EOSAT, 1995). India's new IRS-1C satellite offers panchromatic 5 m imagery (EOSAT, 1996), and France has for several years offered panchromatic 10 m imagery from the SPOT satellite systems (EOSAT, 1993).

Figure 6.4 Black and white rendition of a false colour infrared photograph showing vegetation stress (dark area) around aircraft wreckage in trees.

148

Synthetic aperture radar (SAR)

In situations where scenes are obscured by darkness, haze, smoke, fog or cloud, it may not be possible to acquire data by conventional remote sensing techniques (such as by aerial photography). Synthetic aperture radar (SAR) could play a vital role.

SAR is a powerful microwave instrument which can 'see the ground' through almost any weather condition, day or night. Ground resolutions of less than 1 m can be obtained from aircraft (Godbole et al., 1995) and 10 m resolutions are now commercially available from satellites (Natural Resources Canada, 1996b).

Even in good weather conditions, SAR may be well suited to detecting, plotting and analyzing certain types of detail. It could be used to locate chunks of metal debris that may be partially concealed or camouflaged by features such as light snow or thin vegetation. In a marine environment, it can be effectively used to detect and plot targets on water which may include fuel/oil slicks (Ryerson, 1996b).

Sonar

Water covers more than 70% of the earth's surface. Whether by chance or by the intent of the perpetrator(s), evidence may fall into water and submerge. As in cases where an aircraft disintegrates over land, detailed mapping of large areas may be required. Sonar can be very effectively used as shown in the Air India Boeing 747-237B, VT-EFO occurrence. Miles of ocean floor were mapped in detail at a depth of approximately 2000 m (6600 feet).

Sonar is an acronym for 'sound navigation and ranging'. It operates by transmitting repeating sound pulses through water via a transducer (a device that converts energy from one form to another). Objects and terrain reflect echos which are received by the transducer. The echos are amplified and then displayed. A picture of the bottom and objects on it can be created. In the same way that SAR can 'see the ground' through a variety of obscuring mediums, so too can sonar 'see the ocean, lake, and river beds' in poor water visibility conditions. The feasibility and/or success of conducting diving operations to recover evidence in hazardous and poor visibility conditions could hinge on its use (Foot and Garstang 1985, Garstang, 1987).

Towed side scan sonars offer numerous benefits to investigators when searching large areas. These sonars typically are constructed with two transducers, one installed on each side of a towed body commonly referred to as a 'fish'. A narrow sonar beam is transmitted out of each side (sonar channel) of the fish. A composite image is formed when echos from each channel are combined. The swath covered in a single pass typically is far larger than that which can be inspected by conventional optical means (i.e. divers, manned submersibles, remote operated vehicles).

The use of a tow fish with side scanning transducers offers flexibility and often superior results in difficult operating conditions (Figure 6.5). For example, thermoclines (a permanent or temporary boundary layer formed between warm and cold water masses) can block or interfere with sonar signals. Several thermoclines can be encountered in deep bodies of water. A towed side scan sonar can be deployed below thermocline layers. The fish can also be adjusted in height and range (relative to the bottom) to improve resolution, and to enhance details by allowing the operator to examine features from a variety of vantage points. This increases the probability of target detection and identification (Figure 6.6). Some short range, high resolution systems can provide measurements in the centimetres (inches) range. Investigators should be aware that there are a number of sonar systems that are capable of full ocean depth operations (Wright, 1994).

In order to ensure complete sonar search coverage of a scene, to accurately determine the location of underwater targets, and to facilitate recovery, acoustic positioning (Figures 6.7, 6.8) can be efficiently integrated with sonar and diving activities (Kelland, 1991). This

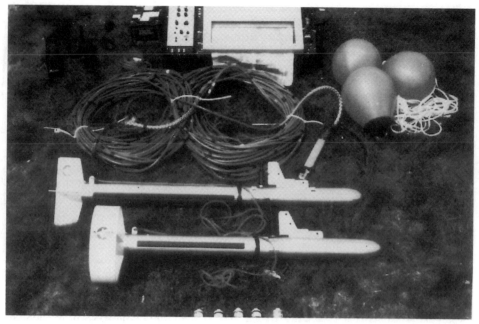

Figure 6.5   General view of a towed side scan sonar system showing the graphic recorder, shallow water tow cable, and two tow fish.

Figure 6.6   Side scan sonar image of a light transport aircraft resting on the bottom of the Arctic Ocean.

Figure 6.7 Underwater acoustic transponders in flotation jackets being deployed to establish a navigation array.

Figure 6.8 Navigation consoles used to integrate surface and subsurface positioning (e.g. vessel and ROV operations).

approach was successfully used by the Royal Canadian Mounted Police during the Air India VT-EFO investigation. The value of accurate positioning in a hostile environment, such as that of the oceans, cannot be overstated. The success or failure of marine operations may depend on it.

Remote operated vehicle (ROV)

Verification of underwater targets and documentation of evidence prior to recovery necessitates visual examination. This can be carried out by manned or unmanned diving operations. Due to a variety of reasons such as safety, human physiological limitations and cost, more and more diving operations are being carried out using remote operated vehicles (ROV). These unmanned robotic type vehicles can be deployed from the surface and guided to the bottom by operator(s) situated on board a surface craft (Figure 6.9). A ROV typically possesses its own propulsion system which is controlled via a communications link to the surface (most often through a tether). Vehicles are built in a variety of sizes, shapes and weights to suit specific mission requirements. A host of remote sensing equipment and tools (including manipulator arms for inspection and recovery work) can be outfitted to them (Remote Operating Vehicles of the World, 1996).

In underwater operations, the need to obtain clear, high quality visual imagery of evidence is paramount. Most ROVs are equipped with at least one remote control visual 'eye' for close-quarter work. This eye usually is a video camera which provides the operator with continuous real-time imagery. Fibre optics technology permits the transmission of signals (e.g. video) over long distances with minimal line loss. Additional cameras (both still and video) can be augmented as required. Selection of the proper camera for the proper job is crucial. Investigators now have at their disposal a wide array of equipment optimized for use in the underwater blue-green spectrum (Mackay, 1995, 1996; Vigil, 1996).

Figure 6.9 View of the SCARAB I ROV being launched on an inspection and recovery mission during one of the Air India VT-EFO diving campaigns.

Underwater laser imaging

In other than pristine diving conditions, darkness is soon encountered as one descends into the watery depths. Artificial light must be provided in order to see. This light is absorbed and scattered by particles and/or organisms suspended in the water column. The backscatter effect is analogous to driving an automobile at night in fog. The more turbid the water, the thicker the fog encountered. As a consequence, conventional optical imaging systems have typically been limited to short range applications.

A recent development that profoundly changes the way in which underwater optical data is gathered is laser imaging. Laser imaging can circumvent underwater backscatter effects. It can produce impressive optical images at considerably greater ranges than conventional underwater cameras. Laser systems can be operated by divers (Schwartz, 1993) and mounted on ROVs (Bonnier et al., 1994) and on towed bodies (Carey et al.,1996). Towed laser systems can be very effectively used to supplement side scan sonar data when surveying large areas. Some targets do not present good sonar signatures (e.g. bodies and fabrics). Lasers can offer clear panoramic views of evidence that are invaluable.

6.4.5 *Photogrammetry*

Investigators are familiar with the common expression 'a picture is worth a thousand words'. Documentation of evidence through photography has long been accepted as a fundamental investigative process. However, few investigators may realize that 'a picture may also contain a thousand measurements'. Three dimensional measurements can be derived from imagery through the application of photogrammetry. The *Manual of Photogrammetry* (Slama, 1980) defines photogrammetry as:

> the art, science and technology of obtaining reliable information about physical objects and the environment through processes of recording, measuring and interpreting photographic images and patterns of electromagnetic radiant energy and other phenomena.

One of the most common applications is the compilation of topographic maps from aerial photography. As noted in the *Manual of Photogrammetry*, measurements from photographs have replaced field surveys to such a degree that the use of photographs and photogrammetry in mapping is often referred to as 'aerial survey' or 'photogrammetric survey'. In this type of survey, stereo photogrammetry is most often used. It involves the capture and use of overlapping pairs of photographs or images. When these pairs are observed through a stereoscopic viewing device, the illusion of a three dimensional scene is created. The ability to view and study scenes in this manner has long been recognized as an excellent tool for acquiring information through image interpretation. Not only can an observer visually explore or probe to locate and classify features or targets, that same observer can also measure and plot those findings in three dimensions by applying photogrammetric techniques. Scale maps of the terrain and engineering drawings of objects can be made.

The quality, accuracy and diversity of information that can be quickly derived is remarkable. For example, fixed-wing aerial mapping aircraft commonly use cameras equipped with a 9×9 inch (23×23 cm) format negative specifically designed for photogrammetry and remote sensing applications. When used with fine grain film, appropriate lens cones and filters, together with a stabilized and motion compensated system, these cameras can produce very high resolution stereo imagery. Targets can be swiftly pinpointed, tentatively identified and precise measurements taken (inches or less) as required. A host of details such as topography, back roads, footpaths and potential helicopter landing zones can all be evaluated and plotted to scale. This type of airborne resource can be rapidly deployed, and used in a flexible manner to collect imagery along various flight lines and at different altitudes.

The use of this technology is not limited to applications over land but can also be applied over bodies of water. High resolution airborne imagery can be used to capture in an instant, fleeting transient evidence. For example, in cases where an in-flight bombing has taken place over water, investigators have a very small window of opportunity to document and inventory specific forms of evidence before they become lost, altered, or mixed up. Evidence such as flotsam (which may include bodies, wreckage, and cargo) can be quickly recorded prior to dispersion (due to wind, wave, and current effects), sinking, or recovery (i.e. during rescue operations). The same concepts apply to scenarios on land. Wind action alone can quickly bury items under sand/or snow in a desert or arctic landscape.

It should be noted that the observer is not restricted to conducting this type of work in the visible spectrum of light commonly portrayed in conventional photography. The process can also be applied to other types of images created from patterns of electromagnetic radiant energy outside of the visible spectrum. It is not limited to aerial mapping applications but can also be utilized terrestrially in a variety of different subjects (Karara, 1989). Just as a scene can be viewed and analyzed in three dimensions remotely from the air, so too can wreckage or pieces of evidence be processed remotely from vantage points underwater or on the ground. The vantage point can be as far away as a spacecraft or as near as the detector of an electron microscope (Wolf, 1983, Greve, 1996).

Maps and charts of the scene

Accurate up-to-date maps and/or hydrographic charts of the scene will be immediately required in large numbers to support field operations early in the investigation. Suitable maps and charts may or may not exist depending on factors such as the location and timing of the occurrence. If investigators are lucky enough to have access to a paper supply, existing stocks could be rapidly consumed due to the tremendous demand created by the large scale of the operations. Some of the advantages of employing electronic based information (i.e. GIS) in lieu of paper based products becomes much more apparent under these circumstances.

Orthophotographs and orthophotographic maps   For land operations where paper maps are in short supply, or in instances where no suitable maps exist (i.e. insufficient scale or detail), raw vertical airborne imagery (i.e. stereo aerial photographic prints) can initially be used for general reference. Images of this type can contain numerous displacements (distortions) created by factors such as relief of terrain and camera tilt. These displacements can be removed by applying photogrammetric techniques. If an aerial photograph is printed with the distortions (e.g. effects of relief and camera tilt) removed, it is referred to as an orthophotograph. It possesses the orthographic geometry of a map and the fine detail of a photograph.

By assembling orthophotographs at a specified uniform scale in a map format, an orthophotographic map is created. It can be appreciated that once either an orthophotograph or an orthophotographic map has been generated, numerous copies can be rapidly printed and distributed to field parties for plotting and navigation purposes. The use of this type of imagery may be a viable option to supplement depleted stocks or to replace inadequate or nonexistent paper map products for the occurrence scene. This imagery can of course be used electronically in an appropriately equipped portable laptop computer in lieu of hard copy products.

It should be noted that orthophotograph and orthophotographic map production capabilities are sometimes incorporated into GIS packages. These packages are also specifically designed to integrate and output different map products that could exist in different scales, projections, vertical and horizontal datums. They can be equipped with extensive visualization routines to present data in three dimensions. Advanced systems can drape photographs onto three dimensional computer models to create realistic terrain, and present it in stereo. These systems can also generate a bird's eye perspective and fly-through, if desired. Images can be output in hard copy in stereo pairs for subsequent three dimensional examination by field parties at remote locations using an appropriate viewing device such as a portable stereoscope.

6.4.6 Global Positioning Systems (GPS)

While carrying out their duties, field teams will face challenging navigation problems in the air, on the land and water. Provided that suitable maps, unique land features, and abundant vantage points exist, conventional map and compass plotting can be effectively carried out in some instances. This may not always be practical or expedient. Poor weather, night operations, and featureless surroundings (i.e. deserts, arctic barren lands, offshore marine environments) could easily impede positioning and degrade accuracies to unacceptable levels. The use of portable Global Positioning System (GPS) receivers overcome most of these problems and offer investigators outstanding capabilities.

GPS is a portable, day/night, all weather, satellite based radio navigation system. An individual with a GPS receiver can instantly determine his or her precise location, elevation, speed and heading from information transmitted from orbiting satellites. The receiver antenna must be within the line of sight of the constellation overhead in order to function. (*Note*: This can pose problems where a significant portion of the sky is obstructed such as by tall trees. Portable antenna masts and receivers designed to operate with auxiliary sensors (Deren et al., 1994; Krakiwsky and McLellan, 1995) have been constructed to overcome some of these problems. Equipment such as electronic distance measuring theodolites (laser transits) can also be used with conventional GPS receivers in areas where the sky is obstructed.)

GPS was built by the United States Department of Defense for military positioning. The military establishment of the Russian Federation of States operates a similar system called the Global Navigation Satellite System (GLONASS). Discussions will be focused on GPS as at present it is used by most international commerical applications.

GPS has been made available for international civilian use with limits placed on the achievable accuracies. Accuracies comparable to military GPS receivers can be obtained by employing special techniques such as differential positioning (Natural Resources Canada, 1993). It is possible to obtain real-time centimetre (inch) accuracies with this equipment (Graham, 1994). Because of its versatility, GPS technology has been successfully applied in numerous fields related to navigation and surveying. For example, it has become an invaluable aid in linking relative positioning devices (i.e. underwater acoustic navigation systems) to real world coordinate systems. It has also been extensively integrated into GIS and remote sensing.

An investigator with a GPS receiver can autonomously plan routes, and conduct searches and surveys. The equipment is relatively easy to use. Some receivers come equipped with graphic displays built into the body of the units to assist operators in visualizing where they are, and to show course deviations. Survey receivers typically are designed to log data and are fitted with an input/output port to exchange information to and from personal computers.

Investigators can interface GPS receivers to portable computers (i.e. laptops) to operate GIS systems in real-time. Detailed static and/or real-time dynamic moving map displays can be shown on computer screens in far greater detail than GPS displays (i.e. orthophotographs, large scale electronic charts and maps). With this equipment, multi-media information can be rapidly collected in the field to build or supplement GIS data files. For example, imagery can be captured on-the-fly using digital cameras for direct input into computers (Figure 6.10). Video footage or selected frames from a conventional video camera can also be input via a

Figure 6.10   General view of a portable multi-media GIS system comprised of a laptop computer (equipped with a removable hard disk and CD-ROM; built-in video capture board, speakers and microphone); interfaced to a mega-pixel digital camera, differential GPS receiver, and radio modem.

computer connection to an internal video capture board. Conventional imagery can of course be taken with the intent of input at a later date.

6.4.7   Cockpit Voice Recorder and Flight Data Recorder Analysis

Although the ICAO documents referenced provide a basic working knowledge of some of the tools of the trade and techniques applied by the accident investigation community, a few topics warrant specific mention due to recent developments. One of these topics pertains to recorders.

Most modern commercial airliners are equipped with a Flight Data Recorder (FDR) and a Cockpit Voice Recorder (CVR). These devices are designed to record and preserve key information about the operation of the aircraft should an accident occur. These same recorders may also provide invaluable evidence should a criminal act or act of unlawful interference take place.

The recorders have been built and installed with crash survival issues in mind. Measures have been taken to harden the units to be resistant to harsh environments that include exposure to impact loads, fire, or immersion in water. The units are quite often installed in the tail section of the aircraft. The recorders are manufactured in different sizes and shapes. Although the media frequently refer to them as the 'black boxes', they are typically painted in a bright colour scheme (such as fluorescent orange) and often exhibit retroreflective tape strips to enhance their conspicuous appearance (Figure 6.11).

Recorders require special playback facilities. Read-out and analysis should only be undertaken by experts and if explosive sabotage is suspected, ideally, by experts with previous experience in examining recorders from such occurrences. A short description of the recorders and the type of data each can provide follows.

Figure 6.11 Oblique view of a CVR and FDR. Note the presence of an underwater acoustic homing beacon near the handle of the FDR.

Cockpit voice recorder

The main function of the CVR is to provide firsthand information from the flight deck. One of the key parts of this recorder system is a cockpit area microphone. It is installed primarily to record voice communications originating from, and directed to, the pilots. This microphone is usually situated on the overhead avionics panel. It is, in effect, an 'electronic bug'. Once aircraft systems are powered up, it operates on a continuous basis, recording all conversation that occurs in its proximity. The CVR also usually records (via other wiring) voice communications transmitted or received on radio, on the aircraft's interphone or intercom, as well as on the aircraft's public address system. Voice or audio signals introduced into headsets or speakers from other sources (such as from approach or navigation aids) are also commonly recorded. Other significant sounds associated with background activities or noise (i.e. aural warning or alarm signals, airflow, and engine sounds) are sometimes picked up by microphones. Early vintage CVRs were typically limited to recording the last 30 continuous minutes. New generation CVRs have in some cases expanded this recording time up to two hours.

In some circumstances, audio analysis of CVR recordings can provide direct evidence of a criminal act or act of unlawful interference. For example, the identity and actions of onboard perpetrator(s) might be intentionally or accidentally recorded onto audio tracks (e.g. a suicide bomber or hijacker may talk directly with the crew on the flight deck, or use the radios or public address system).

In cases where no perpetrator(s) are onboard and an undetected hidden explosive device detonates, it may cut power or input to the CVR causing it to cease operating almost instantly. In these circumstances, the recording may end suddenly with no audible indication of anything having occurred out of the ordinary. In other cases where the destruction of the aircraft occurs over a longer time span and the CVR continues to function, the possibility exists that sounds associated with the event might be recorded (e.g. perforations of the fusel-age caused by a blast might be indicated via a recording of audible alarms triggered as a consequence of loss of cabin pressurization, etc.). Although it is conceivable that the sound of the explosion itself might be recorded, this is not considered to be a likely event. The speed of most explosive reactions and blast effects far exceed the speed of sound. Consequently, the CVR is often rendered inoperative before any useful audio information can be recorded. Only subtle clues, if any, may be provided by conventional audio analysis of the CVR.

New developments with regard to vibration spectrogram analysis of the CVR recordings, show that the CVR also records other information which can be interpreted. Indeed this method, which is described by F.W. Slingerland in Chapter 14, has been used to distinguish structural decompression failures from that caused by a bomb.

Flight data recorder

The main function of the FDR is to provide sufficient information to reconstruct the flight path of the aircraft in three dimensions. Some of the parameters recorded include altitude, airspeed, and heading, all related to time. Other information such as the status or operation of equipment and systems may also be logged, depending on the capability of the unit and of the hardware it is mated to. Old designs were often limited to less than 20 parameters and sometimes employed mechanical foil or wire recording devices. Depending on the design, both old and new units commonly record the last 25 continuous hours of operation. The new generation of FDRs are incorporating solid state technology and many of these systems have the capacity to record hundreds of parameters. One of the newest aircraft, the Boeing 777, is equipped with a standard configuration that logs approximately 750 parameters and this can be expanded if desired. This aircraft also utilizes two complementary FDR/CVR pairs of recorders; one in the front of the aircraft, the other in the rear.

Analysis of FDR recordings can provide evidence of a criminal act or act of unlawful interference. When the destruction of the aircraft occurs over a sufficiently long time span and the FDR continues to function, then changes in some parameters might be able to be associated with an explosive sabotage event. For instance, the onset of a unique vertical acceleration spike has been recorded on some FDRs as a result of the detonation of an onboard explosive device. However, like the CVR, the FDR is also prone to being rendered inoperative almost instantly due to the rapid nature in which damage is inflicted by an explosion. Consequently, its recordings may also suddenly end with no indication of anything having occurred out of the ordinary. In any event, FDR and CVR data can in most instances, at least establish the circumstances immediately prior the disaster.

Modern FDR/CVR playback facilities

Significant improvements have not been limited to the design of the FDR and CVR but have also occurred in other related recorder fields such as data recovery, playback, and analysis. Prior to these new developments, investigators were often forced to compare a multitude of different numbers (often represented on graphs) in order to visualize how different aircraft parameters changed with time. CVR recordings also had to be related to these values. Through this process, attempts were made to reconstruct the sequence of events and analyze the data. Assimilating and analyzing all this fragmented information and conveying it to others in an easily understandable manner was challenging.

The use of new generation computer technology and graphics applications has now made it possible to present the information in a radically different format. State-of-the-art FDR/CVR playback facilities can now graphically portray the aircraft in three dimensions on computer displays, and fly it in a manner that replicates the recorder data in real-time. Chase plane views, cockpit views, instrument displays, and different camera positions and orientations can be created to suit observer requirements (Figure 6.12). The speed and direction of

Figure 6.12 Frame from a computer simulation showing selected data recovered from the FDR.

the animation can also be adjusted. Geographic information such as mountains, lakes, cities, runways, and other landmarks can be added to the scene. The audio tracks of the CVR can be synchronized with the FDR data and played simultaneously to reconstruct movements or actions with sounds or voice communications. Imagery and sounds can be re-recorded onto video cassettes for playback at other locations. The use of a suitably equipped FDR/CVR facility can be an invaluable tool that greatly enhances the reconstruction and analysis capabilities of investigators.

FDR/CVR recovery

Recovery and analysis of the FDR and CVR should be a very high priority, since they might be able to quickly provide investigators with crucial information. The accident investigation community can supply specifics about the recorders (i.e. size, weight, shape, appearance, installation location). If the section of the aircraft where the recorders are installed cannot be found, or if the recorders have been shed in-flight, a search for the units will have to be undertaken.

Special provisions have been made in the event the FDR and CVR are lost in water. One or both of the recorders will typically be equipped with a portable underwater acoustic homing beacon, commonly referred to as a 'pinger'. The beacon is self-activating upon fresh or salt water immersion. It is designed to facilitate underwater recoveries up to depths of 20 000 feet (US Department of Transportation, 1983, 1990; Society of Automotive Engineers, 1988). Specialized underwater acoustic homing equipment is required to locate the beacons (Figure 6.13). Accident investigation agencies usually either possess this equipment or can obtain it on short notice. Some navies, coast guard agencies, oceanographic related organizations, and/or search and rescue units might be suitably equipped and trained to handle searches of this nature.

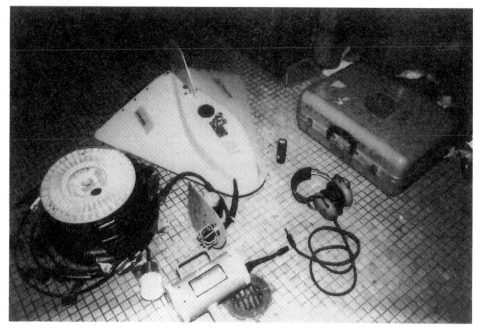

Figure 6.13 Oblique view of towed and hand held underwater acoustic homing equipment.

Disaster response plans should include provisions to handle FDR/CVR water scenarios. These provisions should of course address not only locating the 'pinger(s)' but also deal with different dive recovery options to document and retrieve the recorders, as well as other pieces of evidence. Diving and salvage experts should be consulted. It is recommended that those police officers or representatives responsible for carrying out such liaison work possess, in advance, a basic understanding of topics that will need to be discussed in detail such as: manned diving operations (US Department of Commerce, 1991), underwater police investigation techniques (Teather, 1994), Remote Operating Vehicle capabilities (Remote Operating Vehicles of the World, 1996), and the operating principles of key pieces of modern search equipment such as side scan sonar (Mazel, 1985), and underwater acoustic positioning systems (Kelland, 1991).

6.4.8 Trajectory Analysis of Wreckage

Another specialized tool and/or technique that can be employed to investigate the scattered remains of a disintegrated aircraft is to carry out a trajectory analysis of the fallen items. This process involves analyzing the distribution and location of wreckage, and correlating it with the estimated free-fall paths of objects (Figure 6.14).

Mathematical models are used to describe and predict the interaction of numerous variables. Often, many permutations and combinations are tried in iterative steps, before a best-fit solution is obtained. Scores of calculations may have to be carried out when numerous objects are involved. With the advent of modern powerful computers and the continued refinement of techniques, significant improvements have taken place over the years. Interactive computer programs have been written to provide solutions much more accurately, and in orders of magnitude less in time, than that which could be achieved by calculations carried out by hand (Bergen-Henengouwen, 1970, 1971, 1973; Matteson, 1974a, 1974b; Steele, 1983; Anker and Taylor, 1989; Grainger, 1990).

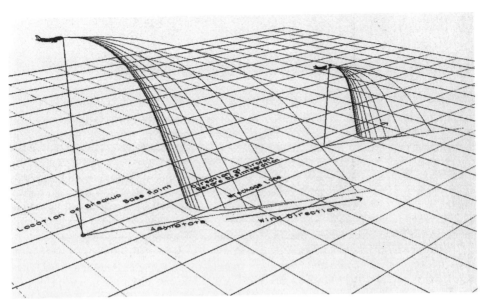

Figure 6.14   Computer rendition of a 3D trajectory analysis for a sequential inflight break-up.

The locations where items will fall, or conversely, the initial in-flight parameters of objects at the time of break-up, can be estimated. This type of information may be of assistance when searching for missing parts or when analyzing the sequence of the in-flight break-up. The FDR and CVR will typically be used in these analyses. As previously noted, this data may be limited to defining the in-flight parameters of the aircraft immediately prior to break-up (the recorders may have abruptly stopped functioning at the moment of the explosion). Additional data, before as well as after the explosion, might be available from military and civilian Air Traffic Control (ATC) facilities. These facilities sometimes record surveillance radar and communications. Meteorological reports together with the particulars of fallen items may all be combined to obtain a solution. Many of the practical problems associated with acquiring sufficient detail in order to conduct a proper trajectory analysis (i.e. determining the position, elevation, angle of impact into terrain and shape of the scattered debris) can often be very effectively addressed by applying some of the subjects previously discussed (i.e. GIS, remote sensing, photogrammetry, GPS).

In general, the closer an item is to an explosive when it detonates, the greater the energy imparted to it by the blast, and the greater the likelihood that it will be propelled away from the center of the blast with significant velocity. The more powerful the explosive charge, the more potent the effect. Thus, the first group of items off the aircraft could well be pieces of the explosive device and objects close to the explosion such as victims, luggage, clothing, cargo containers, and portions of air frame. Such material clearly has high forensic potential in the form of biological, chemical, metallurgical, or structural post-blast evidence (some of which could be transient in nature). Hence, trajectory analysis may provide valuable 'triage' by indicating which areas to search first to optimize the potential of recovering key evidence.

6.4.9   Reconstruction of Wreckage

Wreckage can be reconstructed by using physical and/or computer modelling techniques. A short discussion of both topics follows. A case study will also be presented that will illustrate the practical application of both types of reconstruction techniques, and their integrated use with many of the technologies previously discussed.

Physical reconstruction

Investigators should be prepared for the possibility of having to conduct a fire investigation in concert with the explosion investigation (Clodfelter and Kuchta, 1985; National Fire Protection Association 1989, 1991, 1995). The secondary effects of an explosion can cause both in-flight and post ground impact fires, which can mask and sometimes destroy certain types of evidence. Even if the in-flight fire is of short duration, it can spread rapidly and cause considerable damage, particularly if flames are exposed to the air stream. Rupture of containers such as a fuselage fuel tank, pressurized oxygen bottles or hydraulic lines can release copious quantities of flammable and/or combustible liquids and gases, which can be readily ignited. Similarly, the detonation of an explosive device in a luggage container of a cargo hold could cause a fire to erupt among the large quantities of combustible material present.

Physical reconstruction of wreckage is an effective technique that can be used to assist in analyzing the badly damaged and fragmented remains of a disintegrated airliner, irrespective of whether it has been demolished by fire and/or explosion. The reconstruction can either be two dimensional, such as laying out components on a floor, or three dimensional, where parts are arranged vertically as well as horizontally (i.e. held in place using apparatus such as jigs or scaffolding). During the exemplary Pan Am Boeing 747 Lockerbie investigation, the latter technique was used to painstakingly reconstruct a large portion of the aircraft fuselage, cargo compartment, and selected luggage containers.

The process involves assembling individual pieces into positions that reflect their location prior to failure (see Chapter 9, Figure 9.2). The damage on each piece is systematically assessed and compared to the damage on adjacent parts and elsewhere. Individual pieces are inspected to analyze evidence such as discolorations, deposits, soot and smear marks, fracture and deformation details. The presence of certain types of evidence, its continuity, or lack thereof, is carefully evaluated to determine failure patterns and to develop an appreciation for the magnitude and the direction of the loads that were applied at the time of break-up. This information is combined with data from all other sources (such as trajectory analysis and FDR and CVR results) to focus search and analysis efforts to specific areas. These areas are then further reconstructed and scrutinized in greater detail to detect, document, and collect as many types of post-blast evidence as possible.

(1) Damage to the exterior skin   One type of evidence that can be revealed during reconstruction work is the post-blast rupture and break-up of the exterior skin of the aircraft. The fuselages of most large transport aircraft are of an all metal, semimonocoque design (a framework of vertical and longitudinal members covered with a skin that carries a large percentage of the loads). The skin is typically fabricated from a ductile aluminum alloy. The fuselage is, in effect, a thin skinned pressure vessel when at altitude.

At the time of detonation, solid explosive material is violently converted into a relatively compact volume of high energy gases by a chemical process. Under ideal conditions, these gases expand outward producing a pressure (shock) wave which initially travels at supersonic speeds. At the shock front, the pressure, temperature and density rise almost instantaneously to peak values much greater than that in the ambient atmosphere. As the front passes, these values decay to lower than ambient and eventually return to it. As a consequence, a region of high velocity and high temperature airflow is produced immediately behind the front.

If the explosive device is in close proximity to an interior surface of the fuselage and conditions are right, the shock front could shatter a portion of the skin. Even if the skin is not immediately shattered by the shock front, it may soon fail in overload as the region of high velocity and high temperature airflow impinges on it. Overpressure damage associated with this airflow can sometimes create an irregular-shaped, radial burst pattern in the exterior skin. This fracture pattern characteristically takes place at a location where significant blast overpressures have been concentrated and vented. Portions of the exterior skin are peeled or rolled outward away from the center of the rupture, creating petals or curls in the sheet metal around the periphery (Figures 6.15, 6.16). The outflow of high pressure and high temperature gas through the rupture initially predominates over the outside air stream. Thus, petals or curls in the exterior skin can be created in directions both against and along the aircraft's slipstream, as well as at a variety of angles relative to it. This type of deformation pattern may be accentuated when the explosion takes place at high altitude, due to cabin pressurization.

Most large airliners maintain a constant low altitude cabin pressure (equivalent to approximately 8000 ft above sea level) during cruise flight at high altitude. Maximum pressure differentials between the inside of the cabin and the outside ambient atmosphere could be in the order of 6–9 lb/in^2 depending on the flight level of the aircraft. It is this additional pressure, among other things, that can have an effect on how the aircraft ruptures during explosive overpressure loading.

The shattering effect of the shock front, and the impingement and venting of the high velocity/high temperature airflow onto and through the exterior skin, can leave important forensic evidence. The shock front can potentially create explosive spalling and cladding on the interior and exterior surfaces of the skin, respectively. Micro cratering (small indentations which resemble meteor craters, formed by the impact of high energy particles created by an explosion at close range) might also be generated on the inside surface of the skin near the center of the rupture. Other types of post-blast evidence such as sooting, gas erosion, fissured surfaces, etc. can also be potentially created at this location. For a more in-depth discussion of post-blast evidence of this nature, investigators are referred to Chapter 13 of this book and to other articles published on the subject (Tardif and Sterling, 1967, 1969; Clancey, 1968; Burgoyne and Clancey, 1973; Higgs et al., 1978; Higgs, 1982; Newton, 1968, 1985).

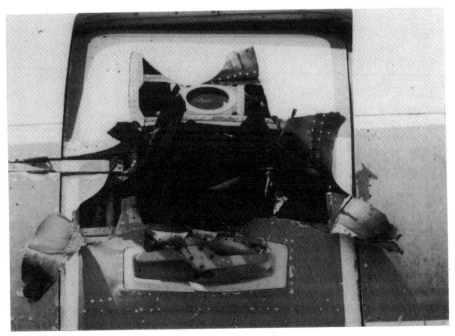

Figure 6.15 Side view of an irregular-shaped radial burst pattern in the exterior skin of a transport aircraft following the detonation of a small improvised explosive device.

Figure 6.16 Side view of a petal or curl created in the exterior skin of an aircraft fuselage following the detonation of an improvised explosive device (fabricated from a high performance plastic explosive). The dark sooted end is near the center of the blow hole created by the blast.

Investigators should note that while petals and curls may be forming in the exterior skin of the aircraft, numerous fractures may be propagating in a variety of directions. In general, tear fractures will spread outward away from the blow hole, and away from other breaches created elsewhere at the time of the explosion. Many of these fractures and others associated with secondary damage generated during the break-up, may intersect with one another. These intersections can further divide and sever the exterior skin of the aircraft into numerous jigsaw pieces. Hence, the full extent and degree of explosive overpressure loading may not be clearly apparent until reconstruction work is carried out. Likewise, the significance of subtle corroborating evidence could easily be overlooked. For example, pieces of exterior skin that were exposed to significant internal overpressure loads may bulge outward in between rivet lines. This type of deformation can occur while the skin is restrained and forced to stretch and distend prior to failure. Similarly, the heads of rivets at these locations will commonly pull through the sheet metal and/or fail in tensile overload, as a consequence of the skin being blown directly off the fuselage (Figures 6.17, 6.18).

(2) Preservation and analysis of fractures   As with other forms of evidence, proper collection, preservation and handling techniques are essential to ensure that fractures are not destroyed, altered or contaminated. Fracture surfaces are fragile, even in metals. Microstructural detail, which is of key importance in fractography (the analysis of fracture surfaces and the processes that create them), can easily be obliterated or badly damaged. This can occur mechanically such as by physically mating the fracture surfaces of broken parts together, and/or probing or cleaning a fracture surface with fingers (which will transfer moisture and salts etc.) or indirect means (e.g. exposing the fracture surface to a harmful environment such as humid air).

Fractures should be analysed soon after they have been made. If this is not practical or if the broken surfaces have been, or are expected to be, exposed to a hostile or uncontrolled

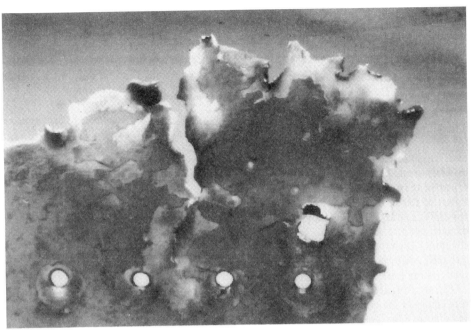

Figure 6.17 Close-up view of the outside surface of the petal or curl shown in Figure 6.16. Note the fracture pattern and how the heads of the rivets have cleanly pulled through the skin along a rivet line.

Figure 6.18 Close-up view of the inside surface of the petal or curl shown previously. Note the micro craters, soot deposits and fracture pattern.

environment during recovery, shipment or while in storage, steps should immediately be taken to protect and preserve the surfaces. It may be necessary to clean and/or apply corrosion prevention coatings (e.g. in cases where metal parts have been immersed in salt water). Items may also have to be cut to facilitate recovery. Before any work of this nature is undertaken, it is important that proper documentation be carried out and that all work is coordinated with other members of the investigation team (see paragraphs dealing with project management software, photogrammetry and GIS).

It is important to stress that the debris and deposits on the fracture surfaces and on adjacent areas could contain key chemical or metallurgical forensic evidence and/or vital information required for failure analysis purposes. The same could be said for the base material of the part(s). This evidence could be destroyed or irreparably altered if items are not properly treated and/or if handled incorrectly. Problems of this nature can be circumvented provided that well thought out processes and procedures are implemented during the early stages of the investigation. For example, it may be possible in some instances to apply a tenacious coating that not only protects a fracture surface from the effects of mechanical damage but also shields it from potentially harmful environmental effects. This same coating could also preserve the precise location of deposits and debris, while at the same time facilitate their removal on a pliable replica of the surface when it is peeled off at a later date. One such coating is cellulose acetate replicating tape.

Use of a plastic coating such as acetate replicating tape may or may not be appropriate depending on the circumstances faced. As with any technique, there are advantages and disadvantages. The use of certain general purpose corrosion prevention coatings might be acceptable and more practical if provisions are made either to first collect surface debris and deposits prior to application, or to do so at a later date by recovering them from spent solvents used to remove the coating (e.g. by filtering and/or chemical analysis). In order to determine the best course of action and to deal with unique situations as they arise, it is recommended that material and/or metallurgical specialists familiar with forensic evidence be part of the investigations team. Their close working liaison with chemists and other technical specialists could be invaluable. Investigators are referred to the distinguished and authoritative *Metals Handbook* series (ASM International, 1984–1996) for guidance and information. The edition of Volumes 11 and 12 entitled *Failure analysis and prevention* and *Fractography* are of particular relevance. These documents provide detailed information on a host of subjects ranging from general procedures, techniques and precautions employed in the investigation and analysis of metallurgical failures, to preparation and preservation of fracture specimens.

Computer reconstruction

Situations may arise where it may not be possible or practical to physically reconstruct wreckage and other forms of evidence by conventional means. The remains of the aircraft and its contents could be widely scattered in deep water such as on an ocean floor. Stereo photography and photogrammetry enable remote (non-contact) inspection and surveying to be conducted. The use of this technology in underwater applications was successfully applied more than 30 years ago to map the ocean floor (Cook and Hale, 1962; Culkin and Keller, 1962) and to map the remains of an ancient ship wreck (Karius et al., 1965).

Current technology can combine the ability to examine and precisely measure features (remotely if required) with Computer Aided Design/Drafting (CAD), Computer Aided Manufacturing (CAM), and Computer Aided Engineering (CAE) techniques to accurately model and reconstruct physical evidence in computers. Reconstruction work can sometimes be carried out more easily and/or more accurately by computer than by manhandling wreckage or evidence to create physical models. Segments of fuselage from wide body aircraft could potentially be enormous in size and weight. Similarly, some items may be relatively small and light but may not be in a condition that facilitates handling and/or manipulation with other pieces of structure (i.e. human remains).

In addition to photogrammetry, the development of high precision electronic distance measuring theodolites integrated with automatic electronic data collectors, and software to bridge the gap between data collection and CAD/CAM/CAE applications, have made computer modelling much easier and more practical. This is particularly so with the new generation of mirror reflectorless electronic theodolites. With these theodolites, measurements can be rapidly taken by projecting a laser beam directly onto the surface of the object being surveyed without the need to place a mirror reflector at this location. CAD/CAM/CAE models can be quickly produced using this equipment.

In some instances, investigators might be able to edit existing computer models rather than create them from scratch. Some aviation maintenance companies have generated detailed computer models for the aircraft they service, and manufacturers are now designing and modelling aircraft in computers for production purposes. There are also commercial companies that make and sell computer models for a variety of industrial applications. In addition, software is commercially available to automate the process of creating anthropometrically correct human body models (taking factors such as age, sex, race, and percentile body type into account).

As in conventional reconstructions, computer modelling work can be carried out in either two or three dimensions. The outline of pieces can be created or entire objects can be reconstructed and rendered showing whatever surface detail is required. Paint schemes, soot and witness marks, perforations, deformations, and so on, can be precisely replicated. Much more than drawings can be produced from the scale models created when using CAD/CAM/CAE techniques. Once objects or parts have been generated, they can be reassembled in their correct relative positions for display, manipulation, and analysis in a computer. Some software products allow operators to move human body models to simulate physiological movements or positions. The products can sometimes be used to analyze vision, reach, or to reconstruct events to analyze injury patterns. A variety of measurements such as distances, angles, areas, volumes, weights (based on material properties such as density), and centre of gravity calculations can be made. Objects and/or surfaces can be manipulated individually or in groups to carry out tasks such as estimating loads (i.e. via finite element analysis) and studying break-up sequences and damage patterns (i.e. mating parts to conduct fit analyses to analyze witness marks).

Complex details or objects can be broken down into more manageable pieces or components and assigned to different layers and groups within CAD/CAM/CAE files, which can be graphic attribute linked to databases. In effect, a GIS, a computerized database management system used for the capture, storage, retrieval, analysis, and display of spatial (e.g. locationally defined) data, can be created. Reconstruction data can be locationally defined with respect to an aircraft coordinate system (i.e. fuselage stations, waterlines, body buttock lines). This information can also be related to geographic map coordinates (where the pieces were found in the field) to tie in and link to any conventional GIS. Databases can be queried and the results can be displayed graphically on computer models (i.e. selectively presented with other information in different colour codes, at various magnifications, and viewing angles). The results can be linked with imagery (i.e. photographs taken of the wreckage accessible from multimedia databases) and can be output in hard-copy whether it be drawings, images, or text reports.

Computer modelling and reconstruction work is not limited to just analyzing external features or characteristics. The internal structure of objects can be examined in three dimensions by stereoscopically viewing overlapping pairs of radiographs. Three dimensional measurements can be made using X-ray photogrammetry techniques and this data can also be used to create CAD/CAM/CAE models. The use of this technology may be of particular value and utility to bomb technicians and investigators. The internal workings of a recovered improvised explosive device, or its components, can be remotely studied in detail through non-destructive stereo X-ray techniques. Similarly, the position and depth of bomb fragments in objects (including human remains) might be accurately located and their trajectory paths determined using this technology. This information could be used to assist in recovering the items and other types of forensic evidence, as well as to assist in determining the probable points of origin (from the trajectory paths) to locate the position and/or possible orientation of the explosive device(s). The same technology could aid pathologists in evaluating internal injury patterns (i.e. bone fracture direction and type) or aid orthodontists in identifying victims (i.e. from jaw and teeth information). The extent of forensic applications of this technology is vast and only a few possibilities have been highlighted.

6.4.10 *Specialized Post-blast Structural Analysis*

Interpretation and evaluation of the response of aircraft structure to blast and fragmentation damage requires specialist knowledge. Unique technical expertise resides with military personnel who design aircraft and equipment to withstand weapons effects. Specialized analysis techniques and computer codes (such as BR-1, BLAST, BR-2) have been developed for the military to model fragmentation, blast pressure effects, and structural response (Avery, 1981). This knowledge and expertise could be of great value when carrying out tasks such as analyzing the type and spread of damage from detonated explosive device(s). It could be used to assist in establishing the post-blast sequence of events, or possibly quantify the size of the explosive charge involved and/or the makeup of the explosive devise (i.e. by working in reverse from the damage patterns observed, blast fragment weights etc., computer programs can be run to try different permutations and combinations until suitable matches are obtained to quantify variables). Contributions from the different branches of the military service could all be of potential value. For example, pertinent bomb fragment penetration data into different materials such as wood (i.e. to potentially analyze damage inflicted to a wooden shipping crate in a cargo hold) could reside with an Army agency such as the United States Army Corps of Engineers (US Department of the Army, 1986; Hyde, 1986).

Within the last few years new initiatives have been undertaken to specifically address issues associated with commercial aircraft hardening and survivability to counter terrorism involving explosive sabotage. The United States Department of Transportation, Federal Aviation Administration, Aviation Security Research and Development Service, is an example of a lead agency sponsoring significant work in this field (US Department of Transportation, 1992). Specialized expertise and unique reference material stemming from this work (some of which is based on full scale destructive tests) could also be of great value to a criminal investigation.

6.4.11 *Documents*

Documents such as luggage tags, shipping labels, cargo manifests and personal effects can sometimes provide valuable evidence in bomb investigations (Blueschke and Kwasny, 1989). A significant portion of documents found at the scene will be in a damaged and/or deteriorated condition from exposure to the elements and/or possibly from post-blast effects. The potential evidential value of this material should not be discounted at first glance because of its poor state or apparent lack of decipherable detail.

Information can sometimes be retrieved from paper documents that have been immersed in water for several years (Garstang, 1989). Even if the water soaked documents may appear to be a congealed mass covered in slime or sediment, when properly handled and preserved (frozen), information can sometimes be recovered through a freeze drying technique (Schmidt,

1985). Often, pages can be easily separated and information can be directly read from the document after treatment (Figure 6.19).

In some instances where ink entries appear faint or illegible, detail can sometimes be deciphered using infrared luminescence and absorption techniques (Creer and Ellen, 1970). This could occur in cases (Garstang, 1992) where the entries become faded or sun bleached (Figures 6.20, 6.21). Similar infrared techniques can also sometimes be used to decipher burnt documents. Fragile, burnt evidence can be preserved and strengthened by applying a very

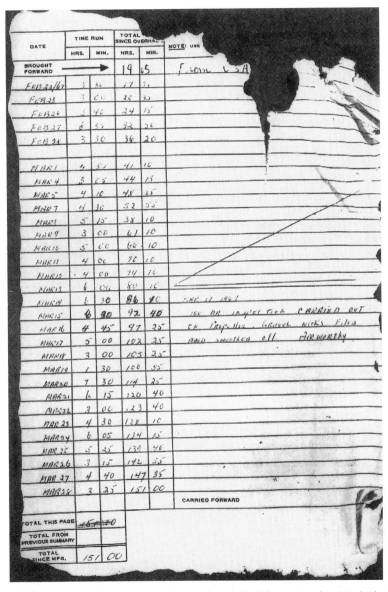

Figure 6.19 General view of a sample page from a freeze dried document showing the legibility of the entries after treatment. The document was recovered from aircraft wreckage which had been submerged in fresh water for approximately $15\frac{1}{2}$ years.

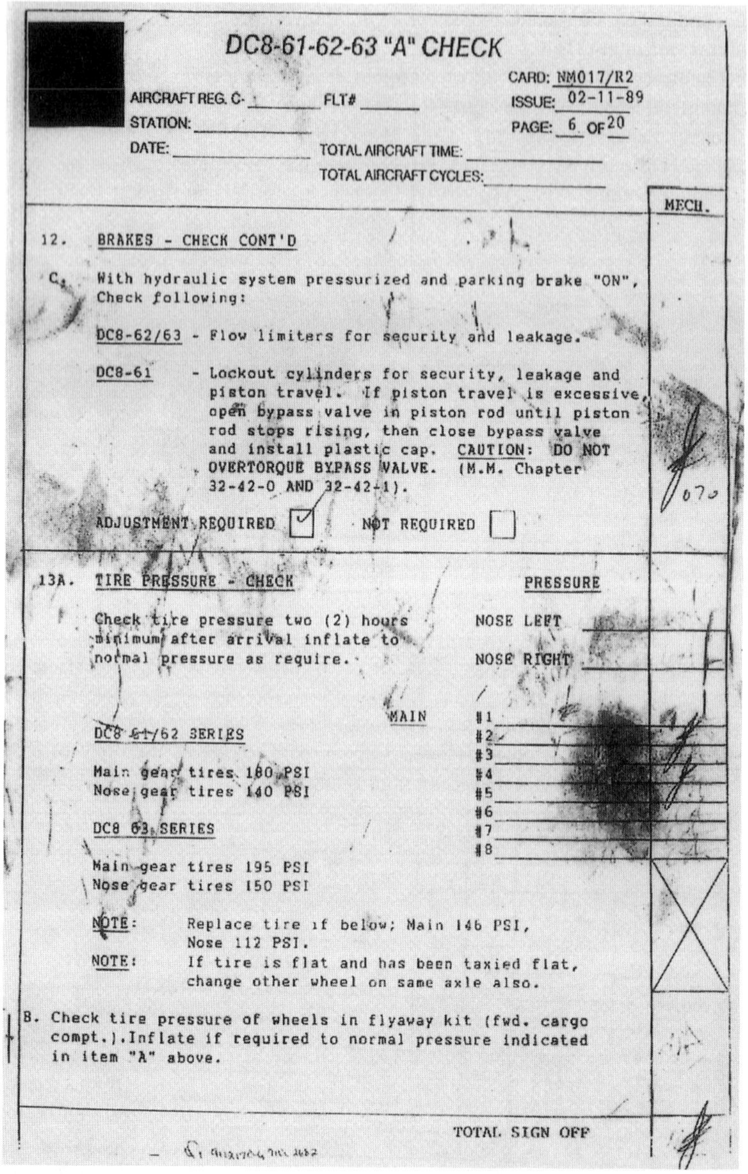

Figure 6.20 General view of a questioned document showing the faded and partially obliterated '13A. TIRE PRESSURE – CHECK' area of interest, situated to the right of the word '<u>MAIN</u>', underlined and numbered from 1 to 8 inclusive.

thin parylene coating (Humphrey, 1984, 1986, 1995). Infrared analysis techniques have been successfully applied using a video spectral comparator to recover information from burnt documents that have been water soaked, freeze dried, and parylene coated (Garstang, 1990). The use of parylene has facilitated the recovery of burnt evidence that would have otherwise been lost due to its fragile nature (Figure 6.22).

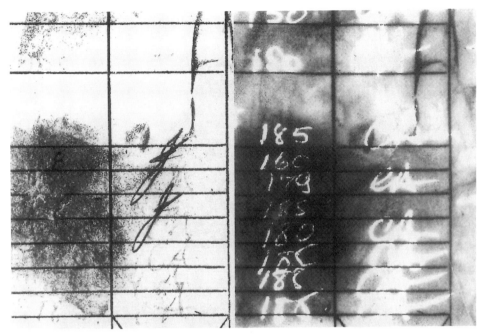

Figure 6.21 Composite, close-up views of the same area of interest in the previous figure showing two different spectral returns. One return highlights what appears to be the number '8' on the second line, and a 'c' shaped entry (with a small dash above it) on the fourth line. The second return highlights the original underlying entries. The over-writing altered the '160' entry on the second line to '180', and appears to have altered the questionable '155' entry on the fourth line to a possible '185' value. This was very important evidence in determining that a fatal DC8 crash was caused by fire originating in underinflated tires at take-off.

6.5 Case Study

The following case study (Garstang et al., 1994) photographically illustrates many practical applications of some of the technology previously discussed. The crash was accidental and did not involve explosives, but the technology is that which could be used in a major criminal case.

A commercial helicopter with a pilot and four passengers on board crashed in a remote mountainous region. All on board died. The helicopter had disintegrated into numerous small pieces which were found to be scattered across a snow-covered plateau on a ridge. A small technical team (comprising four investigators) documented the scene and completed all field work at the site within three days (under difficult working conditions), prior to the onset of a severe winter storm. The data collected was subsequently analyzed in a laboratory environment.

Upon arrival at the site, a satellite GPS was used to determine the geographic coordinates and the elevation of a point to create a survey bench marker. Photogrammetry target blankets were placed around the perimeter of the scene, and survey control data was derived for each blanket location (relative to the bench marker) using an electronic theodolite, commonly referred to as a laser transit (Figure 6.23). All features at the site were documented and inventoried by taking terrestrial and aerial stereophotography (helicopter), using a

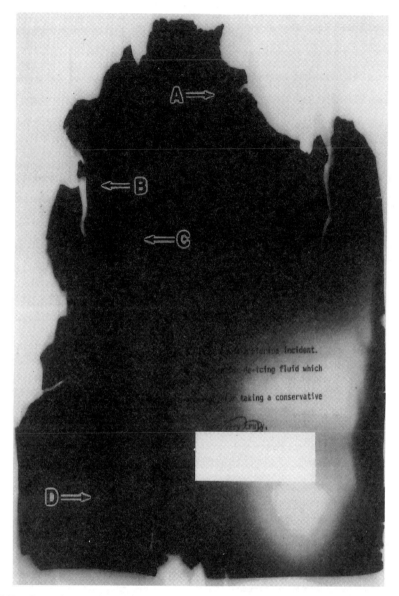

Figure 6.22 General view of a burnt, water soaked, document that was freeze dried and parylene coated. The document was able to be successfully deciphered to recover important evidence at the areas marked using infrared analysis techniques in a video spectral comparator.

conventional off-the-shelf 35 mm standard lens reflex camera for subsequent photogrammetric analysis. The documentation and inventory were supplemented with video imagery, which was integrated with an audio commentary and written notes. Wreckage was recovered and physically reconstructed. Colour-coded, multi-layered CAD files were made for analysis purposes (Figure 6.24). Pre-existing stereo 23 × 23 cm mapping photography was obtained for

Figure 6.23 General view of an electronic theodolite being used to determine the coordinates of target blanket markers placed around the periphery of the scene to establish survey control for photogrammetry.

the area (summer season) from a forestry department. This imagery together with digital map data was used to analyze witness locations and assess terrain features through a GIS approach.

A scale engineering computer model of the helicopter and of the terrain at the site were made. The helicopter model was mated to witness marks on the ground to determine the flight path and attitude of the helicopter at impact (Figure 6.25). Anthropometric computer models of the victims were created and mated to the inside of the helicopter (Figure 6.26). Injuries determined from the autopsy were colour coded and transferred to the human body models for study and analysis (Figure 6.27). A trajectory analysis of the ejected body of the pilot was carried out to estimate the initial launch conditions to assist in quantifying the speed and direction of the helicopter at impact.

6.6   Summary

No matter what the motive of the perpetrator(s), the in-flight destruction of a large commercial civil transport aircraft by explosive sabotage is a blatant, cold blooded act of mass murder. Often, innocent members of the international travelling public fall victim. There are usually far reaching political, psychological, and economic ramifications associated with actions of this type, both nationally and internationally. It is hoped that the information put forward might assist investigators in some way, in achieving a successful conclusion to the daunting investigative challenge faced in these type of cases.

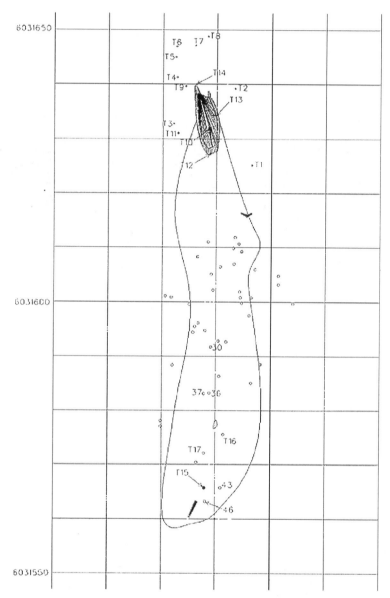

Figure 6.24 Black and white rendition of a colour coded, multi-layered CAD file showing the location of some of the evidence documented at the scene.

Figure 6.25 Computer reconstruction showing the engineering scale helicopter model mated to ground scars on the terrain model.

Figure 6.26 Computer reconstruction showing the engineering scale anthropometric human body models of the victims mated to the helicopter model.

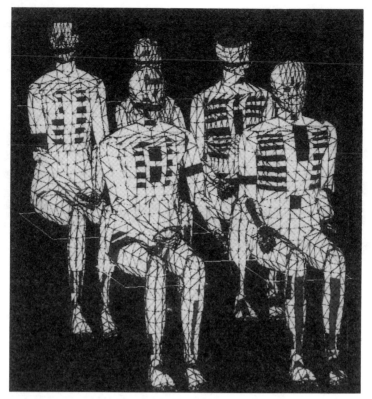

Figure 6.27 Computer reconstruction showing a black and white rendition of the colour coded injury patterns plotted on the engineering scale anthropometric human body models of the victims.

Acknowledgement

Photographs are reproduced by permission of the Transportation Safety Board of Canada. © Transportation Safety Board of Canada, 1996. Published by Taylor and Francis with permission.

This article has been written for the police community from the personal viewpoint of an engineering transportation accident investigator. The opinions expressed are those of the author and do not necessarily reflect, in any way, the opinions or policies of the Transportation Safety Board of Canada. The Transportation Safety Board of Canada accept no responsibility for any use made of this material.

References

ANKER, R. and TAYLOR, A. F., 1989, The trajectories of falling parts following in-flight break-up, *Proceedings of the Twentieth International Seminar of the International Society of Air Safety Investigators, Forum*, ISASI, **22**, 133–140.

ARONOFF, S., 1991, *Geographic information systems: a management perspective*, Ottawa: WDL Publications.

ASM INTERNATIONAL, 1984–1996, 9th and 10th Editions, *Metals handbook* series, Materials Park, OH: ASM International.

AVERY, J. G., 1981, addendum by Jacobson, M. J., 1988, *AGARD-AG-238 design manual for impact damage tolerant aircraft structure*, Nueilly sur Seine: Advisory Group for Aerospace Research & Development, North Atlantic Treaty Organization.

BERGEN-HENENGOUWEN, S. G., 1970, *Wreckage Trajectory Analysis in Aircraft Accident Investigation*, unpublished Master of Engineering Thesis, Carleton University, Ottawa.

BERGEN-HENENGOUWEN, S. G., 1971, Wreckage trajectory analysis in aircraft accident investigation, *Canadian Aeronautics and Space Journal*, **17**, 335–361.

BERGEN-HENENGOUWEN, S. G., 1973, *Aircraft wreckage trajectory analysis user manual*. Author address: Aeronautical and Mechanical Engineering Dept., Southern Alberta Institute of Technology, Calgary, Alberta.

BLUESCHKE, A. and KWASNY, R., 1989, Terrorism and the Document Examiner, *Royal Canadian Mounted Police Gazette*, **51**, 8–12.

BONNIER, D., FORAND, J. L., FOURNIER, G. R. and PACE, P. W., 1994, LUCIE— Laser enhanced underwater camera, *Sea Technology*, **35**, 55–59.

BURGOYNE, J. H. and CLANCEY, V. J., 1973, Annex 13-sabotage and malicious acts against aircraft – practical problems, *Symposium on International Aircraft Accidents Investigation*, London: Royal Aeronautical Society.

Canadian Transportation Accident Investigation and Safety Board Act, Section 7, R.S.C., 1985, C-23.4.

CAREY, D., ESTABROOK, N. and SAADE, E., 1996, Recent applications of laser line scan technology and data processing, *SAIC Science and Technology Trends*, pp 190–195, San Diego: Science Applications Instrument Company (SAIC).

CHARLES, M. M., 1989, The Lockerbie aircraft disaster, *Proceedings of the Twentieth International Seminar of the International Society of Air Safety Investigators, Forum*, ISASI, **22**, 21–25.

CHARLES, M. M., 1990, *The Department of Transport Air Accidents Investigation Branch Aircraft Accident Report 2/90, Report on the Accident to Boeing 747-121, N739PA at Lockerbie, Dumfriesshire, Scotland on 21 December 1988*, London: HMSO.

CLANCEY, V. J., 1968, Explosive evidence in an airplane accident, *Canadian Aeronautics and Space Journal*, **14**, 337–343.

CLODFELTER, R. G. and KUCHTA, J. M., 1985, Aircraft mishap fire pattern investigations, *Report # AFWAL-TR-85-2057*, Wright-Patterson AFB, Ohio: USAF Aero Propulsion Laboratory, AF Wright Aeronautical Laboratories, Air Force Systems Command.

COOK, C. E. and HALE, W. B., 1962, Underwater microcontouring, *Photogrammetric Engineering*, **28**, 96–98.

CREER, K. E. and ELLEN, D. M., 1970, Infra-red luminescence in the examination of documents, *Journal of the Forensic Science Society*, **10**, 159–164.

CULKIN, C. W. and KELLER, G. H., 1962, Mapping of underwater features, *Surveying and Mapping*, **22**, 283–290.

DEEPSEA DEVELOPMENT SERVICES INC., 1992, New underwater towed laser survey system, *Waves*, Spring Valley, CA: Windate Enterprises Inc.

DEL VECCHIO, J. J., EDWARDS, D., GARSTANG, J. H. and RINKER, J., 1989, An air photo analysis of an airplane crash, *Aviation Space and Environmental Medicine*, A6–A15.

DEREN, G., KRAKIWSKY, E., McLINTOCK, D., 1994, DGPS and barometry for seismic surveys, *GPS World*, Feb., pp. 20–26.

DUMFRIES AND GALLOWAY CONSTABULARY, 1989, Official press release — Lockerbie Incident investigation — miscellaneous statistical data.

EOSAT, 1993, *Comparison of satellites and sensors earth resources satellite systems*, Lanham, MA: Earth Observation Satellite Company.

EOSAT, 1995, *Russian satellite photography digitized high resolution products fact sheet*, Lanham, MA: Earth Observation Satellite Company.

EOSAT, 1996, IRS-1C products available, *EOSAT Notes*, **11**, 4–7.

FIORINO, F., 1996, Airline outlook – semper paratus, *Aviation Week & Space Technology*, **144**, 15.

FOOT, J. E. and GARSTANG, J. H., 1985, Sonar search of Lac Findley for Kiowa helicopter CH136 258, *Canadian Aviation Safety Board Engineering Report EP 158/85*, Ottawa: CASB.

GARSTANG, J. H., 1987, Search, Survey, Salvage, Cessna 172N C-GJQZ, Nelson River, Gillam, Manitoba, 26 July 1986, *Canadian Aviation Safety Board Engineering Report EI 173186*, Ottawa: CASB.

GARSTANG, J. H., 1989, Document analysis, Cessna TU206B, CF-VSK, Cameron Lake, British Columbia, 01 March, 1968, *Canadian Aviation Safety Board Engineering Report LP 419/83*, Ottawa: CASB.

GARSTANG, J. H., 1990, Document analysis, Fokker F-28 Mk 1000, C-FONF, Dryden Municipal Airport, Dryden, Ontario, 10 March 1989, *Canadian Aviation Safety Board Engineering Report LP 54/89*, Ottawa: CASB.

GARSTANG, J. H., 1992, Engineering Laboratory internal request for document analysis services, *Appendix 17 of Transportation Safety Board of Canada Engineering Branch Report LP 227/91, Records Group Report on the Investigation of the DC-8-61, C-GMXQ Accident, Jeddah, Saudia Arabia, 11 July 1991*, Ottawa: Transportation Safety Board of Canada.

GARSTANG, J. H., KARAFOTIAS, D. and LANDRIAULT, L.R., 1994, *Transportation Safety Board of Canada Engineering Branch Report LP 28/94, Site Survey, Crashworthiness, Structures, Bell 206B3, C-GRAH Investigation, 2 km North of Houston, British Columbia, 29 January 1994*, Ottawa: Transportation Safety Board of Canada.

GODBOLE, P., HASLAM, G. and VANT, M., 1995, The Spotlight SAR project, *IEEE Canadian Review*, **22**, 5–8.

GRAHAM, D. M., 1994, Real-time, 'on-the-fly' centimeter-level positioning, *Sea Technology*, **36**, 57–59.

GRAHAM, R. W., REED, R. E. and WARNER, W. S., 1996, *Small format aerial photography*, Knoxville, TN: Whittles Publications Services.

GRAINGER, L., 1990, Aircraft wreckage trajectory analysis final report, unpublished undergraduate paper, Princeton University, Department of Mechanical and Aerospace Engineering.

GREVE, C. W., (Ed.), 1996, *Digital photogrammetry: An addendum to the manual of photogrammetry*, Bethesda, MA: The American Society for Photogrammetry and Remote Sensing.

HIGGS, D. G., JONES, P. N., MARKHAM, J. A. and NEWTON, E., 1978, A review of explosives sabotage and its investigation in civil aviation, *Journal of Forensic Science Society*, **18**, 137–160.

HIGGS, D. G., 1982, Explosives sabotage and its investigation in civil aircraft, *Journal of Occupational Accidents*, **3**, 249–258.

HORD, R. M., 1986, *Remote Sensing Methods and Applications*, New York: John Wiley.

HUMPHREY, B. J., 1984, The application of parylene conformal coating technology to archival and artifact conservation, *Studies in Conservation*, **29**, 117–123.

HUMPHREY, B. J., 1986, Vapor phase consolidation of books with the parylene polymers, *Journal of the American Institute for Conservation of Historic and Artistic Works*, **25**, 15–29.

HUMPHREY, B. J., 1995, Document coating helps solve irs tax fraud case, *Law Enforcement Technology*, 52 to 53.

HYDE, D. W., 1986 (revised 1992), *Microcomputer Programs CONWEP/FUNPRO, Applications of TM 5-855-1, Fundamentals of Protective Design for Conventional Weapons*, Vicksburg, MI: Department of the Army, US Army Corps of Engineers, Structures Laboratory, Waterways Experiment Station.

INTERNATIONAL CIVIL AVIATION ORGANIZATION, 1970, *Manual of aircraft accident investigation*, 4th edition, Montreal: ICAO.

INTERNATIONAL CIVIL AVIATION ORGANIZATION, 1985, *Manual of civil aviation medicine*, 2nd Edition, Montreal: ICAO.

INTERNATIONAL CIVIL AVIATION ORGANIZATION, 1994, *International Standards and Recommended Practices – Aircraft Accident and Incident Investigation – Annex 13 to the Convention on International Civil Aviation*, 8th Edition, Montreal: ICAO.

INTERNATIONAL CIVIL AVIATION ORGANIZATION, 1996, *Security Manual for Safeguarding Civil Aviation against Acts of Unlawful Interference*, 5th Edition, Montreal: ICAO.

KARARA, H. M., (Ed.), 1989, *Non-topographic photogrammetry*, 2nd Edition, Bethesda, MA: The American Society for Photogrammetry and Remote Sensing.

KARIUS, R., MERIFIELD, P. and ROSENCRANTZ, D., 1965, Stereo-mapping of underwater terrain from a submarine, *Ocean Science and Ocean Engineering*, **2**, 1167–1177.

KELLAND, N. C, 1991, Acoustics as an aid to salvage location and recovery, *The Hydrographic Journal*, **60**, 27–33.

KRAKIWSKY, E. and McLELLAN, J., 1995, Making GPS even better with auxiliary devices, *GPS World*, 46–53.

MACKAY, D., 1995, A perspective on conventional underwater cameras, *Sea Technology*, **36**, 21–27.

MACKAY, D., 1996, Underwater cameras-SIT or CCD, *Sea Technology*, **37**, 35–38.

MACLEAN, A., 1994, *Remote sensing and GIS: an integration of technologies for resource management*, 1st Edition, Bethesda, MA: American Society for Photogrammetry and Remote Sensing.

MATTESON, F. H., 1974a, *Analysis of in-flight disintegration accidents*, unpublished doctoral dissertation, Stanford University, Ann Arbor, Mich: University Microfilms International.

MATTESON, F. H., 1974b, Analysis of wreckage patterns from in-flight disintegrations, *Journal of Safety Research*, **6**, 60–71.

MAZEL, C., 1985, *Side scan sonar record interpretation*, Salem, NH: Klein Associates Incorporated.

MILLER, T. L. and RICHARDSON, M. J., 1995, Megapixel imaging technology, *Sea Technology*, **36**, 53–57.

NATIONAL FIRE PROTECTION ASSOCIATION, 1989, *Aircraft fire and explosion investigation manual*, Quincy, MA: NFPA.

NATIONAL FIRE PROTECTION ASSOCIATION, 1991, *Fire protection handbook*, 15th Edition, Quincy, MA: NFPA.

NATIONAL FIRE PROTECTION ASSOCIATION, 1995, *Guide for fire and explosion investigations*, Quincy, MA: NFPA.

NATURAL RESOURCES CANADA, 1993, *GPS positioning guide, geodetic survey of Canada*, Ottawa: Natural Resources, Canada.

NATURAL RESOURCES CANADA, 1996a, *Digital chart of the world*, Sales information sheet, Ottawa: Natural Resources Canada, Canada Centre for Geomatics.

NATURAL RESOURCES CANADA, 1996b, *RADARSAT — an overview*, Information Sheet, Ottawa: Natural Resources Canada, Canada Centre for Remote Sensing.

NEWTON, E., 1968, Aircraft damaged or destroyed by deliberate detonation of explosives (sabotage), *Canadian Aeronautics and Space Journal*, **14**, 385–395.

NEWTON, E., 1985, Investigating explosive sabotage in aircraft, *The International Journal of Aviation Safety*, **3**, 43–48.

O'TOOLE, K., 1996, Aviation insurers are still cautious despite quiet year, *Flight International*, p. 16.

PHILIPSON, W., (Ed.), 1996, *The manual of photographic interpretation*, 2nd Edition, Bethesda, MA: The American Society for Photogrammetry and Remote Sensing.

PROCTOR, P., 1996, Boeing homes in on future 747 design, *Aviation Week & Space Technology*, 32–33.

179

REEVES, R. G., (Ed.), 1975, *Manual of remote sensing*, 1st Edition, Falls Church, VA: American Society of Photogrammetry,

Remote operating vehicles of the world, 1996–97 Edition, Ledbury (Eng): Oilfield Publications Limited.

RYERSON, R. A., (Ed.), 1996a, *The manual of remote sensing earth observing platforms & sensors*, 3rd edition, Bethesda, MA: The American Society for Photogrammetry and Remote Sensing.

RYERSON, R. A., (Ed.), 1996b, *The manual of remote sensing: principles & applications of imaging radar*, 3rd Edition, Bethesda, MA: The American Society for Photogrammetry and Remote Sensing.

SCHMIDT, J. D., 1985, Freeze drying of historic/cultural properties: a valuable process in restoration & documentation, *Technology and Conservation*, **9**, 20–26.

SCHWARTZ, B., 1993, *Diver and ROV deployable laser range gate underwater imaging system*, New Orleans, LA: Marine Technology Society.

SLAMA, C., (Ed.), 1980, *Manual of photogrammetry*, 4th Edition, Bethesda, MA: American Society of Photogrammetry and Remote Sensing.

SOCIETY OF AUTOMOTIVE ENGINEERS, 1988, *Aerospace Standard AS8045, Minimum performance standard for underwater locating devices (acoustic) (self-powered)*, Warrendale, PA: Society of Automotive Engineers.

STEELE, R. M. G., 1983, *Trajectory plots of aircraft debris following in-flight break-up*, unpublished MSc thesis, Cranfield Institute of Technology, England.

TARDIF, H. P. and STERLING, T. S., 1967, Explosively produced fractures and fragments in forensic investigations, *Journal of Forensic Sciences*, **12**, 247–272.

TARDIF, H. P. and STERLING, T. S., 1969, Detection of explosive sabotage in aircraft crashes, *Canadian Aeronautics and Space Journal*, **15**, 19–27.

TARDIF, H. P. and STERLING, T. S., 1969, *Report 608/69, some explosive aspects of aircraft crash investigations*, Valcartier: Canadian Defence Research Establishment.

TEATHER, R. G., 1994, *Encyclopaedia of underwater investigations*, Flagstaff, AR: Best Publishing Company, USA.

THE LORD ADVOCATE, 1989, *Note to news – PAN AM 103*, Edinburgh, Scotland.

US DEPARTMENT OF THE ARMY, 1986, *Technical manual TM 5-855-1, Fundamentals of protective design for conventional weapons*, Washington, DC: United States Government Printing Office.

US DEPARTMENT OF COMMERCE, 1991, *NOAA diving manual-diving for science and technology*, Washington, DC: National Oceanic and Atmospheric Administration.

US DEPARTMENT OF TRANSPORTATION, 1983, *Federal Aviation Administration – advisory circular AC 21-10A — flight recorder and cockpit voice recorder underwater locating devices*, Washington, DC: FAA.

US DEPARTMENT OF TRANSPORTATION, 1990, *Federal Aviation Administration – technical standard order TSO-C121 — underwater locating devices (acoustic) (self-powered)*, Washington, DC: FAA.

US DEPARTMENT OF TRANSPORTATION, 1992, *Federal Aviation Administration Aviation Security Research and Development Service Proceedings: FAA Aircraft Hardening and Survivability Symposium*, Atlantic City, NJ: US Department of Transportation.

VIGIL, A. E, 1996, Update on Low-light-level camera technology, *Sea Technology*, **37**, 10–13.

WOLF, P. R., 1983, *Elements of photogrammetry with air photo interpretation and remote sensing*, 2nd Edition, New York: McGraw-Hill.

WRIGHT, A., 1994, Deep-towed side-scan sonars, *Sea Technology*, **35**, 47–57.

Appendix:
ICAO Annex 13 Definitions

When the following terms are used in the Standards and Recommended practices for Aircraft Accident and Incident investigation they have the following meaning:

Accident. An occurrence associated with the operation of an aircraft which takes place between the time any person boards the aircraft with the intention of flight until such time as all such persons have disembarked, in which:

(a) a person is fatally or seriously injured as a result of:
    (i) being in the aircraft; or
    (ii) direct contact with any part of the aircraft, including parts which have become detached from the aircraft; or
    (iii) direct exposure to jet blast;
except when the injuries are from natural causes, self-inflicted or inflicted by other persons, or when the injuries are to stowaways hiding outside the areas normally available to the passengers and crew; or

(b) the aircraft sustains damage or structural failure which:
    (i) adversely affects the structural strength, performance or flight characteristics of the aircraft; and
    (ii) would normally require major repair or replacement of the affected components;
except for engine failure or damage, when the damage is limited to the engine, its cowlings or accessories; or for damage limited to propellers, wing tips, antennas, tires, brakes, fairings, small dents or puncture holes in the aircraft skin; or

(c) the aircraft is missing or is completely inaccessible.

Note 1. For statistical uniformity only, an injury resulting in death within 30 days of the date of the accident is classified as a fatal injury by ICAO.

Note 2. An aircraft is considered to be missing when the official search has been terminated and the wreckage has not been located.

Accredited Representative. A person designated by a State, on the basis of his or her qualifications, for the purpose of participating in an investigation conducted by another State.

Advisor. A person appointed by a State, on the basis of his or her qualifications, for the purpose of assisting its accredited representative in an investigation.

Aircraft. Any machine that can derive support in the atmosphere from the reactions of the air other than the reactions of the air against the earth's surface.

Causes. Actions, omissions, events, conditions, or a combination thereof, which led to the accident or incident.

Flight Recorder. Any type of recorder installed in the aircraft for the purpose of complementing accident/incident investigation.

Note. See Annex 6, Parts I, II and III, for specifications relating to flight recorders.

Incident. An occurrence, other than an accident, associated with the operation of an aircraft which affects or could affect the safety of operation.

Note. The type of incidents which are of main interest to the International Civil Aviation Organization for accident prevention studies are listed in the ICAO *Accident/Incident Reporting Manual (Doc 9156).*

Investigation. A process conducted for the purpose of accident prevention which includes the gathering and analysis of information, the drawing of conclusions, including the determination of causes and, when appropriate, the making of safety recommendations.

Investigator-in-charge. A person charged, on the basis of his or her qualifications, with the responsibility for the organization, conduct and control of an investigation.

Note. Nothing in the above definition is intended to preclude the functions of an investigator-in-charge being assigned to a commission or other body.

Maximum Mass. Maximum certificated take-off mass.

Operator. A person, organization or enterprise engaged in or offering to engage in aircraft operation.

Preliminary Report. The communication used for the prompt dissemination of data obtained during the early stages of the investigation.

Safety Recommendation. A proposal of the accident investigation authority of the State conducting the investigation, based on information derived from the investigation, made with the intention of preventing accidents or incidents.

Serious Incident. An incident involving circumstances indicating that an accident nearly occurred.

> *Note 1.* The difference between an accident and a serious incident lies only in the result.

> *Note 2.* Examples of serious incidents can be found in Attachment D of Annex 13 and in the ICAO *Accident/Incident Reporting Manual* (*Doc 9156*)

Serious Injury. An injury which is sustained by a person in an accident and which:
- (a) requires hospitalization for more than 48 hours, commencing within seven days from the date the injury was received; or
- (b) results in a fracture of any bone (except simple fractures of fingers, toes, or nose); or
- (c) involves lacerations which cause severe haemorrhage, nerve, muscle or tendon damage; or
- (d) involves injury to any internal organ; or
- (e) involves second or third degree burns, or any burns affecting more than 5% of the body surface; or
- (f) involves verified exposure to infectious substances or injurious radiation.

State of Design. The State having jurisdiction over the organization responsible for the type design.

State of Manufacture. The State having jurisdiction over the organization responsible for the final assembly of the aircraft.

State of Occurrence. The State in the territory of which an accident or incident occurs.

State of the Operator. The State in which the operator's principal place of business is located or, if there is no such place of business, the operator's permanent residence.

State of Registry. The State on whose register the aircraft is entered.

> *Note.* In the case of the registration of aircraft of an international operating agency on other than a national basis, the States constituting the agency are jointly and severally bound to assume the obligations which, under the Chicago Convention, attach to a State of Registry. See, in this regard, the Council Resolution of 14 December 1967 on Nationality and Registration of Aircraft Operated by International Operating Agencies (*Doc 8722*).

Investigation of Gas Phase Explosions in Buildings

CHRISTOPHER D. FOSTER

7.1 Introduction

This chapter is concerned principally with the investigation of gas phase explosions in buildings. It is not intended that it should provide a comprehensive treatment on the subject of characterising and evaluating all types of accidental explosion. Such details are contained, for example, in the work of Strehlow and Baker (1976) and Baker et al. (1983).

An explosion is a phenomenon resulting from the sudden release of energy with the potential to produce shock or blast waves, and missiles, both of which can cause remote damage. The ways in which energy can be released suddenly, and generally speaking the type of damage caused as a result, can be broadly divided into three categories as discussed below.

Detonation

Detonations are caused by high explosives, certain unstable solids and liquids, and much less frequently, the rapid oxidation of certain gases taking place in long, narrow ducts or pipes. The detonation of a high explosive is caused by a very rapid reaction process taking place in the material which leads to an almost instantaneous rise in pressure at the source of the reaction. This in turn sets up a shock wave which travels outwards at a speed greater than that of sound in air. The shock wave causes shattering and crushing of objects nearby. Similarly, local effects of heat are usually in evidence. Remote damage may result from the interaction of a blast wave with neighbouring structures.

Deflagration

Combustible gases, vapours, dusts and mists or aerosols of combustible liquids which are mixed with air in the correct proportions will usually propagate a flame. Typically, flames will travel at a speed of a few metres per second, i.e. in the order of 100 times slower than the speed of sound in air. Combustion reactions in the flame front produce a sudden increase in the number of molecular species, and a sharp rise in temperature causing expansion of the gases. When a deflagration occurs in a structure offering a degree of confinement, this expansion will cause the pressure to rise until the weakest element or elements of that structure fail. This will relieve, or partially relieve, the pressure and allow the products of the deflagration to vent to atmosphere or to a neighbouring compartment.

Pressure generation can also accompany deflagrations in the open air. A pressure wave or blast wave of sufficient magnitude to cause damage to structures is dependent on the rate of combustion being sufficiently fast and it was previously believed to be important that the fuel should be released in tonnage quantities. The subject of unconfined vapour cloud explosions (UVCEs) was comprehensively reviewed following the Nypro chemical plant explosion at Flixborough, UK in 1974 (Gugan, 1979). In a recent review, it is emphasised that the presence of obstructions or a degree of confinement of the vapour cloud are necessary to increase the degree of turbulence leading to significant flame acceleration and hence blast damage (Pritchard and Roberts, 1993). Indeed, in experiments using cyclohexane and propane with ignition in a region of the cloud that was partially confined and which then vented into an unconfined region containing repeated arrays of pipes, transition from deflagration to detonation was obtained.

Hydraulic/pneumatic failure of pressurised containers

Stored energy, for example in steam reservoirs, pressurised air containers and chemical reactors containing a liquid under pressure, can be released suddenly if the container bursts. Higher rates of release of energy accompany the failure of structures containing compressed gases and superheated liquids. Blast waves generated by a bursting pressurised vessel are capable of causing serious damage to a building in which the vessel is located and to neighbouring structures.

The majority of explosions in buildings result from a deflagration. In Section 7.2 some important properties which determine the intensity of an explosion of combustible gas/air mixtures, and to a limited extent dust/air mixtures, are discussed. Factors influencing the accumulation of gas from a leak into a compartment are also considered.

7.2 Some Important Properties of Combustible Gases, Vapours and Dusts in Admixture with Air

7.2.1 Flammability Limits

Mixtures of combustible gases, vapours and dusts in air are capable of propagating a flame only when the concentration of fuel in air lies within what is known as the 'flammable range'. The flammable range of a fuel/air mixture is bounded by two well defined limits known as the 'lower explosive limit' (LEL) and the 'upper explosive limit' (UEL), although for dusts the UEL is ill-defined. For gases and vapours, these limits are expressed as the percent concentration by volume of the fuel in air. For dusts, the LEL is expressed in terms of a weight per unit volume. For many dusts this usually lies within the range 40–50 g/m^3, which conveniently also corresponds closely to the LEL concentration of many organic gases and vapours. It follows that fuel/air mixtures whose composition lies below the LEL or above the UEL, will not propagate a flame under normal atmospheric conditions. Fuel/air mixtures within the flammable range lying reasonably close to the LEL are generally referred to as 'lean' mixtures, whereas those lying relatively close to the UEL are referred to as 'rich' mixtures. Some typical flammability limits for hydrogen and several hydrocarbons in air are set out in Table 7.1.

Table 7.1 Maximum laminar burning velocities and flammability limits for hydrogen and some hydrocarbons in air

Fuel	Maximum laminar burning velocity (m/s)	Lower flammability limit (vol%)	Upper flammability limit (vol%)
Hydrogen	3.5	4	75
Methane	0.45	5	15
Propane	0.52	2.2	9.5
Butane	0.50	1.9	8.5
Hexane	0.52	1.2	7.5
Acetylene	1.58	2.5	80
Cyclohexane	0.52	1.3	8.0
Gasoline		1.2	7.0

7.2.2   Stoichiometric Mixtures

The concept of stoichiometric mixtures usually applies to fuels in the gaseous or vapour state. A stoichiometric mixture is one in which there is theoretically just sufficient oxygen to balance the chemical equation so that the hydrocarbon fuel is converted to carbon dioxide and water. As an example, for propane the stoichiometric mixture would be represented by the equation:

$$C_2H_6 + 3.5O_2 = 2CO_2 + 3H_2O$$

For stoichiometric mixtures, the percent concentration by volume of fuel in air usually lies closer to the LEL than to the UEL.

7.2.3   Factors Influencing the Accumulation of Gas Leaking into a Compartment

Harris (1983: 18–35) prepared a detailed review of the way in which leaking gas mixes with air, and how the mixture formed distributes itself in an enclosure. The factors influencing the accumulation of gas in an enclosure are:

- gas density;
- the conditions of leakage;
- ventilation.

Gas density

Gases which are less dense than air, such as natural gas, form mixtures between the point of leakage and the ceiling. Concentration profiles are determined by both the conditions of the leak and ventilation. Conversely, gases which are heavier than air such as the volatile constituents of gasoline and liquefied petroleum gases, form layers at floor level with inverted concentration profiles.

Source of leakage

A series of experiments carried out with natural gas leaking into unventilated enclosures by British Gas (Harris, 1983: 18–35) demonstrated that:

- sources of leak close to the ceiling produce shallow, gas-rich layers between the point of leakage and the ceiling;

- sources of leak close to the floor fill the enclosure from the point of leakage to the ceiling with a mixture of a reasonably uniform gas concentration. Moreover, a uniform gas concentration is established quite quickly;

- downward pointing leaks result in better mixing and deeper layers than occurs with leaks that are either horizontal or upward facing;

- higher volumetric leak rates reduce the time taken to reach a certain gas concentration, and higher leak velocities promote better mixing.

More recently it has been shown that variations in the gas leakage conditions and geometry can be accounted for by relevant dimensionless parameters, leading to the development of a simple mathematical model to predict the build-up of gas in a nominally unventilated enclosure (Cleaver et al., 1994). The predicted maximum gas concentrations in the enclosure generally agreed with experimental results from studies with natural gas to within a factor of two. The model was also found to be applicable to leaks involving other gases such as propane.

Gas leak rate

If circumstances permit, then the rate of leak of gas from a perforation, crack or opening in a pipe forming part of the installation should be measured on-site. This can be done by using compressed air, regulated to the appropriate gas supply pressure, and a flow meter. The resulting measurement is to be corrected for the gas under consideration by applying the factor: (air density/gas density)$^{0.5}$. Alternatively, a U-tube manometer can be attached to the relevant section of pipe of known volume within which the pressure is raised to the normal gas pressure. The time taken for the pressure to drop by a measured amount is then determined and the flow rate through the point of leak can be calculated from the equation below reproduced from Harris (1983: 144).

$$Q = (7.2V/t)[(PiPw)^{0.5} - (PfPw)^{0.5}] \tag{7.1}$$

where Q is the flow rate (m^3/h); V is the volume of the section of the pipe under test (m^3); Pi is the initial test pressure (mbar gauge); Pf is the final test pressure (mbar gauge); Pw is the normal working pressure (mbar gauge); and t is the time taken for the pressure to drop from Pi to Pf(s).

When it is impractical to determine the probable gas leak from measurements on-site, then the approximate leak rate can be calculated (Harris, 1983: 140). If the leak is from the straight portion of an open-ended pipe, the leak rate is given by:

$$Q = 4.03 \times 10^{-3}(\Delta p d^{4.8}/S^{0.8}L)^{0.555} \tag{7.2}$$

where Q is the gas flow rate in m^3/h; Δp is the pressure drop along the length of the pipe (mbar); d is the pipe internal diameter (mm); S is the specific gravity of the gas (0.58 for natural gas); and L is the length of the pipe over which the pressure drop occurs. (A table of the straight pipe length equivalents of fittings such as bends of varying angles and T-pieces, is reproduced in Harris (1983: 141)).

The above equation is applicable if:

- gas flow in the pipe is turbulent;

- the operating pressure is ⩽30 mbar gauge;

- the internal diameter of the pipe is between 12.5 and 150 mm;

- there are no pressure losses caused by expansion and contraction of the pipework.

When the gas has been leaking from a restricted opening such as a crack or perforation, then the approximate leak rate can be calculated from the equation below (Harris, 1983: 141):

$$Q = 5.1 \times 10^{-2} Cd Av(\Delta p/\rho)^{0.5} \tag{7.3}$$

where Q is the gas flow rate in m^3/h; Cd is the discharge coefficient (assumed to be 0.6); Av is the area of the orifice through which the gas was leaking (mm^2); Δp is the pressure drop across the orifice (mbar); and ρ is the gas density (kg/m^3, which for natural gas is 0.75).

Ventilation

Natural ventilation in a building results from temperature differences and the effects of wind. Typical adventitious ventilation rates in buildings are reported to range from about 0.5 to 3.0 volume changes per hour (Harris, 1983: 18–35). Further experimental studies carried out by British Gas investigated the effects of different ventilation patterns on the build up of natural gas in an enclosure (Harris, 1983: 18–35). These revealed that for the same ventilation and gas leakage (mid-height) rates, the effectiveness of the ventilation flow pattern on limiting the concentration of gas in the mixture at steady state conditions increased in the order:

- cross-flow at the bottom of the enclosure caused by wind;
- downward flow caused by wind;
- upward flow, typically caused by temperature differences;
- cross-flow at the top caused by wind.

The gas concentration profile and position of the layer also depends on the ventilation flow pattern. The variation with time of the average gas concentration in an enclosure can be calculated from the perfect gas mixing equation as:

$$Ct = [100 \, Qg/(Qa + Qg)][1 - \exp(-[Qa + Qg]t/v)] \tag{7.4}$$

where Ct is the volume percent of gas in the gas–air mixture after a time, t; Qg is the volume flow rate of the gas (m^3/h); Qa is the flow rate of the ventilation air (m^3/h), typically between 0.5 and 3.0 room changes per hour; t is the time in hours from the onset of the gas leakage; and V is the volume of the enclosure (m^3).

Significantly higher ventilation rates should be used where ventilation grilles or other openings are known to be present in the exterior of the structure. Due account should also be taken of atmospheric conditions and in particular, the effects of a wind blowing against a structure fitted with a ventilation grille. For cross-flow, and upward ventilation patterns in a compartment, and for leaks involving lighter-than-air gases, it may be more appropriate to calculate V as the volume of the compartment above the point of leak (Harris, 1983: 18–35).

7.2.4 Flame Speed and Burning Velocity

The propagation of a flame through a fuel/air mixture takes the form of a thin reaction layer which, under ideal circumstances, is spherical in shape, moving radially outwards from a central point of ignition. Constraints imposed by the compartment in which the deflagration is taking place, obstacles in the path of the expanding flame front and variations in mixture composition will cause the spherical shape to distort. The flame will also be distorted towards a vent opening which forms as a result of the removal of a vent cover, or the failure of a weak element of the enclosure boundary. The flame speed, S_f, of a particular fuel/air mixture is the velocity at which the flame front moves with respect to a stationary observer.

The combustion process in the flame front gives rise to high temperatures which cause expansion of the gases moving behind the flame front. This in turn pushes the flame front forward. The flame speed is related to a fundamental property of fuel/air mixtures known as the 'burning velocity', S_b, which is the velocity with which the flame front moves relative to the unburnt mixture lying before it. The relationship between flame speed and burning velocity is represented approximately by the equation below:

$$S_f = S_b \cdot (T_f/T_i) \tag{7.5}$$

where T_f is the flame temperature (K); and T_i is the initial temperature of the fuel/air mixture (K).

The burning velocity varies with the type of fuel and is an important property of the fuel in that the rate of pressure rise in an enclosure increases with burning velocity. The rate of pressure rise in a vented, or partially vented deflagration is a factor which determines the maximum explosion pressure reached in an enclosure. The burning velocity of a fuel/air mixture increases as the fuel content increases from a composition close to the LEL, generally reaching a maximum on the gas rich side of a stoichiometric mixture, and then falls as the concentration of the mixture approaches the UEL. Table 7.1 provides examples of burning velocities for hydrogen and a number of hydrocarbon fuels. Thus, from equation 7.5, the maximum laminar flame speed of the saturated hydrocarbons will lie approximately in the range 3.5 to 4.1 m/s. Hydrogen and unsaturated hydrocarbons such as acetylene and ethylene have significantly higher burning velocities, and hence laminar flame speeds.

The burning velocity of a particular fuel/air mixture is also susceptible to variations in temperature and pressure. It is generally found that an increase in temperature will cause an increase in burning velocity. Conversely, an increase in pressure will cause a decrease in the burning velocity. The burning velocity will also increase with the degree of turbulence. Thus, the propagation of a flame in a compartment where turbulence can be induced by, for example, the presence of obstacles, will increase the rate of pressure rise within that compartment. When a flame is propagating through a flammable gas/air mixture contained within a pipe or duct having a large ratio of length to diameter (i.e. 50–100 : 1), flame acceleration caused by turbulence can cause a transition from deflagration to detonation. This has been observed in accidental explosions resulting from the ignition of a large spillage of gasoline into underground sewage systems.

7.3   The Development of Pressure

In the preceding paragraphs, reference was made to the expansion of gases caused by a rapid rise in temperature as being the principal reason for the generation of pressure, particularly in a confined explosion. Pressure waves are transmitted in air at the speed of sound in air which is 334 m/s at 20°C. Maximum flame speeds for methane and most commonly available hydrocarbon fuels are very much slower than this, typically in the region of a few metres/second as has already been discussed. Thus, as a flame front moves through a fuel/air mixture and expansion of the gases causes the pressure to rise, the pressure information is transmitted omnidirectionally and can be assumed to be the same throughout the compartment at any instant in time. The exception to this is where higher burning rates result from turbulence induced by obstructions, the geometry of which also provides a degree of confinement.

Pressure rises in combustion chambers caused by the ignition of pockets of gas/air mixtures are reported by Cubbage and Marshall (1972). The maximum overpressure increases with the volume of the gas/air pocket, and for a stoichiometric mixture which completely fills a strong enclosure, the maximum overpressure can exceed 8 bar. Most building structures and industrial plant are neither designed to, nor capable of, withstanding such pressures. Consequently, as the pressure in a compartment rises, the weakest elements fail and provide a vent through which the products of combustion and burning fuel can escape. Windows are usually the weakest elements in buildings. The variation of pressure with time for a vented explosion taking place under idealised conditions (where the gas/air mixture which completely fills the enclosure is ignited at the centre) is illustrated in Figure 7.1. Typically, there are two main pressure peaks. The first peak, P_v, corresponds to the pressure at which the vent is removed and this becomes less distinct as the breaking pressure of the vent increases. The second peak, P_m, corresponds to the maximum final pressure which is dependent on the burning velocity of the gas, the vent area and the speed of removal of the vent. In practice the idealised variation of pressure with time is usually an oversimplification of events, and the following points should be noted:

- the rate of venting increases when hot, burnt gases are discharged through the vent;
- the magnitude of P_m often depends on achieving the maximum flame area in the enclosure;
- the maximum flame area, and hence P_m, can be achieved for gas/air mixtures which do not completely fill an enclosure depending on the volume of the mixture, and the position within the enclosure of both the gas/air mixture and the source of ignition;
- for gas/air mixtures filling an enclosure, a source of ignition remote from the vent will lead to expulsion of some of the unburnt mixture. When the flame emerges from the vent, the unburnt mixture ignites. This may generate pressure outside the vent which reduces the rate of venting and increases the magnitude of P_m.

The fitting of suitably sized explosion relief panels to plant, such as ovens, in which the risk of an explosion is foreseeable, is a means by which the explosion overpressure can be prevented from reaching a magnitude which will cause damage. The sizing of explosion relief vents is a well established practice (Maisey, 1965; Cubbage and Marshall 1973; Butlin and Tonkin 1974; Rashbash et al., 1976; Marshall, 1977; Rogowski 1977; Bartknecht 1981;

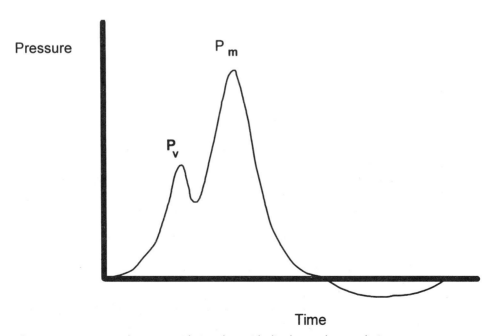

Figure 7.1   Variation of pressure with time for an idealised vented gas explosion.

NFPA 68, 1988), and the equations involved provide a useful means of assessing the maximum likely explosion overpressure which could occur in an enclosure filled with a gas/air mixture. These empirical equations have limitations which are discussed in the references cited. They are applicable to structures similar to those in which the experiments leading to their derivation were carried out. NFPA 68 recommends the use of a model developed by Bartknecht (1981) for structures that can withstand more than a 10 kPa overpressure, in which category lie most building structures, ship's tanks and boilers for example. Using this approach, the maximum pressure developed as a result of a vented deflagration, P_m(bar), is given by:

$$P_m = \left[\frac{d \cdot (V)^f \cdot e^{g \cdot P_v}}{A_v} \right]^h \tag{7.6}$$

where P_v is the vent release pressure (bar); A_v is the vent area (m²); V is the enclosure volume (m³); and d, f, g and h are constants for specific gases as discussed below.

This equation is derived from nomograms (Hart, 1985) produced from extensive studies of vented explosions, in which the vent was a plastic foil bursting disc having a failure pressure ranging from 0.1 to 0.5 bar (10–50 kPa). The disc was somewhat stronger, therefore, than the typical failure for glass which, ranges from about 2 to 7 kPa. Values of the constants in the above equation for methane, propane and hydrogen are:

	d	f	g	h
Methane	0.050	0.770	1.230	1.215
Propane	1.148	0.703	0.942	1.490
Hydrogen	0.279	0.680	0.755	2.545

Alternatively, the method of Cubbage and Marshall (1973) can be used to assess the maximum overpressure achieved in a vented explosion in a compartment by using the formula:

$$P_m = 0.1[P_v + 23(S_T^2 \cdot K \cdot W/V^{0.33}] \tag{7.7}$$

where P_m is the maximum pressure (kPa); P_v is the vent failure pressure (kPa); K is the vent coefficient, usually taken to be $V^{2/3}/A_v$ where V is the vessel volume (m³) and A_v is the vent area (m²); W is the weight per unit area of vent cladding (kg/m²); and S_T is $S_0\beta$ where S_0 is the maximum laminar burning velocity (m/s) and β is a turbulence factor to take account of the influence of turbulence on the deflagration by the presence of internal obstacles such as furniture in the enclosure.

Values of β suggested (Rasbash et al., 1976) in cases where there is no turbulence prior to ignition are:

$\beta = 1.0$ if there are no internal obstacles;

$\beta = 1.5$ for enclosures of room or laboratory size where turbulence is generated by furniture or obstacles on one level;

$\beta = 5.0$ for an explosion propagating through large openings into other sections of an enclosure, or where the obstacles are distributed throughout the entire enclosure volume.

(If significant turbulence may be present prior to ignition, an appropriate value of β might be as high as 8 or 10.)

In most buildings, windows will be the weakest element of the external structure. If these fail completely in a compartment of a building in which a deflagration occurs, then an upper limit on the maximum explosion overpressure can be assessed by inserting appropriate values for the parameters in equations 7.6 and 7.7. If this upper limit exceeds the failure pressure of other structural elements of the building, such as doors, walls and floors which do not in fact fail, then the upper limit will be overestimated. In this situation it may be assumed that the typical failure pressures of these other building elements was not exceeded. The effect of a pressure pulse on a building structure is considered below.

7.3.1   The Effect of a Pressure Pulse on a Building Structure

One of the objectives of an investigation will be to assess the maximum explosion over-pressure within a building or enclosure as this can provide an indication of the amount of fuel involved in the deflagration. It must be emphasised, however, that this is not an exact science. It is usual, therefore, to determine a range within which the likely maximum overpressure lies by using a number of methods as will be discussed below.

A considerable amount of research and experimentation on the effects of a pressure pulse on building structures followed the Ronan Point gas explosion in London in 1968. This explosion resulted in the progressive collapse of one corner of a 22 storey block of flats (Mainstone 1973, 1974, 1976; Taylor and Alexander 1974; Buildings and the hazards of explosions, 1974). A Tribunal was set up to inquire into the cause of the explosion, and the reason for the progressive collapse of the structure. The recommendations of the Tribunal led to an amendment of the then current Building Regulations relating to the construction of a building having five or more storeys, including a basement. The amendment included a requirement that portions of structural elements of a building should be designed to with-stand loads produced by overpressures up to 34.47 kPa (5 psi). No explanation for the choice of this overpressure was provided.

Generally speaking, the duration of an imposed pressure load resulting from a defla-gration in a building is likely to exceed the natural period of vibration of the structure, or structural elements, which are typically in the range 10–40 ms (Harris, 1983: 84). In the circumstances, the loading experienced by the building or structural element will effectively be the same as a static pressure of the same magnitude as the peak pressure generated by the explosion. Exaggerated damage can occur when the natural period of vibration of a building or structural element is similar to the duration of the imposed pressure load. The equivalent static overpressure can then be up to a factor of $\pi/2$ greater than the peak pressure generated by the explosion. The inertia of a structure, or structural element, is also a factor which can have a significant influence on the magnitude of the peak pressure during a deflagration. Higher peak pressures occur for structures which have a high inertia.

Window glazing

Detailed experiments have been carried out to investigate the effects of gaseous explosions in buildings on the failure of window glazing (Mainstone, 1971; Harris et al., 1977). Relation-ships between glass area and thickness, and the breaking pressure for different kinds of glass are shown in Figures 7.2 and 7.3.

Clearly, if window glazing is broken during a deflagration in a building, the only conclu-sion which can be reached is that the breaking pressure of the glass was exceeded. Conversely, if glazing is intact, then this is of assistance in placing an upper limit on the maximum explosion pressure. In the aforementioned experiments the distance of travel of glass windows or window fragments was determined as a function of the explosion overpressure generated, and this is shown in Figure 7.4. Measurements of the maximum distance travelled by glass fragments at the scene of an explosion can, therefore, provide a useful indication of the peak pressure generated by the explosion.

In a limited number of experiments with both single and double glazed units (of dimension 1 m × 1 m, utilising 32 oz glass) it was established that the failure pressure of the double glazed unit was a factor of approximately two times the failure pressure of the single unit, i.e. 6.8 kPa compared with 3.4 kPa (Astbury et al., 1972).

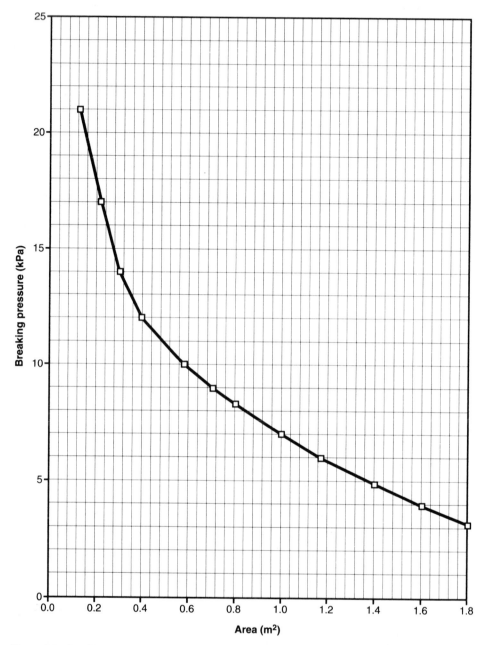

Figure 7.2 Breaking pressure as a function of area for 5-mm thick glass panes (mean curve for plain, patterned and floated glass). (Reproduced with kind permission from Harris, Marshall and Moppett, 1977.)

Brick walls

Hendry et al. (1973) provide useful data on the lateral strength of 10.5 inch (267 mm) cavity brick walls having different lengths, numbers of returns and levels of precompression. The walls were tested with statically applied loads. For walls without returns (another wall

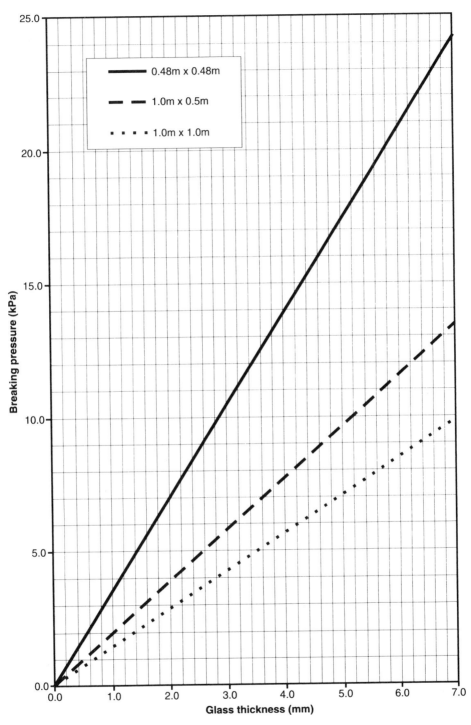

Figure 7.3 Mean breaking pressure of glass panes as a function of glass thickness. (Reproduced with kind permission from Harris, Marshall and Moppett, 1977.)

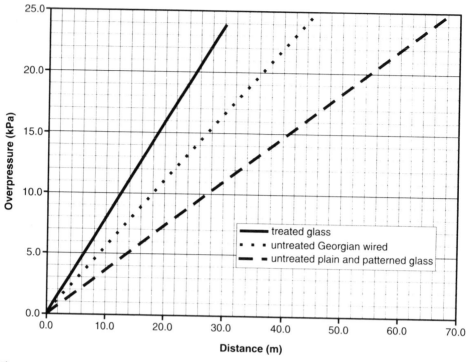

Figure 7.4 Mean distance of travel of fragments of glass as a function of the explosion overpressure. (Reproduced with kind permission from Harris, Marshall and Moppett, 1977.)

perpendicular), it was found that cracking started soon after the tensile bending strain exceeded the precompressive strain. The strength of the walls was increased by the presence of returns. In the case of short walls, the vertical joints failed first followed by the wall buckling outwards. The strengthening effect of the returns reduced as the length of the wall increased. For a wall aspect ratio (length/height) of about unity, the presence of one return increases the failure pressure by a factor of about 1.5, and the presence of two returns increases the failure pressure by a factor of about 3. For an aspect ratio of about 2, these factors reduce to about 1.2 and 1.5 respectively, and some increase in strength is provided up to an aspect ratio of 3.

The Ronan Point gas explosion also stimulated a considerable amount of research on the effects of deflagration overpressures on load-bearing brick structures, both in single and multiple compartments (Rasbash et al., 1970; Astbury et al., 1970, 1972). Pressure–time profiles observed during some of the tests showed that gas explosions are relatively slow conflagrations building up to the first peak in about one hundred to several hundred milliseconds (Astbury et al., 1970). The shape of the pressure profile, however, depends upon the nature of the gas and the venting characteristics of the enclosure. Moreover, an analysis of the response of brick panels to dynamic loading of the type produced during a deflagration has shown that the deflection of a panel may be increased by a factor of two if the rise time to peak pressure is reduced from about 200 ms to 50 ms. Although the dynamics of a deflagration should not be overlooked when interpreting the failure of brick walls, the studies have confirmed that the failure pressure of a brick wall determined from static pressure tests will be representative of a failure caused by a smooth pressure pulse generated during a gas explosion.

During the experiments designed to examine the effect of cascade explosions involving interconnected rooms, pressure spikes or oscillations were observed superimposed on the main pressure pulse (Astbury et al., 1970). Studies have shown that the spikes on these complex pressure pulses can provide a duration of applied load which matches the natural vibrational frequency of a brick structure (Astbury and Vaughan, 1972). In these circumstances there is the potential for greater damage to brickwork than would be expected from static pressure loading conditions, i.e. the equivalent static overpressure may be up to a factor of $\pi/2$ greater than peak pressure generated by the explosion, as discussed earlier. It should be noted, however, that the aforementioned studies were carried out in empty brick enclosures, and in practice bulky objects such as furniture and furnishings can significantly reduce the magnitude of these pressure spikes which are triggered by acoustic waves (Zalosh, 1979).

Some typical ranges of failure pressures of different wall constructions and other structural building elements under the conditions of a gas explosion have been collated by Harris (1983: 94), and these are summarised in Table 7.2.

7.3.2 Blast Waves

The sudden failure of a pressurised enclosure creates a pressure disturbance on the surrounding air. Because of its compressibility the air will be heated locally, thus causing the velocity of sound to increase. The front of the pressure disturbance becomes steeper as it moves through the air and the initial pressure disturbance may soon assume a classical shock wave profile such as that illustrated in Figure 7.5. The transition from the initial pressure disturbance to a shock wave takes place over a distance which is approximately a few times the mean diameter of the enclosure.

The characteristic feature of the shock wave is the shock front. In the shock front of an established ideal shock wave which is moving away from its point of origin, the pressure rises virtually instantaneously to what is called the peak overpressure (P_i^+ in Figure 7.5). The velocity of propagation of the shock front will depend upon this peak over pressure relative to the still air ahead of the shock wave. The positive shock duration is the period of time, P^+, that the pressure remains above atmospheric. In conjunction with the velocity, this gives the duration as a length. The impulse of the shock wave is obtained by integrating the pressure curve with respect to time (i.e. the area beneath the positive pressure curve).

Behind the shock wave the air is moving with a velocity known as the 'streaming velocity'. This is sometimes called the 'blast wind' or 'transient wind'. As the shock wave continues to propagate away from the source, it gradually decays until the overpressure has fallen to about atmospheric pressure, at which point it behaves as an acoustic wave. When the overpressure falls below atmospheric pressure, the movement of air reverses direction and the blast wind is directed towards the source of the blast (sometimes referred to as the rarefaction or suction phase), and the overpressure falls below atmospheric pressure. When the negative pressure phase rises again to atmospheric pressure, the blast wind direction again reverses, and moves away from the source.

Table 7.2 Typical failure pressures of structural building elements exposed to gas explosion

Building element	Typical failure pressure (kPa)
114 mm brick wall	22–35
Unrestrained brick walls	7–15
100 mm thermalite block wall	14–22
Double plasterboard	3–5
50 mm breeze block wall	2–5
Single plasterboard	2–5
Glass windows	2–7
Room doors	2–3

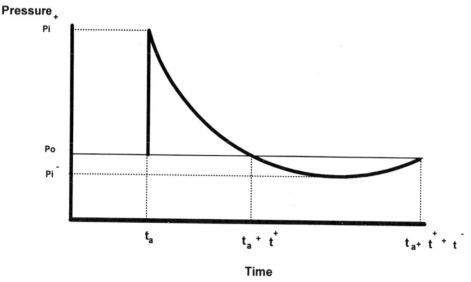

Figure 7.5 Decay of pressure with time for a classic shock wave.

To a stationary target the passage of an intense shock wave may be a catastrophic event. Nothing is known about the approach of the shock wave until it strikes. Its approach is not heralded by a sound wave since it is travelling faster than sound. The pressure around the target suddenly rises to the reflected peak overpressure, or face-on overpressure. The magnitude of the pressure pulse resulting from this reflection is typically a factor of at least two times the peak overpressure of the shock wave when the target is normal to the direction of travel of the shock wave. The magnitude of this factor depends on the peak incident overpressure. This process involves arresting the streaming gases behind the shock wave which in turn produces a directional pressure effect on the target. This sudden application of a directional pressure to the target is the most devastating manifestation of the passage of a shock or blast wave. Rigid structures such as walls and windows tend to break. The extent of the damage is related to the magnitude of the shock wave which is usually assumed to be applied perpendicularly to the direction of travel, i.e. normal to the reflecting surface, to avoid complications arising from reflection. This is often referred to as the 'side-on' shock overpressure.

Behind the shock wave, the gas is streaming forward and this results in the target being subjected to aerodynamic drag forces which further contributes to the net transverse pressure.

Finally, the effect of the rarefaction may also manifest itself. It is typical of explosions involving low energy density sources, such as gas phase explosions in buildings and bursting vessels, that the negative phase of the blast wave is sizable when compared with the positive phase (Baker et al., 1983). The effect is most noticeable with glass window damage. The passage of the shock wave may crack the windows and they will begin to fail inwards. Before they have time to move into the building, however, the pressure drop outside the building due to the passage of the rarefaction will often result in a reversal of the process with the result that a significant amount of glass ends up outside the building. This is particularly noticeable where wired glass is involved. It is often found to be bowed outwards, i.e. towards the blast source.

The Overpressure Working Party (1994), produced a detailed monograph on explosions in the process industries which deals with the interaction of blast waves with building structures. The static strength of a structure alone is not enough to determine the vulnerability of that structure to damage, and it is necessary to take account of pressure, impulse and duration in respect of the blast wave, together with information on the mechanical properties of the structure or object. As a first approximation, however, damage remote from the site of the explosion is usually related to the side on peak overpressure of the blast wave.

Peak pressures in a blast wave required to cause specific levels of damage are reasonably well known from data collected on the effects of blast waves produced by conventional explosives. From these data, it is possible to produce tables such as Table 7.3 (Scilly and High, 1986), which provides a few examples of levels of damage caused by the interaction of shock-wave overpressures with buildings and other structures. It is important to note from Table 7.3 that there is usually a range of peak overpressures which will cause the same degree of damage. This variation arises because explosives of different sizes were used to generate the shockwaves which resulted in different combinations of peak pressure and impulse. It must also be emphasised that at a specified distance from the source of an explosion, the shock-wave generated by a gas explosion in plant or in a building is likely to have a different profile from one generated by the detonation of TNT.

Butlin and Tonkin (1974) determined an empirical relationship between peak pressure resulting from a gas explosion in a horizontally vented enclosure and external pressure at various distances normal to the vent which takes the form:

$$P_m = P_1 \left[\frac{d_1 + d'}{d'} \right] \tag{7.8}$$

where P_m is the maximum overpressure at the site of the explosion; P_1 is the side-on peak overpressure at a distance d_1 from the source of the explosion; and d' is a characteristic dimension of the enclosure usually taken as the cube root of its volume.

Butlin (1976) estimated the likely range of explosion overpressures caused by a gas explosion in the ground floor dwelling of a block of flats in Liverpool from the extent of blast damage caused to windows of neighbouring buildings. The importance of turbulence in creating a higher peak pressure within the enclosure than is predicted from venting theory

Table 7.3 Levels of structural damage associated with blast waves produced by TNT

Approximate side-on shock wave overpressure (kPa)	Damage
0.04	Reports of isolated window damage
0.14	Annoying noise
0.2	Occasional breaking of large glass windows already under strain
0.7–1.0	5% glass windows broken
1.5–2.5	50% glass windows broken
3.7–4.1	90% glass windows broken
3.5–7.0	Large and small windows usually shattered and occasional damage to window frames
2.7–44	Roof tiles displaced
3.7–6.2	Minor structural damage to houses
10	Serious structural damage to houses; cracking of exterior walls; corrugated asbestos type sheeting shattered; panels torn from fastenings; houses may not be repairable
21	Steel framed buildings distorted, cladding torn off.
15.9–27.6	Partial or total collapse of roof of house
16.5–35.8	Telegraph poles snapped, large trees destroyed
34.5–79.3	50 to 70% external house brickwork destroyed or unsafe
60	Reinforced concrete walls cracked

for example the method of Cubbage and Marshall in equation was illustrated in this work.

Scaling laws are frequently used to predict the properties of blastwaves generated by large scale explosions involving explosives, from small scale tests. In Hopkinson's scaling law the scaled distance, z, is defined as:

$$z = \frac{d}{w^{1/3}}$$

(7.9)

where d is the distance from the source of the explosion (m); and w is the weight of the explosive (kg).

This means that the same blastwaves will be generated at an identical scaled distance when two explosive charges having the same geometry and of the same explosive, but of a different size, are detonated in the same atmosphere. The relationship between scaled distance and blastwave peak overpressure has been determined experimentally using different masses of TNT and has been shown to agree well with theoretical predictions. The results of a large number of tests have been used to produce overpressure-scaled distance graphs of the type shown in Figure 7.6. It is important to note that this graph relates to blast waves generated from surface explosions which, on account of reflection from the ground, causes the shock wave overpressure to increase by a factor of about 1.8 when compared with the shock wave generated by a 'free air' explosion, e.g. in an upper storey of a tower block.

Thus, the magnitude of a blast overpressure produced at a certain distance from the source of an explosion involving a known weight of TNT may be predicted from Figure 7.6. Conversely, Figure 7.6 may be used to determine the likely mass of TNT involved in an explosion from an assessment of the peak shockwave overpressure required to cause an observed level of damage in the far field. As the assessment of damage may be somewhat speculative, and the effects of topography on the passage of the blastwave may be unknown, the results of such calculations should be treated with caution.

Scaled distance calculations are also used to assess possible levels of damage in the far field caused by a bursting pressure vessel (Overpressure Working Party, 1994). Two methods are recommended for pressure vessel explosions involving loss of containment of a permanent gas stored under pressure. In the first method, the energy available from the expanding gases may be calculated from the equation below, which is the expansion of $E = \int P \cdot dV$

$$E = \frac{P_m \cdot V}{(\gamma - 1)} \cdot \left\{ \left[1 - \left(\frac{P_a}{P_m}\right)^{\gamma - 1/\gamma} \right] + (\gamma - 1) \cdot \left(\frac{P_a}{P_m}\right) \cdot \left[1 - \left(\frac{P_a}{P_m}\right)^{-1/\gamma} \right] \right\}$$

(7.10)

where E is energy (kJ); P_m is the dynamic enclosure failure pressure (kPa abs.); P_a is the atmospheric pressure (kPa abs.); V is the enclosure volume (m³); and γ is the ratio of specific heats of gas (assumed constant).

It is difficult to know what proportion of the energy released is used in generating a blastwave, but it is suggested (Overpressure Working Party, 1994) that a figure of 50% of available energy is a reasonable approximation. This is probably a conservative estimate because it is believed that efficiencies no greater than 40% have been measured. Once the energy available has been calculated, it is converted to an equivalent mass of TNT by dividing it by the energy content of TNT which is 4.6 MJ/kg^{-1}. Then by using the scaled range – peak overpressure curve in Figure 7.6 the potential peak overpressure at a specified distance from the source of the bursting vessel may be estimated. This procedure takes no account of the difference in the shape of the blastwave produced by a bursting vessel compared with that produced by detonating TNT.

In principle, the procedure may be used in reverse to make some assessment of the stored pressure in the vessel prior to failure. It is also tempting to use the same reverse procedure for estimating the available energy released during a deflagration in an enclosure, and hence the maximum explosion overpressure attained within the enclosure These calculations should,

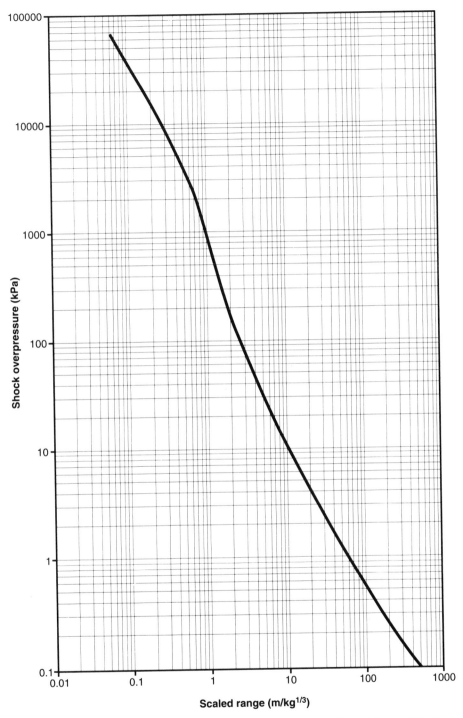

Figure 7.6   Side-on shock overpressure –V– scaled range. (Reproduced from Baker et al. Explosions hazards and evaluation 1983 p. 207, with kind permission from Elsevier Science-NL, Sara Bergerstraat 25, 1055 KV Amsterdam, The Netherlands.)

however, be treated with caution because there are a number of uncertainties which will affect the result, such as:

- P_m will be dependent upon the nature of the vent and the venting characteristics of the enclosure;

- γ needs further consideration to account for the high temperature of the combustion products;

- flame and combustion products may emerge from the vent at very high velocities, and for considerable distances from the enclosure which influences the energy imparted into the blastwave;

- the shape of the pressure pulse in the blastwave is likely to be different from that produced by the detonation of high explosive;

- assessment of the shockwave overpressure from the extent of damage caused to structures at a specific distance from the source of the explosion will be somewhat speculative. Moreover, the blast characteristics of solid phase explosives are only known approximately.

The second method of assessing possible levels of damage in the far field caused by a bursting pressure vessel was developed by Strehlow and Ricker (1976), and this was supplemented and extended by Baker et al. (1975). By using classical shock tube theory, a large number of numerical studies were performed to predict blast wave overpressures from bursting glass spheres. The effects on the results of these calculations of varying the initial sphere pressure, the initial gas temperature and the ratio of the specific heats of the gases in the vessel were also studied. The studies led to a graphical method of predicting shock overpressure at a distance from the bursting sphere. The Overpressure Working Party (1994) and Baker et al. (1983) describe the procedures to be followed in using this method which, as with the TNT equivalence method described above, can be used to determine the maximum pressure developed in an enclosure from an assessment of remote damage. The same caution should be exercised in interpreting the results.

7.4   The Scene Investigation

The main objectives of an investigation into the cause of an explosion may be summarised as:

- to ascertain the general category of the source of energy release, i.e. detonation of a condensed phase explosive, a deflagration or a physical burst;

- if a gas explosion (deflagration) has occurred determine whether the gas was lighter, or heavier than air;

- estimate the likely range within which the maximum explosion overpressure in the enclosure lies;

- identify the probable source and size of the gas leak;

- determine the probable reason for the leak;

- identify the source, or potential sources, of ignition.

A formal approach to the investigation of occurrences such as explosions is described by Burgoyne (1982a, b), and wherever possible it is recommended that this approach is adopted. The outline plan of investigation suggested can be conveniently divided into three parts as follows:

(1)   Carry out a preliminary examination of the scene of the incident without further disturbance, note remote damage on a plan and interview readily available witnesses, possibly leading to an initial view as to the likely area of origin of the explosion.

(2) Carry out a systematic and more detailed examination of the scene, involving disturbance where necessary, and obtain further witness evidence together with relevant documentary information so that a view may be formed as to the source or sources of fuel and the possible sources of ignition.

(3) Test the view formed by the application of a theoretical analysis and calculations or simulation as appropriate and adjust the view to meet the conclusions of such tests. Formulate a final view on the evidence as a whole.

7.4.1 Preliminary Investigation

Before approaching the scene of an explosion, the investigator should ascertain as far as possible the general circumstances of the incident. Witnesses are fallible and their powers of observation and recollections may be deficient. However, such problems can be reduced if the witness is taken to the position on the scene from which their observations were made. The witnesses may also assist by providing information about the state of the scene prior to the incident, routine procedures and methods of operation of plant and machinery, and descriptions of the events leading to the explosion.

It is essential that the safety of the site in its damaged condition is assessed before undertaking an inspection. It is probable that the scene will already have been disturbed to some extent by the Fire Brigade or some other responsible party having a duty to attend. Rescue operations, the clearance of debris and further demolition carried out to make dangerous structures safer will disturb the scene, and it is therefore important for the investigator to commence the scene examination at the earliest possible opportunity:

(1) The whole scene should be systematically photographed in colour, and also if possible recorded on video tape, including those areas which appear not to have been involved in the primary event. This includes peripheral damage to buildings, such as broken windows, which are likely to be repaired quite soon after the incident. Ideally, a written or dictated note should be made of the subject of each photograph, which will provide a useful commentary on the evidence. In this way, the attention of the investigator is also caused to focus on the subject of the photograph in great detail and thereby describe evidence which might otherwise be missed.

(2) Scaled plans, sketches and notes are essential as an aid to memory, and should be used to record areas where walls, doors and partitions have failed, or where they remain intact, or show signs of incipient failure. Readily identifiable patterns of damage distribution should be recorded, such as the failure patterns and displacement of constructional elements of a building, levels of peripheral damage and the distance of travel of glass fragments from windows of the building.

Particular attention should be paid to:

- materials of construction of the building;
- the locations and integrity of the gas main, service pipes and meters;
- the routing of redundant buried gas pipework, which should be tested to ensure that it is no longer live;
- the routing of consumer gas pipe work and the original disposition of appliances;
- the locations of bottled liquefied petroleum gas (lpg), low flashpoint flammable liquids and aerosol cans containing an lpg propellant;
- the presence of external wall vents;
- the presence of flues.

In addition to identifying potential sources of fuel, consideration should be given to possible sources of ignition which may be present, although this aspect of the investigation will assume more importance at a later stage. Moreover, potential sources of leak associated with gas mains, service pipes and consumer's pipe work should be pressure tested systematically for leak tightness. The test involves raising the relevant section of pipe to an appropriate pressure (usually higher than the normal operating pressure), and observing whether the pressure changes, and if so at what rate, by using a U-tube manometer (Figure 7.7). It is especially important to conduct this test on consumer's gas piping before there is any further disturbance of the scene.

7.4.2 Systematic and More Detailed Examination

Detailed examination of the scene falls under the following headings:

- identifying the origin of the explosion;

Figure 7.7 U-tube manometer used to test pipes for leaks.

202

- nature of the fuel;
- identifying the fuel;
- fire associated with an explosion;
- source of leak(s);
- source of ignition.

Identifying the origin of the explosion

Generally speaking, the distribution of damage resulting from an explosion may provide a reliable indication of its origin. Structures such as doors, partitions and walls forming the boundary of an enclosure move away from an area in which pressure is generated provided that the failure pressure of these structures is exceeded. These effects are usually most noticeable where the boundaries form the extremities of a building. Care must be exercised when examining internal structures, such as partition walls, for it will be necessary to take account of interconnected compartments in which the explosion overpressure may rise at similar rates and to similar levels which can significantly reduce the pressure difference across the thickness of the wall at any instant in time.

Due account must also be taken of the possibility of accumulations of a combustible fuel/air mixture in a number of adjacent compartments, in which case the greatest pressure could be generated in a compartment other than the one in which ignition occurred. This is due to an increase in the level of turbulence of the unburnt gases in the secondary compartment into which the primary explosion vents, an increase in the strength of the source of ignition created by flame venting into the secondary compartment and pre-compression of unburnt gases resulting in a higher initial pressure prior to ignition. Combustible dusts lying in various inter-connecting compartments of a building can become roused as they are disturbed by the propagation of an explosion from a neighbouring compartment, so that they too become involved in the combustion event, thereby creating widespread damage. The course of a dust explosion may be traced back to its source by searching for deposits of scorched dust particles, particularly on vertical surfaces.

Nature of the fuel

The next objective of an investigation will be to identify the source of energy release. In the first place, it is necessary to determine whether the explosion was caused by a detonation, the deflagration of a fuel/air mixture, or a bursting vessel.

As mentioned in the introduction, detonations give rise to extremely high rates of energy release accompanied by a very rapid rise in pressure at the source of the explosion. The shock wave produced is characterised by an almost instantaneous rise to the peak pressure, the magnitude of which decays rapidly with increase in distance from the source of the explosion. Shattering, crushing, splintering and scorching of objects close to the source of the explosion will be in evidence, and the degree of such damage will decrease markedly on moving away from the focus of the explosion.

By contrast, a deflagration takes place very much more slowly, and at a rate where the pressure generated by the combustion wave at any instant in time will be equalised throughout the enclosure volume. Similar levels of structural damage to a building may result from a deflagration and a vessel which bursts within an enclosure. The failure under pressure of an electrically operated, instantaneous water heater, and the failure of an electrically heated hot water cylinder in a closed system are examples of bursting vessels which have been found to produce significant structural damage to buildings in which they were installed (Figures 7.8 and 7.9).

Figure 7.8   Structural damage caused by a bursting electrically heated hot water cylinder.

Figure 7.9   Electrically heated hot water cylinder, which had burst.

Thermal damage resulting from the effects of exposure to the transient flame in a deflagration provides a means of distinguishing an explosion caused by a deflagration from one caused by a bursting vessel. This may involve blistering and scorching of large areas of painted and varnished surfaces, wallpaper, fabrics and smaller items having a low thermal inertia such as splinters and slivers of wood, typical examples of which are shown in Figures 7.10 and 7.11. The presence of thermal damage is of particular assistance in defining whereabouts a deflagration occurred, and hence where gas accumulated, in a building which has suffered extensive structural damage.

For explosions involving natural gas, it is generally found that extensive thermal damage results for approximately stoichiometric gas concentrations and higher (Harris, 1983: 135). Thermal damage resulting from deflagrations involving lean fuel mixtures may be difficult to find. In such circumstances evidence of superficial melting or fusing of thin plastic sheets and fibres which have a low fusing temperature will serve to confirm the passage of a transient flame.

Figure 7.10 Thermal damage to a wall caused by deflagration of a natural gas/air mixture.

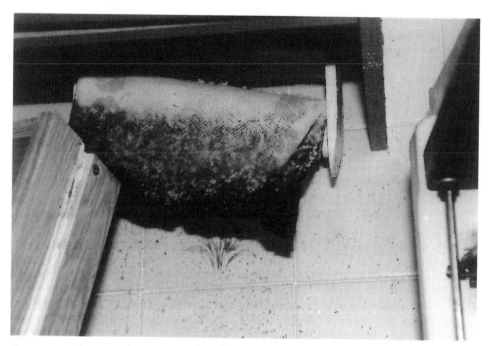

Figure 7.11   Thermal damage to a paper towel caused by deflagration of a natural gas/air mixture.

Identifying the fuel

If a detonation and a bursting pressure vessel are eliminated as the source of the explosion, the next objective will be to identity the fuel involved, and determine whether it is less or more dense than air. In the case of dust explosions, there is a pre-requisite for a combustible dust to be finely divided and of a flour-like composition, and to be sufficiently dispersed in air so as to form a dense cloud. Such situations normally arise in industrial plant, such as cyclones used in dust extraction facilities, loading chutes and during the shaking of bag filters. It should be evident from enquiries made during the preliminary stages of an investigation whether ignition of a dust cloud is a possible cause of the explosion.

Natural gas is lighter than air and will tend to form gas/air mixtures between the source of leak and the ceiling of a compartment in which the leak is taking place. Scorch and blister marks found predominantly at a high level can provide a useful indication that the fuel involved was natural gas.

Conversely, liquefied petroleum gases such as propane and butane, and vapours produced by evaporation from a spillage of volatile flammable liquids such as gasoline, are heavier than air. Mixing of the vapours with air by natural movement of gases within a compartment and by density differences will tend to cause the formation of flammable mixtures close to floor level, and therefore blistering or scorch marks found predominantly at a low level are more consistent with the involvement of heavier than air fuels (Figure 7.12).

Explosions resulting from the evaporation of volatile components of fuels such as gasoline, are often followed by a fire involving initially the less volatile components of the fuel in the area where it was spilled. The perpetrators of deliberately started fires who resort to the use of gasoline sometimes underestimate the possible consequences of delaying ignition, which can lead to the generation of pressure rises sufficient to cause serious structural damage to a building, and serious burn injuries to the person involved.

Figure 7.12 Scorching to the bottom of a door and refrigerator caused by deflagration of a
gasoline/air mixture.

Fire associated with an explosion

Isolated fires caused by explosions involving natural gas and liquefied petroleum gases may
also occur where the original source of gas leak becomes ignited during the deflagration, or
damage to an installation results in a fracture leading to an escape of gas which becomes
ignited, as shown in Figure 7.13. Solid materials which have a very low thermal inertia can be
raised to their autoignition temperatures during the course of a deflagration and continue to
burn causing isolated areas of fire damage. Such items include, for example, lamp shades and
pieces of paper. The incidence of secondary fires following an explosion involving permanent
gases is likely to increase with the concentration of gas in the flammable mixture owing to
greater radiant heat effects from the flame front (Figure 7.14).

Explosions occurring during the course of a pre-existing fire have been reported (Woolley
and Ames, 1975) giving rise to the breakage of windows and the violent emission of flame.
Such incidents are the subject of a review (Croft, 1980/81) and are thought to be caused by

Figure 7.13 Fire damage due to ignition of gas leaking from a broken pipe.

Figure 7.14 Secondary fire initiated by the deflagration of a gas/air mixture.

the formation of vapours and mists of combustible materials at a concentration above the upper explosive limit during the course of a smouldering fire. When air is admitted to the compartment in which the fire is located, the fuel becomes diluted to the extent that a flammable mixture is formed. This mixture becomes ignited by the smouldering fire, or flame produced as the smoulder undergoes transition to flame which is facilitated by improved ventilation.

A serious fire which is preceded by an explosion can mask evidence created by the deflagration of the original fuel/air mixture. An examination of fragments of window glass which have travelled some distance from the building in which the explosion occurred should assist in determining whether the fire or the explosion came first. The generation of pressure within a compartment in which a deflagration is the primary event is quite likely to cause windows to break before they are reached by the flame front. Moreover, if there is any soot deposition resulting from the momentary contact between window glass and a flame front propagating through a fuel/air mixture, this will be very slight when compared with soot deposited by a sustained smouldering fire which precedes an explosion. Thus, fragments of glass blown clear of a building which show no signs of soot or smoke deposition are usually a reliable indication that the explosion preceded the fire.

Conversely, smouldering fires invariably produce pyrolysis products, a proportion of which condense on smooth cool surfaces, such as windows and tiles, in the form of an oily brown film. Such deposits or smoke found on one side of fragments of glass blown clear of a building demonstrate that the explosion was preceded by a fire.

Source of leak

Once the type of gas involved in the incident has been established, the task of identifying the source of fuel is often straightforward. A leak of gas from a fractured or damaged pipe in a building is likely to ignite and continue to burn after a deflagration. As mentioned earlier, localised fire damage created in the vicinity of such leaks greatly assists in determining their location. It may then be necessary to carry out a close visual and metallurgical examination of the suspect point of failure to determine whether the damage developed prior to the explosion or occurred as a consequence of the incident.

Leaks from buried gas mains, or service pipes may lead to an escape of gas through the soil into a basement or an underfloor space of a building. Preferential routes for such leakages are to be found where there are cable ducts and other service entry points. Direct leakage into a floor void may result from damaged consumer's pipe work which is routed through the void. The leakage of methane into buildings may also occur from the fermentation of sewage or from land-fill sites, and this possibility should not be overlooked when conducting a scene examination. Deflagrations resulting from the ignition of gas which has accumulated principally in an underfloor space will invariably cause the floor structure to be damaged in a manner consistent with the development of an overpressure underneath (Figure 7.15), and there will be signs of scorching on the underside of floorboards. In floor voids of older buildings, the absence of cobwebs which are very susceptible to thermal damage may provide a useful indication of the passage of a flame in the underfloor space.

Where the physical evidence is most consistent with the involvement of liquefied petroleum gas, a careful search should be made for an appropriate cylinder or container along with any regulator, fixed or flexible piping, hose clips, etc. and an appliance. Any such items uncovered in debris should be examined and photographed carefully in situ before they are removed for further examination.

Finally, serious explosions have resulted from leaks of liquefied petroleum gas used as a propellant in domestic aerosol cans. Corrosion of the can wall may result in the formation of a pinhole which allows the propellant to leak at a relatively high rate, especially if the pinhole is below the liquid surface in the can. Such leaks should be investigated carefully at the scene, especially if it appears that other more usual sources of leak can be eliminated.

Figure 7.15   Chairs overturned by the upwards displacement of the floor.

Source of ignition

A source of ignition should be capable of releasing sufficient energy into the fuel/air mixture to initiate a combustion reaction which can propagate a flame independently away from the source of ignition.

Possible sources of ignition include:

- naked flames on pilot lights, burners in cooker and heating equipment, cigarette lighters and matches;
- electrical discharges from lighting and appliance switches, room and appliance thermo-stats, and control apparatus such as contactors;
- surfaces at a very high temperature.

In some situations an identification of the source of ignition can be quite straightforward. Pilot lights or frequently operating electrical contacts, for example, may be found in an area where a flammable fuel/air mixture was likely to have accumulated. A consideration of whether the gas involved was heavier or lighter than air, and therefore whether it was likely to have formed fuel/air mixtures at a high or low level in an enclosure will assist in elimi-nating or implicating possible sources of ignition.

Gas–air mixtures can be caused to ignite when they are exposed to hot surfaces. The phenomenon is usually termed autoignition. Minimum autoignition temperatures of gases and vapours are usually determined from experiments carried out in spherical or cylindrical reaction vessels which are sufficiently large to minimise the surface effects which may quench a combustion reaction, both chemically and thermally. The geometry and surface area of hot surfaces are in practice likely to be very different from the aforementioned reaction vessels. For this reason hot surfaces are most unlikely to be viable sources of ignition unless they are raised to a temperature significantly in excess of the minimum autoignition temperature quoted in the literature for the gas under consideration.

In a recent extensive review on the ignition of flammable gases and liquids by cigarettes,

Holleyhead (1996) concludes that whereas carbon disulphide, flowing hydrogen/air mixtures, acetylene, ethylene oxide and diethyl ether can be ignited, most other flammable gases cannot. Moreover, in the experiments which achieved ignition of the gases referred to above, it was often necessary to puff the cigarettes to establish a glow. It is emphasised that there is no evidence to indicate that fuels such as gasoline vapour, propane and butane are likely to be ignited by cigarettes unless exceptional circumstances arise such as momentary flaming of the cigarette paper, or if the glowing end is disturbed while in contact with a flammable gas.

7.5   The Final Stage of the Investigation

The final stage of the investigation involves the application of theoretical analysis and calculations to the results of the scene examination to assist in the formulation of a final view on the evidence as a whole.

The pressure generated at the source of the explosion can be estimated from a consideration of the extent and degree of failure of structural elements of the building, such as window glass, doors, partitions and brick walls. A lower limit on the overpressure developed can be estimated from an examination of window glass which did not fail in the explosion. Incipient failure of a structure provides one of the more useful indications of the maximum explosion overpressure in a compartment of which it forms a part.

Measurements of the maximum distance of travel of glass fragments provides an independent method of estimating the explosion overpressure. A second independent method of calculating the overpressure at the source of the explosion is based upon an observation of blast damage caused to neighbouring building structures by using equations 7.8 and 7.10 in conjunction with Figure 7.6.

A consideration of the scorch and blister patterns in the building together with the calculated overpressure generated by the explosion can be useful in confirming where the deflagration took place, and determining the likely concentration of gas present prior to the explosion. It will be recalled that maximum overpressures are generated for gas-air mixtures exhibiting the highest rates of pressure rise. Rates of pressure rise are in turn dependent upon the burning velocity which is at maximum when the fuel concentration lies on the rich side of the stoichiometric ratio. It follows, therefore, that minimum heat effects and low explosion overpressures will be produced by lean fuel/air mixtures.

Conversely, extensive blistering and scorching accompanied by low overpressures will be produced by fuel rich mixtures. Relatively high explosion overpressures accompanied by quite extensive heat damage will be produced by mixtures having a composition close to the stoichiometric ratio.

If the source of gas leakage is known, it should be possible to measure the leak rate, or alternatively calculate it by using equations 7.2 and 7.3. By making assumptions about the ventilation of the building, the time taken to reach the concentration of gas required to account for observed damage can be estimated from the perfect mixing equation, equation 7.4, or its modified form in which the volume of the enclosure is that which lies between the point of leakage and the ceiling in the case of gases less dense than air.

7.6   Case Studies

7.6.1   *Case 1*

A four-storey end of terrace building with a basement was demolished by an explosion. There were three fatalities. The building was of conventional, brick cavity-wall construction with timber floors. The ground floor was used as a grocery shop, and this had access via a trap door in the floor to timber stairs leading into the basement storage area. A sketch plan of the basement and ground floor is shown in Figure 7.16. The remaining three floors

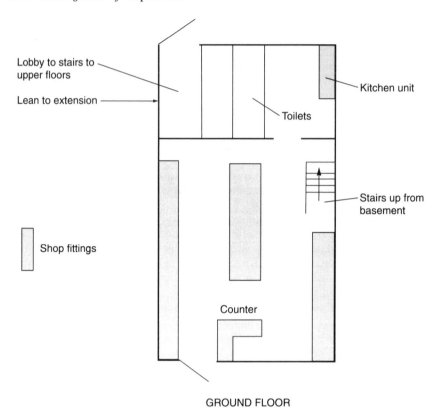

Lobby to stairs to upper floors

Lean to extension

Kitchen unit

Toilets

Stairs up from basement

Shop fittings

Counter

GROUND FLOOR

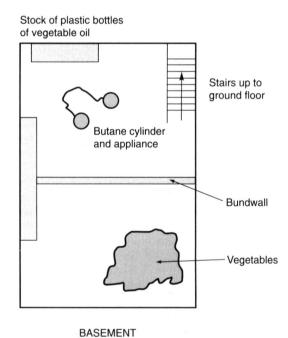

Stock of plastic bottles of vegetable oil

Stairs up to ground floor

Butane cylinder and appliance

Bundwall

Vegetables

BASEMENT

Figure 7.16 Plan of the basement and ground floor of the destroyed building.

of the building were used as residential flats. Several years before the explosion the original gas supply to the premises, which was provided by a service pipe below ground level entering the front of the basement, was capped off and replaced. A new polyethylene, 1 inch service pipe was routed below ground level to the rear, lean-to extension of the shop indicated in Figure 7.16. The service pipe emerged from the ground and via a manifold, was connected to three separate meters associated with the gas supply to each of the three residential flats. The ground floor and basement were not provided with a natural gas supply.

The proprietor of the grocery shop did not work there, but employed an assistant who was responsible for the day to day operation of the premises. Occasionally, the proprietor's father visited the premises. Approximately once a fortnight he decanted palm oil from a large, 45 gallon drum into smaller retail size containers. This was done in the basement of the premises. Because palm oil is viscous, it was necessary to warm the product in the drum and this was done by heating it on a gas ring connected by flexible hosing to a 15 kg sized butane cylinder. The process normally took about 30–60 min.

There is some evidence that during the morning of the explosion, a local street market operator visited the shop assistant and helped him by moving some stock from the basement area. At about 10:30 hours the proprietor took his father to the premises and left him there. It is not known what happened between then and the explosion at 11:15 hours, but the proprietor believes it to be unlikely that his father would have undertaken any palm oil decanting operations since these were carried out earlier during the week.

At approximately 11:15 hours, the street market operator took the shop assistant a cup of tea. As he was leaving the shop the explosion occurred. At this time, the retail premises were occupied by the shop assistant, the proprietor's father and a customer. All three died of their injuries. Residents of the second and third floor flats were in occupation at the time and described the building collapsing beneath them. They were treated for their injuries in hospital and survived. Evidence obtained from the fire brigade suggests that the proprietor's father may have been in the basement when the explosion occurred, whereas the shop assistant and customer were in the shop, the customer being closer to the rear of the premises. Both the customer and the proprietor's father died of substantial burn injuries, whereas the shop assistant suffered a broken neck. He also received flash burns and pathology subsequently revealed that his blood contained a trace of butane.

Scene examination

The explosion resulted in complete collapse of the building (Figure 7.17). Rescue operations took priority. Projectiles were deposited over a wide area at the front of the shop and a number of vehicles in the street were damaged. Debris was strewn across the street, and the frontage of shops on the opposite side of the street had sustained minor structural damage. A few of the windows of the multi-storey, terraced building at the rear of the premises were broken, as shown in Figure 7.17 (right rear). Whereas fragments of glass presumed to have been from the shop were found at a distance of about 55 m from the premises, the majority of the external glazing on the upper floors of the noted building had remained intact, indicating that the force of the explosion had been relieved through the front of the building.

At ground floor level, evidence of scorching and minor charring of torn wallpaper was found on the partitioning of the rear, lean-to extension. The butane cylinder and burner ring were recovered during excavation within the basement. This equipment is shown in Figure 7.18 and the original position in which it was found is marked on Figure 7.16. Whereas the switch on the cylinder regulator valve was found in the 'off' position, the needle valve on the burner ring was open by about 1.5 turns. Other items removed from the basement were slightly heat damaged. For example, there were heat blister patterns on the paintwork of a chair and shrink wrapping around a box containing bottles of vegetable oil had partially melted (Figure 7.19).

Figure 7.17   The scene of the explosion.

After the basement had been cleared, the stairs were repositioned as shown in Figure 7.20. Splinters on the underside of the treads and risers of the staircase were found to be scorched (Figure 7.21), and similar evidence of scorching was found on fractured woodwork around the trap door opening to the basement. The routings of the original gas main and service pipes, and the new service pipes was determined from excavations around the property, and these are shown in Figure 7.22. The new service pipe, manifold and the three meters originally located in the lean-to section at the rear of the shop were recovered and reconstructed, as shown in Figure 7.23. A T-piece from the manifold had evidentially fractured during the course of the explosion.

Analysis

The directional evidence of the blast provided by the position of projectiles, and minor structural damage to other premises in the street indicated that the overpressure within the

214

Figure 7.18 Butane cylinder and gas ring recovered from basement.

Figure 7.19 Partially melted 'shrink-wrap'.

215

Figure 7.20 Reconstruction of basement stairs.

Figure 7.21 Scorched splinters on underside of staircase treads and risers.

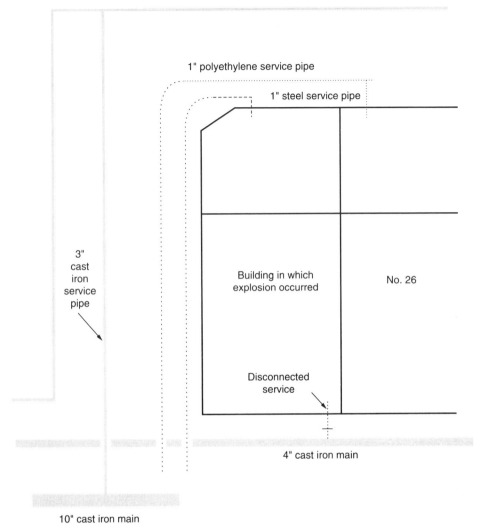

1" polyethylene service pipe

1" steel service pipe

3"
cast
iron
service
pipe

Building in which
explosion occurred

No. 26

Disconnected
service

4" cast iron main

10" cast iron main

Figure 7.22   Routings of gas main and service pipes.

building was predominantly vented through the frontage of the ground floor shop. The total collapse of the building structure is most consistent with loss of structural integrity at a low level.

The presence of scorch marks both in the basement, within the access hatch to the basement stairs and at ground floor level, confirms that the overpressure was generated by a deflagration, there being no evidence to indicate that an explosive device had operated. Moreover, two of the fatalities had suffered severe flash burns, and the third had suffered minor flash burns. The more severe burn injuries were sustained by those occupants who were believed to be in the basement, and at the rear of the shop which would have been close to the trap door leading down to the basement.

Figure 7.23 Reconstruction of new service pipe manifold and three gas meters.

The following potential sources of gas were investigated:

(a)   a leak of natural gas from the underground main or service pipework;

(b)   a leak of natural gas from the installation gas pipework within the building;

(c)   methane gas from sewers;

(d)   butane from the calor gas cylinder and/or appliance.

As far as (a) is concerned, no significant levels of gas leakage were detected when the area was monitored by engineers in the hours following the incident. Moreover, subsequent pressure decay tests confirmed that the gas main and service pipework were in satisfactory order. Furthermore, as natural gas is lighter than air, any leakage of gas which seeped into the basement would tend to rise through the trap door and into the shop. The evidence of low level heat effects in the basement were inconsistent with this scenario.

If there had been a leak of natural gas from the installation pipework, then any evidence of this was destroyed by the explosion. Only the residential flats were served with gas and it is significant that the heat effects were confined to the ground floor and basement. Having regard to the positioning of the gas pipework at the rear of the lean-to extension, the presence of flash burns at low level in the basement could not be explained as natural gas, being lighter than air, would tend to rise. Moreover, it is to be expected that a smell of gas would have been very noticeable to the shop assistant and customers significantly before the explosion occurred.

A study of the routing of drains and sewers in the area did not produce any evidence to support the possibility that methane seeped into the basement via this route.

The available physical evidence was most consistent with the source of gas being provided by the butane cylinder in the basement.

That being the case, it is necessary to explain how the gas escaped, bearing in mind that when the cylinder and gas ring had been removed from the basement by the fire brigade, the regulator valve was found switched to the "off" position. Questioning the fire officers who were carrying out excavations in the basement during clearance and search operations, indicated that it was unlikely they had interfered with the settings on the valve, although the evidence was not conclusive in this regard. This begs the question that the proprietor's father, having gone down into the basement shortly before the explosion occurred, discovered that there was a leak of butane gas from the equipment. It is then postulated that he isolated the leak by turning the regulator valve to the closed position. Although the source of ignition was not identified, it is assumed that this was in some way associated with the activities of the proprietor's father, e.g. using an electrical switch in an area where a flammable gas/air mixture had accumulated.

Maximum pressure   The explosion overpressure was developed in the basement in which the walls and floor were very strong, but the timber ceiling (shop floor) was relatively weak. Venting would initially have been through the small area of trap door opening to the basement until the shop floor failed. The travel distance of glass fragments from the shop frontage (55 m) suggests a maximum pressure of about 20 kPa, which is unlikely to cause failure of the cavity brick external walls. It is probable, therefore, that collapse of the walls was associated with the floor of the shop breaking up, either through impact or leverage forces.

If it is assumed that all of the butane which leaked into the basement burnt there, and that only air vented from the compartment as it failed, then the minimum quantity of gas which leaked before ignition occurred can be roughly estimated from the equation below:

$$\frac{\text{explosion overpressure (abs)} \times \text{compartment volume (m}^3)}{\text{maximum theoretical overpressure (abs)}}$$

If a stoichiometric butane/air mixture had completely filled the basement, of volume 60 m^3, this could theoretically have produced a maximum overpressure of about 8 bar. Hence, from the equation above, the volume of a stoichiometric mixture which generates a pressure of about 20 kPa will be 7.98 m^3.

A stoichiometric butane/air mixture contains 3.1% by volume butane. The volume of butane which leaked prior to ignition was, therefore, approximately 0.248 m^3. The duration of the leak required to account for the explosion damage is given by the equation:

$$\frac{\text{calorific value (MJ/m}^3) \times \text{gas volume (m}^3)}{\text{heat output (MJ/h)}}$$

The gas ring was of commercial hot plate size, which typically have a heat input of about 40 MJ/h. As the calorific value of butane is 112 MJ/m^3, the minimum volume of butane required to account for the explosion damage could have leaked during a period of 0.69 h.

Clearly this is consistent with the circumstantial evidence, as would still be the case if the leak rate was somewhat less than the full input rating of the appliance, and/or a somewhat greater volume of butane was required to account for the observed damage than the minimum volume used in the calculations.

7.6.2 Case 2

This case concerns an explosion in a domestic, mid-terraced dwelling. A street plan of the area is shown in Figure 7.24; the explosion occurred in 'Dwelling B' close to the centre. Figure 7.25 is a sketch plan of dwelling B. The external walls of the building were constructed of potted (lightweight) brick, and the internal partition walls were constructed of breeze (lightweight) block. Party walls were constructed from a double thickness of conventional

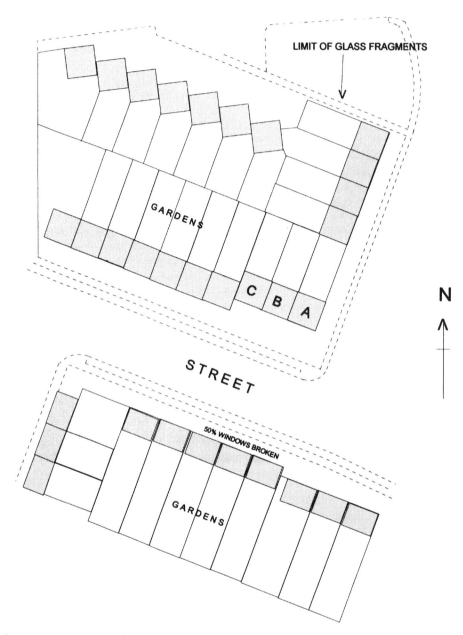

Figure 7.24 Street plan of area of explosion; the explosion occurred in dwelling B.

house brick. The external walls were not keyed to the party walls. The faces of the different brick types were simply bonded together with mortar, leaving the external walls effectively unrestrained. Windows were single glazed in timber frames. The roof was of conventional timber frame construction clad with tiles.

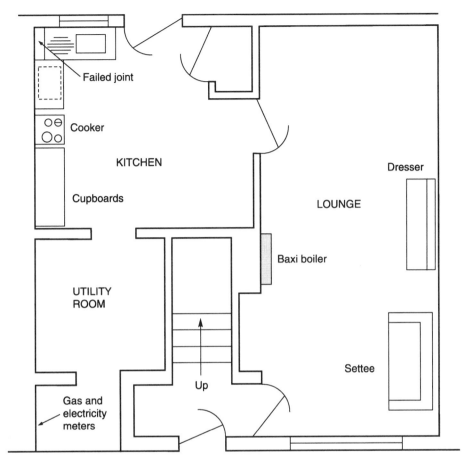

Figure 7.25   Ground floor plan of dwelling B.

During the afternoon on the day of the explosion, dwelling B was occupied briefly at 14:00 hours during which time the occupants visited the lounge and the kitchen. At that time there was no smell of gas. All windows and external doors were closed. The internal door between the kitchen and lounge was left open. The pilot light of a 'Baxi' boiler (a gas appliance which provides radiant heat and also heats water) installed in the lounge was permanently alight.

On the day of the explosion, a short garden wall was being rebuilt against the rear of dwelling C at its boundary with dwelling B (Figure 7.24). The builder was away from the premises from about 11:00 hours until shortly before 16:00 hours, when he returned and laid fresh concrete footings. About 17:00 hours, he noticed a smell of gas, but took no action. At about 17:45 hours, the smell of gas became stronger and at that moment the explosion occurred.

Scene examination

The external walls of dwelling B had been virtually demolished as shown in Figure 7.26. External walls of adjacent properties were cracked and the windows broken. Fragments of

Figure 7.26 Rear view of the damaged building (dwelling B).

glass had been blown into these adjacent properties. Flame venting from the rear windows of dwelling B had scorched plants and shrubbery up to a distance of about 11 m from the building.

In the kitchen of dwelling B, there was clear evidence of the passage of a transient flame, where splinters of a fractured cupboard were charred. In the sitting room, the wallpaper was extensively scorched from floor to ceiling and a coffee table and various other items which had been on the floor at the time of the explosion were also scorched (Figure 7.27). The covering of some polyurethane foam-upholstered furniture in the lounge had ignited and burned for a limited period of time before being extinguished by the fire brigade (see Figure 7.14). In the kitchen, a 15 mm diameter copper pipe had come adrift from a compression fitting by means of which it was connected to a floor point forming part of the consumer's gas carcass system (gas fixtures within the property of the consumer and which are the responsibility of the consumer) (Figure 7.28). Although the nut attached to the floor point could not be released by finger pressure, it was released by minimal force with a spanner. This copper pipe was connected at the other end to a 9 mm copper, plastic-sleeved pipe via reducing couplings. This pipe passed through a hole in the external wall from where it was routed below ground level in the area where fresh concrete had been laid (Figure 7.29). A detailed examination of the failed compression fitting revealed that although it had left a witness mark on the pipe wall, it had not been compressed sufficiently to create a mechanical impression (Figure 7.30).

The party wall dividing dwelling B from dwelling C was cracked indicating that it had been very close to failure as a result of exposure to the explosion overpressure developed in dwelling B. A general survey of the neighbouring area revealed that glass fragments of both reeded and plain 3 mm glass from dwelling B had been propelled at least 40 m from the rear of the building (Figure 7.24). Approximately 50% of windows were broken in the terraced houses directly opposite and at a distance of about 28 m from building B.

Figure 7.27 View of lounge showing scorched wallpaper, coffee table and carpet, and the 'Baxi' boiler.

Pressure tests carried out on the street gas main and service pipe demonstrated that these were leak-tight. Moreover, the gas carcass pipe work in dwelling B was sound with the exception of the open-ended coupling found in the kitchen.

Analysis

There is no doubt that this explosion resulted from the accidental ignition of natural gas which had escaped from a defective compression joint in the kitchen of dwelling B. No other plausible source was found. Although the compression fittings had been tightened sufficiently to make the joint leak tight, the olive (compression ring) had clearly not been compressed on the pipe to provide a sufficiently strong mechanical joint. Thus, any disturbance of the pipe created a risk that it would pull free from the floor point. It is probable that the 9 mm pipe was disturbed by the builder as he was laying fresh cement so that the pipe moved outwards from the floor point to an extent sufficient to provide a serious gas leak. The pipe may not

Figure 7.28 The failed compression fitting.

have parted company completely with the floor point at this stage because the 9 mm pipe had been further disturbed by falling masonry during the explosion. Thus the original gas leak may not, necessarily, have been from an open-ended pipe but from a partial opening caused by displacement of the pipe as a result of the initial disturbance.

(1) Maximum pressure The maximum explosion overpressure developed was estimated from consideration of the following evidence:

(a) Glass fragments had travelled to a distance of up to 40 m from the building. From Figure 7.4, the over-pressure is estimated to have been about 15 kPa.

(b) Approximately 50% of windows were broken by a blast wave in houses situated 28 m from, and immediately opposite, dwelling B. From Table 7.3, it is estimated that the reflected shock overpressure was between 1.5 and 2.5 kPa. By using equation 7.8 and setting the characteristic dimension of the building to be the cube root of the volume of the ground floor room (i.e. 4.59 m^3), the maximum overpressure is calculated to be between 10 and 17 kPa.

(c) From Table 7.2, the typical failure pressure of an unrestrained brick wall lies between 7 and 15 kPa. As the external walls had failed completely, the explosion pressure must have been in the upper end of the range.

(2) Quantity and concentration of gas Extensive thermal damage to furniture and furnishings together with severe damage to the structure of the building indicate that there had been a near stoichiometric concentration of natural gas. Moreover, a significant quantity of gas burnt outside the building as witnessed by the extent to which shrubbery had been scorched.

The source of the leak was at floor level in the kitchen and the most likely source of ignition was the pilot light on the Baxi boiler in the adjacent lounge, the interconnecting door

Figure 7.29 Basic 9 mm sleeved pipe emerging from the external wall of the kitchen.

to which was open. As discussed in Section 7.2.3 under 'Source of leakage', floor level leaks tend to fill the enclosure with a gas mixture of reasonably uniform gas concentration. As the source of the leak was partially enclosed by a kitchen unit, this will probably have influenced the concentration profile to some extent. In particular, the apparent source of the leak into the enclosure may have been somewhat higher than at floor level.

It is unlikely that the gas was escaping from a completely open-ended pipe, as this would have caused a pressure drop in the pipe sufficient to extinguish the pilot light in the Baxi boiler and hence remove the source of ignition. If, for the sake of argument, the leak had been from an open-ended pipe 5 m long and internal diameter 12.5 mm, then by using equation 7.2, the leak rate would have been approximately 9.5 m^3/h.

It is more likely that the pipe had only partially withdrawn from the floor point, thus producing a circumferential crack-like leak. For crack widths of 0.5 mm and 1.5 mm the corresponding areas through which gas can escape are about 20 mm^2 and 60 mm^2. By inserting these leak areas into equation 7.3, gas leak rates of 2.98 m^3/h and 6.6 m^3/h are calculated.

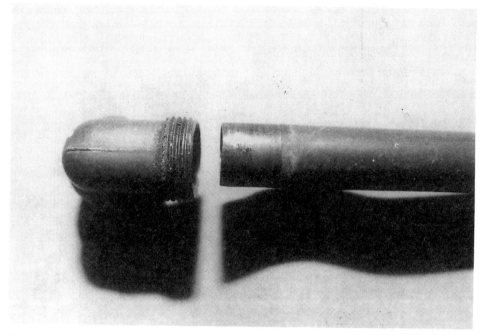

Figure 7.30 Close-up view of the defective compression joint.

The rate of build-up of gas concentration can now be estimated from equation 7.4 for assumed ventilation rates of one and two air changes per hour. The volume into which the gas is leaking is assumed to be the combined volume of the kitchen and lounge and the leak is assumed to be at floor level. The results of these calculations are shown graphically in Figure 7.31 from which it will be seen that a gas leak of 2.98 m^3/h produces a steady state concentration of gas below the lower explosive concentration of 5% by volume. The size of the leak must, therefore, have been larger than this.

A leak rate of 6.6 m^3/h could, however, have produced a near-stoichiometric gas concentration in the 1.75 h which elapsed between the builder resuming his work when he started laying cement, and the explosion. Moreover, if a leak from an open-ended pipe had not caused the pilot light on the Baxi burner to fail, then the corresponding gas leak would also have provided a near stoichiometric gas concentration at the time of the explosion. The factual observations are, therefore, entirely consistent with the circumstantial evidence relating to the leak.

References

ASTBURY, N. F. and VAUGHAN, G. N., 1972, Motion of a brickwork structure under certain assumed conditions, *Technical Note No. 19*, London: The British Ceramic Research Association.

ASTBURY, N. F., WEST, H. W. H. and HODGKINSON, H. R., 1972, *Experimental gas explosions: report of further tests at Potters Marston*, London: British Ceramic Research Association.

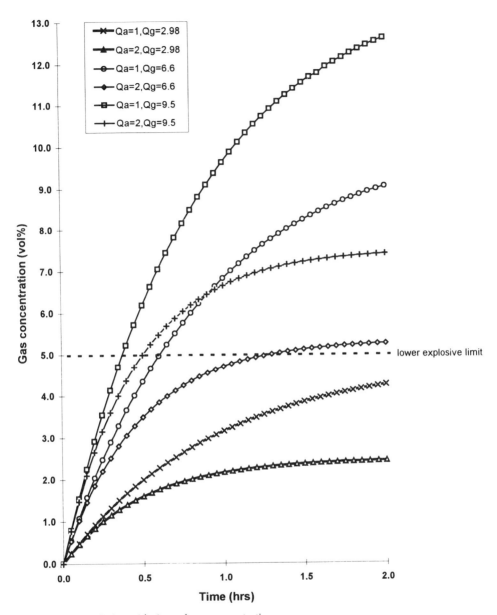

Figure 7.31 Variation with time of gas concentration.

ASTBURY, N. F., WEST, H. W. H., HODGKINSON, H. R., CUBBAGE, E. A. and CLARE, R., 1970, Gas explosions in load bearing brick structures, *British Ceramic Research Association Special Publication No. 68*, London: British Ceramic Research Association.

BAKER, W. E., COX, P. A., WESTINE, P. A., KULESZ, J. J. and STREHLOW, R. A., 1983, *Explosion hazards and evaluation*, New York: Elsevier.

BAKER, W. E., KULESZ, J. J., RICKER, R. E., BESSEY, R. L., WESTINE, P. S., PARR, V. B. and OLDHAM, G. A., 1975, *Workbook for predicting pressure wave and fragment effects of exploding propellant tanks and gas storage vessels*, NASA CR-134906, NASA Lewis Research Centre.

BARTKNECHT, W., 1981, *Explosions: Course, Prevention, Protection*, Berlin: Springer-Verlag.

BUILDINGS AND THE HAZARD OF EXPLOSION, 1974, *Proceedings of a Symposium at the Building Research Establishment* (BRE Report No. 81), Garston: Building Research Establishment.

BURGOYNE, J. H., 1982a, Accident investigation, *Journal of Occupational Accidents*, **3**, 289–297.

BURGOYNE, J. H., 1982b, The scientific investigation of the occurrences of fires, *Fire Safety Journal*, **4**, 159–162.

BUTLIN, R. N. and TONKIN, P. F., 1974, Pressures produced by gas explosions in a vented compartment, *Fire Research Note No. 1019*, Borehamwood: Fire Research Station.

BUTLIN, R. N., 1976, Estimation of maximum explosion pressure from damage to surrounding buildings, *Fire Research Note No. 1054*, Borehamwood: Fire Research Station.

CLEAVER, R. P., MARSHALL, M. R. and LINDEN, P. F., 1994, The build-up of concentration within a single enclosed volume following a release of natural gas, *Journal of Hazardous Materials*, **36**, 209–226.

CROFT, W. M., 1980/81, Fires involving explosions: a literature review, *Fire Safety Journal*, **3**, 3–24.

CUBBAGE, P. A. and MARSHALL, M. R., 1972, Pressures generated in combustion chambers by the ignition of air-gas mixtures, *Institute of Chemical Engineers Symposium Series No. 33*, London: Institute of Gas Engineers.

CUBBAGE, P. A. and MARSHALL, M. R., 1973, Pressures generated by explosions of gas–air mixtures in vented enclosures, *Institute of Gas Engineers Communications No. 926*, London: Institute of Gas Engineers.

GUGAN, K., 1979, *Unconfined vapour cloud explosions*, London: Godwin/Institute of Chemical Engineers.

HARRIS, R. J., 1983, *The investigation and control of gas explosions in buildings and heating plant*, London: Spon/British Gas Corporation.

HARRIS, R. J., MARSHALL, M. R. and MOPPETT, D. J., 1977, The response of glass windows to explosion pressures, *Symposium Series No. 49*, London: Institute of Chemical Engineers.

HART, R. L., 1985, Formula for sizing explosion vents, *Plant/Operation Progress*, **4**, 1–4.

HENDRY, A. W., SINHA, B. P. and MAURENBRECHER, A. H. P., 1973, Full scale tests on the lateral strength of brick cavity walls with precompression, *Proceedings of the British Ceramic Society*, **21**, 165–179.

HOLLEYHEAD, R., 1996, Ignition of flammable gases and liquids by cigarettes, *Science and Justice*, **36**, 257–266.

MAINSTONE, R. J., 1971, The breakage of glass windows by gas explosions, *Building Research Establishment Current Paper No. CP 26/71*, Garston: Building Research Establishment.

MAINSTONE, R. J., 1973, The hazard of internal blast in buildings, *Building Research Establishment Current Paper No. CP 11/73*, Garston: Building Research Establishment.

MAINSTONE, R. J., 1974, The hazards of explosion, impact and other random loadings on tall buildings, *Building Research Establishment Current Paper No. CP 64/74*, Garston: Building Research Establishment.

MAINSTONE, R. J., 1976, The response of buildings to accidental explosions, *Building Research Establishment Current Paper No. 24/76*, Garston: Building Research Establishment.

MAISEY, H. R., 1965, Gaseous and dust explosion venting, Part 1, *Chemical Process Engineering*, **46**, 527–535.

MARSHALL, M. R., 1977, Calculations of gas explosion relief requirements: the use of empirical equations, *Symposium Series no. 49*, London: Institute of Chemical Engineers.

NFPA 68, 1988, *Venting of deflagrations*, Boston: National Fire Protection Association.

OVER-PRESSURE WORKING PARTY, 1994, Explosions in the process industry, *Major Hazards Monograph*, Rugby: Institute of Chemical Engineers.

PRITCHARD, D. K. and ROBERTS, A. F., 1993, Blast effects from vapour cloud explosions: a decade of progress, *Safety Science*, **16**, 527–548.

RASBASH, D. J., PALMER, K. M., ROGOWSKI, Z. W. and AMES, S., 1970, Gas explosions in multiple compartments, *Fire Research Note No. 847*, Borehamwood: Fire Research Station.

RASBASH, D. J., DRYSDALE, D. D. and KEMP, N., 1976, Design of explosion relief systems for handling liquified fuel gases, *Symposium Series No. 47*, London: Institute of Chemical Engineers.

ROGOWSKI, Z. W., 1977, Explosion protection methods by reliefs, *Institute of Chemical Engineers Symposium Series No. 49*, London: Institute of Mechanical Engineers.

SCILLY, N. F. and HIGH, W. G., 1986, The blast effect of explosions, *5th International Symposium on Loss Prevention and Safety in the Process Industries*, pp 39-1–39-15, London, Institute of Chemical Engineers.

STREHLOW, R. A. and BAKER, W. E., 1976, The characterization and evaluation of accidental explosions. *Proceedings of Energy Combustion Science*, **2**, 27–60.

STREHLOW, R. A. and RICKER, R. E., 1976, The blast waves from a bursting sphere, *Association of Industrial Chemical Engineers*, **10**, 115–121.

TAYLOR, N. and ALEXANDER, S. J., 1974, Structural damage in buildings caused by gaseous explosions and other accidental loadings, *Building Research Establishment Current Paper No. CP 45/74*, Garston: Building Research Establishment.

WOOLLEY, W. D. and AMES, S. A., 1975, The explosion risk of stored foam rubber, *Building Research Establishment Current Paper No. 36/75*, Garston: Building Research Establishment.

ZALOSH, R. G., 1979, Gas explosion tests in room-like vented enclosures, *Thirteenth Loss Prevention Symposium*, London: Association of Industrial Chemical Engineers.

8

Chromatography of Explosives

BRUCE McCORD AND EDWARD C. BENDER

8.1 Introduction

Chromatography permits the separation, detection and quantitation of both the explosive and non-explosive components of energetic materials, allowing them to be identified and compared. During the investigation of an explosives incident, chromatography is an essential procedure used to separate trace amounts of unreacted explosives from a wide variety of sample matrices. In this chapter, the application of the following chromatographic techniques to explosives analysis is discussed:

- gas chromatography;
- high performance liquid chromatography;
- thin layer chromatography;
- size exclusion chromatography;
- ion chromatography;
- capillary electrophoresis;
- supercritical fluid chromatography.

The discussion centres on current applications of the above techniques and is not intended to be a comprehensive review. Instead, the advantages and disadvantages of the various techniques are discussed in the context of their use in forensic laboratories which carry out explosives investigations.

8.2 Gas Chromatography (GC)

A gas chromatograph is an instrument which separates compounds in the gas phase. The sample to be analysed is usually injected with a needle into a heated injection port where it is volatilized into a gas. The analyte is swept by an inert carrier gas onto an analytical column in a temperature-controlled oven. As the mixture of compounds passes through the column, it is separated into its components by selective adsorption and/or partitioning between the carrier gas and the column phase. The separated components are swept through the column to a detector which monitors their elution. GC is a useful technique for the analysis of a wide variety of materials and its use is limited only by the volatility of the sample (McNair and Bonelli, 1968). The elements of the gas chromatographic system which may be specifically modified for use in explosives analysis are discussed in detail below. .

GC has an advantage over other forms of chromatography in its superior resolution and ease of interfacing to different detection systems. However, because the analyte must be in the gas phase, the analysis of thermally unstable explosive materials has caused significant problems and has required the development of rigorous analytical conditions. The keys to successful GC analysis of explosives are to volatilize the explosive at the lowest temperature possible, elute the explosive from the column at the lowest temperature possible and keep everything that the explosive contacts as inert as possible.

8.2.1 Injection Systems

Injection systems are normally of two types: (1) split–splitless; and (2) on-column.

Each type has advantages and deficiencies in use with explosive compounds. With either injector, temperature is a critical parameter in the success of the analysis. Injectors may be operated at one temperature (isothermal) or at a temperature which is increased very rapidly (temperature program). The choice of injector and temperature is based on the concentration and thermal stability of the sample.

Split–splitless injectors

A split–splitless injector may deliver the majority of a sample to the analytical column or only a small percentage of it. The user selects the injection format based on the sample concentration and type of column used. Splitless injections are used in the injection of very dilute samples. Split injections deliver a smaller portion of the sample which prevents overloading of capillary columns.

Isothermal split–splitless injectors should be set at a temperature where the least volatile sample component can be introduced onto the analytical column. This temperature can be high enough to cause a thermally labile, volatile explosive to decompose. Thus, a temperature programmed injector is preferred in explosives analysis. With this device, the injector temperature is increased rapidly while it is continuously swept by carrier gas. This process permits the analytes to be removed from the injector at the moment of volatilization and does not expose them to unnecessarily high temperatures.

On-column injectors

With on-column injectors, the sample is injected directly into the analytical column using a specially designed narrow bore needle. The advantage of this apparatus is that the explosive can be injected at a low temperature. As the temperature of the column oven is increased the explosives will start their migration at their volatilization temperature. These injectors also have a drawback. All material is deposited on the column – including non-volatile components. The column can become quickly blocked and contaminated (Figure 8.1), necessitating removal of a section of it. Thus in practice, a pre-column (guard column) is used and replaced once contaminated.

The pre-column (guard column)

A pre-column consists of about 1 m of empty silica capillary tubing which should be replaced as soon as any change in resolution or response time is noted for a calibration standard. This prevents degradation and contamination of the analytical column (Lloyd, 1991). A pre-column is unnecessary when using a split–splitless injector; the injector liner performs a similar function of accumulating nonvolatile constituents of injected mixtures. The liner must be cleaned regularly to maintain inertness. Douse (1985) successfully analyzed a number of different explosives using splitless injection. The injector was maintained at 175°C and the liner was cleaned daily.

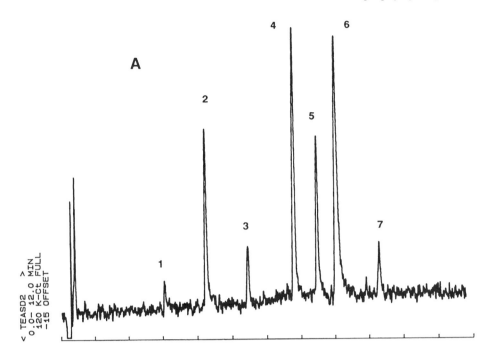

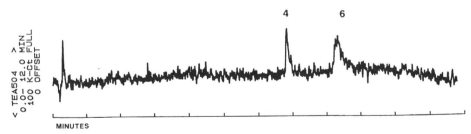

Figure 8.1 Standard explosives mixture separated by GC/TEA on a clean capillary column (A) and on an unprotected capillary column contaminated by a post-blast extract (B). Components are: (1) ethylene glycol dinitrate; (2) nitroglycerine; (3) 2,4-dinitrotoluene; (4) trinitrotoluene; (5) pentaerythritol tetranitrate; (6) RDX; (7) tetryl. Conditions: on-column injection onto a 15 m, 0.32 mm id DB-5 capillary column (5% phenylmethylsilicone, J&W Inc.). Temperature: 50–180°C at 15°C/min. Chemiluminescence (TEA) detection.

Figure 8.1 illustrates how quickly a capillary column can be rendered useless for explosive analysis. The first chromatogram shows a standard mix of explosives on a capillary column using a chemiluminescence (TEA) detector (see below). The second chromatogram is the same mixture after a single injection of a post-blast extract of debris. The column efficiency has been reduced to the point that only two of the seven peaks are observed. After replacement of the pre-column, the column recovered its resolution. This also illustrates the importance of sample clean-up (see 8.2.4. below).

8.2.2 Columns

Fused silica capillary columns have been used for explosives analysis since Douse's work in the early 1980s (Douse, 1981, 1983). The stability of the wall coating in these capillaries is important in maintaining reproducibility. Coatings used in explosives analysis usually are either methylsilicone or 5% phenylmethylsilicone and are bonded to the capillary walls. The thinner the stationary phase, the more rapidly the explosives will elute; however, the column capacity is lower with a thin stationary phase. Column diameter choice is a trade-off between efficiency and capacity. While most separations may be performed on standard diameter capillaries (0.25 mm, 0.32 mm) (Douse, 1981, 1983; Fine et al., 1984), some widebore (0.53 mm) capillary work has been done (Hable et al., 1991). The column length should be as short as possible without degrading the separation. This will allow the explosive to spend the least amount of time on the column, minimizing chances for thermal degradation. Most explosives can easily be resolved on a 15 m column (Figure 8.2).

The chromatogram in Figure 8.2 shows the separation of commonly encountered explosives on a 15 m DB-5 (5% phenylmethylsilicone, J&W, Inc.) 0.32 mm internal diameter capillary in less than 10 min. The temperature of the column oven was ramped from 50°C to 180°C at a rate of 15°C/min. The injection was on-column and the detector was chemiluminescence (TEA) based (see below).

8.2.3 Detectors

Detectors for GC analysis of explosives must be selective and sensitive. Chemiluminescence (TEA) and mass spectrometer detectors are widely used. Other detectors used include flame ionization and electron capture.

Chemiluminescence (TEA)

Chemiluminescence detectors are the most selective of the GC detectors available for explosives (Douse, 1983; Fine et al., 1984). One version of this detector manufactured by Thermedics Corporation, Chelmsford, MA, is referred to as a *thermal energy analyzer* (TEA) which is a registered trademark. These detectors have sensitivity in the low picogram range and give responses only to nitrate or nitroso moieties, making them ideal for post-blast screening of explosives (Douse, 1983, 1985, 1987; Goff et al., 1983). Within the detector the GC eluent is passed through a furnace which pyrolyses it. If the eluent contains nitrate or nitroso groups, the pyrolysis produces nitrogen oxide, which is then reacted with ozone to form excited nitrogen dioxide. This complex breaks down to unexcited nitrogen dioxide by emitting light. The detector contains a photomultiplier tube preceded by a filter designed to pass only light of the wavelength emitted by excited nitrogen dioxide (see Chapter 10, Figure 10.3).

Nitrated organic explosives are among the very few compounds which give a TEA response. The temperature of the pyrolyser is important. Nitrate esters and nitramines are readily pyrolysed at about 550°C, but nitroaromatics like TNT require a temperature of about 800°C. Calibration is typically checked with a mixture of nitrated organic explosives as shown in Figure 8.2.

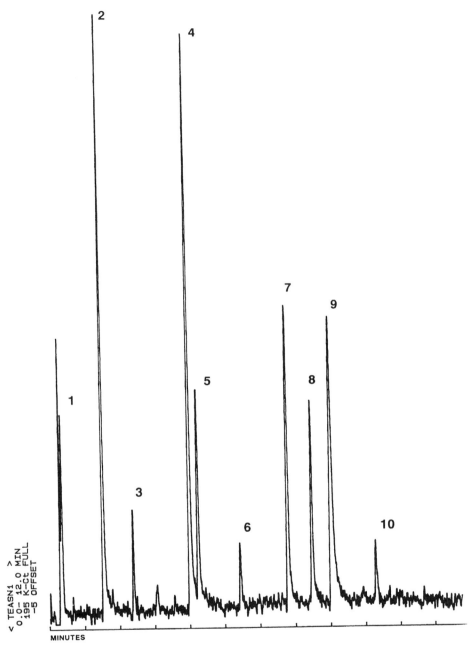

Figure 8.2   GC/TEA analysis of a standard explosive mixture. Compounds are: (1) nitromethane; (2) ethylene glycol dinitrate; (3) o-mononitrotoluene; (4) nitroglycerine; (5) diethylene glycol dinitrate; (6) 2,4-dinitrotoluene; (7) trinitrotoluene; (8) pentaerythritol tetranitrate; (9) RDX; (10) tetryl. Conditions in para. 8.2.2.

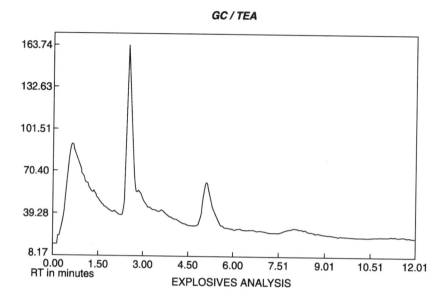

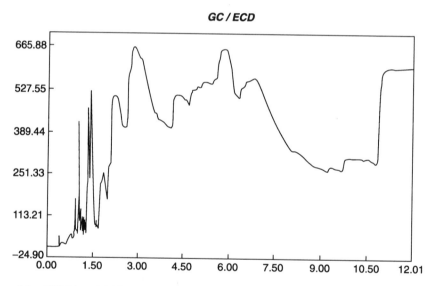

Figure 8.3 GC/TEA and GC/ECD analysis of an acetone extract of debris from a dynamite explosion. Conditions: 15 m DB5 columns, 32 mm od, 25 μm film thickness; cold on-column injection; oven program: 50–180°C at 20°C/min; injector program: 50–200°C at 60°C/min; ECD detector: 8 mCi Ni63 at 200°C; TEA detector: Thermedics TEA Model 510. Peaks identified in GC/TEA: EGDN at 2.50 min; NG at 4.90 min; peaks identified in GC/ECD: none.

GC/TEA is the most widely used screening technique for high explosives. Due to its high selectivity, many non-explosive compounds which produce a signal on other detectors such as flame ionization and electron capture, do not respond to the TEA. As an example, Figure 8.3 illustrates chromatograms from GC/TEA and GC/ECD analyses of the same acetone extract of clothing debris from a dynamite explosion. The GC/TEA chromatogram shows the peaks due to nitroglycerine and ethylene glycol dinitrate, whereas neither is discernable in the GC/ECD. The extract had been partially cleaned-up by passing it through a silica gel column.

Unfortunately, the newer generation of commercial blasting high explosives, such as emulsions based on ammonium nitrate and hydrocarbon fuel, are not detected by the TEA as they contain no nitrated organic compounds. Notwithstanding the selectivity of the TEA, samples must still be properly prepared via a clean-up step to avoid capillary degradation (Douse, 1982, 1985; Douse and Smith, 1986; Kolla, 1991a; Yinon and Zitrin, 1993; Hiley, Chapter 10). This is especially true if an on-column injector is used (see 8.4.4 below).

In Chapter 10, Hiley describes in detail the quality control issues pertaining to GC/TEA.

Mass spectrometer

The mass spectrometer is a selective detector for GC, and is an excellent complement to the TEA due to its ability to confirm the identity of explosive materials previously screened by the TEA.

The technique itself, and its application to analysis of explosives is comprehensively discussed by Zitrin in Chapter 9 and has been reviewed by Yinon (1991) and Fetterolf (1995).

Electron capture detector (ECD)

Electron capture is a sensitive means of detecting explosives, although it is less selective than the TEA or mass spectrometer. These devices operate by passing a beam of electrons between two electrodes; when a sample elutes from the GC column, organic nitrate groups and other electronegative moieties 'capture' these electrons, interrupting the current and thereby producing a signal. ECD are useful in analyzing trace amounts of explosives but are less appropriate for dirty post-blast samples as these tend to produce complex chromatograms (Douse, 1981, 1983). This relative lack of selectivity has resulted in the TEA largely replacing the ECD for explosives residue analysis (Douse, 1983; Fine et al., 1984).

The ECD has been employed successfully for the analysis of trace explosives on hand swabs, but only after sample clean-up (Douse, 1981).

Flame ionization (FID)

Flame ionization detectors operate by passing the eluent from the GC column through a hydrogen flame which ionizes organic compounds. This results in a detector which responds to most organic compounds (McNair and Bonelli, 1968). The FID is useful in the pre-blast analysis of explosives and explosive components, but it has limited application in detecting traces of nitrated organic explosives after an explosion. Because it is not a selective detector, the FID was replaced first by the more selective ECD and then by the much more selective TEA. However, the relatively new detonator-sensitive emulsion explosives and the widely used ammonium nitrate/fuel oil (ANFO) blasting agent do not contain nitrated organic compounds, and thus the FID is now being used in many routine protocols for the detection of the hydrocarbon fuels (Midkiff and Walters, 1993; Bender et al., 1993).

The GC/FID has many applications for the analysis of non-explosive components in bulk explosives. For example, plastic bonded explosives (PBX) contain oils and plasticizers which are amenable to GC analysis. These materials are easily removed from the PBX by the use of non-polar solvents such as pentane and hexane. The extract is then simply injected into the instrument. However, this analysis does require high temperatures for volatilization and elution of the sample components (Keto, 1989). Hydrocarbon analysis can provide valuable sourcing information due to differences in formulations of particular explosives. Such differences can even occur between different lots of the same product. Two chromatograms of oils from different samples of the explosive Semtex 'H' are shown in Figure 8.4. The samples were

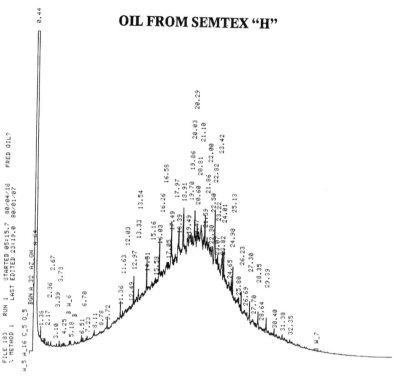

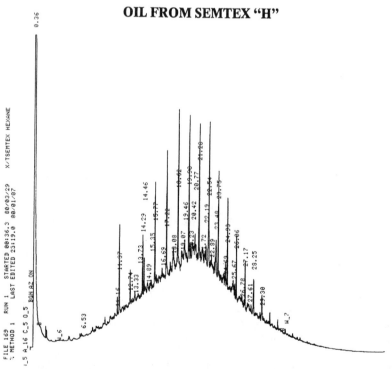

Figure 8.4 GC/FID analysis of two oil samples from Semtex-H. Conditions in para. 8.2.3 'Flame ionization'.

analyzed using an aluminum-clad, methylsilicone column specially designed for high temperature analyses. The GC was equipped with a programmed temperature vaporizer (PTV) injector which was ballistically heated to 450°C. The column oven was programmed from 100°C–430°C at 10°C/min, and the FID detector was held at 450°C. GC/FID clearly shows that the oils have different elution profiles. This analysis provides a valuable point of comparison between samples since the oils used in PBX formulations vary in composition. Emulsion explosives also contain oils which show some variance between samples (Bender et al., 1993).

Petrolatums (Vaseline®) are sometimes incorporated into improvised plastic explosives, e.g. mixed with potassium chlorate. Figure 8.5 shows chromatograms of petrolatums from different manufacturers and different lots of the Vaseline® product. The analysis was done on a 30 m thin phase (0.1 μm) methylsilicone capillary with a standard split–splitless injector (350°C) and FID detector (350°C). The column was ramped from 100°C to 320°C at 15°C/min.

Smokeless powder additive packages have also been analyzed by GC/FID (Andrasko, 1992; Selavka et al., 1989). Effective comparisons and identifications of these components can be made. A problem with using GC for smokeless powder analysis is that certain stabilizer decomposition products are difficult to analyze. These materials are too thermally unstable or involatile for GC analysis.

8.2.4 A Sample Clean-up Procedure for GC Analysis

As illustrated in Figure 8.1, improperly prepared samples can contaminate and render an analytical column useless. Solid phase extraction followed by thin-layer chromatography (TLC) is an effective clean-up procedure which may be carried out as follows. A sample is first dissolved in a non-polar solvent (iso-octane, hexane). The solution is passed through a (silica) solid phase extraction cartridge. Additional solvent is passed through the cartridge to remove the non-polar contaminants and then discarded. Any extracted explosive is retained on the column. The original sample is then extracted with methylene chloride and passed through the cartridge. This fraction is collected and the cartridge, which now contains only polar contaminants, is discarded.

The collected fraction is reduced in volume under a stream of nitrogen and spotted onto a silica TLC plate. An explosive mixture is spotted on another area of the plate. The plate is then developed with a 4:1 mixture of trichloroethylene and acetone. The area of the TLC plate containing the explosive mixture is visualized with 5% diphenylamine in ethanol and UV exposure. Care is taken to cover the sample lane of the plate with a clean sheet of white paper to prevent contact between the sample and the visualization solution. The undeveloped area of the plate corresponding to the explosive of interest is scraped off, extracted with solvent (acetone), and reinjected into the GC.

This method was applied to a contaminated sample of the type illustrated in Figure 8.1(B) in which poorly resolved peaks were observed for RDX and PETN. The TLC spot corresponding to the retention factor of RDX was extracted and injected into the GC. The resultant chromatogram illustrated in Figure 8.6 demonstrates that the above procedure produces significantly improved peak shape and signal-to-noise ratio. The analytical conditions were the same as those used to generate Figure 8.1.

Hiley details an alternative column clean-up method in Chapter 10 (see 10.3.3). Other clean-up procedures have been reviewed by Yinon and Zitrin (1993).

8.3 High Performance Liquid Chromatography (HPLC)

HPLC is an ideal choice for instrumental analysis of explosives. The problem of thermal stability of the explosive is eliminated because the analysis takes place at room temperature,

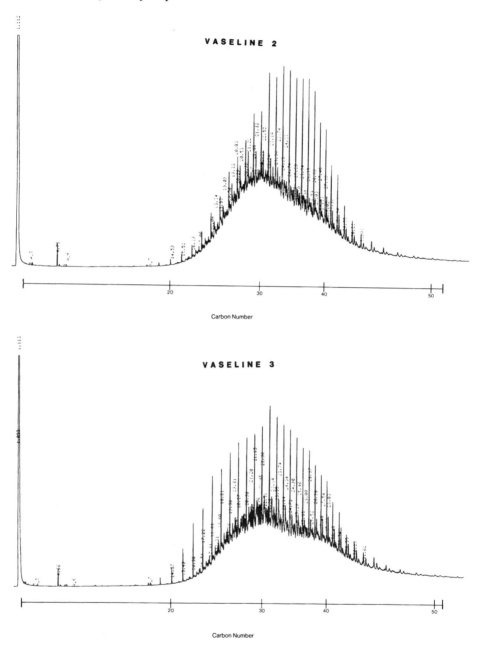

Figure 8.5 GC/FID analysis of two different Vaseline® samples. Conditions in para 8.2.3 'Flame ionization'.

and the formation of active sites on the column from dirty samples is not as great a problem as in gas chromatography (Lloyd, 1992: 179–181). This permits analysis of samples with minimum sample cleanup. Detection systems are available which are sensitive and selective. The only real deficiency has been the lack of an efficient interface between HPLC and mass spectrometry, but on-going research and development are being conducted in this area

RDX

Figure 8.6 GC/TEA analysis of a sample of RDX cleaned-up by TLC (cf. Figure 8.1 (B); same conditions).

(Fetterolf 1995: 233–9). In general, the capacity of HPLC is much greater than that of GC. Sample injections may range from 10–100 μl.

The application of HPLC to analysis of explosives has been reviewed in great detail by Lloyd (1992: 174–261).

8.3.1 *Injectors*

The injection system for a liquid chromatograph is very simple. The sample is injected into a sample loop of fixed capacity; a 20 μl loop is typical. Injection is achieved by switching a valve which places the sample in the eluent stream. The sample then proceeds to the column.

8.3.2 *Solvents*

The most commonly used HPLC separation system for explosives analysis is reversed phase, in which the explosives are separated in mixtures of methanol or acetonitrile and water. The separations can be carried out in an isocratic mode or by using gradient elution in which the eluent composition is changed by increasing the concentration of the organic solvent as the run progresses. Reversed phase HPLC has many advantages for explosives analysis including system stability, low toxicity and transparency to ultraviolet light. Unfortunately, reverse phase solvent systems cannot be used with the TEA detector because water freezes and blocks the cold trap of the pyrolysis unit.

Normal phase solvent systems such as isooctane/methylene chloride/methanol are used with TEA detectors for explosive residue analysis (Fine et al., 1984; Lloyd, 1992: 234).

Smokeless powder additives have been separated using methylene chloride/isooctane gradients with THF or 1,4-dioxane modifiers (E. C. Bender, unpublished; referenced in Wallace and Midkiff, 1993). Care should be used when handling these solvents because of their toxicity and/or flammability.

8.3.3 Columns

The column packing of choice for the reversed-phase analysis of explosives is 'C18' which is also referred to as 'ODS' (octadecylsilane). This packing is appropriate for use with ultraviolet or electrochemical detection systems. The packing size is normally 5 μm, but 3 μm can be used on shorter columns to decrease solvent consumption. Smaller diameter packings (3 μm) have less sample capacity and do not perform well for post-blast residue analysis as they are easily clogged.

The TEA requires normal phase operation. Most explosives which respond to the TEA can be separated using cyano or silica packings. A 25 cm × 4.6 mm CN column with 5 μm packing (Supelco, Inc.) was used to perform the separation illustrated below in Figure 8.1. For smokeless powder analysis, 3 μm silica packings are used in order to maximize resolution. Column capacity is not a restriction in this case.

8.3.4 Detectors

The three detectors most widely used for explosives analysis are:

- ultraviolet (UV);
- chemiluminescence (TEA);
- pendant mercury dropping electrode (PMDE).

UV

The most common means of detecting explosives separated by HPLC is UV absorption. Reverse phase solvents are more transparent to ultraviolet radiation than normal phase solvents. This becomes particularly important when analyzing nitrated esters, such as nitroglycerin and PETN. These compounds produce acceptable signals only at wavelengths of less than 215 nm (Lyter, 1983). However, operation of a UV detector at low wavelengths makes it non-selective, since almost every compound will respond. For this reason, HPLC with UV detection is used more often for quantitative analysis of bulk explosives than for explosive residues. Another application of HPLC with UV detection is the quantitative analysis of explosives in aqueous extracts (Bauer et al., 1990). For example, reversed-phase HPLC is the Environmental Protection Agency (EPA) specified method for the quantitation of explosives in ground water (EPA Method 8330).

With the introduction of diode array detectors, HPLC/UV has become a more selective technique. Not only can the explosive be detected but also an identification with an acceptable degree of certainty can be obtained since the UV spectrum of components can be recorded to supplement retention data (DeBruyne et al., 1989; Bouvier and Oerhle, 1995). Detection limits using this technique, however, are reduced with post-blast samples because of the poor sensitivity of the diode array technique, and due to alteration of UV spectra by background interferences. Such interferences can be substantially reduced using clean-up techniques such as solid phase extraction (see 8.2.4).

Alkyl amine nitrates, which are found in some water gels and emulsions, can be analyzed as trinitrobenzene sulphonic acid (TNBS) derivatives by HPLC (Bender et al., 1993). The derivatization process makes the system very selective even in post-blast analysis. The amine nitrate is extracted into water, then derivatized with a 1% TNBS solution in 10% $NaHCO_3$. An insoluble product is formed which is extracted from the aqueous solution with benzene. The derivative is yellow which allows selective detection in the ultraviolet region (340 nm). A chromatogram of the alkyl amine nitrate derivatives used in explosives is illustrated in Figure 8.7. The solvent program was 0–100% B in 15 min @ 1.0 ml/min using solvents A

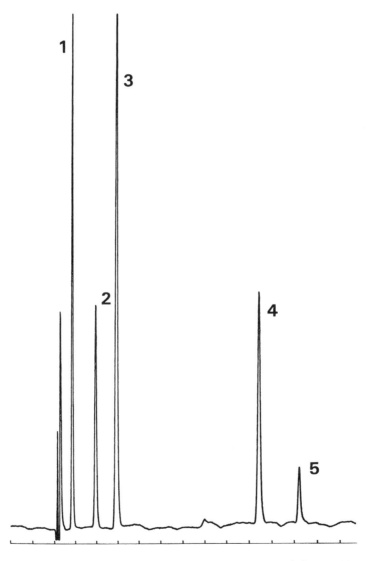

Figure 8.7 HPLC/UV analysis of TNBS derivatives (1) Ethylamine; (2) methylamine; (3) ammonia; (4) ethylenediamine; (5) ethanolamine. Conditions: Whatman PAC column, 25 × 4.6 mm, 5 μm packing; solvent program as described in para. 8.3.4 'UV'.

and B as follows:

	Solvent A	Solvent B
Methylene chloride	90%	75%
Isooctane	9%	20%
2-Propanol	1%	5%

TEA

HPLC/TEA detectors operate similarly to those used in GC (see 8.2.3). The flow from the HPLC column is led to a pyrolysis oven which is maintained at about 550°C. This temperature is required to avoid solvent breakdown products and is lower than that used in GC/TEA (Goff et al., 1983). In HPLC, the excess solvent is removed by a cold trap (about −80°C) which restricts the system to normal phase solvents, because aqueous solvents freeze and block the trap.

The TEA is a very sensitive, selective detector but only for nitrate esters and nitramines. An example of an analysis of a mixture of explosives by this technique is illustrated in Figure 8.8. Unfortunately, aromatic nitro compounds and any other explosives containing the $C-NO_2$ group cannot be analyzed directly by this technique, because the temperature of the pyrolysis oven is not sufficient to release NO. However, raising the temperature creates unacceptable interferences by the solvent system. One technique to partially by-pass this problem is to use an ultraviolet post-column reactor to break the $C-NO_2$ bond prior to TEA analysis (Selavka et al., 1987).

PMDE

The PMDE is a reductive mode electrochemical detector which was first applied by Lloyd (1983) for use in trace explosives analysis. The PDME can detect all the organic explosives of interest at picogram levels. The PDME is superior to thin film detectors because a renewable surface is created through the formation of each mercury drop, limiting problems with electrode contamination. It does require rigorous sample preparation and removal of oxygen from the system before analysis. The detector must be employed in reversed phase systems which is not a severe restriction.

The technology has not been commonly employed because of the toxicity of mercury and the fragility of the system. It is, however, used routinely in the Forensic Science Agency of Northern Ireland (Murray, 1993; see also Chapter 12), in the UK Home Office (King, 1993) and in France (Pokrzywa, 1993).

8.3.5 Application of HPLC to Smokeless Powders

HPLC is valuable for the analysis of non-explosive components in explosive systems and for the comparison and sourcing of explosives. One of the most important analyses is the characterization and comparison of smokeless propellants. Unlike GC, this technique is ideal for the quantitation of thermally unstable decomposition products. An abundance of work has been done in this field by both military and forensic laboratories.

In military laboratories, product performance and stability are monitored by a quantitative analysis of the stabilizer content. The analysis is usually performed on an isocratic reversed phase system with UV detection. The concentration of the stabilizer (diphenylamine (DPA) or centralite) and its decomposition products are determined from stored master lots. These data are compared to acceptable values. If the lot does not meet specifications, it is recalled. In addition, the presence of decomposition products, such as 2,4′-diphenylamine, can cause a

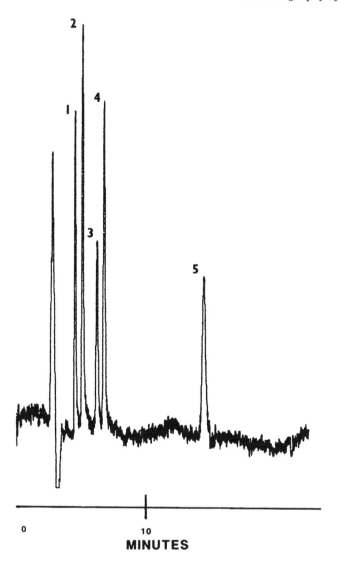

Figure 8.8 HPLC/TEA analysis of a mixture of nitrate ester and nitramine explosives. Compounds are: (1) nitroglycerine; (2) pentaerythritol tetranitrate; (3) tetryl; (4) RDX; (5) HMX. Conditions: 25 cm × 4.6 mm Supelco CN column; 5 μm packing; solvent 20% acetone/isooctane at a flow of 1 ml/min.

propellant to be scrapped (Stine, 1991). Other additives, such as plasticizers, surface coatings, and other incidental materials, are not usually evaluated.

Forensic scientists, however, are not interested in the performance of a smokeless powder but in the identity of its source. Small amounts of intact powders can often be recovered following an explosion and connected to a suspect. Thus, the characterization of all components is important, including additives and incidental materials (Wallace and Midkiff, 1993). Other important factors are the potential contributions from reworked powder added to the batch and changes in powder composition due to environmental conditions. HPLC is well suited for determining these issues. Most separations proposed for smokeless powder analysis are reversed phase. Stine (1991) used an acetonitrile/water (65:35) isocratic mobile phase on a C18 column to quantitate decomposition products of diphenylamine. Robertson and Kansas (1990) used acetonitrile and water to separate the decomposition products of DPA and 2-nitro DPA.

Normal phase chromatography on silica columns has been used both isocratically and with gradients to separate smokeless powder constituents. Isocratic systems have not been able to analyze all the compounds of interest, but gradients using modifiers, such as THF (Curtis and Berry, 1989) and 1,4-dioxane (E. C. Bender, unpublished), have been more successful. The normal phase adsorption mechanism tends to be better at resolving closely related compounds, and the reproducibility of these separations on modern silica columns is quite good. Figure 8.9 illustrates a normal phase gradient separation using a 1,4-dioxane modifier on silica. The column was 3 μm silica (Supelco, Inc.), 15 cm × 4.6 mm. Solvent A

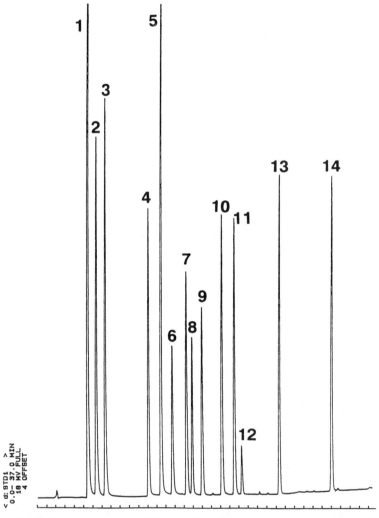

Figure 8.9 HPLC/UV separation of smokeless powder additives and decomposition products. Compounds are: (1) 2-nitrodiphenylamine; (2) diphenylamine; (3) *N*-nitrosodiphenylamine; (4) 2,6-dinitrotoluene; (5) 2,4-dinitrotoluene; (6) O-NBA (IS); (7) di-n-octylphthalate; (8) trinitrotoluene; (9) di-n-butyl phthalate; (10) di-n-ethyl phthalate; (11) di-n-methyl phthalate; (12) nitroglycerine; (13) ethyl centralite; (14) methyl centralite.

was 1000 ml of isooctane with 3 ml of dioxane. Solvent B was 500 ml methylene chloride + 500 ml isooctane + 50 ml dioxane. The gradient profile was as follows:

Time (min)	%A	%B
0.0	100	0
4.0	100	0
15.0	85	15
40.0	0	100

The flow was 1.0 ml/min, and the eluent was monitored by an ultraviolet detector set at 254 nm.

Most of the components of interest in smokeless powder analysis can be analyzed using this separation.

8.4 Thin Layer Chromatography (TLC)

TLC was one of the first techniques used for explosive analysis (Jenkins and Yallop, 1970; Parker et al., 1975; Archer, 1975) and is still used today as an analytical method to indicate the presence of an explosive or as a preparative or clean-up technique (see Hiley, Chapter 10). The advantages of this separation system are low cost, reasonably fast analysis times, simultaneous analysis of several samples, and low solvent consumption. The sensitivity of TLC is generally in the low microgram to high nanogram range, making the technique too insensitive for post-blast analysis of RDX and PETN, which typically are present in trace quantities. The selectivity of the technique is increased by the choice of visualization reagents as described below.

In Chapter 10, Hiley discusses the use of TLC in the analysis of explosives and their residues, and details quality control procedures. Specific application to smokeless powder analysis is discussed by Bender in Chapter 11.

8.4.1 The Thin-layer Chromatographic Process

TLC systems consist of a glass or aluminum plate which is coated with a thin layer of adsorbent; most commonly silica gel, with alumina a distant second. Reversed phase plates containing a non-polar adsorbent layer are also available but are rarely needed for explosive analysis. The sample is spotted on one end of the plate and placed in a tank which contains a developing solvent. The solvent front proceeds up the plate by capillary action to some predetermined level, then the plate is removed from the tank.

The position of the analytes after development is their retention factor, R_f, the ratio of the distance travelled by the spot to that of the solvent front. In practice, known standards are always analyzed on the same plate as unknown samples.

Solvent systems

Solvent systems that have been proposed for TLC of explosives are numerous and have been discussed in detail by Yinon and Zitrin (1981). For the post-blast analysis of commonly encountered explosives, Hiley (Chapter 10, see 10.4.1), reports favourably on use of a 9:1 mixture of toluene/ethyl acetate on silica gel, because it separates nitrocellulose, nitroglycerin, TNT, tetryl, PETN, HMX and RDX, which make up the majority of organic explosives (Chapter 10, Figure 10.2).

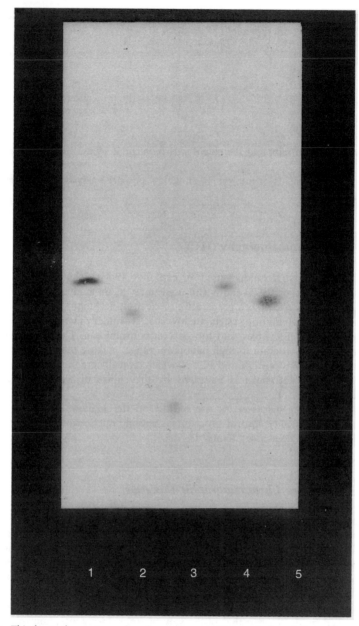

Figure 8.10 Thin layer chromatography of some organic high explosives (left to right: (1) TNT; (2) tetryl; (3) RDX; (4) PETN; (5) NG). Experimental conditions are described in para. 8.4.1 'Solvent systems'.

A 4:1 mixture of trichloroethylene:acetone also is effective at separating explosives (Yinon and Zitrin, 1981, and references therein) as illustrated in Figure 8.10 for the separation of TNT, tetryl, RDX, PETN and NG using KOH/ethanol + Griess reagent for visualization. DPA with UV irradiation could also have been used.

A more non-polar solvent system is necessary to separate less polar explosives such as EGDN and nitromethane. Douse (1982) used a 7:3 mixture of toluene and cyclohexane to separate nitrate esters. This system did not effectively separate the nitramines, but 2:1 chloroform:acetone was successful.

In most reported solvent systems, nitrocellulose (NC) has an R_f of zero. Douse (1982) has reported that a solvent system consisting of 3:2 acetone:methanol system allows the separation of nitrocellulose with an R_f of 0.64. Peak (1980) has also reported the separation of certain nitrocelluloses, however, propellant grade NC remained at the origin.

Separation of various inorganic and ionic components of explosives has been proposed (Beveridge et al., 1975), but instrumental techniques such as ion chromatography and capillary electrophoresis (see below), infrared spectroscopy, and elemental analysis (see Chapters 10 and 11) perform these examinations with greater speed and reliability.

Visualization

The TLC plate can be screened under UV light to determine the position of absorbent materials. The aromatic explosives (e.g. TNT) and the nitramines (e.g. HMX, RDX) absorb UV light and can be seen, but the nitrated esters such as nitroglycerin do not, and must be visualized by a chemical reaction. UV-absorbing explosives should also be treated, as the characteristic colors produced by this method, combined with the R_f, give the technique its selectivity.

The most common visualization process is the Griess reaction. This is a two step process in which the nitro group of the explosive is first hydrolyzed to form nitrite ions by spraying the plate with a solution of 1 M base and then heating it at about 100°C for about 10 min (Lloyd, 1967). Nitroaromatic explosives form a colored complex at this stage. Once nitrite has been formed, the plate is sprayed with modified Griess reagent to visualize nitramines and nitrate esters.

There are several formulations for modified Griess reagent (Zeichner and Glattstein, 1986). In part this is due to the realization that some chemicals used in earlier formulations may be carcinogenic. The reagent typically consists of two parts. One current formulation is a 1% aqueous solution of sulfanilic acid and a 1% N-naphthylethylene diamine in a 70/30 mix of ethanol/acetic acid. The reagents are sequentially sprayed on the plate. A positive reaction produces pink spots. An alternative procedure is to spray the plate with a freshly prepared 50/50 mixture of 0.5 g α-napthol in 165 ml 30% acetic acid and 1 g of sulfanilic acid in 100 ml 30% acetic acid. This version produces orange spots. Yet another formulation is described by Hiley in Chapter 10 which produces magenta spots.

Douse and Smith (1986) have reported that 3,3′-iminobispropylamine is a selective and sensitive visualization reagent for nitroaromatic compounds. It forms stable colored complexes which are tabulated in the original paper.

Clean-up

TLC is valuable as a clean-up technique for other analytical methods – which also serve to confirm the identity inferred from the TLC result. Post-blast extraction of debris frequently produces complex mixtures. If such samples are injected into gas chromatographs they can quickly contaminate the analytical column. The high backgrounds produced by these contaminants may not permit the analysis of trace quantities of explosives (see 8.2.4 above).

Post-blast extracts can be spotted or streaked along the bottom of the TLC plate. A non-polar solvent (pentane, isooctane) should first be used to develop the plate. This will remove oils and non-polar contaminants from the sample which can affect the migration times (R_f). Most explosives will remain at the origin; but some of the volatile nitrated esters such as EGDN will migrate.

Next the plate is developed by the primary solvent with only one edge of the plate sprayed with the visualization solution. The unvisualized portion of plate is then scraped off at the R_f of interest and extracted.

8.5   Size Exclusion Chromatography

Size exclusion (SEC) or gel permeation chromatography (GPC) has been used primarily for the analysis of non-explosive components such as polymeric binders present in explosive mixtures. The SEC system consists of an HPLC pump, injector, and detector. The difference between SEC and HPLC is the separation mechanism. With GPC, samples are separated because of differences in apparent size in solution.

8.5.1   Columns and Separation

The SEC column contains silica or polymeric packings which have known pore sizes. Analytes are separated based on their ability to permeate into these pores. Smaller molecules are entrapped within the pores longer and are preferentially retained on the column. If the molecule is larger than the typical pore size total exclusion takes place. On the other hand if the molecule can fit into every pore there is total permeation. Separation takes place between the time (volume) of these two events. The advantage of SEC is that everything injected into the chromatograph will elute before a known time. A series of columns are normally used to cover the range of molecular sizes present in the sample. Larger pore columns (i.e. 10^3–10^6 Å) will separate the polymer and the small pore columns (i.e. 100 Å, 60 Å) the additives and oligimers (Lloyd, 1984).

8.5.2   Detectors

The refractive index (RI) detector is commonly used in SEC because of its universal response. Many polymers do not have any appreciable chromophores that would allow use of an UV detector. UV detectors, however, are commonly used in series with RI detectors to monitor the effluent for the low molecular weight components.

   The TEA detector and an electrical conductivity detector may also be used for specific applications (see 8.5.4 below).

8.5.3   Solvents

The solvents used in size exclusion is mostly dependent on polymer solubility. Tetra-hydrofuran, toluene and acetone are frequently used as SEC solvents.

8.5.4   Applications

A major application of SEC is the analysis of plastic-bonded explosives. All components present, including the polymer, plasticizer, oil and explosive, can be separated and quanti-tated (Keto, 1989). Because this technique has a larger sample capacity than standard HPLC and because all samples elute in a known time frame, it is also useful as a clean-up technique. Fractions can be removed and analyzed by other methods such as infrared spec-troscopy and mass spectroscopy.

   The most common polymeric explosive, nitrocellulose (NC), has been analyzed by SEC at nanogram levels of detection (Lloyd, 1984, 1986). This analysis was performed using a silica size exclusion column with an electrochemical detector. The electrochemical detector per-mitted a selective determination of the NC species. It was also demonstrated by Lloyd that different kinds of NC could be distinguished. This is very important because NC is used in nail polish, varnish, paint, glue, collodion and film. Finding traces of NC in solvent extracts of post-blast debris may be of little significance unless this distinction can be made. The analysis

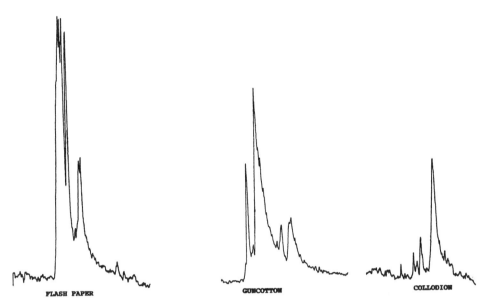

Figure 8.11 Size exclusion chromatography of different nitrocelluloses using a TEA detector. Conditions: three columns were used in series: two Supelco LC-3-Si and one Supelco LC-Si; all columns were 4.6 × 6.2 mm, 5 μm packing; acetone solvent at 1 ml/min.

of NC is also important in detection of gunshot residue, because it is the major component of firearm propellants (Lloyd, 1986).

Nitrostarch can also be analyzed by SEC if a selective detector is used. The TEA is an ideal detector for this analysis and also for the analysis of NC. The latter is illustrated in Figure 8.11. The separation was performed on a silica size exclusion column using an acetone mobile phase. Nanogram amounts of NC could be detected, and gun cotton could be distinguished from the other commonly encountered NCs.

8.6 Ion Chromatography (IC)

In many criminal cases, energetic materials such as black powder, flash powder, ammonium nitrate, and homemade energetic mixtures are used in improvised explosive devices such as pipe bombs (Bureau of Alcohol, Tobacco and Firearms, 1994). These devices typically contain an oxidizer such as potassium nitrate mixed with a fuel. Because such mixtures can leave up to 60% of their original weight behind as inorganic residue upon deflagration, the inorganic anions and cations resulting from the explosive reaction are readily available evidence used to determine the explosive composition (Conkling, 1985).

For analysis of the inorganic ions left following an explosive blast, ion chromatography (IC) is perhaps the most specific and sensitive method that can be used (Reutter et al., 1983; Rudolph and Bender, 1983). This technique, developed in the early 1970s, has the ability to detect parts per million levels of the anions and cations left behind in the residue (Small et al., 1975; Buechele and Reutter, 1983). While the complex chemical reactions that take place during the explosive reaction cannot be completely characterized, enough is known to allow a trained investigator to determine the composition of the explosive used in the device.

IC has the ability to detect volatile ions such as ammonium. Such ions can be lost by analytical techniques such as X-ray diffraction which require evaporation of the aqueous extracts. One disadvantage of the IC technique is the paucity of sensitive techniques necessary

251

to confirm the presence of low levels of inorganic anions and organic cations. Fortunately newer methods such as capillary electrophoresis and liquid chromatography/mass spectrometry are able to fill this void (Hargadon and McCord, 1992; Stewart and Horlick, 1996).

Table 8.1 illustrates some typical commercial and homemade preparations and the particular ions which might be expected to be seen using an ion chromatographic technique.

8.6.1 Anion Analysis by IC

There are two types of IC systems: (1) suppressed IC; and (2) unsuppressed IC.

Early instrumentation for IC consisted solely of suppressed or dual column chromatographic methods (Small, 1989). Suppressed IC utilizes a strong ion exchanger followed by a second column or membrane which neutralizes the ionic eluent, allowing just the conductivity of the sample ions to be detected. However, this method was inadequate for explosives analysis due to the wide variety of charge densities present in residue (Barsotti et al., 1983). As a result, no single combination of column and eluent could achieve separation of all relevant ions.

To help alleviate this and similar problems, gradient techniques have been developed (Smith, 1988; Jandik et al., 1990). At the cost of increased complexity these methods allow a much wider range of ions to be detected. Alternatively, techniques utilizing weaker ion exchange media with electronic compensation or inverse photometric detection are available (Small and Miller, 1982; Sherman and Danielson, 1987; Bender, 1989; Verweij et al., 1986). A third type of ion analysis utilizing capillary electrophoresis will be discussed later.

The two best approaches to analysis of explosive residues with IC are:

- gradient elution with eluent conductivity suppression;
- isocratic elution with indirect photometric detection.

Either of these techniques allow the full range of anions or cations present in explosive residue to be analyzed using a single column.

Table 8.1 Ions present in explosive residue

Explosive	Components	Ions Present
Black powder	KNO_3, S, C	K^+, NO_3^-, NO_2^-, HS^-, SO_4^{2-}, OCN^-, SCN^-, HCO_3^-, CO_3^{2-}
Pyrodex	KNO_3, S, C, $KClO_4$, NaOBz, dicyandiamide	K^+, Na^+, NO_2^-, NO_3^-, HS^-, SO_4^{2-}, OCN^-, SCN^-, HCO_3^-, CO_3^{2-}, Cl^-, ClO_4^-,
Flash powder	$KClO_4$, S, Al	K^+, HS^-, SO_4^{2-}, Cl^-, ClO_4^-, ClO_3^-
Smokeless powder	Nitrocellulose, NG, KNO_3, K_2SO_4, NaCl*	K^+, Cl^-, NO_3^-, SO_4^{2-}
ANFO	NH_4NO_3, Fuel Oil	NH_4^+, NO_3^-, HCO_3^-
Tovex	NH_4NO_3, Monomethylamine Nitrate, $Ca(NO_3)_2$	NH_4^+, NO_3^-, MMA^+

NaOBz, sodium benzoate; MMA, monomethyl amine.
* Ionic compounds in smokeless powder provide flash suppression.

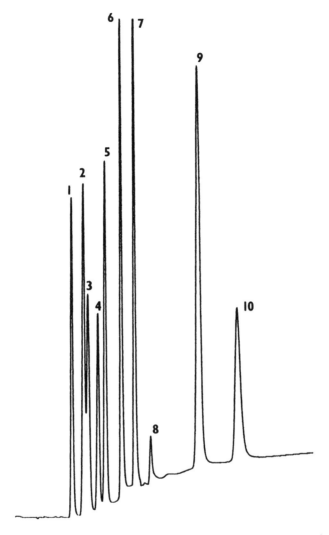

Figure 8.12 Gradient elution of anions present in explosives. Conditions as in Smith (1988). Peak identification: (1)-Cl^-; (2)-NO_2^-, (3)-OCN^-, (4)-ClO_3^-, (5)-NO_3^-, (6)-SO_4^{2-}, (7)-HS^-, (8)-unk, (9)-SCN^-, (10)-ClO_4^-.

Gradient elution with eluent conductivity suppression

The gradient elution technique requires eluent suppression to eliminate the large background produced by an eluent of gradually increasing ionic strength. This procedure, developed by Dionex Inc., can be used for both cation and anion analysis and is capable of resolving a wide variety of ions (Smith, 1988). For anion analysis, a gradually increasing concentration of carbonate ions is used to elute ions of interest off an ion exchange resin. Upon exiting the separation column, the carbonate eluent is neutralized to carbonic acid by means of an ion exchange membrane. The charge of the eluted anions is unaffected by this process and they are detected by conductance. The neutralized carbonic acid produces minimal interference in the conductivity cell. Figure 8.12 shows the separation of a number of ions of interest using this technique.

Isocratic elution with indirect photometric detection

With indirect photometric detection, a UV absorbent eluent is used in combination with a weak ion exchange resin. Instead of a conductivity detector (as was the case above), a standard HPLC/UV absorbance detector can be used. When the sample ions reach the vicinity of the detector cell, the eluent ions are displaced in order to maintain solution neutrality. The result is a decrease in absorbance at the detector in proportion to the amount of eluent displaced. An advantage of this technique is that weaker eluents can be used with low-capacity ion exchange resins. This allows a wider range of sample ions to be analyzed on the same column and eliminates the necessity for gradient elution and eluent suppression (Sherman and Danielson, 1987; Bender, 1989; Glatz and Girad, 1992).

Systems consisting of Vydac anion columns with phthalate eluents have been used to achieve separation of a number of different mixtures of ions present in explosives (Bender, 1989; Verweij et al., 1986). Figure 8.13 shows the separation obtained using a Vydac 3000 column with an eluent consisting of 1.5 mM isophthalic acid at a pH of 4.6. The perchlorate and carbonate peaks are difficult to analyze on standard IC systems, but generally show up in under 15 min with this column.

Utilizing isocratic elution with indirect photometric detection offers certain advantages over the gradient system. Because the method utilizes inverse photometric detection at 280 nm, no specialized equipment such as a conductivity detector is required. This is particularly important for laboratories with limited budgets or areas where such cases occur infrequently. An additional advantage of the technique is its simplicity. Gradient generators and membrane suppressors are not required. The disadvantage of the isocratic technique is that late eluting peaks are broad which limits the sensitivity of the procedure.

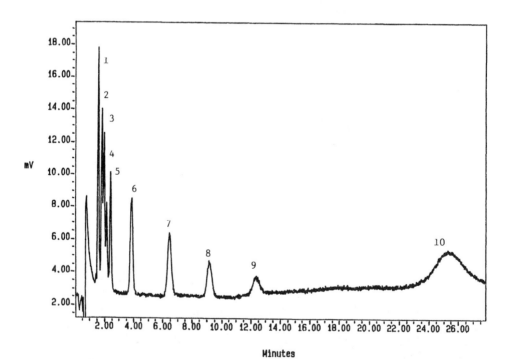

Figure 8.13 Inverse photometric detection of anions present in explosives. Conditions: Vydac 3000 column 1.5 mM isophthalic acid, pH 4.6 flow rate 1.5 ml/min. Peak identification: (1) Cl; (2) NO_2^-, (3) OCN^-; (4) ClO_3^-; (5) NO_3^- (6) SO_4^{2-}; (7) HS^-; (8) SCN^-; (9) ClO_4^-; (10) HCO_3^-.

Table 8.2 Methods for IC of anions in explosive residue

Column	Eluent	Ions found	Reference
Hamilton PRPX100	4 mM Phthalate	Cl^-, NO_2^-, NO_3^-, ClO_3^-, SO_4^{2-}	Kolla (1991b)
Vydac 302IC4.6	4 mM Isophthalate	Cl^-, NO_2^-, ClO_3^-, SO_4^{2-}, HS^-, SCN^- ClO_4^-, HCO_3^-	Bender (1989); Verweij et al. (1986); McCord et al. (1994)
Vydac 3000	4 mM Isophthalate	Cl^-, NO_2^-, OCN^- ClO_3^-, NO_3^-, SO_4^{2-}, SCN^-, ClO_4^-, HCO_3^-	B. R. McCord, unpublished
Dionex As9	Carbonate	Cl^-, NO_2^-, ClO_3^- NO_3^-, SO_4^{2-}	Abramovich-Bar et al. (1993)
Capillary electrophoresis	Chromate/ Borate/DETA	Cl^-, NO_2^-, OCN^- ClO_3^-, NO_3^-, SO_4^{2-}, SCN^-, ClO_4^-, HCO_3^- HS^-	McCord et al. (1994); Kishi et al. (1993); Hargadon and McCord (1992)
LiCrospher RP-18	7 mM Phosphate, 4 mM BTA, 0.14 mM HSA	NO_2^-, ClO_3^-, NO_3^-, SO_4^{2-}, SCN^-, ClO_4^-,	Woolfson-Bartfield et al. (1990)

Table 8.2 summarizes methods for IC of ions in explosive residue.

8.6.2 Cation Analysis by IC

Cation analysis can also be performed on explosives using ion chromatographic techniques (Henderson and Saari-Nordhaus, 1991; Hall and McCord, 1993; McCord et al., 1994). The primary reason for performing cation analysis of explosive residues is that other methods of analysis such as atomic spectroscopy are incapable of determining cations such as ammonium (NH_4^+) and monomethylamine ($CH_3NH_3^+$). These ions are important in the analysis of explosives such as emulsions and slurries used in the mining industry and which have appeared in certain recent terrorist bombings (Henderson and Saari-Nordhaus, 1991; Hall and McCord, 1993; Wallace and Midkiff, 1993).

Analysis of metal ions can also be useful in residue determinations. For example, potassium salts are preferred over sodium salts in explosive formulations as they are less hygroscopic. Another example is calcium nitrate which is a component in certain commercial preparations. Thus the ideal analytical method for the analysis of cationic residue should be capable of determining Group I and II metals as well as amines in a single run.

For analysis of cations in explosives both gradient or single column techniques can be used. A useful single-column technique involves a 0.05 mM cerium III sulfate eluent used with an Interaction Ion 210 column (McCord et al., 1994). Indirect detection occurs at a wavelength of 254 nm. With this procedure sodium, ammonium, potassium, magnesium and calcium may be determined. If a conductivity detector is available, a second type of column such as the Alltech Universal Cation or the Waters Cation M/D is capable of analyzing the above ions as well as monomethylamine (Henderson and Saari-Nordhaus, 1991; Hall and McCord, 1993). These columns combine ion exchange with complexation to achieve analysis of both singly and doubly charged cations in a single run.

255

8.6.3   Confirmation of Identity

One problem which is characteristic of all chromatographic ion analysis techniques is the effect matrix and sample concentration can have on elution. In particular, high ionic-strength samples can result in more rapid elution of individual ions then would be expected. This phenomena can be compensated by means of spiking individual ions to confirm elution order. A particularly useful technique is to add a small amount of a standard containing a mixture of the most common ions to the unknown. Alternatively, sample identity may be ascertained through the use of an orthogonal separation/analysis technique such as capillary electrophoresis, X-ray diffraction, infrared or mass spectrometry.

8.7   Capillary Electrophoresis

The development of capillary electrophoresis (CE) for explosive analysis has provided alternative methodology for analyses involving liquid chromatography and IC (Hargadon and McCord, 1992). This technique provides 10 to 20 times the chromatographic efficiency of HPLC producing GC-like separations in liquid media. To perform the analysis, a small quantity of sample is injected onto a thin 50–100 μm fused silica capillary filled with a conductive buffer. Separation takes place at high voltages based on mass to charge differences between the different ions in solution. An induced flow created by a potential difference of 10–25 kV sweeps the ions past the detector, which typically is on-column and based on UV absorbance.

CE has a number of advantages over traditional HPLC and IC techniques in addition to its higher efficiency. These include ease of automation, low solvent consumption, low injection volumes and fewer moving parts. Because the analysis occurs within an open capillary tube, problems such as efficiency loss and poisoning of the stationary phase that can occur in packed columns are minimized. The disadvantage of the technique is the reduction in dynamic range resulting from the relatively small injection size. As with IC, sample matrix effects are possible and confirmation of peak elution order through sample spiking is often necessary.

There are two modes of capillary electrophoresis of interest to the explosives analyst:

- free zone capillary electrophoresis for the analysis of ions in solution;
- micellar electrokinetic chromatography (MEKC) for the analysis of neutral species.

Free zone capillary electrophoresis has been applied to the analysis of anions and cations from explosive residue, and MEKC has been applied to the analysis of gunshot residue (GSR), pipe bombs, and high-explosive residue in soil and ground water. Ion analysis by CE will be reviewed below and organic analysis by MEKC will be discussed in the section which follows.

8.7.1   Ion Analysis by Capillary Electrophoresis

In capillary electrophoresis of anions, separation occurs due to the difference in charge to mass ratio of each solvated ion. The ions are injected into the capillary and swept past the detector by means of an induced or electroosmotic flow produced by the high voltages (20 000 V) used in the separation. For proper anion analysis, this flow must be oriented in the direction of the positive electrode where the detector is located (Jones and Jandik, 1991).

Normally, electroosmotic flow is induced by a layer of solution cations attracted to the negatively charged capillary wall. This results in a net flow towards the negative electrode. By the addition of quaternary ammonium ions to the CE buffer, the capillary walls are coated with positively charged ions, effectively reversing the flow and allowing the anion separation to be performed quickly and efficiently. Minimum detectable concentrations are on the order of 0.5 ppm, and a wide variety of ions can be separated and detected (Jones and Jandik, 1991).

The main reason capillary electrophoresis is of interest in explosive residue analysis is its ability to obtain highly efficient separations of all ions of interest. Late eluting peaks in IC suffer from band broadening which results in poor sensitivity. A second problem in IC analysis is the poisoning of the stationary phase with organic contaminants. These problems do not occur in capillary electrophoresis because the separation is electromotive and does not rely on any interaction with the column or buffer (Jones and Jandik, 1991). As a result, peak elution order is completely different in the two techniques, and comparison of the same sample analyzed using both CE and IC is an excellent way to confirm peak identity (Hargadon and McCord, 1992). Contaminants in the CE capillary are easily rinsed out at the conclusion of an analysis.

Selectivity in CE can be further enhanced by performing the analysis at wavelengths below 220 nm (Hargadon and McCord, 1992). Under these conditions ions such as nitrate, nitrite and thiocyanate absorb light giving a positive response. Figure 8.14 shows an example of the analysis of a Pyrodex pipe bomb residue with a UV detector set at both 280 and 205 nm. Peak elution order is easy to determine under these conditions by analysis of the differential response of each ion.

Capillary electrophoresis has also shown value in the confirmation of IC peaks such as fluoride, perchlorate and cyanate (McCord and Hargadon, 1993). Similarly, IC can be used to help elucidate peak identity in CE results (Hargadon and McCord, 1992).

While the chromatographic profile can be used to determine the type of explosive utilized, it must be emphasized that variations in the type of containment device, in the heat of the

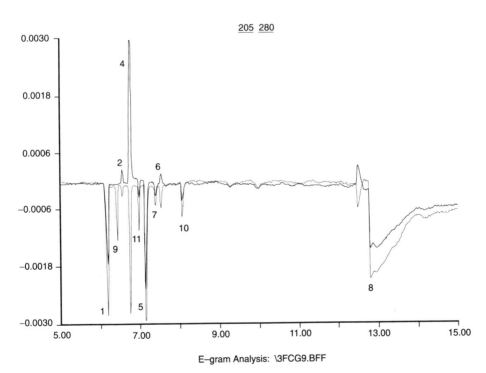

Figure 8.14 Capillary electropheogram of Pyrodex® pipe bomb residue anions. Conditions as in McCord et al. (1994). Analysis performed at 205 and 280 nm. Peak identification: (1) Cl^-; (2) NO_2^-; (4) NO_3^-; (5) SO_4^{-2}; (6) SCN^-; (7) ClO_4^-; (8) HCO_3^-; (9) HS^-; (10) OCN^-; (11) unknown.

blast, and in the amount of unburned material extracted can greatly affect the relative amounts of ions found in a residue extract (McCord et al., 1994). Among the threads and end caps of the pipes, a larger proportion of the starting mixture is left unburned, while fragments from the pipe center have correspondingly less unburned material. Other ions such as nitrite are intermediates, and their amounts vary depending on factors such as burn rates and the heat of the blast. Due to this problem, the combination of sensitivity and specificity that CE techniques can achieve can make the difference between a properly analyzed sample and an indeterminate result.

8.7.2 Capillary Electrophoresis of Organic Explosives: Micellular Electrokinetic Chromatography

Even though most organic explosives are uncharged molecules, it is possible to perform analyses of these materials via capillary electrophoresis using a technique known as micellar electrokinetic chromatography (MEKC) (Northrop et al., 1991). This highly efficient technique is particularly appealing in situations such as the analysis of smokeless powder residue in which complex mixtures occur. The MEKC technique allows a wide variety of compounds to be analyzed, and does not require extensive setup or equilibration times which are necessary for gradient HPLC (Weinberger, 1993).

In MEKC the capillary is filled with a buffer containing a detergent such as sodium dodecyl sulfate at a concentration sufficient to form aggregates or micelles. The micelles are charged clusters with hydrophobic interiors and allow the organic analytes to partition between the buffer solution and the interior of the micelle. Due to their negative charge, the micelles move counter to the flow in the column. As the analysis proceeds, the more hydrophobic molecules elute later as they are more strongly retained in the micelle. The result is a highly efficient separation (Weinberger, 1993). Figure 8.15 gives an example of a standard used in the analysis of smokeless powders. In the figure, the more polar materials such as nitroglycerine elute first followed by other more hydrophobic compounds such as TNT, diphenyl amine and phthalates.

At present MEKC has been used in two main areas in explosives analysis:

- organic gunshot residue (GSR) analysis;
- the analysis of explosives in soils.

These two choices point to the two main advantages of the technique, low sample requirements and high resolving power. In both situations a wide variety of similar compounds must be resolved and thermal instability of these compounds discourages the use of gas chromatography. A potential advantage of CE is the low sample volumes utilized which permit very small samples to be analyzed.

Organic gunshot residue analysis

Application of MEKC in the analysis of organic GSR has been developed by Northrop et al. (Northrop et al., 1991; Northrop, 1993; Northrop and MacCrehan, 1992). They note that increasing concern about toxic compounds in bullet primers has driven manufacturers to offer lead-free primers. These materials cannot be detected by present GSR analysis methods which detect lead, barium and antimony. They have proposed an alternative procedure using CE to look for organic constituents resulting from the smokeless powder propellants and from primers. While not yet an established technique for this analysis, the CE technique shows some promise. In a series of firing range studies using 9 mm and 45 caliber ammunition, traces of nitroglycerine and other components were detected using masking tape lifts

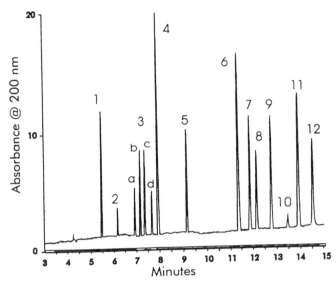

Figure 8.15 Standard solution containing constituents present in smokeless powders. Conditions: 15 mM sodium phosphate pH 7 with 25 mM SDS, 75 μm fused silica capillary at 25 kV, 200 nm. Peak identification: (1) nitroglycerine; (2) 2,4,6 TNT; (3) dinitrotolunene isomers: (a) 2,4; (b) 2,6, (c) 3,4, (d) 2,3; (4) dimethylphthlate; (5) 2-napthol; (6) diethylphthalate; (7) diphenylamine; (8) N-nitroso-DPA; (9) methylcentralite; (10) 2-nitro-DPA; (11) ethyl centralite; (12) dibutylphthalate.

and CE detection. The components were separated using a 2.5 mM borate buffer with 25 mM SDS as the micelle forming agent. This same technique has also been applied to the analysis of smokeless powder and its residues collected from pipe bombs (Smith et al., 1995).

Analysis of explosives in soils

In another application of the technique, MEKC has been used to separate explosive compounds extracted from contaminated soils (Kleibohmer et al., 1993). The technique used an SDS/borate buffer to resolve 24 different explosive compounds including the isomers of aminonitrotoluene, dinitrotoluene, and nitrotoluene. Total analysis time was under 10 min. This technique holds great promise for the analysis of environmental samples and mixtures due to its ability to rapidly screen for explosive components.

8.8 Supercritical Fluid Chromatography (SFC)

Supercritical fluid chromatography is another recently developed technique which has applications in explosives analysis. In SFC, mobile phases such as carbon dioxide (or other gases) are used at a pressure and temperature above their critical point (Lee and Markides, 1990). For carbon dioxide the pressure required is 72 atm at 32°C. Under these conditions certain molecules become very soluble in the carbon dioxide and solubility can be controlled by varying the temperature and pressure of the mobile phase.

Unlike gas chromatography, injection can be made onto a packed column or capillary without requiring vaporization. This is one reason the technique shows promise as a mechanism for the analysis of thermally unstable explosives. Supercritical carbon dioxide can also be used as an efficient media for selective extraction (SFE). By placing the sample in a high-pressure cell, selective extractions can be performed by varying the density and temperature of the carbon dioxide as well as by adding solvent modifiers (Lee and Markides, 1990).

Initial work in the analysis of high explosives by SFC was performed by Douse using a TEA detector (Douse, 1988). Detection levels using this technique were 20–60 pg for a variety of explosives. Further work in this area was performed by Francis et al. (1994), who used a *p,p*-cyano-biphenyl stationary phase to obtain improved detection of HMX. Figure 8.16 shows the result of this analysis. Most of the analyses performed using SFC however, utilize UV or FID detectors where detection levels are two to three orders less favorable (Griest et al., 1989; Munder et al., 1990).

There has been very little work done on the SFC/MS of explosives mainly due to problems of sensitivity with capillary systems and high gas loads produced with packed columns (Via and Taylor, 1994). Detection limits in the picogram range have been reported for nitro-

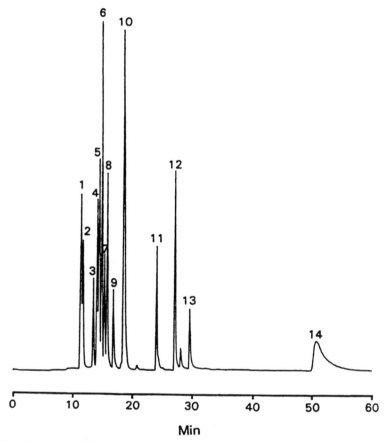

Figure 8.16 SFC–TEA chromatogram of a synthetic mixture of explosives. Conditions as in Francis et al. (1994). Peak identification: (1) nitroglycerine; (2) 2,6 DNT; (3) *N*-nitrosodiphenylamine; (4) 2,4 DNT; (5) 1-nitronaphthalene; (6) PETN; (7) 2-nitronaphthalene; (8) TNT; (9) 2-nitrodiphenylamine; (10) 1,3,5-TNB; (1) tetryl; (12) RDX; (13) 1-nitronaphthalene; (14) HMX. ©Francis, E. S., Eatough, D. J. and Lee, M. L., 1994, Capillary supercritical fluid chromatography with nitro- and nitroso-specific chemiluminescence detection, *Journal of Microcolumn Separations*, **6**, 395–401. Reprinted by permission of John Wiley & Sons Ltd.

aromatics using capillary SFC and negative ion chemical ionization (Lee and Markides, 1990). Via and Taylor (1994) also had some success using a thermospray interface with packed column SFC. Detection limits in the femtogram range were reported for a mixture of dinitrotoluene and nitrodiphenyl amines using negative ion chemical ionization.

Another area of interest has been the SFC analysis of degradation products in propellants in an effort to monitor shelf life (Asraf-Khorassani and Taylor, 1989a, 1989b; Via and Taylor, 1994). As a propellant ages, diphenyl amine stabilizers in the powder become nitrated. Nitro- and nitrosodiphenyl amine by-products are readily analyzed by SFC without the degradation that can sometimes occur in GC analysis (Lee and Markides, 1990). A number of laboratories have also begun to examine supercritical fluid extraction followed by some other analytical technique for the analysis of post-blast explosives in difficult matrices such as soils or swabs (Kolla et al., 1993; Bunte et al., 1994). Successful extractions have also been performed on mixtures such as smokeless powders or plastic explosives (Slack et al. 1992; Taylor, 1993).

References

ARCHER, A. W., 1975, Separation and identification of minor components in smokeless powders by thin layer chromatography, *Journal of Chromatography*, **108**, 401–404.

ABRAMOVICH-BAR, S., BAMBERGER, Y., RAVREBY, M. and LEVY, S., 1993, Applications of ion chromatography for determination and identification of chlorite, nitrite and nitrate in explosives and explosive residues, in Yinon, J. (Ed.) *Advances in the Analysis and Detection of Explosives*, pp. 41–54, Dordrecht: Kluwer Academic Publishers.

ANDRASKO, J., 1992, Characterization of smokeless powder flakes from fired cartridge cases and from discharge patterns on clothing, *Journal of Forensic Sciences*, 37, 1030–1047.

ASRAF-KHORASSANI, M. and TAYLOR, L. T., 1989a, Analysis of propellant stabilizer components via packed and capillary supercritical fluid chromatography/fourier transform infrared chromatography, *Journal of High Resolution Chromatography*, **12**, 40–44.

ASRAF-KHORASSANI M. and TAYLOR, L. T., 1989b, Quantitative supercritical fluid chromatography/fourier transform infrared spectroscopy study of methylene chloride and supercritical carbon dioxide extracts of double base propellant, *Analytical Chemistry*, **61**, 145–148.

BARSOTTI, D. J., HOFFMAN, R. M. and WENGER, R. F., 1983, The use of ion chromatography in the analysis of water gel explosives, *Proceedings of the International Symposium on the Analysis and Detection of Explosives*, pp. 209–212, Washington, DC: US Government Printing Office.

BAUER, C. F., KOZA, S. M. and JENKINS, T. F., 1990, Liquid chromatographic method for determination of explosives residues in soil: collaborative study, *Journal of the Association of Official Analytical Chemists*, **73**, 541–552.

BECK, W. and ENGLEHARDT, H., 1992, Capillary electrophoresis of organic and inorganic cations with indirect UV detection, *Chromatographia*, **33**, 313–316.

BENDER, E. C., 1989, Indirect photometric detection of anions for the analysis of low explosives, *Crime Laboratory Digest*, **16**, 78–83.

BENDER, E. C., CRUMP, J. and MIDKIFF, C. R., 1993, The instrumental analysis of intact and post-blast water gel and emulsion explosives, in Yinon, J. (Ed.) *Advances in the Analysis and Detection of Explosives*, pp. 178–188, Dordrecht: Kluwer Academic Publishers.

BEVERIDGE, A. D., PAYTON, S. F., AUDETTE, R. J., LAMBERTUS, A. J. and SHADDICK, R. C., 1975, Systematic analysis of explosives residues, *Journal of Forensic Sciences*, **20**, 431–454.

BOUVIER, E. S. P. and OEHRLE, S. A., 1995, Analysis and identification of nitroaromatic and nitramine explosives in water using HPLC and photodiode-array detection, *LC-GC*, **13-2**, 120–130.

BUECHELE, R. C. and Reutter, D. J., 1983, Ion chromatography in bombing investigations, *Analytical Chemistry*, **55**, 1468A–1472A.

BUNTE, L. G., KROLL, A., HIRTH, T. and KRAUSE, H., 1994, Extraction of explosives from soils with supercritical carbon dioxide, *25th International Annual Conference ICT, Energetic Materials – Analysis, Characterization, and Test Techniques*, pp. 74-1–74-16.

BUREAU OF ALCOHOL, TOBACCO, AND FIREARMS, 1994, Arson and Explosives Incidents Report, Department of the Treasury, Bureau of ATF, Explosives Division, 650 Mass Ave NW, Washington, DC 20226.

CAMMANN, K., KLEIBOHMER, W. and MUSSENBROCK, E., 1994, Micellar electrokinetic chromatography, a concurrence for HPLC, *GIT Fachz. Lab.*, **3**, 162–170.

CONKLING, J. A., 1985, *Chemistry of Pyrotechnics*, New York: Marcel Dekker.

CURTIS, N. J., and BERRY, P., 1989, Derivatives of ethyl centralite in Australian gun propellants, *Propellants, Explosives, Pyrotechnics*, **14**, 260–265.

DE BRUYNE, P. A. M., ARIJS, J., VERGAUWE, D. A. G. and DE BISSCHOP, H. C. J. V., 1989, The HPLC determination of some propellant additives, *Proceedings Third Symposium on Analysis and Detection of Explosives*, pp. 27-1–27-15, Berghausen: Fraunhofer Institut für Chemische Technologie (ICT).

DOUSE, J. M. F., 1981, Trace analysis of explosives at the low picogram level by silica capillary column gas–liquid chromatography with electron-capture detection, *Journal of Chromatography*, **208**, 83–88.

DOUSE, J. M. F., 1982, Trace analysis of explosives in handswab extracts using Amberlite XAD-7 porous polymer beads, silica capillary column gas chromatography with electron capture detection and thin layer chromatography, *Journal of Chromatography*, **234**, 415–425.

DOUSE, J. M. F., 1983, Trace analysis of explosives at the low picogram level using silica capillary column gas chromatography with thermal energy analyser detection, *Journal of Chromatography*, **256**, 359–362.

DOUSE, J. M. F., 1985, Trace analysis of explosives at the low nanogram level in handswab extracts using columns of Amberlite XAD-7 porous polymer beads and silica capillary column gas chromatography with thermal energy analysis and electron capture detection, *Journal of Chromatography*, **328**, 155–165.

DOUSE, J. M. F., 1987, Improved method for the trace analysis of explosives by silica capillary column gas chromatography with thermal energy analysis detection, *Journal of Chromatography*, **410**, 181–189.

DOUSE, J. M. F., 1988, Trace analysis of explosives by capillary supercritical fluid chromatography, *Journal of Chromatography*, **445**, 244–250.

DOUSE, J. M. F. and SMITH, R. N., 1986, The analysis of explosives and firearm discharge residue in the Metropolitan Police Forensic Science Laboratory, *Journal of Energetic Materials*, **4**, 169–186.

FETTEROLF, D. D., 1995, Detection and identification of explosives by mass spectrometry, in Yinon. J. (Ed.), *Forensic Applications of Mass Spectrometry*, pp. 215–260, Ann Arbor: CRC Press, Inc.

FINE, D. H., YU, W. C., GOFF, U. E., BENDER, E. C. and REUTTER, D. J., 1984, Picogram analysis of explosive residues using the thermal energy analyzer (TEA), *Journal of Forensic Sciences*, **29**, 732–746.

FRANCIS, E. S., EATOUGH, D. J. and LEE, M. L., 1994, Capillary supercritical fluid chromatography with nitro- and nitroso-specific chemiluminescence detection, *Journal of Microcolumn Separations*, **6**, 395–401.

GLATZ, J. A. and GIRARD, J. E., 1982, Factors affecting the resolution and detectability of inorganic anions by nonsuppressed ion chromatography, *Journal of Chromatographic Science*, **20**, 266–273.

GOFF, E. U., YU, W. C., and FINE, D., 1983, Applications of the nitro/nitroso specific detector to explosive residue analysis, *Proceedings of the International Symposium on the*

Analysis and Detection of Explosives, pp. 169–181, Washington, DC: US Government Printing Office.

GREIST, W. H., GUZMAN, C. and DEKKER, M., 1989, Packed column supercritical fluid chromatographic separation of highly explosive compounds, *Journal of Chromatography*, **467**, 423–429.

HABLE, L., STERN, C., ASOWATA, C. and WILLIAMS, K., 1991, The determination of nitroaromatics and nitramines in ground drinking water by wide-bore capillary gas chromatography, *Journal of Chromatographic Science*, **29**, 131–135.

HALL, K. E. and McCORD, B. R., 1993, The analysis of mono- and divalent cations present in explosive residues using ion chromatography with conductivity detection, *Journal of Forensic Sciences*, **38**, 928–934.

HARGADON K. A. and McCORD, B. R., 1992, Explosive residue analysis by capillary electrophoresis and ion chromatography, *Journal of Chromatography*, **602**, 241–247.

HENDERSON, I. K. and SAARI-NORDHAUS, R., 1991, Analysis of commercial explosives by single column ion chromatography, *Journal of Chromatography*, **602**, 149–154.

JANDIK, P., LI, J. B., JONES, W. R. and GJERDE, D. T., 1990, New method of background eluent conductivity elimination in gradient ion chromatography, *Chromatographia*, **30**, 509–517.

JENKINS, R. and YALLOP, H. J., 1970, The identification of explosives in trace quantities on objects near an explosion, *Explosivestoffe*, **18**, 139–141.

JONES, W. R. and JANDIK, P., 1991, Controlled changes of selectivity in the separation of ions by capillary electrophoresis, *Journal of Chromatography*, **546**, 445–458.

KETO, R. O., 1989, Analysis of the eastern bloc explosive Semtex-H, *Proceedings of the Third Symposium on Analysis and Detection of Explosives*, pp. 11-1–11-20, Berghausen: Fraunhofer-Institut für Chemische Technologie (ICT).

KING, R. M., 1993, The work of the explosives & gunshot residues unit of the Forensic Science Service (UK), in Yinon, J. (Ed.) *Advances in the Analysis and Detection of Explosives*, pp. 91–100, Dordrecht: Kluwer Academic Publishers.

KISHI, T., NAKAMURA, J., KOMO-OKA, Y. and FUKUDA, H., 1993, A scheme for the analysis of explosives and explosive residues in Japan, in Yinon, J. (Ed.) *Advances in the Analysis and Detection of Explosives*, pp. 11–17, Dordrecht: Kluwer Academic Publishers.

KLEIBOHMER, W., CAMMANN, K., ROBERT, J. and MUSSENBACH, E., 1993, Determination of explosives residues in soils by micellar electrokinetic capillary chromatography and high-performance liquid chromatography, a comparative study, *Journal of Chromatography*, **638**, 349–356.

KOLLA, P., 1991a, Trace analysis of explosives from complex mixtures with sample pretreatment and selective detection, *Journal of Forensic Sciences*, **36**, 1342–1359.

KOLLA, P., 1991b, Trace analysis of salt based explosives by ion chromatography, *Forensic Science International*, **50**, 217–226.

KOLLA, P., ENGLEHARDT, H. and ZAPP, J., 1993, Sample preparation by supercritical fluid extraction in explosives trace analysis, in Yinon, J. (Ed.) *Advances in the Analysis and Detection of Explosives*, pp. 55–65, Dordrecht: Kluwer Academic Publishers.

LEE, M. L. and MARKIDES, K. E., (Eds), 1990, *Analytical Supercritical Fluid Chromatography and Extraction*, Provo, Utah: Chromatography Conferences Inc.

LLOYD, J. B. F., 1967, Detection of microgram amounts of nitroglycerin and related compounds, *Journal of the Forensic Science Society*, **7**, 198.

LLOYD, J. B. F., 1983, High-performance liquid chromatography of organic explosives components with electrochemical detection at a pendant mercury drop electrode, *Journal of Chromatography*, **257**, 227–236.

LLOYD, J. B. F., 1984, Detection and differentiation of nitrocellulose traces of forensic science interest with reductive mode electrochemical detection at a pendant mercury drop electrode coupled with size exclusion chromatography, *Analytical Chemistry*, **56**, 1907–1912.

LLOYD, J. B. F., 1986, Liquid chromatography of firearms propellant traces, *Journal of Energetic Materials*, **4**, 239–271.

LLOYD, J. B. F., 1991, Forensic explosive and firearms traces: trapping of HPLC peaks for gas chromatography, *Journal of Energetic Materials*, **9**, 1–17.

LLOYD, J. B. F., 1992, HPLC of explosives materials, in Giddings, J. C., Gruska, E. and Brown, P. R. (Eds.), *Advances in Chromatography*, **32**, 179–181.

LYTER, A. H., 1983, A high performance liquid chromatographic (HPLC) study of seven common explosive materials, *Journal of Forensic Sciences*, **28**, 446–450.

McCORD, B. R. and HARGADON, K. A., 1993, Explosives analysis by capillary electrophoresis, in Yinon, J. (Ed.) *Advances in the Analysis and Detection of Explosives*, pp. 133–144, Dordrecht: Kluwer Academic Publishers.

McCORD, B. R., HARGADON, K. A., HALL, K. E. and BURMEISTER, S. G., 1994, Forensic analysis of explosives using ion chromatographic methods, *Analytica Chimica Acta*, **288**, 43–56.

McNAIR, H. M. and BONELLI, E. J., 1968, *Basic Gas Chromatography*, Berkeley, CA: Consolidated Printers.

MIDKIFF, C. R. and WALTERS, A. N., 1993, Slurry and emulsion explosives: new tools for terrorists, new challenges for detection and identification, in Yinon, J. (Ed.) *Advances in the Analysis and Detection of Explosives*, pp. 77–90, Dordrecht: Kluwer Academic Publishers.

MUNDER, A., CHESLER, S. N. and WISE, S. A., 1990, Capillary supercritical fluid chromatography of explosives, *Journal of Chromatography*, **521**, 63–70.

MURRAY, G., 1993, Explosive residue analysis in Northern Ireland, unpublished presentation, an *International Seminar and Workshop on Explosive Residue Analysis Protocols*, FBI Academy, Quantico, Vancouver, June.

NORTHROP, D. M., 1993, in Guzman, N. A. (Ed.), *Capillary Electrophoresis Technology*, pp. 673–691, New York: Marcel Dekker.

NORTHROP, D. M. and MacCREHAN, W. A., 1992, Sample collection, preparation, and quantitation in the micellar electrokinetic capillary electrophoresis of gunshot residue, *Journal of Liquid Chromatography*, **15**, 1041–1063.

NORTHROP, D. M., MARTIRE, D. E. and MacCREHAN, W. A., 1991, Separation and identification of organic gunshot and explosive constituents by micellar electrokinetic capillary electrophoresis, *Analytical Chemistry*, **63**, 1038–1042.

PARKER, R. G., McOWEN, J. M., and CHEROLIS, J. A., 1975, Analysis of explosives and explosives residues, Part 2: thin layer chromatography, *Journal of Forensic Sciences*, **20**, 254–256.

PEAK, S. A., 1980, A thin layer chromatographic procedure for confirming the presence and identity of smokeless powder flakes, *Journal of Forensic Sciences*, **25**, 675–681.

POKRZYWA, G., 1993, Explosive residue analysis in France, unpublished presentation, an *International Seminar and Workshop on Explosive Residue Analysis Protocols*, FBI Academy, Quantico, Vancouver, June.

REUTTER D. J., BUECHELE, R. C. and RUDOLPH, T. L., 1983, Ion chromatography in bombing investigations, *Analytical Chemistry*, **55**, 1468A–1472A.

ROBERTSON, D. and KANSAS, L., 1990, Surveillance of the Army's Propellant Stockpile: Analysis of Stabilizer Content by High Performance Liquid Chromatography, *Technical Report ARAED-TR-90020*, New Jersey: Picatinney Arsenal.

RUDOLPH, T. L. and BENDER, E. C., 1983, A scheme for the analysis of explosives and explosive residues, *Proceedings of the International Symposium on the Analysis and Detection of Explosives*, pp. 71–78, Washington, DC: US Government Printing Office.

SELAVKA, C. M., TONTARSKI, R. E. and STROBEL, R. A., 1987, Improved determination of nitrotoluenes using liquid chromatography with photolytically assisted thermal energy analysis, *Journal of Forensic Sciences*, **32**, 941–952.

SELAVKA C. M., STROBEL, R. A. and TONTARSKI, R. E., 1989, The systematic identification of smokeless powders: an update, *Proceedings of the Third Symposium on Analysis and Detection of Explosives*, pp. 11-1-11-20, Berghausen: Fraunhofer-Institut für Chemische Technologie (ICT).

SHERMAN, J. H. and DANIELSON, N. D., 1987, Comparison of mobile phase counterions for cationic indirect photometric chromatography, *Analytical Chemistry*, **59**, 490.

SLACK, G. C., MCNAIR, H. M. and WASSERZUG, L., 1992, Characterization of Semtex by supercritical fluid extraction and off-line GC–ECD and GC–MS, *Journal of High Resolution Chromatography*, **15**, 102–104.

SMALL, H., 1989, *Ion Chromatography*, pp. 57–74, New York: Plenum Press.

SMALL, H. and MILLER, T. E., 1982, Indirect photometric chromatography, *Analytical Chemistry*, **54**, 462–469.

SMALL, H., STEVENS, T. S. and BOWMAN, W. C., 1975, Novel ion exchange chromatographic method using conductimetric detection, *Analytical Chemistry*, **47**, 1801–1809.

SMITH, K. D., MCCORD, B. R., MACCREHAN, W. A., MOUNT, K. and ROWE, W. F., 1995, Detection of smokeless powder residue on pipe bombs by micellar electrokinetic capillary chromatography, *Proceedings of the Fifth International Symposium on the Analysis and Detection of Explosives*, Bureau of Alcohol, Tobacco, and Firearms, Rockville, MD; in press.

SMITH, R. E., 1988, *Ion Chromatographic Applications*, pp. 29–37, Boca Raton, FL: CRC Press.

STEWART, I. I. and HORLICK, G., 1996, Developments in the electrospray mass spectrometry of inorganic species, *Trends in Analytical Chemistry*, **15**, 80–90.

STINE, G. Y., 1991, An investigation into propellant stability, *Analytical Chemistry*, **63**, 475A–478A.

TAYLOR, L., 1993, The supercritical fluid extraction and analysis of aged single-base propellants, *American Laboratory*, **25**, 8, 22–26.

VERWEIJ, A. M. A., DE BRUYNE M. M. A. V. and KLOOSTER, N. T. M., 1986, Anionenaustauschchromatographie in der analyse von Bombenruckstanden, *Archiv fur Kriminologie*, **177**, 91–94.

VIA, J. and TAYLOR, L. T., 1994, Packed column supercritical fluid chromatography/chemical ionization mass spectrometry of energetic material extracts using a thermospray interface, *Analytical Chemistry*, **66**, 1385–1395.

WALLACE, C. L. and MIDKIFF, C. R., 1993, Smokeless powder characterization, an investigative tool in pipe bombings, in Yinon, J. (Ed.) *Advances in the Analysis and Detection of Explosives*, pp. 29–39, Dordrecht: Kluwer Academic Publishers.

WEINBERGER, R., 1993, *Practical Capillary Electrophoresis*, pp. 147–188, New York: Academic Press.

WOOLFSON-BARTFIELD, D., GRUSHKA, E., ABRAMOVICH-BAR, S., LEVY, S. and BAMBERGER, Y., 1990, Reversed phase ion chromatography with indirect photometric detection of inorganic anions from residues of low explosives, *Journal of Chromatography*, **517**, 305–315.

YINON, J., 1991, Forensic identification of explosives by mass spectrometry and allied techniques, *Forensic Science Review*, **3**, 17–27.

YINON, J. and ZITRIN, S., 1981, *The Analysis of Explosives*, Chapter 5, Oxford: Pergamon.

YINON, J. and ZITRIN, S., 1993, *Modern Methods and Applications in Analysis of Explosives*, Chapters 2 and 4, Chichester: John Wiley.

ZEICHNER, A. and GLATTSTEIN, B., 1986, Improved reagents for firing distance determination, *Journal of Energetic Materials*, **4**, 187–197.

Analysis of Explosives by Infrared Spectrometry and Mass Spectrometry

SHMUEL ZITRIN

9.1 Introduction

The identification of explosives is an important aspect of many police investigations. Circumstances range from possession of explosive substances through pranks and experimentation to terrorist bombings. The analytical results may dictate the direction of an investigation – was the explosion intentionally caused or accidental; if intentional, was it a 'criminal' act or a political/terrorist act; is there a link to earlier occurrences; is there sufficient evidence to lay charges?

These and many other questions arise during explosion investigations. From the scientist's perspective, identification of an explosive can be a most significant analytical challenge. This chapter discusses the advantages and disadvantages of infrared spectrometry and mass spectrometry in analytical protocols for analysis of unreacted explosives and their residues.

9.2 Criteria for Identification

A wide range of methods has been used for the identification of unexploded explosives and of post-explosion residues (Yinon and Zitrin, 1981, 1993; Washington and Midkiff, 1986; Beveridge, 1992). The methods include:

- chemical tests (usually spot tests based on colour reactions);
- chromatographic methods (mainly thin layer chromatography (TLC), gas chromatography (GC), high performance liquid chromatography (HPLC) and ion chromatography (IC));
- spectrometric methods (mainly infrared (IR) spectroscopy and mass spectrometry (MS); occasionally nuclear magnetic resonance (NMR) spectrometry).

On-line combinations of chromatographic and spectrometric methods, mainly GC/MS and HPLC/MS, have also been used.

The criteria for positive identification of a single compound differ between various forensic laboratories. Some laboratories still use colour reactions (spot tests) as part of the analysis, but it is generally accepted that the identification of an organic compound should not be based on colour reactions only. Exceptions are sometimes made for inorganic ions.

Reliance on chromatographic methods for the identification of an organic compound is quite common in forensic laboratories, especially when several different chromatographic methods are applied. Although chromatographic methods are basically intended for separation, the incorporation of highly specific detection of the separated compounds often enables a relatively safe identification by these methods only. Examples of specific detection methods are certain colour reagents for visualization of TLC plates (Yinon and Zitrin, 1981: 59–85), electron-capture detection (ECD) and thermal-energy analyser (TEA) detection in GC (Yinon and Zitrin, 1993: 46–66) or HPLC (Yinon and Zitrin, 1993: 83–85).

Such detection is specific for one or more groups of explosives but not for a single compound. Thus the Griess reagent used for spraying TLC plates (following an alkali spray) gives specific colour reactions for nitrate esters and nitramines but does not distinguish between individual compounds within these classes. Individual compounds can be distinguished by their different rates of migration on the TLC plates. Similarly, TEA detection is highly specific for groups of compounds containing nitro or nitroso groups (depending on the experimental conditions) but individual compounds must be distinguished by having different retention times. Although the combination of chromatographic retention data and group-specific detectors enhances the reliability of the identification of explosives by chromatographic methods only, such an identification cannot be considered foolproof. Therefore, use of a spectrometric method such as IR, MS or NMR as a criterion for identification is highly recommended (Zitrin, 1986). The reason is that the information obtained from these spectrometric methods is more directly related to the molecular structures of the analytes than the information obtained from chromatographic data.

9.2.1 Advantages and Applicability of IR Spectrometry and Mass Spectrometry

A complete IR spectrum of a pure explosive often constitutes a reliable basis for its identification. For organic molecules, the IR spectrum not only identifies functional groups by their characteristic absorption but often, due to its 'fingerprint' value, can identify the whole molecule. Sometimes, however, it is difficult to distinguish between individual compounds of the same class (i.e. nitrate ester explosives) solely by their IR spectra. In these cases, the combination of retention data from a chromatographic method and the IR spectrum gives a safe (unambiguous) identification of the explosive. Inorganic ions which appear in explosive mixtures (e.g. nitrate, chlorate, ammonium) have strong characteristic absorptions in the IR region.

Explosives are often mixtures of several explosive and non-explosive compounds. IR spectrometry can supply reliable information about the identity of the components present in high concentration in simple mixtures. If the IR spectrum of the mixture does not permit unambiguous identification, the spectrum of a compound (or compounds) known to be present in the mixture can be artificially subtracted to enhance that of other components. This capability of spectral substraction is an integral part of modern FTIR (Fourier Transform Infrared) instruments (see 9.3). Obviously, the success of subtraction in producing a meaningful spectrum depends on the composition of the mixture.

Another approach is to separate the mixture prior to IR analysis. The separation is usually based on different solubilities of the components in one or more solvents. After the separation, the IR spectrum of the relevant component is run and the forensic chemist decides if the quality of the spectrum enables unambiguous identification. The commercially available on-line GC/IR, in which components are eluted from the GC column directly to the IR spectrometer, has not yet found a place in the analysis of explosives.

HPLC was tried as a preceding technique for separating explosives prior to their IR analysis. In an off-line procedure (Riddell and Mills, 1983), the explosives eluted from the LC column were collected and analysed by FTIR. An on-line procedure, where the FTIR served as a detector to the LC, was also evaluated (Cantu et al., 1983; Riddell and Mills, 1983), but the results suffered from rather incomplete IR spectra of the explosives.

It is especially difficult to base the identification of explosives on IR spectrometry in cases of post-explosion residues. The amount of the original explosive in the debris is usually very small and is often mixed with large amounts of contaminants. The chances to get a meaningful IR spectrum on which identification of the explosive can be based are therefore very slim. Exceptions are inorganic ions, whose characteristic strong absorption bands are sometimes discernible in the IR spectra obtained from the aqueous extract of the post-explosion debris.

Mass spectrometry (see 9.4.1) is probably the most reliable method for the identification of many organic compounds. In its usual mode – positive ions formed by electron ionization (EI) – the mass spectrum of a pure compound is generally accepted as a proof of identity. Explosives from the same chemical class, such as nitrate esters, may have similar mass spectra in the EI mode but it is then possible to use chemical ionization (CI) mass spectrometry, which complements the EI mode by giving additional information, mainly in the molecular weight region. The combination of EI/MS and CI/MS constitutes a very reliable method for a complete identification of most organic explosives. Other techniques, such as negative-ion mass spectrometry and tandem mass spectrometry (MS/MS) can be useful for specific problems.

For explosive mixtures, the components of the mixture should be separated prior to the mass spectrometric analysis. The most common separation is carried out by gas chromatography combined on-line with mass spectrometry (GC/MS). This combination, which has been commercially available for more than twenty years, has reached a very advanced state. Following separation in the GC column, each of the eluted components enters directly into the ion source of the mass spectrometer. Ideally, the result is a full mass spectrum for each of the separated components. On-line chromatographic separation prior to the mass spectrometric analysis can also be carried out by liquid chromatography (LC/MS). LC/MS may be especially suitable for explosives which do not elute easily in GC (e.g. involatile or thermally-labile explosives).

Mass spectrometry is more suitable than IR spectrometry for post-explosion analysis of explosives. Not only is GC/MS technologically better established than GC/IR, but it is more sensitive, at least by three orders of magnitude, being capable of analysing explosives in the nanogram range. This sensitivity is very important in post-explosion analysis, where only small amounts of the explosive are present.

For forensic laboratories which adhere to the criterion that the identification of a single organic compound should not be based on chromatographic methods only, GC/MS is the method of choice (Zitrin, 1986). It is a good policy to include GC/MS in the analytical scheme for the analysis of explosives in post-explosion situations.

9.3 IR Spectrometry

The IR region of the electromagnetic spectrum lies in the energy range of most molecular vibrations. Thus, irradiation of molecules with IR light will result in absorption at the vibrational frequencies of the molecule. Since the absorption pattern of each compounds depends on its molecular structure, it is possible to identify the compound by its absorption pattern: its IR spectrum.

In the 'classical' dispersive spectrometer the polychromatic light passes through the sample, which absorbs some wavelengths and transmits the rest. The transmitted light is then dispersed by a prism or grating before being detected. The resulting spectrum usually consists of downward absorption peaks, plotted against the wavelength or wavenumber. The wavenumber, which is the reciprocal of the wavelength, is expressed in reciprocal centimetres (cm^{-1}).

In FTIR the IR beam is not separated to discrete wavelengths before its detection. Instead,

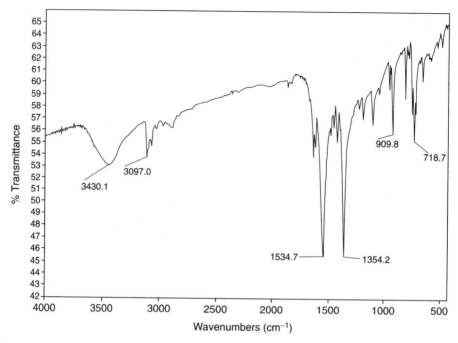

Figure 9.1 IR spectrum of trinitrotoluene (TNT) (experimental conditions: KBr; 32 scans; resolution 4.0 cm^{-1}).

an interferometer is used which sends simultaneously all the transmitted wavelengths to the detector. The resulting interference pattern (interferogram) is transformed by computer software into an IR spectrum, using a mathematical procedure called Fourier Transform. Most modern IR spectrometers are FTIR instruments, which have several advantages over dispersive instruments, including higher sensitivity and better resolution. The IR spectra illustrated below were all recorded on FTIR instruments. The experimental conditions are given in the Appendix.

From an analytical point of view it is convenient to divide the vibrations of an organic molecule into two types: vibrations associated with the molecule as a whole and vibrations associated with specific functional groups. The usual range of the IR spectrum is 4000 cm^{-1} to 250 cm^{-1}. Vibrations characteristic of the molecule as a whole usually give rise to absorption bands below 1300 cm^{-1}, and are very useful for the identification of an unknown molecule by comparing its spectrum to that of an authentic known molecule ('standard'). This region is therefore often called the 'fingerprint' region. Vibrations associated with specific functional groups usually lead to absorption bands above 1500 cm^{-1}. They are useful in the analysis of unknown samples, by indicating the presence or absence of specific functional groups in the molecule. For example, a strong absorption band near 1700 cm^{-1} could come from a carbonyl group; absorption in the 2200 cm^{-1} region can often be correlated to a carbon—carbon or carbon—nitrogen triple bond; aromatic rings absorb in the 1600–1500 cm^{-1} region. Such correlations, are recorded in many books (e.g. Bellamy, 1962; Williams and Fleming, 1980).

9.3.1 Explosives Containing the Nitro Group

The symmetric (v_s) and asymmetric (v_{as}) stretching vibrations of the NO$_2$ group give rise to two strong absorption bands at wavelengths which depend on the type of atom to which the nitro group is directly attached (Yinon and Zitrin, 1981: 154–6 and references). These two stretching vibrations of the NO_2 group have an important diagnostic value. Their presence or absence should be the first thing to notice when an unknown sample is suspected to be an explosive.

Nitroaromatic compounds

In nitro compounds, in which the nitro group is directly bonded to a carbon atom ($C-NO_2$), the two stretching vibrations of the NO_2 group appear in the ranges 1590–1510 cm^{-1} (v_{as}) and 1390–1320 cm^{-1} (v_s). Nitroaromatic compounds are nitro compounds in which the carbon directly attached to the nitro group is part of an aromatic ring. Their two bands appear at 1560–1520 cm^{-1} (v_{as}) and 1370–1340 cm^{-1} (v_s) (Alm et al., 1978). Dinitroaromatic and trinitroaromatic compounds, many of which are explosives, were reported (Conduit, 1959) to have their v_{as} absorption bands at 1552–1539 cm^{-1} and at 1567–1554 cm^{-1} respectively. The IR spectrum of 2,4,6-trinitrotoluene (TNT) is shown in Figure 9.1. The two strong absorption bands at 1534 cm^{-1} (v_{as}) and 1354 cm^{-1} (v_s) are clearly observed.

Nitrate esters

The two NO_2 stretching vibrations in nitrate esters, which contain the $C-O-NO_2$ bond, appear as sharp and intense bands at 1660–1640 cm^{-1} (v_{as}) and 1285–1270 cm^{-1} (v_s) (Pristera et al., 1960; Urbanski and Witanowski, 1963). The absorption bands due to the NO_2 stretching vibrations in polynitrate esters are often split into two or more peaks. The splitting is normally attributed to the existence of rotational isomers, due to hindered rotation around the C-C bond (Urbanski and Witanowski, 1963). The splitting of the v_s (NO_2) and v_{as} (NO_2) in the IR spectrum of pentaerythritol tetranitrate (PETN, ($C[CH_2ONO_2]_4$)) shown in Figure 9.2 (Abramovich-Bar), was attributed to crystal-lattice effects. This was proved by

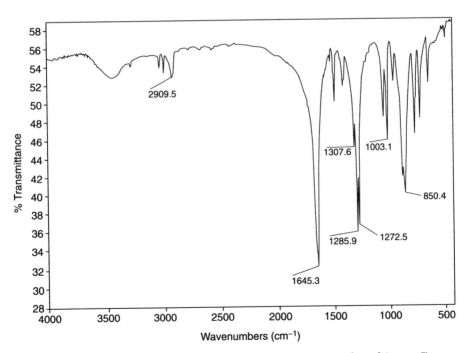

Figure 9.2 IR spectrum of pentaerythritol tetranitrate (PETN) (experimental conditions as Figure 9.1).

271

recording its spectrum as a supercooled liquid; the splitting disappeared and single symmetric peaks were obtained for the two bands (Urbanski and Witanowski (1963). The IR spectra of most nitrate ester explosives are very similar to each other, so the identification of an individual nitrate ester by comparing its spectrum to that of a known one is often very difficult. Figures 9.3, 9.4, 9.5 and 9.6 illustrate the spectra for ethylene glycol dinitrate (EGDN; $C_2H_4(ONO_2)_2$), nitroglycerine (NG; $C_3H_5(ONO_2)_3$), nitrocellulose (NC; $C_6H_7(OH)_x(ONO_2)_y$) and metriol trinitrate (MTN; 1,1,1-trimethylolethane trinitrate (H_3C . $C[CH_2ONO_2]_3$)) respectively.

Nitramines

The two NO stretching vibrations of the $N-NO_2$ bond in nitramines were reported (Alm et al., 1978) to absorb in the ranges 1590–1530 cm^{-1} and 1310–1270 cm^{-1} for the asymmetric and symmetric vibrations respectively. Figures 9.7 and 9.8 show the IR spectra of the heterocyclic nitramines RDX (tricyclomethylene trinitramine; $C_3H_6N_3(NO_2)_3$ and HMX (tetracyclomethylene tetranitramine $C_4H_8N_4(NO_2)_4$). Several peaks which were attributed to the NO_2 stretching vibrations (Werbin 1957; Alm et al., 1978) appear in the above listed ranges.

Other

There are other absorption bands which are characteristic of certain classes of explosives such as 2,4,6-trinitroaromatic (e.g. 1081 cm^{-1}), 2,4-dinitroaromatic (e.g. 913–922 cm^{-1}) and others

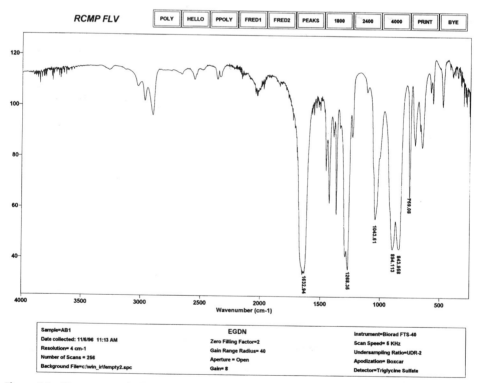

Figure 9.3 IR spectrum of ethylene glycol dinitrate (EGDN) (experimental conditions: diamond cell, 6 × beam condenser; Biorad FTS-40 FTIR spectrometer; 256 scans; resolution 4.0 cm^{-1}).

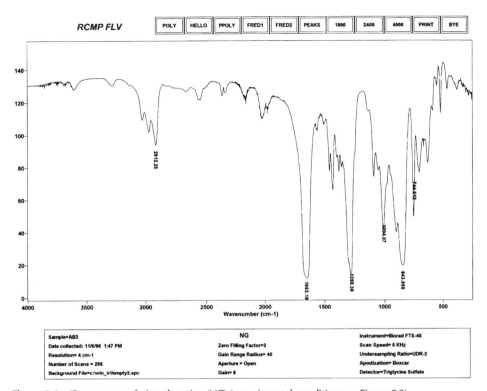

Figure 9.4 IR spectrum of nitroglycerine (NG) (experimental conditions as Figure 9.3).

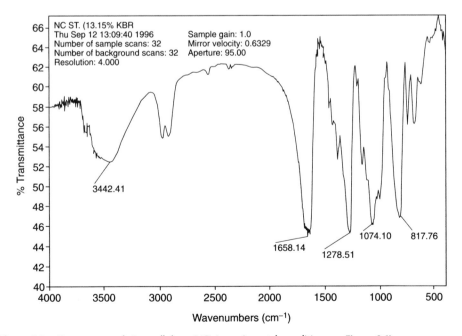

Figure 9.5 IR spectrum of nitrocellulose (NC) (experimental conditions as Figure 9.1).

273

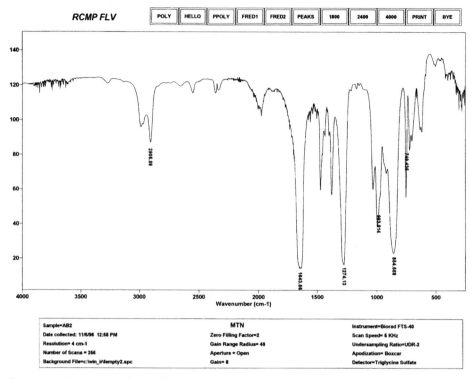

Figure 9.6   IR spectrum of metriol trinitrate (MTN) (experimental conditions as Figure 9.3).

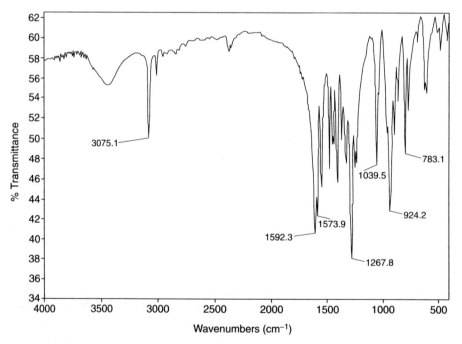

Figure 9.7   IR spectrum of RDX (experimental conditions as Figure 9.1).

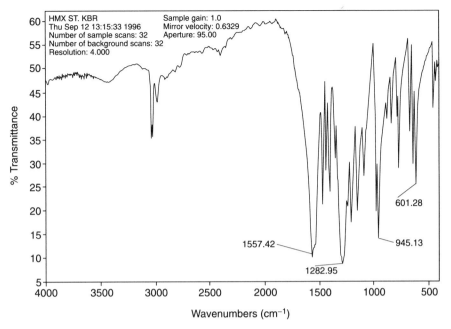

Figure 9.8 IR spectrum of HMX (experimental conditions as Figure 9.1).

(Yinon and Zitrin, 1981: 159). Since most appear in the 'fingerprint' region of the spectrum, they are not very useful for the identification of unknown samples.

Some primary explosives, such as lead styphnate or diazodinitrophenol (DDNP, Dinol), have the nitroaromatic structure and can be easily identified by their IR spectra. In addition to its nitroaromatic nucleus, DDNP contains a diazo group, which has a characteristic absorption band at 2211 cm^{-1} as shown in Figure 9.9.

The aliphatic nitrate explosive urea nitrate ($H_2N-CO-NH_3^+NO_3^-$), has been suspected as having been used in a terrorist bombing. The IR spectrum of urea nitrate is illustrated below in Figure 9.10.

9.3.2 Organic Peroxides

Organic peroxides, often powerful explosives, are usually unstable and highly sensitive to shock, friction or heat (Urbanski, 1964). They are not considered safe enough to be used as standard primary explosives (initiators). Triacetonetriperoxide (TATP), **1**, first prepared in the late 19th century by German chemists (Wolffenstein, 1895), has reappeared in recent years in terrorist cases. In this context it was first identified in Israel (Zitrin et al., 1983), where it has become very popular among terrorists. It has been used as a main charge rather than as an initiating explosive. Subsequent papers (Evans et al., 1986; White, 1992) reported its appearance in the United States. Figure 9.11 shows the IR spectrum of TATP. When first encountered in a terrorist case (Zitrin et al., 1983), the absence of the characteristic bands of the NO_2 stretching vibrations was very helpful in its identification. It was unambiguously identified by combination of IR, NMR and mass spectrometry. The band at 872 cm^{-1} was attributed to the O—O stretching vibration in peroxides (Bellamy, 1962: 343–9), but its diagnostic value is

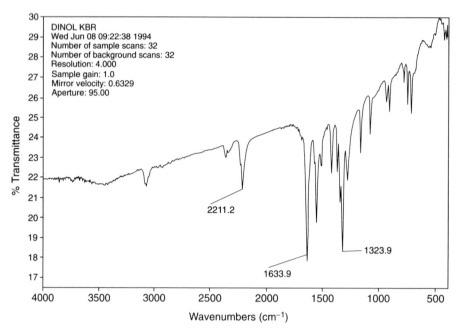

Figure 9.9 IR spectrum of diazodinitrophenol (DDNP) (experimental conditions as Figure 9.1).

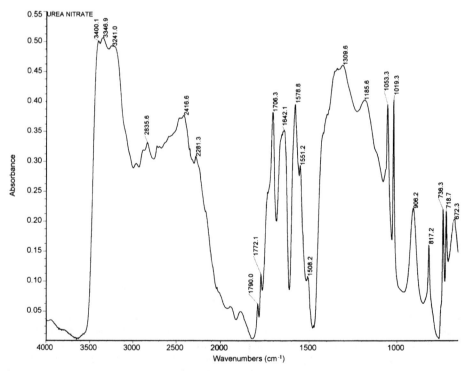

Figure 9.10 IR spectrum of urea nitrate (transmission spectrum; Nic-Plan IR microscope on a Nicolet 710 bench, 256 scans, 4 cm^{-1} resolution).

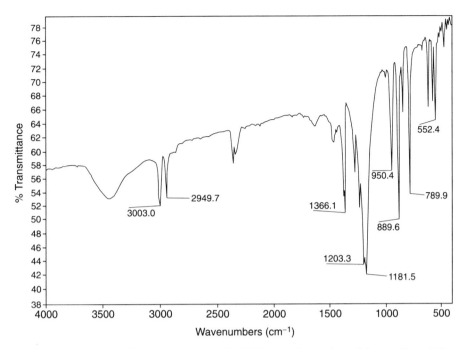

Figure 9.11   IR spectrum of triacetone triperoxide (TATP) (experimental conditions as Figure 9.1).

limited when identification of an unknown sample is required.

Another organic peroxide, hexamethylenetriperoxide diamine (HMTD), **2**, was also encountered in cases related to terrorist activity (Reutter et al., 1983; Zitrin et al., 1983). Its potential danger is that, unlike common primary explosives such as lead azide or mercury fulminate, it does not contain a heavy metal so it cannot be detected by X-ray screening. It was indeed found as a white powder inside a detonator made of plastic material (Zitrin et al., 1983). Its identification was based on IR, NMR, mass spectrometry and melting point determination (Reutter et al., 1983; Zitrin et al., 1983). Peak assignments in the IR spectrum of HMTD were suggested, in a study which included also its Raman and NMR spectra (Sulzle and Klaeboe, 1988). The spectrum (Sulzle and Klaeboe, 1988) is illustrated in Figure 9.12.

1   **2**

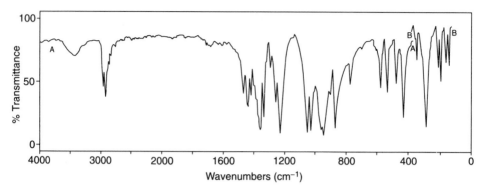

Figure 9.12 IR spectrum of hexamethylene triperoxide diamine (HMTD) (from Sulzle and Klaeboe, 1988, with permission).

9.3.3 Inorganic Explosives

Inorganic anions appear as oxidizing agents in explosive mixtures. Nitrates (NO_3^-), are used in traditional black powders, in many commercial explosives such as dynamites slurries, emulsions and ammonium nitrate–fuel oil (ANFO). Nitrates have found extensive use in home-made, improvised explosive mixtures. The common nitrate salts are ammonium (NH_4NO_3), potassium (KNO_3) and sodium ($NaNO_3$). Ammonium nitrate is the most commonly used commercial explosive and has recently been mentioned in connection to major terrorist bombings. Chlorates (ClO_3^-) have usually been avoided in commercial explosives, being so unpredictably dangerous. Their mixtures with reducing agents such as sugars or aluminium powder have, however, been used as improvised explosives (often with disastrous results to the users). Nitrates, chlorates and perchlorates are constituents of many flash powder compositions (Meyers, 1978).

Other anions, such as carbonates (CO_3^{2-}) or sulphates (SO_4^{2-}), though having no role in the explosion process, are encountered in some explosive mixtures and also appear as explosion products. All the anions mentioned above have strong characteristic absorption bands in the IR region, which are shown in Table 9.1 (Bellamy, 1962: 9).

IR spectrometry is therefore very important for the identification of inorganic ions, mainly in unexploded mixtures but sometimes in post-explosion residues (Beveridge et al., 1975, 1983; Washington et al., 1977; Kaplan and Zitrin, 1977).

The residues from black powder confined in pipes and exploded by flame initiation contained mainly sulphates, carbonates and nitrates, all of which could be identified by IR (Beveridge et al., 1975; Washington et al., 1977). A highly diagnostic ion found in the acetone extract of black powder was thiocyanate, which was identified by its IR spectrum (Beveridge

Table 9.1 IR absorption bands of inorganic ions

Ions	Absorption bands	
Carbonates	1450–1410 cm^{-1} (very strong)	880–860 cm^{-1} (medium)
Sulphates	1130–1080 cm^{-1} (very strong)	680–610 cm^{-1} (medium-weak)
Nitrates	1380–1350 cm^{-1} (very strong)	840–815 cm^{-1} (medium)
Chlorates*	980–940 cm^{-1} (very strong)	

* Miller and Wilkins, 1952; Nyquist and Kagel, 1971.

et al., 1975). A characteristic absorption band of the thiocyanate anion appears at 2020 cm^{-1} (Miller and Wilkins, 1952; Nyquist and Kagel, 1971; Beveridge et al., 1975).

IR was the basis for the analysis of anions in flash powders, containing $KClO_4$, $KClO_3$ or KNO_3 as oxidizing agents (Meyers, 1978).

The IR spectra of sodium, potassium and ammonium nitrate are shown in Figures 9.13, 9.14 and 9.15 respectively. The spectra of sodium and potassium chlorate and potassium perchlorate ($KClO_4$) are shown in Figures 9.16, 9.17 and 9.18 respectively. The IR spectrum of potassium sulphate is shown in Figure 9.19. It is interesting to compare it to potassium perchlorate. The similarity underlines the need for caution in interpretation. However, any potential problems which this might cause are readily overcome by combining IR with elemental analysis in systematic analytical protocols as discussed in Chapter 11 for analysis of low explosive chemical mixtures and their residues.

Extensive compilations of IR spectra of inorganic salts (Miller and Wilkins, 1952; Nyquist and Kagel, 1971) and of compounds related to explosives (Chasan and Norwitz, 1972) have been published. The latter includes IR spectra of the primary explosives lead azide, mercury fulminate and lead styphnate all of which have characteristic absorption bands in the IR region.

9.3.4 Non-explosive Additives

IR spectra of the common stabilizers in smokeless powders have been published (Pristera, 1953). They include diphenylamine, 2-nitrodiphenylamine, methyl centralite (*N,N′*-dimethyl-*N,N′*-diphenylurea) and ethyl centralite (*N,N′*-diethyl-*N,N′*-diphenylurea). Because of the relatively low concentrations of stabilizers in smokeless powder formulations, analysis by IR must be preceded by separation from the major components of the powder, usually by extraction or column chromatography.

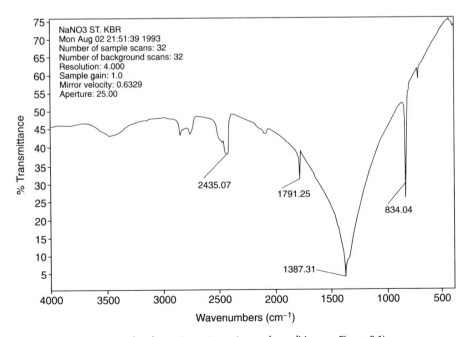

Figure 9.13   IR spectrum of sodium nitrate (experimental conditions as Figure 9.1).

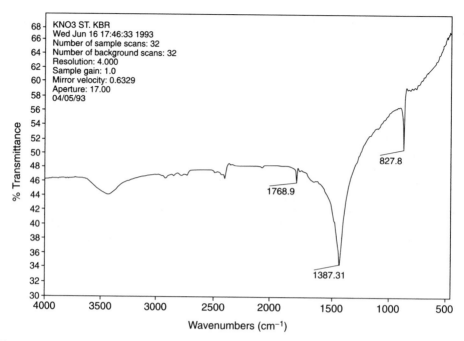

Figure 9.14 IR spectrum of potassium nitrate (experimental conditions as Figure 9.1).

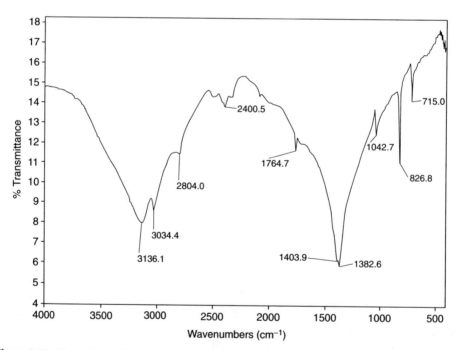

Figure 9.15 IR spectrum of ammonium nitrate (experimental conditions as Figure 9.3).

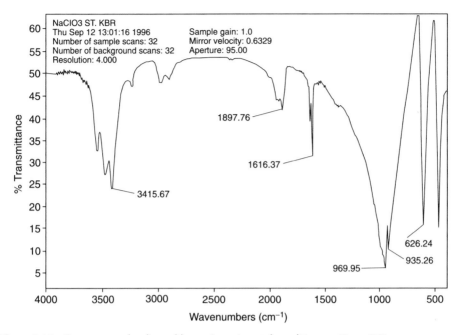

Figure 9.16   IR spectrum of sodium chlorate (experimental conditions as Figure 9.1).

Plasticizers are essential components of plastic explosives. Many plastic explosives are based on the same high explosive (e.g. RDX, PETN), but contain different plasticizers. Thus, the individualization of a plastic explosive cannot be based on the analysis of the explosive alone and the plasticizers must also be identified. Common plasticizers are usually esters of

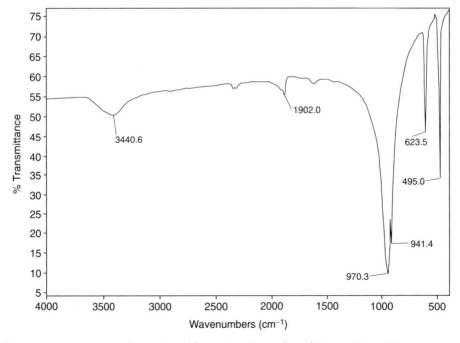

Figure 9.17   IR spectrum of potassium chlorate (experimental conditions as Figure 9.1).

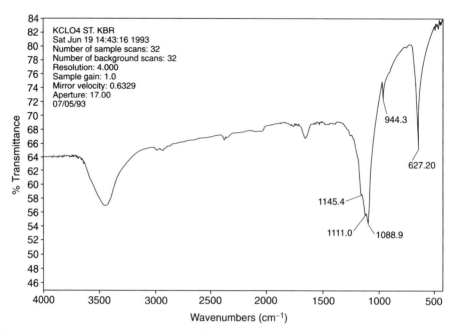

Figure 9.18 IR spectrum of potassium perchlorate (experimental conditions as Figure 9.1).

dicarboxylic acids such as phthalic, adipic or sebacic acids. Their IR spectra are characterized by a strong absorption band at 1750–1735 cm^{-1}, due to the C=O stretching vibrations of esters (Bellamy, 1962c). The US-made military explosive 'Composition C-4' has been analysed by IR spectrometry in the normal (transmission) mode (Peimer et al., 1980) and in the

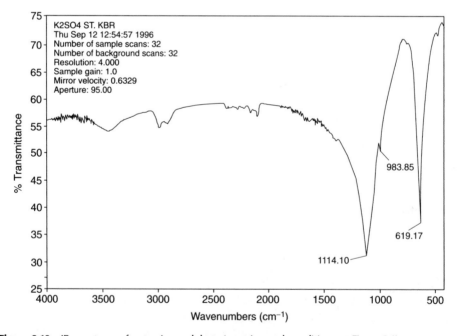

Figure 9.19 IR spectrum of potassium sulphate (experimental conditions as Figure 9.1).

multiple internal reflectance (MIR) mode (Keto, 1986). The two plasticizers whose IR spectra were recorded were octyl(2-ethylhexyl) esters of adipic acid (DOA) and sebacic acid (DOS). Polyisobutylene, another component of C-4, was also analysed by IR spectrometry (Keto, 1986).

Some other reported applications of IR spectrometry follow. It was used to identify poly-ethylene in explosive compositions of Russian missiles (Solomonovici, 1986). Non-explosive components of Semtex-H, a Czechoslovakian-made plastic explosive which has been widely used by terrorists (Feraday, 1993), have been analysed by FTIR following separation by gel permeation chromatography (Keto, 1989). They were identified as styrene-butadiene copoly-mer (binder) and dilaurylthiodipropionate (antioxidant). Wax, used as a desensitizer in the RDX-based military explosives Composition A-3 and Composition A-4, was analysed by IR (Hong et al., 1993). Trioxane, an unusual adulterant in a double-base smokeless powder found in an improvised incendiary device was identified by its IR spectrum (Cabiness and Blackledge, 1983).

9.4 Mass Spectrometry

In mass spectrometry, molecules are introduced into the ion source of the mass spectrometer, where they obtain an electrical charge, i.e. are converted into ions. The ions formed include molecular ions (the charged molecules) and fragments, produced by decomposition of the molecular ions. All the ions then enter the analyser part of the mass spectrometer, where they are separated according to their masses. The common analysers are magnetic and the quad-rupole. Their description, which is beyond the scope of this book, can be found in many textbooks (e.g. Chapman, 1993). Following their separation, the ions pass through a slit onto a detector, which is connected to a data acquisition system. The result is a 'mass spectrum', in which masses of different ions are plotted against their relative abundances. (More accurately, it is the mass/charge (m/z) ratio which is plotted, but since the charge is usually 1 electron charge it is convenient to refer to 'masses'. Throughout this chapter, the notation 'm/z' is used to note the masses of the ions.)

9.4.1 Ionization

The two types of ionization are: (1) electron ionization (electron impact, EI); and (2) chemical ionization (CI).

Electron impact

The most common ionization mode is electron ionization in which electrons emitted from a filament, after acquiring an energy of 70 eV, hit the sample molecules. This leads to a removal of an electron from the bombarded molecule, producing a positive ion-radical called the *molecular ion*:

$$M + e^- \rightarrow M^{\cdot+} + 2e^-$$

The molecular ion acquires (from the impinging electrons) enough energy to further decompose into fragment ions (see 9.4.2).

Chemical ionization

Another ionization mode is chemical ionization (CI), in which a molecule is ionized by receiv-ing a proton from a 'reagent ion'. The reagent ion is formed when a gas (the 'reagent gas') is introduced into the ion source at relatively high pressure (1 torr instead of 10^{-6} torr in EI). At such pressures, the gas molecules, after being ionized by the usual electron bombardment, collide with neutral gas molecules to produce new ions. With methane as reagent gas, the following reactions occur:

$$CH_4 + e^- \rightarrow CH_4^{\cdot+} + 2e^-$$

$$CH_4^{\cdot+} + CH_4 \rightarrow CH_5^+ + CH_3^{\cdot}$$

The CH_5^+ reagent ion will transfer a proton to most organic molecules M:

$$CH_5^+ + M \rightarrow CH_4 + MH^+$$

The MH^+ ions are also capable of decomposing into fragment ions, but this fragmentation usually occurs to a lesser extent than that of the M^+ ions in EI. The degree of fragmentation in CI/MS can be controlled by using different reagent gases (Yinon and Zitrin, 1981: 192–4). For example, the reactant ions in hydrogen (H_3^+) or methane (CH_5^+) are much stronger Brønsted acids than the reactant ions in ammonia (NH_4^+) or isobutane ($C_4H_9^+$). Thus, with hydrogen or methane, the proton is transferred to the sample molecule with more energy than with ammonia or isobutane. As a result, the MH^+ ions tend to fragment more when hydrogen or methane are used as reagents, while when ammonia or isobutane are used the fragmentation occurs to a lesser extent and more undecomposed MH^+ ions are observed.

9.4.2 Interpretation

The information in a mass spectrum includes masses of fragments of the molecule and often its molecular weight. This information, especially in an EI mass spectrum, is often specific enough to be considered a 'fingerprint' of the compound. By comparing this spectrum to a published spectrum of a known compound (often following a computerized library search), a highly reliable identification can be made. In addition to the 'fingerprint' value of a mass spectrum, the molecular weight (often obtained by using the CI mode) and the masses of the fragments can be correlated by the experienced expert to the molecular structure, hence the important role of mass spectrometry in structure elucidation of unknown compounds.

An example of a mass spectrum (in the EI mode), that of 2,4-dinitrotoluene (2,4-DNT), is shown in Figure 9.20. The molecular ion $M^{\cdot+}$ appears at m/z 182. In EI mass spectra the *molecular ion* is usually the ion with the highest mass. Ions at lower masses are formed by fragmentation of molecular ions into ions ('fragment ions') and neutral species. Obviously, the neutral species are not recorded by the mass spectrometer.

The peak formed from the most abundant ion in the mass spectrum is called '*base peak*'. It is customary to arbitrarily assign to it an abundance value of 100% so abundances of other ions are relative to the base peak.

The base peak may be the molecular ion peak but may also be a fragment ion as in the spectrum of 2,4-DNT, whose base peak at m/z 165 is due to loss of a hydroxyl radical from the molecular ion. It is therefore designated $(M - OH)^+$ ion [M *minus* OH]. Similarly, the ion at m/z 119 corresponds to $(M - OH - NO_2)^{\cdot+}$ [M *minus* OH *minus* NO_2].

In some ions in Figure 9.20 one can observe low abundant ions at one mass unit higher than the ions with higher abundance. These are 'isotope peaks', which are formed because of the natural abundance of ^{13}C (1.1%). Thus the ion at m/z 183 is the isotope peak of the molecular ion. Ions containing chlorine or sulphur give rise to typical isotope peaks.

9.4.3 Negative-ion Mass Spectrometry

In negative-ion mass spectrometry, when the ion source is at low pressures as in normal EI mode, negative ions are produced by several mechanisms (Melton, 1970): resonance-capture

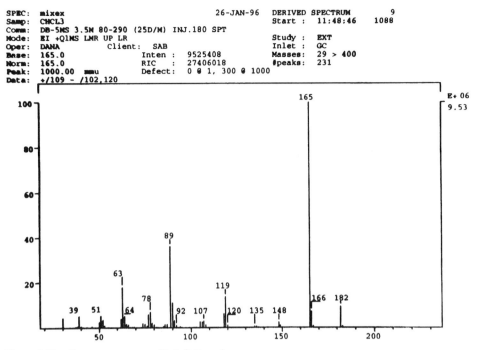

SPEC: mixex   26-JAN-96   DERIVED SPECTRUM   9
Samp: CHCL3   Start : 11:48:46   1088
Comm: DB-5MS 3.5M 80-290 (25D/M) INJ.180 SPT
Mode: EI +Q1MS LMR UP LR   Study : EXT
Oper: DANA   Client: SAB   Inlet : GC
Base: 165.0   Inten : 9525408   Masses: 29 > 400
Norm: 165.0   RIC   : 27406018   #peaks: 231
Peak: 1000.00 mmu   Defect: 0 @ 1, 300 @ 1000
Data: +/109 - /102,120

Figure 9.20   EI mass spectrum of 2,4-dinitrotoluene.

of the electron followed by collisional stabilization, dissociative capture of the electron and ion-pair formation. The probability of these processes occurring is less than the positive EI ionization by at least two orders of magnitude so its application to the trace analysis of organic compounds has been limited. Negative-ion mass spectrometry is strongly dependent on the electron energies and on the molecular structure of the analyte. Its sensitivity increases when the molecule contains electronegative groups.

A significant increase in sensitivity was obtained by using high-pressure negative-ion mass spectrometry, either at normal CI source pressure (1 torr) or at atmospheric-pressure ionization (API). In negative-ion chemical ionization (NCI) ions are produced by resonance-capture of low-energy electrons (formed by collision of the higher-energy electrons with the moderator gas) and also by a variety of ion–molecule reactions between the sample molecules and the negative ions of the reagent gas (Yinon and Zitrin, 1981: 201–5; Yinon and Zitrin, 1993: 105–17).

9.4.4   Gas Chromatography/Mass Spectrometry (GC/MS)

The upgrading in the status of mass spectrometry to become a leading method in organic analysis was mainly due to the technical achievement of combining, on-line, gas chromatography with mass spectrometry (GC/MS) (Karasek and Clement, 1988). In modern GC/MS instruments, complex mixtures are separated by the GC capillary column, and each of the separated compounds enters directly into the ion source of the mass spectrometer. This capability allows the rapid identification of every single component in a complex mixture, a task that in the past could take many years. The combination of liquid chromatography–mass spectrometry (LC/MS) (Yergev et al., 1990) is also commercially available so compounds which are involatile or thermally labile can be separated and identified by MS.

In practice, in addition to its principal role as identification tool, the mass spectrometer serves as a detector for the gas and liquid chromatographic system. As the mass spectrometer ionizes the compounds eluted from the chromatographic system, it can measure the total ion current, which is proportional to the concentration of the eluted compound in the mixture (in a way, this is similar to the flame-ionization detector in conventional GC). The resulting chromatogram, which is identical to a conventional GC trace, is termed *total-ion chromatogram* (TIC) or *reconstructed ion chromatogram* (RIC).

Figure 9.21 shows the RIC of a mixture of several explosives analysed by GC/MS. NG is eluted first, after 62 s, followed by 2,4-DNT (109 s), TNT (159 s), PETN (183 s) RDX (218 s) and tetryl (262 s) (see also 9.4.6 'Nitramines'). The similarity to a conventional GC trace is clearly observed. A more refined option is to reconstruct a chromatogram, termed '*mass chromatogram*', which includes only compounds whose mass spectra contain a pre-selected mass. The mass chromatogram is very useful in locating a suspected compound in a complex mixture, as is often the case in post-explosion analysis (see examples in 9.5.1).

9.4.5 Mass Spectrometry/Mass Spectrometry (MS/MS)

A recent and most promising technique, also commercially available, is tandem mass spectrometry or MS/MS (Busch et al., 1988), in which two mass spectrometers are combined in series. In a typical MS/MS operation, ions emerging from the first mass spectrometer (after normal separation) are selected to enter a collision cell. In this cell the selected ions (called parent ions or primary ions) collide with atoms of an inert gas such as argon. As a result of this collision the ions undergo a fragmentation process called *collision-induced dissociation* (CID). The resulting fragment ions (called daughter ions or secondary ions) are then mass-analysed by the second mass spectrometer to produce a daughter-ion mass spectrum (also

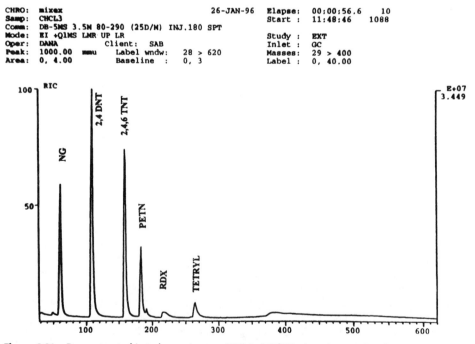

Figure 9.21 Reconstructed ion chromatogram (RIC) in GC/MS of a mixture of explosives.

called CID spectrum), which is often characteristic of the pre-selected parent ion. MS/MS has been extensively used for structure elucidation of ions. The CID spectrum of the ion in question is compared to the CID spectrum of an ion having a known structure (which can be produced from another molecule) in the same m/z value. If the two CID spectra are identical, it can be assumed that the two ions have the same structures.

MS/MS has the potential to be a suitable analytical method for the identification of small amounts of an analyte in a complex mixture, in cases where the analyst looks for a specific compound. An example is a post-explosion situation, where a certain explosive is suspected to have been used. An ion from the mass spectrum of this explosive is selected, and is subsequently identified by its CID spectrum.

It is possible that in this way the need for clean-up of the debris prior to chromatography can be eliminated.

9.4.6 Explosives Containing Nitro Groups

Any organic compound containing a nitro group can be analysed by mass spectrometry. The most important explosive compounds are the nitroaromatics ($Ar-NO_2$), the nitrate esters ($O-NO_2$) and the nitramines ($N-NO_2$). This section discusses the mass spectra of the most commonly encountered explosives, and explains how the major fragments are formed. Section 9.5 gives examples of how mass spectral information has been applied in casework.

Nitroaromatic explosives

(1) *EIMS* The EIMS of nitroaromatic compounds is characterized by loss of NO_2 radicals, with retention of the charge on the aromatic nucleus (Yinon and Zitrin, 1981: 181–6). The alternative formation of an NO_2^+ ion at m/z 46, which is the major process in nitrate esters (see 9.4.6 Nitrate esters, EIMS), is negligible in nitroaromatic compounds. Thus the EI mass spectra of 1,3,5-trinitrobenzene (TNB) (Meyerson et al., 1972; Volk and Schubert, 1968) and 2,4,6-trinitrotoluene (TNT) (Volk and Schubert, 1968) contain abundant $(M - 3NO_2)^+$ ions at m/z 75 and 89, respectively, but only low abundant ions at m/z 46.

Another process in EIMS of nitroaromatic compounds is the loss of NO from the molecular ion to produce $(M - NO)^+$ ions, especially in compounds like picric acid, where the resulting $(M - NO)^+$ ions are stabilized by resonance (Zitrin and Yinon, 1978). This ion does not appear in the EI mass spectrum of TNT. In TNT, as in most nitroaromatic compounds it is the complementary process – formation of NO^+ ion at m/z 30 – which predominates. The loss of NO must obviously involve some sort of rearrangement of the nitroaromatic $ArNO_2$ to an aromatic nitrite ArONO (Beynon et al., 1964). The EI mass spectrum of TNT (T. Tamiri, unpublished data) is shown in Figure 9.22.

In addition to the $(M - 3NO_2)^+$ ion at m/z 89 [molecular ion minus 3 NO_2 groups; i.e. $227 - (3 \times 46) = 89$] and the NO^+ ion at m/z 30, the base peak appears at m/z 210, corresponding to the loss of an OH radical from the molecular ion.

This loss is an example of an 'ortho effect', where interaction between two ortho substituents causes the ortho isomer to fragment differently from the meta or para isomers. Thus 2-nitrotoluene (ortho-nitrotoluene), having a hydrogen-containing substituent (methyl) in an adjacent position to the nitro group, readily loses OH from its molecular ion while no such loss is observed in the 3 (meta) or 4 (para) nitrotoluenes (Beynon et al., 1964; Carper et al., 1984). Similarly, the dinitrotoluene (DNT) isomers with nitro groups ortho to the methyl (2,3-; 2,4-; 2,5- and 2,6-) show an intense peak at m/z 165, corresponding to $(M - OH)^+$ (see for example Figure 9.20) while the 3,4- and 3,5-DNT do not (Beynon et al., 1964). This process is so pronounced in 2,4,6-TNT, which contains two nitro groups ortho to its methyl group,

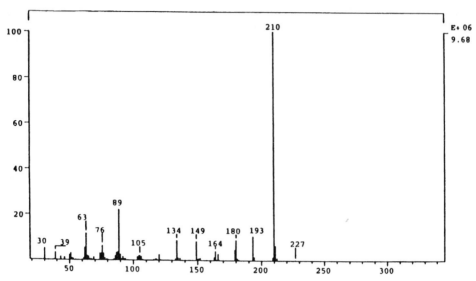

Figure 9.22 EI mass spectrum of TNT.

that the molecular ion of TNT at m/z 227 is hardly observed. The ions at m/z 193, 180 and 164 in the EI mass spectrum of TNT are formed by the loss of a second OH radical, an NO radical and an NO_2 radical, respectively, from the $(M - OH)^+$ ion.

The structures of fragment ions and the fragmentation pathways in the EIMS of TNT were extensively studied by high-resolution mass spectrometry (Zitrin and Yinon, 1978; Carper et al., 1984), metastable scanning of precursors (Bulusu and Axenrod, 1979) and MS/MS techniques (Carper et al., 1984; McLuckey et al., 1985; Yinon, 1987; Yinon et al., 1991b). Some of these studies (Carper et al., 1984; Bulusu and Axenrod, 1979) employed isotopic labelling of TNT.

MS/MS was employed in studies of EI fragmentation pathways of several nitroaromatic compounds which are related to explosives such as aminonitrobenzenes (Yinon, 1990) and dinitroaromatic compounds (Yinon, 1992).

The low-pressure negative-ion mass spectra of TNT (Yinon and Boettger, 1972/73), as well as of other nitroaromatic compounds (Brown and Weber, 1970), using low-energy electrons (10 eV or less), gave rise to a highly abundant NO_2^- ion as the base peak, in contrast to the positive-ion EI/MS of nitroaromatics, where the abundance of the NO_2^+ ion is low.

(2) *CIMS* The CI mass spectra of nitroaromatic explosives (Gillis et al., 1974; Saferstein et al., 1975; Zitrin and Yinon, 1976a, 1976b; Pate and Mach, 1978) contain intense MH^+ ions and usually very few fragment ions. TNT, whose EI mass spectrum does not contain a molecular ion, has an abundant MH^+ ion in its CI spectrum, at m/z 228 (Figure 9.23).

CI mass spectra of nitroaromatic compounds have often included ions which corresponded to loss of 30 mass units from the MH^+ ions. Their existence and relative abundances depended considerably on the type of the reagent gas and on experimental parameters such as the type of instrument, the ion source temperature and the presence of water in the system (Harrison and Kallury, 1980). At first they had been mistakenly assumed (Gillis et al., 1974; Zitrin and Yinon, 1976b) to be formed by loss of NO from the MH^+ ions. It was later found (Maquestiau et al., 1979; Brophy et al., 1979) that their formation involved a reduction of the nitroaromatic compound to an aromatic amine:

$$ArNO_2 \rightarrow ArNH_2$$

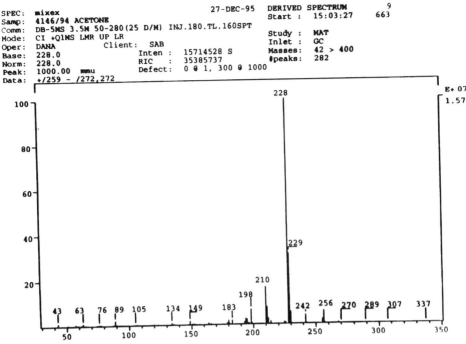

Figure 9.23 CI mass spectrum of TNT, using methane as reagent gas.

These protonated amines were especially abundant when water was used as the reagent gas (Yinon and Laschever, 1981). The ion $(MH - 30)^+$ in the CI mass spectrum of TNT when GC/MS (Gielsdorf, 1981; Weinberg and Hsu, 1983) and LC/MS (Yinon, 1983; Beberich et al., 1988) were used was probably due to this reduction process.

The NCI mass spectrum of TNT, using isobutane as reagent gas was carried out at source pressures up to 0.4 torr (Yinon, 1980). The abundant ions were $(M - H)^-$ at m/z 226 (base peak), the molecular anion $M^{\cdot -}$ at m/z 227 and the $(M - OH)^-$ ion at m/z 210. The detection limit of TNT by NCI, monitoring the molecular anion, was found to be below the nanogram range, while in positive-ion CI the detection limit (monitoring the MH^+ ion) was a few nanograms, which was higher by more than an order of magnitude (Lee et al., 1988). Negative ions of nitroaromatic compounds were formed by API using corona-discharge ionization (Asselin and Pare, 1981) and glow-discharge ionization (McLuckey et al., 1988; Eckenrode et al., 1990; Yinon, 1990, 1992). The details of these ionization techniques are described elsewhere (Yinon and Zitrin, 1993: 113–17).

Using ambient air, in which $O_2^{\cdot -}$ acted as a major reactant ion, significant differences were observed between 2,4-DNT and other DNT isomers (Asselin and Pare, 1981): 2,4-DNT was the only isomer which produced an intense $(M - H)^-$ ion (at m/z 181) while in the other isomers $M^{\cdot -}$ (at m/z 182) was the major ion in the high mass region. This was explained by the acidity of the hydrogens in the methyl group of the 2,4-isomer (where the two nitro groups in the ortho and para position act most efficiently as electron-withdrawing groups), which enabled proton transfer to the basic $O_2^{\cdot -}$ reagent ion. In the other isomers the hydrogens are not acidic enough so the competing reaction of electron transfer from the $O_2^{\cdot -}$ ion predominates. In 2,6-DNT, it is a steric hindrance which interferes with the resonance effect of the nitro groups, diminishing the acidity of the methyl hydrogens. The picture in TNT, where the methyl is also flanked by two nitro groups, was similar: $M^{\cdot -}$ (at m/z 227) was most abundant when $O_2^{\cdot -}$ was the major reagent ion (McLuckey et al., 1988; Eckenrode et al.,

1990) while with NO_2^- as the reagent ion, the base peak was the $(M - H)^-$ ion at m/z 226 (Eckenrode et al., 1990).

MS/MS was employed in studies of fragmentation pathways of hexanitrostilbene in the CI and NCI modes (Yinon et al., 1991a).

Nitrate esters

(1) EIMS The EI mass spectra of nitrate ester explosives such as EGDN, NG (Figure 9.24) and PETN are very similar, containing the ions NO_2^+ at m/z 46 (base peak), NO^+ at m/z 30 and $CH_2ONO_2^+$ at m/z 76 (Yinon and Zitrin, 1981: 187–8). The formation of an abundant NO_2^+ ion is in contrast to the EIMS of nitroaromatic compounds, in which formation of NO_2^+ is negligible and the charge is retained on the complementary ion $(M - NO_2)^+$.

The similarity in the EI mass spectra of nitrate esters, combined with the fact that molecular ions are absent and the ions are in the low-mass region (Fraser and Paul, 1968) does not permit reliance on EIMS only for the identification of a specific nitrate ester.

Using GC/MS, the differentiation between nitrate esters can be based on their different GC retention times but the recommended approach is to analyse the nitrate ester by GC/MS in both ionization modes: EI and CI (Zitrin, 1986; Tamiri and Zitrin, 1986).

(2) CIMS Unlike their EI mass spectra, the CI mass spectra of nitrate esters (Yinon, 1974; Gillis et al., 1974; Pate and Mach, 1978; Tamiri and Zitrin, 1986) include peaks in the molecular weight region, corresponding to MH^+ and $(MH - HONO_2)^+$ ions. Cellulose nitrate ('nitrocellulose', NC) is an exception: being an involatile polymer, its mass spectrum cannot

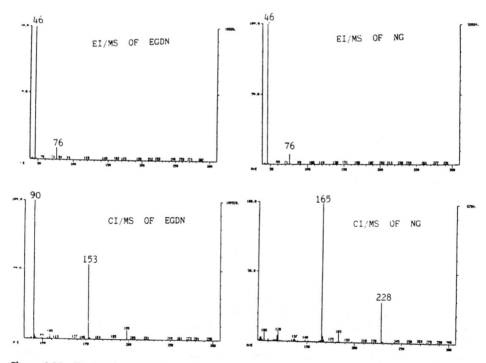

Figure 9.24 EIMS and CIMS (using methane as reagent gas) of NG and EGDN.

be obtained under normal conditions. However, a CI mass spectrum of NC containing some structurally-related ions was reported (Chen and Campbell, 1989), using a direct-evaporation probe mass spectrometric technique (DEMPS).

Figure 9.24 (Tamiri and Zitrin, 1986) shows the EI and CI mass spectra of EGDN and NG. The similarity between the EI spectra is clearly seen, while the CI spectra are very different.

Hydrolysis of polynitrate esters to lower nitrate esters is an important phenomenon which, if unrecognized, can interfere with interpretation. Hydrolysis products of NG (Helie-Calmet and Forestier, 1979; Tamiri et al., 1993) and PETN (Tamiri et al., 1993; Basch et al., 1986; Bamberger et al., 1989) have been observed in post-explosion extracts. These included the mono- and dinitrates of glycerol and the mono-, di- and trinitrates of pentaerythrytol. The identification in post-explosion samples was based on EIMS and CIMS of the free esters (Basch et al., 1986) and of their silylated derivatives (Tamiri et al., 1993), as well as on TLC (Helie-Calmet and Forestier, 1979; Basch et al., 1986) and NMR (Basch et al., 1986; Bamberger et al., 1989). This hydrolysis appeared as an artifact in the CI analysis of nitrate esters (Yinon, 1980b; Yelton, 1982), especially in LC/MS (Beberich et al., 1988; Yinon and Hwang, 1983; Voyksner and Yinon, 1986), where the aqueous LC eluent served as CI reagent gas in the subsequent on-line mass spectrometric analysis. Thus, the highly abundant ion at m/z 183, which appeared in the CI spectrum of NG (Yinon, 1980b; Yinon and Hwang, 1983), was most probably the MH^+ ion of glycerol dinitrate and not, as assumed, a rearrangement product of the MH^+ of NG. The CI-ammonia spectrum of PETN included, in addition to the $(M + NH_4)^+$ ion of PETN, the $(M + NH_4)^+$ ions of the lower nitrate esters of pentaerythrytol (Yelton, 1982). Using an LC eluent containing ammonium acetate as buffer, the NCI mass spectrum of PETN was recorded (Beberich et al., 1988). PETN produced two major peaks at m/z 315 and 375, corresponding to $(M - H)^-$ and $(M + CH_3COO)^-$, respectively. Other peaks, at m/z 330, 285, 240 and 195, which were assumed to be complex rearrangement products of PETN, were most probably due to the $(M + CH_3COO)^-$ ions of pentaerythrytol trinitrate, dinitrate, mononitrate and pentaerythrytol itself, respectively. The structural formulae of pentaerythrytol, **3**, and its nitrate esters: **4** (mononitrate), **5** (dinitrate), **6** (trinitrate) and **7** (PETN), together with their molecular weights are shown below.

The CI mass spectra of nitrate ester explosives such as EGDN, NG and PETN often include ions at m/z values $(M + 30)$ and $(M + 46)$, corresponding to $(M + NO)^+$ and $(M + NO_2)^+$ respectively (Gillis et al., 1974; Yinon, 1974; Saferstein et al., 1975; Pate and Mach, 1978; Gielsdorf, 1981; Tamiri and Zitrin, 1986; Voyksner and Yinon, 1986; Chen and Campbell, 1989). These unusual adduct ions are probably formed by ion–molecule reactions

3 (M=136)   **4** (M=181)   **5** (M=226)

6 (M=271)   **7** (M=316)

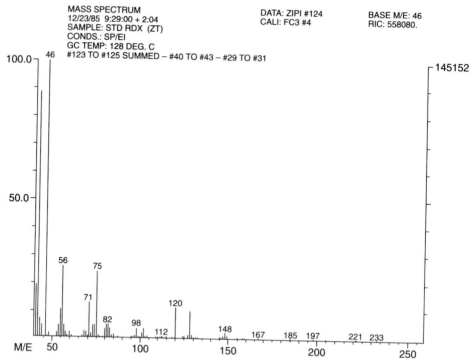

Figure 9.25 EIMS of RDX.

between the NO^+ and NO_2^+ ions and the neutral molecules of the explosives (Pate and Mach, 1978). Their formation therefore depends on the sample pressure, causing their relative abundances in the CI mass spectrum to be highly irreproducible. Sometimes, with high sample pressures in the ion source (especially when using a relatively 'tight' ion source, such as dual EI/CI source) these adduct ions are abundant even in the EIMS of nitrate ester explosives (Lee et al., 1989).

The NCI mass spectra of nitrate esters contain two major ions in the low-mass region (Yinon, 1980a; Bouma and Jennings, 1981; Ottoila et al., 1982; Parker et al., 1982; Lee et al., 1989): ONO_2^- at m/z 62 and NO_2^- at m/z 46. The molecular weight region (Yinon, 1980a; Bouma and Jennings, 1981; Voyksner and Yinon, 1986; Beberich et al, 1988; Lee et al., 1989) is characterized by $M^{\cdot-}$ and $(M + ONO_2)^-$ ions. Other adduct ions, containing species from the CI reagent gas or (in LC/MS) from the LC mobile phase, can also be observed (Bouma and Jennings, 1981; Yinon and Hwang, 1983; Voyksner and Yinon, 1986; Beberich et al., 1988; Lee et al., 1989).

Nitramines

(1) EIMS The EIMS of the two heterocyclic nitramines, RDX and HMX, have been extensively studied (Volk and Schubert, 1968; Bulusu et al., 1970; Stals, 1971; Gielsdorf, 1981; Inoue et al., 1989), including the use of high resolution (Bulusu et al., 1970; Yinon et al., 1982), isotopic labelling (Bulusu et al., 1970) and metastable transitions (Stals, 1971). Collision-induced dissociation (CID) was also used to study the fragmentation pathways in

RDX and HMX (Yinon et al., 1982; McLuckey et al., 1985; Yinon and Zitrin, 1993) and some other heterocyclic nitramines (Yinon et al., 1990) under EI conditions.

The EI mass spectra of RDX (Figure 9.25) and HMX (Figure 9.26) are very similar; in both spectra the molecular ions cannot be observed and the most abundant ions are the NO_2^+ ion at m/z 46 and the NO^+ ion at m/z 30. More diagnostic ions appear at m/z 120, 128 and 148. Based on high-resolution measurements (Bulusu et al., 1970; Yinon et al., 1982), the elemental composition of the ion at m/z 120 was found to be $CH_2N_3O_4$, which corresponded to $CH_2N(NO_2)_2^+$. In order to explain the presence of two nitro groups in one CH_2N moiety (a situation which does not exist in the molecule of RDX), an unusual migration of a nitro group had to be assumed (Bulusu et al., 1970). CID studies (Yinon et al., 1982) suggested another possibility: the precursor ion of the ion at m/z 120 was, at least partly, the adduct ion $(M + NO_2)^+$ at m/z 268. As mentioned earlier and below, $(M + NO)^+$ and $(M + NO_2)^+$ adduct ions were observed in CIMS of nitrate esters and nitramines Their appearance in the EI mass spectra of RDX and HMX is somewhat unusual, and could be the result of a high sample pressure in the ion source.

(2) CIMS   The CI mass spectra of RDX have been recorded, using a variety of reagent gases (Gillis et al., 1974; Yinon, 1974, 1980b; Saferstein et al., 1975; Zitrin and Yinon, 1976; Vouros et al., 1977; Pate and Mach, 1978; Zitrin, 1982; Chen, 1993). The spectra usually contain a low abundant MH^+ ion at m/z 223, adducts ions related to the reagent gas, such as the $(M + NH_4)^+$ ion in the $CI - NH_3$ spectrum (Vouros et al., 1977) or $(M + H_3O)^+$ ion in the $CI - H_2O$ spectrum (Yinon, 1974) and sometimes $(M + NO)^+$ ions (Gillis et al., 1974; Zitrin and Yinon, 1976a; Yinon, 1980b). $(M + NO_2)^+$ ions were reported in CID studies of the CIMS of RDX (McLuckey et al., 1985; Yinon et al., 1982). It is well established in CIMS that different reagent gases, according to the acid strength of their protonating agents, influence

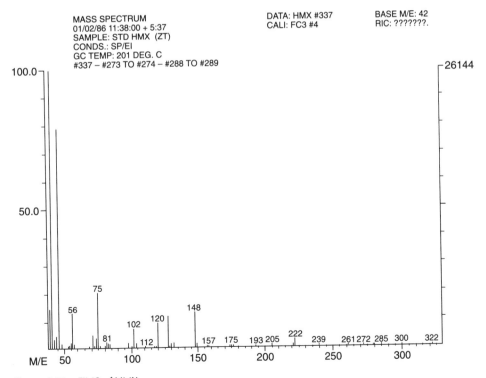

Figure 9.26   EIMS of HMX.

the extent of fragmentation of the MH^+ ions (see 9.4.1), but normally the same fragments ions are produced. This is not the case with RDX, where different fragment ions are produced with different reagent gases (Zitrin, 1982). Using gases such as ammonia or isobutane, whose reactant ions are weak Brønsted acids (NH_4^+ and $C_4H_9^+$, respectively), a set of fragment ions at *m/z* 84, 131 and 176 is observed. Using hydrogen or methane, whose reactant ions are strong Brønsted acids (H_3^+ and CH_5^+), the major ions appear at *m/z* 75 and 149, while the former set of ions is completely absent. The appearance of these two sets of ions was explained in terms of the energetics involved (Zitrin, 1982). Protonation of RDX by weak Brønsted acids (NH_4^+ and $C_4H_9^+$) is accompanied by transfer of relatively small amounts of energy. The resulting MH^+ ions would not have enough energy for simple cleavage reactions (leading to the ions at *m/z* 75 and 149) but only for rearrangement reactions (leading to the ions at *m/z* 84, 131 and 176). Protonation of RDX by strong Brønsted acids (H_3^+ and CH_5^+) is accompanied by transfer of enough energy for the occurrence of simple cleavage reactions. The CI mass spectra of RDX with methane and isobutane as reagent gases are shown in Figure 9.27.

The CI mass spectra of HMX (Yinon, 1974; Gillis et al., 1974; Saferstein et al., 1975; Vouros et al., 1977; Gielsdorf, 1981; Moncur and Sharp, 1982) usually included MH^+ ions at *m/z* 297, whose abundance varied according to the experimental conditions. Using ammonia as reagent gas (Vouros et al., 1977; Moncur and Sharp, 1982) the base peak appeared at *m/z* 314, corresponding to the $(M + NH_4)^+$ adduct ion.

HMX is highly involatile so its mass spectra were recorded by using a direct probe (Volk and Schubert, 1968; Gillis et al., 1974; Yinon, 1974; Saferstein et al., 1975; Zitrin and Yinon, 1976a; Vouros et al., 1977; Gielsdorf, 1981; Moncur and Sharp, 1982; Parker et al., 1982; Chen, 1993) or LC/MS (Voyksner and Yinon, 1986; Beberich et al., 1988; Chen, 1993) but not GC/MS, where the elution of HMX through the GC column was unsuccessful (Gielsdorf, 1981; Zitrin, 1986). The analysis of HMX by capillary column GC was reported (Douse,

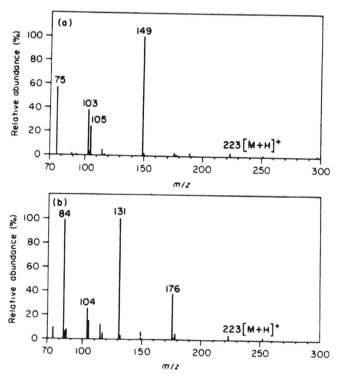

Figure 9.27 CIMS of RDX, using methane and isobutane as reagent gases.

1981; Hable et al., 1991), but no data were given as to the identity of the eluted compound, so a decomposition product could not be excluded. The identification of the eluted compound is probably not straightforward because of the similarity between the mass spectra of RDX and HMX. Even in CI, where different MH^+ ions are expected, the results may not be unequivocal due to formation of cluster ions of the general formula $(CH_2NNO_2)_n H^+$ from both RDX (Yinon, 1974, 1980b; Doyle and Campana, 1985) and HMX (Yinon, 1974; Gillis et al., 1974; Gielsdorf, 1981).

NCI mass spectra of RDX and HMX, using isobutane as a moderator gas were recorded (Yinon, 1980b; McLuckey et al., 1985). $(M + 102)^-$ ions, which appeared in the spectra of both nitramines (McLuckey et al., 1985), could be due to loss of NO_2 from the $(M + 2CH_2NNO_2)^-$ adduct ions. NO_2^- ions at m/z 46 (Yinon, 1980; McLuckey et al., 1985), $H(NO_2)_2^-$ ions at m/z 93 (Yinon, 1980) and many other fragment ions and adduct ions were also observed under these conditions (Yinon, 1980; McLuckey et al., 1985).

Using LC/MS with the LC eluent containing ammonium acetate as a buffer, the NCI spectra of RDX and HMX were characterized by major $(M + CH_3COO)^-$ adduct ions, at m/z 281 and 355, respectively (Voyksner and Yinon, 1986; Berberich et al., 1988). In another LC/MS analysis of RDX and HMX (Chen, 1993), their NCI spectra were dominated by ions at m/z 102, probably $(CH_2NNO_2CH_2N)^-$ and included also the adduct ions $(M + NO)^-$, $(M + NO_2)^-$ and the $(M + 102)^-$ ion mentioned above.

CID studies of RDX and HMX in the CI mode, using isobutane (McLuckey et al., 1985) and methane (Yinon et al., 1982) as reagent gases, were reported. The precursors of some fragment ions in the CI spectra of RDX and HMX were found to be – in addition to the MH^+ ions – the adduct ions $(M + NO)^+$ and $(M + NO_2)^+$ (Yinon et al., 1982; McLuckey et al., 1985).

Tetryl, a trinitroaromatic explosive with a nitramine group, produced different mass spectra according to the way in which the sample was introduced into the mass spectrometer. Using a direct insertion probe, the EI spectrum (Volk and Schubert, 1968; Zitrin and Yinon, 1976, 1978) had its base peak at m/z 241, corresponding to the $(M - NO_2)^+$ ion. This ion was completely absent in the GC/MS of tetryl, where the EI mass spectrum included a highly abundant ion at m/z 242 (Gielsdorf, 1981; Tamiri and Zitrin, 1986; Inoue et al., 1989). This phenomenon was explained (Tamiri and Zitrin, 1986) by the hydrolytic decomposition of tetryl, **8**, during its GC analysis, producing *N*-methylpicramide, **9** (first published in *Journal of Energetic Materials*, **4**, 215–237).

This hydrolysis accounts for the ion at m/z 243 in the CI mass spectrum of tetryl (Saferstein et al., 1975; Zitrin and Yinon, 1976, 1978; Yinon, 1980b; Gielsdorf, 1981; Tamiri and Zitrin, 1986) which corresponds to the MH^+ ion of *N*-methylpicramide (Tamiri and Zitrin, 1986). This is supported by high-resolution measurements (Rowley, 1989), CID studies

8   **9**

(Yinon et al., 1993) and the use of CD_4 as a reagent gas (Yinon et al., 1993). It should be noted that under CI conditions the hydrolysis of tetryl occurs, at least partly, also when the sample is introduced directly into the mass spectrometer (Saferstein et al., 1975; Zitrin and Yinon, 1976, 1978; Pate and Mach, 1978; Yinon, 1980) although the MH^+ ion of tetryl, at m/z 288 can also be observed (Saferstein et al., 1975; Zitrin and Yinon, 1978; Yinon, 1980). Obviously, in GC/MS the hydrolysis is complete and only the CI mass spectrum of *N*-methylpicramide is obtained (Gielsdorf, 1981). The ion at m/z 212, observed in some NCI spectra of tetryl (Gielsdorf, 1981; Parker et al., 1982), could thus be formed from *N*-methylpicramide by the well known reduction of nitroaromatic compounds (see 9.4.6 'Nitro-aromatic explosives, CIMS').

9.4.7 Organic Peroxides

Along with IR (see 9.3.3) and NMR (Zitrin et al., 1983; Reutter et al., 1983), mass spectro-metry played an important role in the identification of organic peroxides when first encoun-tered in a forensic laboratory (Zitrin et al., 1983). The mass spectrometry of organic peroxides has been extensively reviewed (Schwarz and Schiebel, 1983).

Triacetonetriperoxide (TATP)

The EI mass spectrum of triacetonetriperoxide (TATP), **1** (see 9.3.2), is hardly informative, being very similar to the EI mass spectrum of acetone. Its CI spectrum, however, using iso-butane (Zitrin et al., 1983) or methane (Evans et al., 1986; White, 1992) as reagent gases, produced a distinct peak at m/z 223, corresponding to the MH^+ ion of TATP. In addition, the CI-isobutane spectrum contained a series of low abundant ions which differed by 16 mass units, corresponding to the mass of an oxygen atom. When first analysed as 'completely unknown' (Zitrin et al., 1983), this series of ions had been the first clue that a peroxide might be involved.

Hexamethylenetriperoxide diamine (HMTD)

The EI mass spectrum of hexamethylenetriperoxide diamine (HMTD), (structure **2**, 9.3.2), contains a molecular ion at m/z 208 and fragment ions at m/z 176 and 88, corresponding to $(M - O_2)^{\cdot +}$ and $N(CH_2)_3O_2^+$ respectively (Zitrin et al., 1983; Reutter et al., 1983). The CI mass spectra of HMTD, using isobutane (Zitrin et al., 1983) or methane (Reutter et al., 1983) as reagent gases, had their base peak at m/z 209, which corresponded to the MH^+ ion.

9.4.8 Some Case Reports and Analyses Based on Mass Spectrometry

Semtex

Semtex, a Czechoslovakian-made plastic explosive, has been used in many terrorist bombings since the early 1970s (Feraday, 1993). It first appeared in a forensic laboratory in the early 1970s in letter bombs (Zitrin, 1971/72). Its explosive components, RDX and PETN, were then identified by IR spectrometry and TLC.

Subsequent mass spectrometric analysis by direct-insertion probe CIMS, used isobutane (Zitrin and Yinon, 1976) and water (Yinon, 1974) as reagent gases. The CIMS of Semtex, using isobutane as a reagent gas was studied by MS/MS (Yinon, 1988).

Both RDX and PETN have relatively low vapour pressures (Yinon and Zitrin, 1981: 243; Yinon and Zitrin, 1993: 5, 10). The positive response of some Semtex samples to explosive vapour detectors based on electron-capture detection (ECD) was therefore studied, using

GC/MS, by Hobbs and Conde (1989), who attributed the positive responses to the unexpected presence of EGDN. Slack et al. (1992) identified EGDN together with some hydrocarbons in other Semtex samples which had been extracted by supercritical fluid extraction (SFE) and then analysed by GC/MS. They concluded that the EGDN was responsible for the positive response to Semtex of an ECD-based explosives 'sniffing' device.

GC/MS was used for the identification of some non-explosive components in Semtex explosives (Hobbs, 1993). These included two plasticizers, bis(2-ethylhexyl)phthalate (dioctyl phthalate; DOP) and butyl citrate, an antioxidant (N-phenyl-2-naphthylamine) and the dyes Sudan I (in Semtex H) and Sudan IV (in Semtex A).

Colour tests

The use of field tests as preliminary presumptive tests for explosives is well established, but so is the caution which should be exercised in interpreting positive results. A field test based on the Griess test was crucial in the 1975 conviction of the 'Birmingham Six', which was overturned in 1991 (Scaplehorn, 1993). It is now generally accepted that positive results of field tests based on colour reactions should be subsequently confirmed by reliable laboratory methods such as GC/MS.

An unknown powder, tested by an explosives-testing kit ('ETK') (Almog et al., 1986), produced a positive result in a reaction between polynitroaromatic compounds and alkalis (Yinon and Zitrin, 1981: 30–6). Using TLC, IR, GC/MS (EI and CI modes) and NMR, the powder was later identified in the laboratory as 1-*tert*-butyl-2-methoxy-3,5-dinitro-4-methylbenzene (Bamberger et al., 1989). This compound, known as 'musk ambrette' is one of several nitroaromatic compounds used as artificial musk fragrances in perfumes, soaps and cosmetic preparations. In its analysis, the CI mass spectrum indicated a molecular weight of 268 and the EI mass spectrum matched a previously published spectrum of musk ambrette (Yurawecz and Puma, 1983). The 'nitro musks' are polynitroaromatic compounds, and therefore produce typical colours with alkalis, but not with some organic bases (Douse and Smith, 1986).

Unusual explosive mixtures

Mass spectrometry has been used in the identification of some unusual explosive mixtures.

(1) A mixture of diethylene glycol dinitrate (DEGDN) and metriol trinitrate (MTN; 1,1,1-trimethylolethane trinitrate) was identified in post-explosion debris (Reutter et al., 1983). Due to the rather similar EI mass spectra of nitrate esters, which do not include molecular weight information (see 9.4.6 'Nitrate esters'), the two compounds were identified mainly by their CI mass spectra, using methane as a reagent gas. In CIMS, the MH^+ ions of these compounds were easily observed. The CI mass spectrum of MTN (Figure 9.28, from Reutter et al., 1983), along with CI spectra of other polynitrate explosives, have been published (Lee et al., 1989).

(2) A mixture of potassium chlorate, 2,4-DNT and 2,6-DNT was identified in an old explosive, which had been found in bottles as a yellow granular substance (McDermott, 1994). The $KClO_3$ was identified by spot tests and IR spectrometry, and the dinitrotoluene isomers were identified by GC/MS.

(3) GC/MS was used to identify trace amounts of impurities in crude TNT including DNT and TNT isomers, *p*- and *m*-dinitrobenzenes, 1,3,5-trinitrobenzene and 1-nitronaphthalene (Chang, 1971).

Polynitropolycyclic 'cage' compounds

Much effort has been put in recent years to synthesize polynitropolycyclic 'cage' compounds in order to study their potential as high energy density explosives and propellants

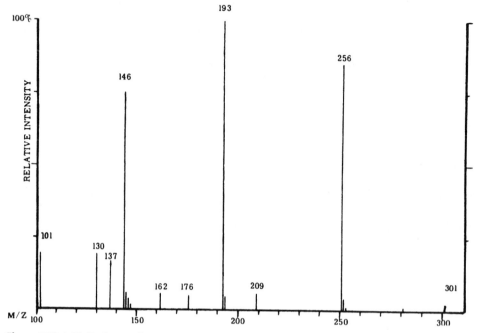

Figure 9.28   CIMS of metriol trinitrate (from Reutter et al. (1983), with permission).

(Marchand, 1993). MS/MS analysis was carried out in order to study fragmentation processes in nitroadamantanes (Yinon and Bulusu, 1986a) and polynitrobishomocubanes (Yinon and Bulusu, 1986b). Direct-exposure probe mass spectrometry (DEPMS) technique in the CI mode was applied to some nitrocage compounds (Chen and Campbell, 1989, 1993).

9.4.9   Stabilizers

Smokeless powders contain stabilizers whose function is to slow the decomposition of the nitrate esters (especially nitrocellulose) and prolong shelf-life. During storage, the NC decomposes, liberating nitrogen oxides which catalyse further decomposition. The stabilizers react with the nitrogen oxides, slowing the decomposition of the nitrate ester. The stabilizers themselves are converted into their nitro- or nitroso- derivatives (Yinon and Zitrin, 1981: 78–80). Common stabilizers are diphenylamine (DPA) or some of its derivatives and substituted ureas (e.g. ethyl and methyl centralites: *N,N'*-diethyl-*N,N'*-diphenylurea, and *N,N'*-dimethyl-*N,N'*-diphenylurea, respectively).

The EI mass spectra of DPA and its nitro- and nitroso- derivatives usually contain highly abundant molecular ions, except for the *N*-nitroso derivatives, where molecular ions are absent (Alm, 1969). The EI spectra of the *N*-nitroso derivatives are very similar to the spectra of the parent compounds. Thus one can distinguish between DPA and its *N*-nitroso derivative only by the presence of the NO^+ ion at *m/z* 30 in the *N*-nitroso derivative. Except this, the spectra are practically identical.

The EI mass spectrum of ethyl centralite (EC) was studied, using deuterium labelling and different electron energies (Hanus et al., 1982). The spectrum is very simple, containing a molecular ion at *m/z* 268 and some fragment ions formed by simple cleavage reactions. The ion at *m/z* 164 was attributed to a rearrangement in which styrene was lost from the molecular ion.

9.4.10 Gunshot Residues

The characterization of particles as gunshot residues is normally done by their elemental composition, analysed by a scanning electron microscope with an energy-dispersive X-ray analyser. The combined presence of certain elements, such as antimony, barium and lead, can characterize a particle as a gunshot-residue particle.

Another, less common approach is to characterize propellant residues on the basis of their organic constituents.

Mach et al. (1978a) analysed 33 smokeless powders using GC/MS in EI and CI modes, for NG, the two stabilizers DPA and EC, the plasticizer dibutyl phthalate (DBP) and the additive 2,4-DNT. The CI mass spectra of DPA and EC, with methane as a reagent gas, contained mainly peaks corresponding to their respective MH^+ ions at m/z 170 and 269. The CI mass spectrum of DBP included peaks corresponding to its MH^+ ion at m/z 279 and to two fragment ions: $(M - C_4H_9OH)^+$ at m/z 205 and the protonated phthalic anhydride at m/z 149. The latter ion is the base peak in the EI mass spectra of phthalate esters (Yinon and Zitrin, 1981: 190). The study was repeated on flakes found on suspects' hands immediately after shooting (Mach et al., 1978b).

IR spectrometry and GC/MS were used to analyse the organic components in an isolated propellant grain (Kee et al., 1990). The involatile NC was identified by its IR spectrum and NG, DPA, EC and DNT were analysed by GC/MS, in the EI mode. In some propellant brands different profiles were observed for unburnt and burnt particles. For example, NG, DPA and EC were identified in an unburnt particle from a certain double-base smokeless powder, but only NG and DPA were identified in burnt particles from the same powder.

Martz and Laswell (1983) set up a library of about 100 smokeless powders based on GC/MS in the EI mode. Each entry was a 'composite' spectrum, formed by artificially summing the mass spectra of the individual organic components such as NG, DPA and EC.

9.4.11 Sugars

Sugars have been used as reducing agents in improvised explosive mixtures, together with oxidizing agents such as potassium chlorate. IR and X-ray diffraction (XRD) are suitable for their identification, but they may not be successful in post-explosion analysis since the residues are often syrups. GC/MS has been employed for the post-explosion identification of sugars, as their trimethylsilyl (TMS) derivatives (Beveridge et al., 1983; Nowicki and Pauling, 1988). EI mass spectra of fructose, dextrose, sucrose, maltose, lactose and mannitol, a related alcohol, have been reported and some of these sugars were identified in aqueous extracts of residues from explosion of their mixture with $KClO_3$ (Nowicki and Pauling, 1988).

9.4.12 Other Reducing Agents (Sulphur, Hydrocarbons)

Other reducing agents which may appear in improvised explosive mixtures are sulphur and hydrocarbons.

Sulphur, a constituent of black powder, was identified by a direct-probe EIMS of a fraction collected from the LC analysis of a smokeless powder (Rudolph and Bender, 1983). Its mass spectrum includes a peak corresponding to the molecular ion of S_8 at m/z 256 (accompanied by the typical 'isotopic' ions) and also peaks corresponding to lower aggregates of sulphur atoms with differences of 32 m/z units between them.

A series of hydrocarbons was analysed by GC/MS in fuel oil which was the reducing component in ammonium nitrate–fuel oil (ANFO) explosive mixture (Rudolph and Bender, 1983). Saturated hydrocarbons from $C_{12}H_{26}$ to $C_{24}H_{50}$ were identified.

9.5 Mass Spectrometry in Post-explosion Analysis

The main problems associated with the post-explosion analysis of explosives are choosing the 'right' exhibits from the scene (i.e. those from which the chances to recover the explosive are best), the minute amounts of the relevant analyte (usually traces of the original explosives which had not been detonated) and the large amounts of contaminants. The complex matrix requires appropriate cleaning procedures prior to the analysis. A comprehensive review of these cleaning methods was published (Yinon and Zitrin, 1993: 170–90). Following the cleaning, usually made by column chromatography on commercially available columns, chromatographic methods such as TLC, GC and LC have been widely used.

Although chromatographic methods are basically intended for separation, the introduction of specific detectors enhances their specificity and enables the analyst to rely on them for the identification of individual organic compounds. The incorporation of the thermal-energy analyser (TEA) in GC and LC (see 9.2.1 and Chapters 8 and 10) makes these methods specific to explosives containing nitro groups. The sensitivity of the TEA detector is often compatible with that required in post-explosion analysis (Yinon and Zitrin, 1993: 190–4). These developments allow many forensic laboratories to identify explosives in post-explosion debris by chromatographic methods only, with a high degree of reliability (Kolla, 1991, 1994). At the same time, the advances in mass spectrometric methods and in their application to analytical problems, make it easier for the analyst to adhere to the criterion that the identification of individual organic compounds should not be based on chromatographic methods only.

GC/MS, LC/MS and MS/MS have the potential to play an important role in post-explosion analysis, possibly by being complementary to each other. Of these methods, GC/MS was reported to have been employed in a large number of real-world cases (Reutter et al., 1983; Basch et al., 1986; Tamiri and Zitrin, 1986; Zitrin, 1986; Beveridge, 1992; Tamiri et al., 1993).

9.5.1 Case Examples

Case 1

The total-ion chromatogram (the normal GC trace based on the total-ion current produced from compounds emerging from the GC column; see 9.4.4) obtained in the GC/MS analysis of an acetone extract from the remains of a bomb which had been detonated on a roadside is shown in Figure 9.29 (Tamiri and Zitrin, 1986). Three chromatographic peaks were identified as TNT, RDX and *N*-methylpicramide, which indicated the original presence of tetryl (see 9.4.6 'Nitramines').

Case 2

In many post-explosion analyses it is difficult to find the peaks originating from the explosives in the total-ion chromatogram. When there is an *a-priori* indication – often based on TLC – about the suspected explosive, then the mass chromatogram is very helpful (see 9.4.4). The advantage in using the mass chromatogram to locate the relevant peak in the total-ion chromatogram is demonstrated in Figure 9.30 (Tamiri and Zitrin, 1986), which shows the results of a GC/MS analysis from an actual case. The peak attributed to RDX can hardly be observed in the total-ion chromatogram, but the mass chromatograms of the ions at m/z 30 and 46 located its position. RDX was then identified by its EI mass spectrum.

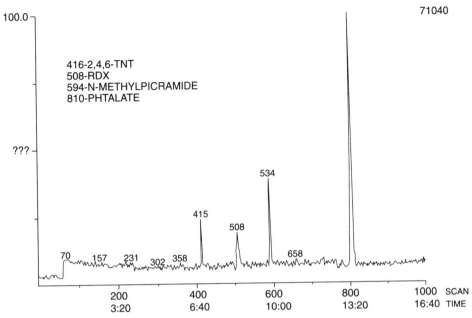

416-2,4,6-TNT
508-RDX
594-N-METHYLPICRAMIDE
810-PHTALATE

Figure 9.29 The total-ion chromatogram in the GC/MS analysis of an acetone extract from the remains of a bomb which had detonated on a roadside.

Case 3

The elution of PETN under normal GC conditions was reported to occur with loss of sensitivity (Kolla, 1991, 1994). In addition, PETN, as well as NG, often decomposed to lower nitrate esters during the explosion process (Basch et al., 1986; Tamiri et al., 1993). The use of very short capillary columns and relatively low injector and transfer-line temperatures (Tamiri et al., 1993) minimizes the loss of sensitivity.

The lower part of Figure 9.31 (Tamiri et al., 1993) shows the total-ion and mass chromatograms of a post-explosion extract from an actual case, where the GC/MS analysis (in the CI mode, with methane as a reagent gas) was carried out using the 'gentle' GC conditions mentioned above. The mass chromatograms, recorded at m/z 317 and 254, corresponding to the MH^+ and $(MH - HONO_2^+)$ ions of PETN in CIMS, respectively, allowed to single out the chromatographic peak of PETN. Locating this chromatographic peak in the total-ion chromatogram would have been practically impossible without the mass chromatograms. The upper part of Figure 9.31 shows the resulting CI mass spectrum of the PETN chromatographic peak.

Case 4

Although the formation of hydrolysis products (see 9.4.6 'Nitrate esters') complicates the chromatographic profiles in the post-explosion analysis of PETN and NG, a procedure which takes advantage of their formation was suggested (Tamiri et al., 1993). As the hydrolysis products contain free hydroxyl groups, they can be easily converted to their trimethylsilyl (TMS) derivatives. These derivatives can then be successfully analysed by GC/MS.

In one post-explosion case (Tamiri et al., 1993), the TLC analysis showed several Griess-positive spots, in addition to the spot corresponding to PETN. The organic extract was

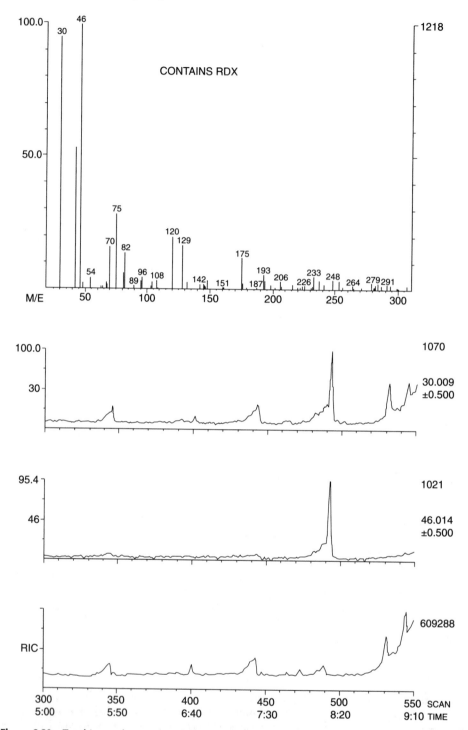

Figure 9.30 Total-ion and mass chromatograms of a post-explosion extract and EI mass spectrum of the peak emerging after 492 seconds, identified as RDX.

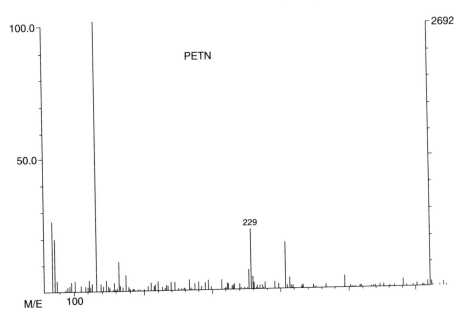

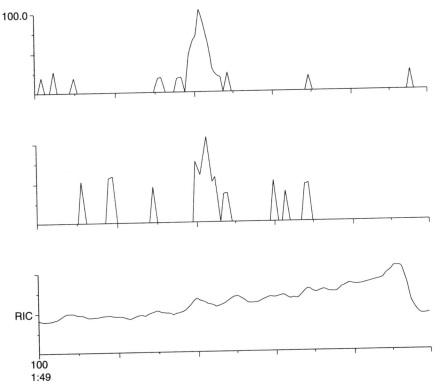

Figure 9.31 Total-ion and mass chromatograms and CI mass spectrum, using methane as the reagent gas of an extract from an actual case involving PETN.

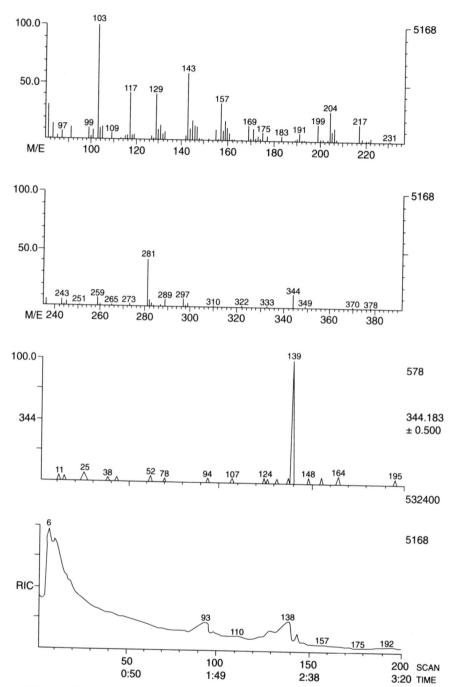

Figure 9.32 Total-ion and mass chromatogram and EI mass spectrum of a silylated extract from an actual case involving PETN.

silylated and then analysed by GC/MS. Figure 9.32 shows the total-ion and mass chromatogram and the CI mass spectrum of this silylated extract. The TMS derivative of pentaerythritol trinitrate (see 9.4.6, structural formula **6**) was identified. This approach was found to be superior to a previously reported one (Basch et al., 1986), where the hydrolysis products were collected by preparative LC, and the collected LC fractions were analysed by CIMS, using a direct-insertion probe and methane as the reagent gas. Naturally, on-line LC/MS of the post-explosion extract could also be employed.

Case 5

Beveridge (1991) used *selective ion monitoring* of three abundant ions in the CIMS of EGDN (m/z 153, 90, 46) and NG (m/z 228, 165, 46) using methane reagent gas to confirm their identity in 5-year old exhibits. This technique, in which the mass spectrometer is set to measure only specific masses, is well suited to trace sample analysis where contamination may prevent obtaining a total-ion chromatogram.

In addition to GC/MS and LC/MS, tandem mass spectrometry (MS/MS) may be potentially suitable in certain situations of post-explosion analysis (Zitrin, 1986). If the presence of a specific explosive is suspected (e.g. following a TLC analysis) then ions known to be formed from the suspected explosive are selected after their separation in the first mass spectrometer. They are allowed to enter the collisionally-induced activation (CID) cell where collisions with a neutral gas (e.g. argon) take place. The resulting ions are then analysed by the second mass spectrometer and the spectra obtained may characterize the explosive. In theory, such a procedure could save tedious cleaning of the post-explosion extract. The method, which requires expensive instrumentation, has yet to be tested in actual cases.

9.6 Inorganic Ions

The trace analysis of inorganic ions, especially anions, is important in cases where commercial (e.g. ammonium nitrate, dynamites) or home-made improvised explosives (mainly mixtures of oxidizing anions with reducing agents) are involved. Spot tests (Yinon and Zitrin, 1981: ch. 2), ion chromatography (Yinon and Zitrin, 1993: ch. 2, sect. 2.5) and capillary electrophoresis (McCord and Hargadon, 1993), have been employed for the identification of trace amounts of these anions.

IR spectrometry and XRD, though included in some schemes of post-explosion analysis (Beveridge et al., 1975, 1983; Rudolph and Bender, 1983) are not sensitive enough when only traces of the anions are present. Since GC/MS cannot be applied directly to inorganic anions, alternative methods for their post-explosion analysis are spot tests or chromatographic methods. A possible way to rectify this would be to find a way to apply GC/MS to these anions.

An attractive idea is to incorporate the inorganic ions into organic molecules, which then would be identified by GC/MS. A method to identify trace amounts of nitrate ions in water was based on nitration of an activated aromatic ring, followed by GC/MS analysis of the nitroaromatic product (Tan, 1977). As nitration of an aromatic ring involves an electrophilic attack by the nitronium ion NO_2^+, an activated aromatic ring should include electron-donating groups. 1,3,5-trimethoxybenzene, which was chosen as the aromatic compound, facilitated the nitration to such an extent that it could be carried out quantitatively under relatively mild conditions: the medium was sulphuric acid–water (2:1) and the reaction was carried out at room temperature. Surprisingly, the reported product was not 2,4,6-trimethoxynitrobenzene but nitrobenzene itself, which was identified by GC/MS, in both EI and CI modes. No convincing explanation was given for this unexpected product.

Nitrate ions, after being converted to NO_2^+ ions, were identified in post-explosion debris from actual cases, by their reaction with *t*-butylbenzene (Glattstein et al., 1995). Mono- and dinitro- derivatives of *t*-butylbenzene were identified by GC/MS. Naturally, this method is not suitable for other anions which are not capable of an electrophilic attack on an aromatic ring.

Another approach is to use a reagent with a functional group which can be replaced by the inorganic anion through nucleophilic substitution. A method based on this approach was developed for a GC analysis of the substitution products of pentafluorobenzyl bromide (Chen et al., 1987, 1990). Among the ions analysed were cyanide, nitrite, sulphide and thiocyanate. The same approach, using *n*-butyl *p*-toluenesulphonate (Funazo et al., 1985a) and pentafluorobenzyl *p*-toluenesulphonate (Funazo et al., 1985b) as derivatizing agents, was used successfully for the GC analysis of nitrates (in addition to the above mentioned ions). In all these analyses, the inorganic anions had to be transferred from their natural aqueous solutions into a non-aqueous phase, where the substitution reactions took place. This was done by various phase-transfer catalysts which included quarternary ammonium salts (Chen et al., 1987; Funazo et al., 1985a, 1985b) or the commercially available polymer Kryptofix 222 B (Chen et al., 1990).

The phase-transfer technique was also used to selectively extract oxidative anions present in explosive mixtures (e.g. nitrate, chlorate) into an organic phase, prior to their analysis by spot tests and IR spectrometry (Glattstein and Kraus, 1986).

The method of nucleophilic substitution of pentafluorobenzyl bromide by anions present in the aqueous extract of post-explosion debris was used in some real-world cases (Glattstein et al., 1995). The anions were transferred into the organic phase by phase-transfer, using a quarternary ammonium salt as the phase-transfer catalyst. Figure 9.33 shows the EI mass spectrum of pentafluorobenzyl thiocyanate formed in an explosion involving black powder. The peak at *m/z* 239 corresponds to the molecular ion and is accompanied by the expected 'isotopic' ions due to the presence of sulphur. Thiocyanate is a known product of the burning of black powder. IR spectrometry had been used in its post-explosion analysis (Beveridge et al., 1975), but this was successful only in a limited number of cases, where the amounts of the thiocyanate were not too small.

The possibility of using GC/MS to identify inorganic ions in post-explosion cases may be considered a breakthrough in post-explosion analysis of commercial and improvised explo-

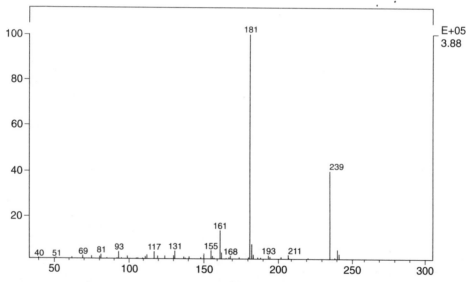

Figure 9.33 EI mass spectrum of pentafluorobenzyl thiocyanate, identified in an actual case involving black powder.

sives. It should be noted that the pentafluorobenzyl derivatives had been chosen mainly because their high response to electron-capture detection in GC. Other reagents may be tried if the analysis of the substitution products is to be carried out by GC/MS.

Acknowledgements

I am grateful to the following for permission to publish spectra from unpublished collections: Sara Abramovitch-Bar, Israeli National Police, for Figures 9.1, 9.2, 9.5, 9.7, 9.8, 9.9, 9.13 through 9.19; Tsippy Tamiri, Israel National Police, for figures 9.20 to 9.26; Sandy Beveridge, Royal Canadian Mounted Police, for Figures 9.3, 9.4 and 9.6, and Mary Tungol, Federal Bureau of Investigation, for Figure 9.10.

References

ALM, A., 1969, The mass spectra of diphenylamine compounds with nitro and nitroso substituents and of tetraphenyl hydrazine, *Explosivstoffe*, **17**, 156–164.

ALM, A., DALMAN, O., FROLEN-LINDGREN, I., HULTEN, F., KARLSSON, T. and KOWALSKA, M., 1978, Analysis of explosives, *FOA Report C 20267-D1*, National Defence Research Institute, S-104 50, Stockholm.

ALMOG, J., KRAUS, S. and GLATTSTEIN, B., 1986, ETK – an operational explosive testing kit, *Journal of Energetic Materials*, **4**, 159–165.

ASSELIN, M. J. F. and PARE, J. J. R., 1981, Atmospheric pressure chemical ionization negative mass spectra of the dinitrotoluene isomers, *Organic Mass Spectrometry*, **16**, 275–278.

BAMBERGER, Y., LEVY, S., TAMIRI, T. and ZITRIN, S., 1989, The identification of musk ambrette during a routine test for explosives, *Proceedings 3rd International Symposium on Analysis and Detection of Explosives*, pp. 4.1–4.9, Berghausen, Germany: Fraunhofer Institute für Chemische Technologie.

BAMBERGER, Y., MARGALIT, Y. and ZITRIN, S., 1989, Post-explosion analysis by NMR spectrometry, *Proceedings 3rd International Symposium on Analysis and Detection of Explosives*, pp. 26.1–26.24, Berghausen, Germany: Fraunhofer Institute für Chemische Technologie.

BASCH, A., MARGALIT, Y., ABRAMOVICH-BAR, S., BAMBERGER, Y., DAPHNA, D., TAMIRI, T. and ZITRIN, S., 1986, Decomposition products of PETN in post-explosion analysis, *Journal of Energetic Materials*, **4**, 77–91.

BEBERICH, D. B., YOST, R. A. and FETTEROLF, D. D., 1988, Analysis of explosives by liquid chromatography/thermospray/mass spectrometry, *Journal of Forensic Sciences*, **33**, 946–959.

BELLAMY, L. J., 1962, *The infrared spectra of complex molecules*, London: Methuen & Co.

BEVERIDGE, A. D., 1991, Analysis of explosives, *Proceedings of the International Symposium on the Forensic Aspects of Trace Evidence*, pp. 177–190, Washington, DC: US Government Printing Office.

BEVERIDGE, A. D., 1992, Developments in the detection and identification of explosive residues, *Forensic Science Review*, **4**, 17–49.

BEVERIDGE, A. D., GREENLAY, W. R. A. and SHADDICK, R. C., 1983, Identification of reaction products in residues from explosives, *Proceedings 1st International Symposium on Analysis and Detection of Explosives*, pp. 53–58, Washington, DC: US Government Printing Office.

BEVERIDGE, A. D., PAYTON, S. F., AUDETTE, R. J., LAMBERTUS, A. J. and SHADDICK, R. C., 1975, Systematic analysis of explosive residues, *Journal of Forensic Sciences*, **20**, 431–454.

BEYNON, J. H., SAUNDERS, R. A. and WILLIAMS, A. E., 1964, The combined use of low and high resolution mass spectrometry: organic nitro-compounds, *Industrie Chimique Belge*, **29**, 311–328.

BOUMA, W. J. and JENNINGS, K. R., 1981, Negative chemical ionization mass spectrometry of explosives, *Organic Mass Spectrometry*, **16**, 331–335.

BROPHY, J. J., DIAKIW, V., GOLDSACK, R. J., NELSON, D. and SHANNON, J. S., 1979, Anomalous ions in the chemical ionization mass spectra of aromatic nitro and nitroso compounds, *Organic Mass Spectrometry*, **14**, 201–203.

BROWN, C. L. and WEBER, W. P., 1970, Negative-ion mass spectrometry of polynitroaromatics, *Journal of the American Chemical Society*, **92**, 5775–5777.

BULUSU, S., AXENROD, T. and MILNE, G. W. A., 1970, Electron-impact fragmentation of some secondary aliphatic nitramines. Migration of the nitro group in heterocyclic nitramines, *Organic Mass Spectrometry*, **3**, 13–21.

BULUSU, S. and AXENROD, T., 1979, Electron impact fragmentation mechanisms of 2,4,6-trinitrotoluene derived from metastable transitions and isotopic labelling, *Organic Mass Spectrometry*, **14**, 585–592.

BUSCH, K. L., GLISH, G. L. and McLUCKEY, S. A., 1988, *Mass spectrometry/mass spectrometry: techniques and applications of tandem mass spectrometry*, New York: VCH.

CABINESS, L. R. and BLACKLEDGE, R. D., 1983, Trioxane – an unusual component in an improvised explosive-actuated incendiary device, *Journal of Forensic Sciences*, **28**, 282–284.

CANTU, A. A., WASHINGTON, W. D., STROBEL, R. A. and TONTARSKI, R. E., 1983, Evaluation of FTIR as a detector for the HPLC analysis of explosives, *Proceedings 1st International Symposium on Analysis and Detection of Explosives*, pp. 349–363, Washington: US Government Printing Office.

CARPER, W. R., DOREY, R. C., TOMER, K. B. and CROW, F. W., 1984, Mass spectral fragmentation pathways in 2,4,6-trinitrotoluene derived from a MS/MS unimolecular and collisionally activated dissociation study, *Organic Mass Spectrometry*, **19**, 623–626.

CHANG, T-L., 1971, Identification of impurities in crude TNT by tandem GC–MS technique, *Analytica Chimica Acta*, **53**, 445–448.

CHAPMAN, J. R., 1993 (Ed.), *Practical organic mass spectrometry*, 2nd ed., Chichester: John Wiley & Sons.

CHASAN, D. E. and NORWITZ, G., 1972, Qualitative analysis of primers, tracers, igniters, incendiaries, boosters and delay compositions on a microscale by use of infrared spectroscopy, *Microchemical Journal*, **17**, 31–60.

CHEN, S-H., WU, H-L., TANAKA, M., SHONO, T. and FUNAZO, K., 1987, Simultaneous gas chromatographic determination of iodide, nitrite, sulphide and thiocyanate anions by derivatization with pentafluorobenzyl bromide, *Journal of Chromatography*, **396**, 129–137.

CHEN, S-H., WU, H-L., TANAKA, M., SHONO, T. and FUNAZO, K., 1990, Simultaneous gas chromatographic determination of cyanide, iodide, nitrite, sulphide and thiocyanate anions by derivatization with pentafluorobenzyl bromide and using a kryptand as phase-transfer catalyst, *Journal of Chromatography*, **502**, 257–264.

CHEN, T. H., 1993, Comparative study of RDX and HMX by DEPMS and TSLC/MS, in Yinon, J. (Ed.) *Advances in analysis and detection of explosives*, pp. 309–321, Dordrecht: Kluwer Academic Publishers.

CHEN, T. H. and CAMPBELL, C., 1989, Identification and confirmation of some nitrocage compounds and explosives by DEMPS, *Proceedings 3rd International Symposium on Analysis and Detection of Explosives*, pp. 26.1–26.24, Berghausen, Germany: Fraunhofer Institute für Chemische Technologie.

CHEN, T. H. and CAMPBELL, C., 1993, Diagnostic scheme for polynitrocage compounds, in Yinon, J. (Ed.) *Advances in analysis and detection of explosives*, pp. 265–269, Dordrecht: Kluwer Academic Publishers.

CONDUIT, C. P., 1959, Ultraviolet and infrared spectra of some aromatic nitro-compounds, *Journal of the Chemical Society*, **665**, 3273–3277.

DOUSE, J. M. F., 1981, Trace analysis of explosives at the low picogram level by silica capillary column gas-liquid chromatography with electron-capture detection, *Journal of Chromatography*, **208**, 83–88.

DOUSE, J. M. F. and SMITH, R. N., 1986, Trace analysis of explosives and firearm discharge residues in the Metropolitan Police forensic science laboratory, *Journal of Energetic Materials*, **4**, 169–186.

DOYLE, R. J. and CAMPANA, J. E., 1985, Nitramine cluster ions from hexahydro-1,3,5-trinitro-1,3,5-triazine, *Journal of Physical Chemistry*, **89**, 4251–4256.

ECKENRODE, B. A., GLISH, G. L. and McLUCKEY, S. A., 1990, Negative-ion chemical ionization in a quadrupole ion trap using reagent ions injected from an external ion source, *International Journal of Mass Spectrometry Ion Processes*, **99**, 151–167.

EVANS, H. K., TULLENERS, F. A. J., SANCHEZ, B. L. and RASMUSSEN, C. A., 1986, An unusual explosive, triacetonetriperoxide (TATP), *Journal of Forensic Sciences*, **31**, 1119–1125.

FERADAY, A. W., 1993, The Semtex-H story, in Yinon, J. (Ed.) *Advances in analysis and detection of explosives*, pp. 67–72, Dordrecht: Kluwer Academic Publishers.

FRASER, R. T. M. and PAUL, N. C., 1968, The mass spectrometry of nitrate esters and related compounds, *Journal of the Chemical Society (B)*, 659–669.

FUNAZO, K., WU, H-L., MORITA, K., TANAKA, M. and SHONO, T., 1985a, Butylation of inorganic anions for simultaneous determination by gas chromatography, *Journal of Chromatography*, **319**, 143–152.

FUNAZO, K., TANAKA, M., MORITA, K., KAMINO, M., SHONO, T. and WU, H-L., 1985b, Pentafluorobenzyl p-toluenesulphonate as a new derivatizing reagent for gas chromatographic determination of anions, *Journal of Chromatography*, **346**, 215–225.

GIELSDORF, W., 1981, Identification of some explosives by special GC/MS techniques, particularly the PPNICI method (in German), *Fresenius Zeitung Analytisches Chemie*, **308**, 123–128.

GILLIS, R. G., LACEY, M. J. and SHANNON, J. S., 1974, Chemical ionization mass spectra of explosives, *Organic Mass Spectrometry*, **9**, 359–364.

GLATTSTEIN, B. and KRAUS, S., 1986, Determination of oxidizing anions in explosive mixtures by phase transfer, *Journal of Energetic Materials*, **4**, 149–157.

GLATTSTEIN, B., ABRAMOVICH-BAR, S., TAMIRI, T. and ZITRIN, S., 1995, A new approach to the post-explosion analysis of inorganic anions, presentation at the *5th International Symposium on Analysis and Detection of Explosives*, Washington, DC (*Proceedings* in preparation).

HABLE, M., STERN, C., ASOWATA, C. and WILLIAMS, K., 1991, The determination of nitroaromatics and nitramines in ground and drinking water by wide-bore capillary gas chromatography, *Journal of Chromatographic Science*, **292**, 131–135.

HANUS, V., HORAK, Z. and TOLMAN, V., 1982, Unusual mass spectrometric fragmentation of N,N′-diethyl-N,N′-diphenylurea, *Organic Mass Spectrometry*, **17**, 49–52.

HARRISON, A. G. and KALLURY, R. K. M. R., 1980, Chemical ionization mass spectra of mononitroarenes, *Organic Mass Spectrometry*, **15**, 284–288.

HELIE-CALMET, J. and FORESTIER, H., 1979, Characterization of explosives traces after an explosion, *International Criminal Police Review*, 38–47.

HOBBS, J. R., 1993, Analysis of Semtex explosives, in Yinon, J. (Ed.) *Advances in analysis and detection of explosives*, pp. 409–427, Dordrecht: Kluwer Academic Publishers.

HOBBS, J. R. and CONDE, E., 1989, Techniques for the detection of explosives, *Proceedings 3rd International Symposium on Analysis and Detection of Explosives*, pp. 41.4–41.18, Berghausen, Germany: Fraunhofer Institute für Chemische Technologie.

HONG, T. Z., TANG, C. P. and LIN, K., 1993, The analysis of explosives of the paper detonator, in Yinon, J. (Ed.) *Advances in analysis and detection of explosives*, pp. 145–152, Dordrecht: Kluwer Academic Publishers.

INOUE, Y., ARAKAWA, S., UEDA, N., YAMAMOTO, Y. and NAKASHIMA, R., 1989,

Rapid and sensitive analysis of explosives by high performance liquid chromatography and gas chromatography/mass spectrometry (in Japanese), *Tottori Daigaku*, **20**, 97–104.

KAPLAN, M. A. and ZITRIN, S., 1977, Identification of post-explosion residues, *Journal of the Association of Official Analytical Chemists*, **60**, 619–624.

KARASEK, F. W. and CLEMENT, R. E., 1988, *Basic gas chromatography-mass spectrometry. Principles and techniques*, New York: Elsevier.

KEE, T. G., HOLMES, D. M., DOOLAN, K., HAMIL, J. A. and GRIFFIN, R. M. E., 1990, The identification of individual propellant particles, *Journal of the Forensic Science Society*, **30**, 285–292.

KETO, R.O., 1986, Improved methods for the analysis of the military explosive Composition C-4, *Journal of Forensic Sciences*, **31**, 241–249.

KETO, R. O., 1989, Analysis of the Eastern bloc explosive Semtex-H, *Proceedings 3rd International Symposium on Analysis and Detection of Explosives*, pp. 11.1–11.20, Berghausen, Germany: Fraunhofer Institute für Chemische Technologie.

KOLLA, P., 1991, Trace analysis of explosives from complex mixtures with sample pretreatment and selective detection, *Journal of Forensic Sciences*, **36**, 1342–1359.

KOLLA, P., 1994, Gas chromatography, liquid chromatography and ion chromatography adapted to the trace analysis of explosives, *Journal of Chromatography A*, **674**, 309–318.

LEE, M. R., HWANG, D-G. and TANG, C-P., 1989, Trace analysis of explosives by mass spectrometry, *Proceedings 3rd International Symposium on Analysis and Detection of Explosives*, pp. 5.1–5.12, Berghausen, Germany: Fraunhofer Institute für Chemische Technologie.

LEE, M. R., CHANG, S. C., KAO, T. S. and TANG, C. P., 1988, Studies of limit of detection on 2,4,6-trinitrotoluene (TNT) by mass spectrometry, *Journal of Research National Bureau of Standards*, **93**, 428–431.

MACH, M. H., PALLOS, A. and JONES, P. F., 1978a, Feasibility of gunshot residues detection via its organic constituents. Part 1: analysis of smokeless powders by combined gas chromatography-chemical ionization mass spectrometry, *Journal of Forensic Sciences*, **23**, 433–445.

MACH, M. H., PALLOS, A. and JONES, P. F., 1978b, Feasibility of gunshot residues detection via its organic constituents. Part II: a gas chromatography–mass spectrometry method, *Journal of Forensic Sciences*, **23**, 446–455.

MAQUESTIAU, A., VAN HAVERBEKE, Y., FLAMMANG, R., MISPREUVE, H. and ELGUERO, J., 1979, Ionisation chimique de composes aromatiques nitres, *Organic Mass Spectrometry*, **14**, 117–118.

MARCHAND, A. P., ANNAPURNA, P., ARNEY, B. E., GADGIL, V. R., RAJAPAKSA, D., SHARMA, G. V. M., SHARMA, R. and ZOPE, U. R., 1993, Synthesis and explosive performance characteristics of polynitropolycyclic cage explosives, in Yinon, J. (Ed.) *Advances in analysis and detection of explosives*, pp. 241–263, Dordrecht: Kluwer Academic Publishers.

MARTZ, R. M. and LASSWELL, L. D., 1983, Smokeless powder identification, *Proceedings 1st International Symposium on Analysis and Detection of Explosives*, pp. 245–258, Washington, DC: US Government Printing Office.

MCCORD, B. R. and HARGADON, K. A., 1993, Explosive analysis by capillary electrophoresis, in Yinon, J. (Ed.) *Advances in analysis and detection of explosives*, pp. 133–144, Dordrecht: Kluwer Academic Publishers, Dordrecht.

MCLUCKEY, S. A., GLISH, G. L. and CARTER, C. A., 1985, The analysis of explosives by tandem mass spectrometry, *Journal of Forensic Sciences*, **30**, 773–788.

MCLUCKEY, S. A., GLISH, G. L., ASANO, K. G. and GRANT, B. C., 1988, Atmospheric sampling glow discharge ionization source for the determination of trace organic compounds in ambient air, *Analytical Chemistry*, **60**, 2220–2227.

MELTON, C. E., 1970, *Principles of mass spectrometry and negative ions*, New York: Marcel Dekker, Chapter 7.

MEYERS, R. E., 1978, A systematic approach to the forensic examination of flash powders, *Journal of Forensic Sciences*, **23**, 66–73.

MEYERSON, S., VAN DER HAAR, R. W. and FIELDS, E. K., 1972, Organic ions in the gas phase. XXVI. Decomposition of 1,3,5-trinitrobenzene under electron impact, *Journal of Organic Chemistry*, **37**, 4114–4119.

MILLER, F. A. and WILKINS, C. H., 1952, Infrared spectra and characteristic frequencies of inorganic ions; their use in qualitative analysis, *Analytical Chemistry*, **24**, 1253–1294.

MONCUR, J. G. and SHARP, T. E., 1982, Mass spectrometric analysis and thermal decomposition study of 1,3,5,7-tetranitro-1,3,5,7-tetrazacyclooctane (HMX), pp. 122–123, *Proceedings 30th Annual Conference on Mass Spectrometry, Honolulu, HI*.

NOWICKI, J. and PAULING, S., 1988, Identification of sugars in explosive residues by gas chromatography–mass spectrometry, *Journal of Forensic Sciences*, **33**, 1254–1261.

NYQUIST, R. A. and KAGEL, R.O., 1971, Infrared spectra of inorganic compounds, New York: Academic Press, p. 439.

OTTOILA, P., TASKINEN, J. and SOTHMANN, A., 1982, Quantitative determination of nitroglycerin in human plasma by capillary gas chromatography negative ion chemical ionization mass spectrometry, *Biomedical Mass Spectrometry*, **9**, 108–110.

PARKER, C. E., VOYKSNER, R. D., TONDEUR, Y., HENION, J. D., HARVAN, D. J., HASS, J. R. and YINON, J., 1982, Analysis of explosives by liquid chromatography–negative ion chemical ionization mass spectrometry, *Journal of Forensic Sciences*, **27**, 495–505.

PATE, C. T. and MACH, M. H., 1978, Analysis of explosives using chemical ionization mass spectroscopy, *International Journal of Mass Spectrometry and Ion Physics*, **26**, 267–277.

PEIMER, R. E., WASHINGTON, W. D. and SNOW, K. B., 1980, On the examination of the military explosive, C-4, *Journal of Forensic Sciences*, **25**, 398–400.

PRISTERA, F., 1953, Analysis of propellants by infrared spectroscopy, *Analytical Chemistry*, **25**, 844–856.

PRISTERA, F., HALIK, M., CASTELLI, A. and FREDERICKS, W., 1960, Analysis of explosives using infrared spectroscopy, *Analytical Chemistry*, **32**, 495–508.

REUTTER, D. J., BENDER, E. C. and RUDOLPH, T. L., 1983, Analysis of an unusual explosive: methods used and conclusion drawn from two cases, *Proceedings 1st International Symposium on Analysis and Detection of Explosives*, pp. 149–153, Washington, DC: US Government Printing Office.

RIDDELL, R. H. and MILLS, T. M., 1983, Analysis of explosives by HPLC–FTIR, *Proceedings 1st International Symposium on Analysis and Detection of Explosives*, pp. 287–307, Washington, DC: US Government Printing Office.

ROWLEY, J. A., 1989, Positive electron impact and chemical ionization mass spectra of some nitramine nitrates, *Organic Mass Spectrometry*, **24**, 997–1000.

RUDOLPH, T. L. and BENDER, E. C., 1983, A scheme for the analysis of explosives and explosive residues, *Proceedings 1st International Symposium on Analysis and Detection of Explosives*, pp. 71–78, Washington, DC: US Government Printing Office.

SAFERSTEIN, R., CHAO, J-M. and MANURA, J. J., 1975, Isobutane chemical ionization mass spectrographic examination of explosives, *Journal of the Association of Official Analytical Chemists*, **58**, 734–742.

SCAPLEHORN, A. W., 1993, Birmingham six pub bombing case, in Yinon, J. (Ed.) *Advances in analysis and detection of explosives*, pp. 1–10, Dordrecht: Kluwer Academic Publishers.

SCHWARZ, H. and SCHIEBEL, H-M., 1983, Mass spectrometry of organic peroxides, in Patai, S. (Ed.) *The chemistry of functional groups, peroxides*, pp. 105–127, Chichester: John Wiley & Sons.

SLACK, G. C., McNAIR, H. M. and WASSERZUG, L., 1992, Characterization of Semtex by supercritical fluid extraction and off-line GC-ECD and GC-MS, *Journal of High Resolution Chromatography*, **15**, 102–104.

SOLOMONOVICI, A., 1986, Identification of explosives in Soviet weapons, *Journal of Energetic Materials*, **4**, 315–324.

STALS, J., 1971, Chemistry of aliphatic unconjugated nitramines. Part 7. Interrelations between the thermal, photochemical and mass spectral fragmentation of RDX, *Transaction of the Faraday Society*, **67**, 1768–1775.

SULZLE, D. and KLAEBOE, P., 1988, The infrared, Raman and NMR spectra of hexamethylene triperoxide diamine, *Acta Chemica Scandanavia*, **A42**, 165–170.

TAMIRI, T. and ZITRIN, S., 1986, Capillary column gas chromatography/mass spectrometry of explosives, *Journal of Energetic Materials*, **4**, 215–237.

TAMIRI, T., ZITRIN, S., ABRAMOVICH-BAR, S., BAMBERGER, Y. and STERLING, J., 1993, GC/MS analysis of PETN and NG in post-explosion residues, in Yinon, J. (Ed.) *Advances in analysis and detection of explosives*, pp. 323–324, Dordrecht: Kluwer Academic Publishers.

TAN, Y.L., 1977, Microdetermination of nitrate by gas chromatography–mass spectrometry technique with multiple ion detector, *Journal of Chromatography*, **140**, 41–46.

URBANSKI, T., 1964, *Chemistry and technology of explosives*, Vol. 3, p. 225, Oxford: Pergamon Press.

URBANSKI, T. and WITANOWSKI, M., 1963, Infrared spectra of nitric esters. Part 2. Rotational isomerism of some esters, *Transactions of the Faraday Society*, 1046–1054.

VOLK, F. and SCHUBERT, H., 1968, Massenspektrometrische untersuchungen von explosivstoffen, *Explosivstoffe*, **16**, 2–10.

VOYKSNER, R. D. and YINON, J., 1986, Trace analysis of explosives by thermospray high-performance liquid chromatography–mass spectrometry, *Journal of Chromatography*, **354**, 393–405.

VOUROS, P., PETERSEN, B. A., COLWELL, L. and KARGER, B. L., 1977, Analysis of explosives by high performance liquid chromatography and chemical ionization mass spectrometry, *Analytical Chemistry*, **49**, 1039–1044.

WASHINGTON, W. D. and MIDKIFF, C. R., 1986, Explosive residues in bombing-scene investigations. New technology applied to their detection and identification, in Davies, (Ed.) *Forensic Science*, Chapter 17, pp. 245–264, Washington, DC: American Chemical Society.

WASHINGTON, W. D., KOPEC, R. J. and MIDKIFF, C. R., 1977, Systematic approach to the detection of explosive residues. V. Black powders, *Journal of the Association of Official Analytical Chemists*, **60**, 1331–1340.

WEINBERG, D. S. and HSU, J. P., 1983, Comparison of gas chromatographic and gas chromatographic/mass spectrometric techniques for the analysis of TNT and related nitroaromatic compounds, *Journal of High Resolution Chromatography: Chromatography Communications*, **6**, 404–408.

WERBIN, A., 1957, The infrared spectra of HMX and RDX, *Report No. UCRL-5078*, University of California, Radiation Laboratory, Livermore Site, Livermore, California.

WHITE, G. M., 1992, An explosive drug case, *Journal of Forensic Sciences*, **37**, 652–656.

WILLIAMS, D. H. and FLEMING, I., 1980, *Spectroscopic methods in organic chemistry*, Maidenhead: McGraw-Hill.

WOLFFENSTEIN, R., 1895, Uber die einwirkung von wasserstoffsuperoxyd auf aceton und mesityloxyd, *Berichte*, **28**, 2265–2269.

YELTON, R. O., 1982, Ammonia chemical ionization of the explosives pentaerythritol tetranitrate and dipentaerythritol hexanitrate, *Proceedings 30th Annual Conference on Mass Spectrometry, Honolulu, HI*, pp. 665–666.

YERGEV, A. L., EDMONDS, C. G., LEWIS, I. A. S. and VESTAL, M. L., 1990, *Liquid chromatography/mass spectrometry. Techniques and applications*, New York: Plenum Press.

YINON, J., 1974, Identification of explosives by chemical ionization mass spectrometry using water as reagent, *Biomedical Mass Spectrometry*, **1**, 393–396.

YINON, J., 1980a, Analysis of explosives by negative-ion chemical ionization mass spectrometry, *Journal of Forensic Sciences*, **25**, 401–407.

YINON, J., 1980b, Direct exposure chemical ionization mass spectra of explosives, *Organic Mass Spectrometry*, **15**, 637–639.

YINON, J., 1983, Forensic applications of LC/MS, *International Journal of Mass Spectrometry and Ion Physics*, **48**, 253–256.

YINON, J., 1987, Mass spectral fragmentation pathways in 2,4,6-trinitroaromatic compounds. A tandem mass spectrometric collision induced dissociation study, *Organic Mass Spectrometry*, **22**, 501–505.

YINON, J., 1988, Identification of explosives mixtures by tandem mass spectrometry (MS/MS), *Canadian Society of Forensic Science Journal*, **21**, 46–53.

YINON, J., 1990, Mass spectral fragmentation pathways in aminodinitrobenzenes. A mass spectrometry/mass spectrometry collision-induced dissociation study, *Organic Mass Spectrometry*, **25**, 599–604.

YINON, J., 1992, Mass spectral fragmentation pathways in some dinitroaromatic compounds studied by collision-induced dissociation and tandem mass spectrometry, *Organic Mass Spectrometry*, **27**, 689–694.

YINON, J. and BOETTGER, H. G., 1972/73, Modification of a high resolution mass spectrometer for negative ionization, *International Journal of Mass Spectrometry and Ion Physics*, **10**, 161–168.

YINON, J. and BULUSU, S., 1986a, Mass spectral fragmentation pathways in nitro-adamantanes. A tandem mass spectrometric collisionally induced dissociation study, *Organic Mass Spectrometry*, **21**, 529–533.

YINON, J. and BULUSU, S., 1986b, MS/MS of energetic compounds. A collisional induced dissociation study of some polynitrobishomocubanes, *Journal of Energetic Materials*, **4**, 115–131.

YINON, J. and HWANG, D-G., 1983, High-performance liquid chromatography–mass spectrometry of explosives, *Journal of Chromatography*, **268**, 45–53.

YINON, J. and LASCHEVER, M., 1981, Reduction of trinitroaromatic compounds in water by chemical ionization mass spectrometry, *Organic Mass Spectrometry*, **16**, 264–266.

YINON, J. and ZITRIN, S., 1981, *The analysis of explosives*, Oxford: Pergamon Press.

YINON, J. and ZITRIN, S., 1993, *Modern methods and applications in the analysis of explosives*, Chichester: John Wiley & Sons.

YINON, J., FRAISSE, D. and DAGLEY, I. J., 1991a, Mass spectral fragmentation patterns of deuterated hexanitrobibenzil and hexanitrostilbene, *Rapid Communications in Mass Spectrometry*, **5**, 164–168.

YINON, J., FRAISSE, D. and DAGLEY, I. J., 1991b, Electron impact, chemical ionization and negative chemical ionization mass spectra, and mass-analysed ion kinetic energy spectrometry – collision-induced dissociation fragmentation pathways of some deuterated 2,4,6-trinitrotoluene (TNT) derivatives, *Organic Mass Spectrometry*, **26**, 867–874.

YINON, J., HARVAN, D. J. and HASS, J. R., 1982, Mass spectral fragmentation pathways in RDX and HMX. A mass analyzed ion kinetic energy spectrometric/collisional induced dissociation study, *Organic Mass Spectrometry*, **17**, 321–326.

YINON, J., ZITRIN, S. and TAMIRI, T., 1993, Reactions in the mass spectrometry of 2,4,6-N-tetranitro-N-methylaniline (tetryl), *Rapid Communications in Mass Spectrometry*, **7**, 1051–1054.

YINON, J., BRUMLEY, W. C., BRILIS, G. M. and BULUSU, S., 1990, Mass spectral fragmentation pathways in nitramines. A collision-induced dissociation study, *Organic Mass Spectrometry*, **25**, 14–20.

YURAWECZ, M. P. and PUMA, B. J., 1983, Nitro musks fragrances as potential contaminants in pesticides residues analysis, *Journal of the Association of Official Analytical Chemists*, **66**, 241–247.

ZITRIN, S., 1971/72, Unpublished Israel Police files 3555/71, 18/72 and 30/72.

ZITRIN, S., 1982, The chemical ionization mass spectrometry of RDX, *Organic Mass Spectrometry*, **17**, 74–78.

ZITRIN, S., 1986, Post explosion analysis of explosives by mass spectrometric methods, *Journal of Energetic Materials*, **4**, 199–214.

ZITRIN, S. and YINON, J., 1976a, Chemical ionization mass spectrometry of explosives, in Frigerio, A. and Castagnoli, N. (Eds.), *Advances in mass spectrometry in biochemistry and medicine*, Vol. 1, pp. 369–381, Holliswood: Spectrum Publications.

ZITRIN, S. and YINON, J., 1976b, Chemical ionization mass spectra of 2,4,6-trinitro-aromatic compounds, *Organic Mass Spectrometry*, **11**, 388–393.

ZITRIN, S. and YINON, J., 1978, Mass spectrometry studies of trinitroaromatic compounds, *Advances in Mass Spectrometry*, **7**, 1457–1464.

ZITRIN, S., KRAUS, S. and GLATTSTEIN, B., 1983, Identification of two rare explosives, *Proceedings of the 1st International Symposium on Analysis and Detection of Explosives*, pp. 137–141, Washington, DC: US Government Printing Office.

Appendix: Experimental conditions, IR spectra

Figures 9.1, 9.2, 9.5, 9.7, 9.8, 9.9 and 9.13 through 9.19 were recorded on an FTIR instrument as KBr pellets, 32 scans, 4 cm^{-1} resolution.

Figures 9.3, 9.4 and 9.6 were recorded on a Biorad FTS-40 FTIR spectrometer using a diamond cell in a 6 × beam condenser, 256 scans, 4 cm^{-1} resolution.

Figure 9.10 was recorded using a Nic-Plan IR microscope on a Nicolet 710 bench, 256 scans, 4 cm^{-1} resolution.

Quality Control in the Detection and Identification of Traces of Organic High Explosives

ROBIN W. HILEY

10.1 Introduction

The number of organic compounds which have few uses other than as high explosives, and which are widely used in large quantities, can be counted on the fingers of one hand. They are ethylene glycol dinitrate (EGDN), nitroglycerine (NG), trinitrotoluene (TNT), pentaerythritol tetranitrate (PETN) and cyclotrimethylene trinitramine (RDX, hexogen, cyclonite). A few other organic compounds such as cyclotetramethylene tetranitramine (HMX, octogen) and hexanitrostilbene, find limited or specialist use.

These simple facts make it both practical and worthwhile to search for traces of these substances in many investigations of explosives crime. The detection of one of these substances, even in minute amounts, in a sample taken from a surface, constitutes powerful (but not conclusive) evidence of involvement between the surface and a high explosive. Where the surface is at the scene of an unexplained explosion or other destructive event such as an aircraft crash, the detection will be a major piece of evidence in deciding the cause of the event. Perhaps of even greater potential value is the finding of one of these substances on an apparently innocent surface away from any explosion scene; for example, on the seat of a car, the floor of a cupboard, or upon hands. Such findings immediately help investigating police officers, and when stated with due regard to their limitations may constitute important evidence in court proceedings.

The power of modern analytical methods and the considerable rarity of these few high explosives in the general environment makes it practical to search for them at very low levels, referred to here as 'traces'. Detection limits of a few tens of picograms are readily achieved. A complete high explosives trace analysis, including initial detection and two or more stages of confirmation, may thus be carried out on a sample containing perhaps 10 nanograms of RDX.

Until comparatively recent times it would have been true to state that the overwhelming majority of high explosives contain one or more of these compounds. Whilst this is still true of military explosives it is no longer true of explosives manufactured for civilian purposes, which now are most commonly gelled or emulsified mixtures of ammonium nitrate and a fuel. Unfortunately, military explosives, particularly plastic explosives, have found their way into criminal hands on many occasions and there is no indication that this will cease in the foreseeable future. The value of high explosives trace analysis thus remains.

Good quality control, whether formal or informal, is essential in all forensic science but the impact of explosives trace evidence and the serious consequences of error, make good quality control of overwhelming importance in this field. Without quality control, the evidence may at best be effectively challenged, or at worst lead to a wrongful conviction. Quality assurance in explosives and arson cases has been well considered by Brunelle et al. (1982), but they did not specifically address explosives trace analysis.

This chapter considers quality control in the whole process of trace analysis of organic explosives, from sample collection through to confirmation of identification. It does not cover the broader aspects of forensic quality control, such as the review of case reports. The subject would tend to be difficult to follow and rather sterile if treated in the abstract. The main text will therefore be based around a series of examples and practical illustrations.

10.2   Issues and Principles

Three objections which commonly may be raised against the reported finding of an explosives trace in a sample are that:

- the substance detected was not in fact the explosive claimed;
- the sample was contaminated at some stage;
- the trace came to be where it was found by innocent means.

The latter objection is a very important issue which is addressed in Chapter 12, but not one which is influenced by quality control in an analytical laboratory. The principles upon which effective answers to the first two issues may be founded are considered below.

10.2.1   Certainty of Identification

Quite commonly there is no physical evidence to suggest that high explosives are present in the sample presented for trace analysis (e.g. no post-blast damage). In these circumstances, the entire weight of the trace detection and identification must be borne by the analytical process. Thus it must be possible to conclude with a practical certainty that the explosive trace has been correctly identified.

The emphasis upon certainty of identification, rather than upon quantification, is very different from that which characterises most analytical chemistry. Most analysts are concerned with the accurate quantification of a substance known (or assumed) to be present in a comparatively predictable matrix. In explosives trace analysis the original samples are collected from unpredictable and diverse surfaces, and bring with them similarly unpredictable matrices. Quantitative recovery of traces and subsequent quantitative processing and analysis might be possible in principle but would greatly slow the process of evidence collection and enormously increase its cost.

Mathematical procedures for the expression of degree of quantitative certainty are highly developed and well understood, but the converse is true for the degree of certainty in identification. Mathematical treatments exist for expressing simpler aspects, such as the resolving power of a chromatographic system, but full statements of the risk of misidentification in a multi-stage analytical system have not yet been developed.

The two principles which underlie certainty of identification are: (1) comparison; and (2) multiple confirmatory analysis.

Comparison

The first principle is the use of direct comparison with a previously authenticated sample of the compound. This reduces to an absolute minimum any uncertainties associated with calibration of the analytical method ('calibration' is meant in its broadest sense including, for example, the generation of comparison mass spectra).

Multiple confirmatory analysis

The second principle is multiple confirmatory analysis. The initial detection will of necessity be made using a highly specific technique (otherwise countless false alarms will arise) but the necessary degree of certainty cannot be provided by a single analysis. At least one confirmatory analysis must be made, using an analytical system which differs from the initial system in a significant respect. Thus, for example, an initial detection by thin layer chromatography might be confirmed by a second analysis using a different eluent system.

10.2.2 Avoidance of Contamination

In forensic work it is not sufficient merely to avoid contamination. Positive evidence must be available which can be put forward in response to suggestions that contamination took place. It does not seem possible to provide absolute proof of the absence of contamination; some hypothetical route can always be suggested, however unlikely. The best response is therefore to provide clear evidence of the precautions taken, and their effectiveness, against which such suggestions can fairly be judged.

A simple calculation reveals that 10 ng of TNT is contained within a single crystal of about 30 μm in size, and one such particle could provide a confirmed positive detection of TNT. Thus, simple visual standards of cleanliness are insufficient to provide control of contamination in explosives trace work.

Three principles can be added to that of visual cleanliness in the control of contamination:

- separation in time, space or both, between work involving bulk explosives and trace work;
- measurement of contamination levels, both before particular trace work is undertaken and routinely;
- parallel processing of control samples.

If all three principles are applied simultaneously, this forms multi-layered protection against sample contamination, and provides the required clear evidence of contamination control.

10.2.3 Formal Quality Systems and External Accreditation

Quality control in forensic work used to be largely informal. Each forensic scientist took responsibility for quality control of his or her own work, and answered for it if necessary in the courts. This system is quite workable and has the advantages of flexibility and efficiency. The drawbacks are that it is heavily reliant upon the ability and credibility of the individual, and that flexibility may be portrayed as inconsistency when the work of different scientists is compared. There is also a strong movement towards formal quality systems in many areas of science and industry, and in the United Kingdom and elsewhere forensic science is moving with this trend.

Formal quality systems vary but many include the following basic elements:

(1) a written 'Quality Manual' which lays down the structure of the organisation, individual responsibilities within it, and overall policies for the conduct of the work;

(2) detailed written instructions for the various procedures used in the work. An example of such instructions is given in the Appendix;

(3)   a scheme of regular audits to ensure that the system is operating as intended.

When, as in forensic science, the work includes physical or chemical measurements of some kind which should yield a predictable result then a fourth element may be added:

(4)   proficiency testing.

Proficiency testing is the presentation of samples for testing which are known to the presenter but unknown to the tester. The result obtained should lie within the predicted range.

The basic aim of a formal quality system is to make visible to an outsider, in a logically consistent documentary form, all of the processes employed to ensure that the work is conducted to an appropriate standard. In industrial work that standard may, for good financial reasons, be modest but in forensic work a very high standard must inevitably be set.

The value of a formal quality system is greatly increased by subjecting it to external audit and there are now various organisations which specialise in providing such a service. In the United Kingdom, for example, the United Kingdom Accreditation Service (UKAS) will audit the measurement aspects of forensic science, and has produced a guidance document dealing specifically with forensic analysis and examination (Anon, 1992). In Canada, the Standards Council of Canada provides a similar service and in the United States it is provided by various bodies including the American Society of Crime Laboratory Directors and the National Institute of Standards and Technology. If the quality system meets a range of requirements, these bodies accredit the measurement processes which have been audited. Following initial accreditation the external surveillance is repeated at regular intervals, ensuring that standards are maintained.

A second form of external scrutiny is participation within an external proficiency test scheme. In a typical scheme a number of laboratories cooperate in the setting of proficiency tests for each other on a regular basis.

There is a strong link between the requirements of forensic trace explosives analysis and formal quality systems. The adoption of a formal quality system allows all of the principles outlined above to be invoked in a planned and consistent way. The acceptance of external audit and proficiency tests provides an impartial scrutiny of the methods employed, which enhances the credibility of the results obtained.

10.3   Specific Aspects of Quality Control

10.3.1   Standard Reference Materials

A set of standard reference explosives is required in order to carry out identifications by direct comparison. The most important of these are the five explosives noted above, but a wider range is desirable. In many fields of chemical analysis suitable reference materials are commercially available; samples can be purchased together with a certificate of authenticity and purity. However, this is not the case with explosives and the following scheme is a means of generating reference explosives of established authenticity:

- obtain from known and well-established explosives suppliers samples of the materials required in as pure a form as is commercially available; retain the documentation recording the supply and any batch or quality assurance certificates;
- estimate the purity of the samples, using generally accepted methods such as melting point determination and chromatographic separation;
- collect mass spectra, infra-red (IR) spectra and nuclear magnetic resonance (NMR) spectra of the samples and compare systematically with those in the literature; retain the spectra.

For explosives trace identification, standards of very high purity are not essential. Exact quantification is seldom of high evidential significance, and the techniques used for identifica-

Table 10.1 Reference explosives

Explosive	Description	Source	Estimated purity (%)
Nitroglycerine	Liquid	Synthesised on site	99.5
2,4,6-TNT	80.2 flake	Royal Ordnance, Bridgwater, UK	99.7
PETN	Crystalline	Nobel's Explosives, UK	99.9
RDX	Crystalline	Royal Ordnance, Bridgwater, UK	99.9
HMX	Crystalline (Type A)	Royal Ordnance, Bridgwater, UK	99.9

tion must of necessity be so highly selective that minor impurities are unlikely to be of significance. A minimum estimated purity of 98% is suggested as acceptable, provided that no single impurity exceeds 1% of the whole.

Table 10.1 lists the reference samples of NG, PETN, TNT, RDX and HMX held at Fort Halstead.

PETN, TNT, RDX and HMX are chemically stable solids (this property is important for military use), and so once supplies are established as references they rarely need to be renewed. Reference explosives must be stored in properly constructed magazines or lockers conforming with local explosives safety legislation, and extremes of temperature should be avoided. NG and EGDN are not indefinitely stable in the pure state, and reference stocks of these explosives should be examined regularly for signs of degradation. If such signs are observed, the explosives are safely destroyed (see below).

Dilute solutions of reference explosives are used for comparison with forensic trace samples, and we have found such solutions of the common explosives in ethyl acetate to be stable for several months when stored in a refrigerator. A series of quantitative measurements of a 1% solution of NG in ethyl acetate similarly stored revealed no loss of NG within 6 months. Stock solutions at about the 1% level are therefore prepared and used as the basis for further dilutions and are assigned a lifetime of 6 months. More dilute solutions are replaced more frequently, and the very dilute bench solutions used for actual comparisons are replaced daily. Solutions prepared according to this scheme are regarded as satisfactory for forensic identification purposes, but would not be used for precise quantification should this ever be required. Freshly prepared solutions would be used for this purpose.

If a formal quality system is in use, the procedures for establishment and use of reference materials is documented. Points addressed include the preparation and recording of standard solutions, storage conditions and lifetimes.

Destruction of reference standards

When no longer required, reference standards and other explosives should be safely destroyed according to local regulations. We employ the normal process for destruction of secondary high explosives, which is by burning under controlled conditions on a licensed burning ground. Such destruction should not be attempted without appropriate training.

10.3.2 Sampling Kits

Materials used for the collection of explosives trace samples must be demonstrably free of explosives contamination. In the United Kingdom this aim has for many years been achieved through the manufacture of quality-controlled sampling kits used by trained police officers or forensic scientists (Wallace and McKeown, 1993).

The kits, within a secure and impermeable container, contain solvent-moistened swabs and other materials suitable for collecting samples from various surfaces. The exact materials may vary according to local needs and according to the analytical methods which will be applied, but the overriding requirement is that all should be free of explosives contamination at detectable levels.

Figure 10.1 shows the contents of a kit used by the Forensic Science Agency of Northern Ireland (FSANI). The swabs are of acrilan fibre and are moistened with pure isopropanol. To minimise the possibilities of contamination the kit includes detailed instructions and a disposable oversuit in the outer packing, and even a pen for recording the samples.

Whether sampling materials are in the form of pre-manufactured kits or produced as the need arises, they should be prepared in accordance with the principles for contamination avoidance laid out above. Detailed consideration of contamination control measures is given in Section 10.4 below, but essentially, the place of preparation should be physically segregated from work involving bulk explosives, and its freedom from detectable contamination established. The personnel involved should be similarly segregated. Ideally, one room should be dedicated to this use. When preparation is complete, one kit in twenty is selected at random and all of the contents tested for explosives traces. If any traces were to be found the entire

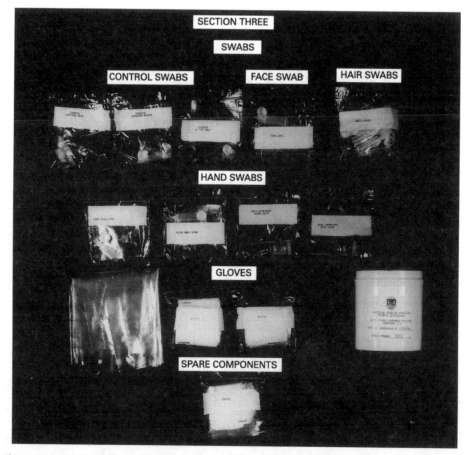

Figure 10.1 Forensic Science Agency of Northern Ireland (FSANI) swab kit.

batch manufactured at the same time would be discarded and the source of contamination investigated.

The testing of sampling materials serves another important purpose. There may be substances within the sampling materials which interfere in analytical methods; either giving responses which can be confused with explosives or reducing the sensitivity of explosives detection. Prior testing of materials enables such problems to be addressed and overcome in advance.

10.3.3 Sample Processing/Clean-up

Before analysis, the original samples usually must be processed. The most common processing steps are extraction into a volatile solvent and concentration of the extract by evaporation. Because the resulting sample will frequently contain much co-extracted material which may cause difficulties in the analyses, various 'clean-up' processes have been devised.

Thin layer chromatography (preparative)

If a particular explosive is sought, preparative thin layer chromatography (TLC) may be used to good effect. We have, for example, separated swab extracts using TLC (see 10.4.1 below) and scraped silica from the plates over the R_f zone in which NG would be found. Subsequent elution of the recovered silica gave a relatively clean solution in which NG traces could readily be detected. Even where analytical TLC plates have been visualised with Griess reagent, sufficient unreacted explosive for further analysis can often be recovered by scraping off the spots and eluting the silica. For more general screening some form of solid phase adsorption is frequently employed.

Column chromatography: solid adsorbents

Strobel and Tontarski (1983) explored the use of bonded phase sorbents to 'clean-up' extracts from explosive debris. Lloyd and King (1990) have described a complete method which both extracts explosives traces and provides a degree of 'clean-up' (separation of the traces from the much larger amount of other material inevitably collected by the sampling process). In this method, the swab sample is extracted using a mixture of propan-2-ol and water. The extract is then shaken with Chromosorb 104, a polymeric solid phase adsorbent having some selectivity for explosives. The adsorbent is separated from the spent supernatant, packed into a very small column and eluted with a small volume of acetonitrile, ready for analysis by liquid chromatography.

For many years, we simply extracted swab samples into diethyl ether and concentrated the extract but for the last 7 years we have used a solid adsorbent clean-up process (Douse, 1985). We prepare disposable adsorption tubes by plugging Pasteur pipettes with small disks of glass fibre filter paper (Whatman Grade GF/A is suitable) at the base of the tapered part and adding a slurry of Amberlite XAD-7 adsorbent in ethyl acetate, sufficient to a give a bed depth of 4 mm above the tapered part of the tube. A further small disk of glass fibre paper retains the adsorbent. We keep a supply of such tubes in readiness, but wash them through with ethyl acetate, methyl-*tert*-butyl ether (MTBE) and finally a mixture of 15%/85% by volume MTBE/pentane immediately before use.

Our method commences by extracting the swab with MTBE. The initial extract is concentrated to a small volume (not to dryness) and rediluted into 5 ml of 15/85 MTBE/pentane. The rediluted extract is passed through an adsorption tube, and the adsorbent washed with a further 1 ml of 15/85 MTBE/pentane. Finally any adsorbed explosives are eluted from the tube with 0.8 ml of ethyl acetate, and the volume of the final extract reduced to approximately 0.1 ml for analysis. We have found that in most instances this procedure produces solutions sufficiently clean for gas chromatographic analysis, but for particularly dirty samples the proportion of MTBE can be increased to 50% or even 100%. With the 15/85 mixture, recoveries of the common explosives are 40% or better, but these fall as the MTBE proportion increases. One troublesome practical point is the tendency of air to accumulate in the adsorbent bed, stopping the flow. We overcome this by applying gentle pressure at the top of the tube with a disposable pipette teat.

Quality control aspects

The two primary quality aspects of sample processing are: (1) to avoid contamination; and (2) to establish and monitor the efficiency of recovery of explosives traces.

Recovery efficiency need not necessarily be very high, but its approximate value should be known for a range of explosives. High recovery efficiencies increase the detection sensitivity of the overall sampling/analysis process, but some sacrifice of processing efficiency may be beneficial if a degree of 'clean-up' is achieved. Cleaner samples produce clearer and more reliable identification data, which is likely to be of greater forensic value than confused data of higher sensitivity. We consider recovery efficiencies of around 30–50% to be quite acceptable if the resulting samples are clean.

The efficiency of recovery is measured by 'spiking' the sample – i.e. adding a known compound. There are two ways to do this. Probably the best system is to spike the original samples with an internal standard similar in chemical characteristics to the common high explosives. For example, Lloyd and King (1990) added dinitrobenzene. The internal standard is recovered and analysed together with any explosives traces, and provides a direct measure of recovery efficiency for every sample.

An alternative system is to spike samples with known amounts of explosives, either when case samples are processed or as a regular routine. This system has the advantage that recoveries of a range of explosives may be checked, but it does not provide direct information for every sample processed. One drawback is that the potential for confusion exists, but if the spike contains many explosives any accidental confusion would be readily apparent.

10.4 Analytical Techniques for Identification of Organic High Explosives

Many analytical methods have been developed for the identification of explosives traces, many of which rely upon a combination of chromatographic separation with selective detection. An excellent recent review is that by Beveridge (1992). The books by Yinon and Zitrin (1981, 1993) cover the ground comprehensively. The use of HPLC has been reviewed in great detail by Lloyd (1992).

Four important aspects may be identified in the quality control of analytical methods for explosives traces. These are:

- demonstrable absence of contamination;
- demonstration that the method performed as expected;
- proper and consistent interpretation of the results;
- clear and adequate documentation of the analyses.

Three of the most popular analytical methods employed for explosives trace detection will now be discussed, and the application of these factors considered in each case. The methods are:

- thin layer chromatography (TLC);

- gas chromatography/chemiluminescence (GC/TEA);
- gas chromatography/mass spectrometry (GC/MS).

10.4.1 *Thin Layer Chromatography*

The earliest methods for explosives trace analysis were simple spot tests (Amas and Yallop, 1966, 1969), but these cannot nowadays be regarded as providing a sufficient degree of selectivity. They can be used for initial screening or for field testing, but confirmation by at least two further methods is desirable if the identifications are to be used for forensic purposes. The value of spot tests was greatly improved when their colour reactions were combined with the separating power of TLC (Lloyd, 1967; Jenkins and Yallop, 1970; Midkiff and Washington, 1974; Beveridge et al., 1975; Midkiff and Washington, 1976; Douse, 1982). Although of somewhat limited sensitivity compared to many instrumental methods, TLC remains an extremely valuable method for trace explosives identification and will be described in some detail here. It became very clear at the International Symposium and Workshop on Explosive Residue Analysis Protocols (unreported) held at the FBI Academy in 1993, that many laboratories cannot justify expenditure upon specialist analytical instruments for explosives trace analysis and in such laboratories the great superiority of TLC over spot tests makes the case for its use very strong. It requires no expensive specialised instrumentation, but by the use of selective visualisation reagents and multiple analyses, can provide a high degree of certainty in identification. It is also quite rapid and several samples may be analysed simultaneously.

Eluent systems

Separation of the common high explosives is comparatively straightforward. With the exception of nitrocellulose, which is a poorly characterised polymer of variable composition, all of the common explosives behave well in normal phase TLC on silica gel plates. 'Normal' phase means that the stationary phase (silica gel layer) is polar and the mobile phase (solvent) is relatively non-polar. There are some common conflicts in R_f value, such as that between NG and PETN with benzene or toluene eluent, but by the use of at least two eluent systems all conflicts between the common explosives can be resolved. We use the eluents shown in Table 10.2 with plain silica gel plates.

The toluene/ethyl acetate mixture is particularly appealing because it provides good separation of all of the explosives shown in the table. Figure 10.2 is a photograph of a TLC plate eluted with this eluent and visualised with Griess reagent (see below).

Table 10.2 Eluents for thin layer chromatography

Eluent system	Typical R_f values						
	NC	HMX	RDX	CE/Tetryl	NG	PETN	TNT
Toluene	0.00	0.01	0.07	0.43	0.61	0.64	0.74
Petroleum ether 40–60°C 40% Petroleum ether 60–80°C 40% Ethyl acetate 20%	0.00	0.02	0.03	0.18	0.32	0.44	0.52
Chloroform 90% Methanol 10%	0.00	0.25	0.53	0.83	0.84	0.88	0.93
Toluene 90% Ethyl acetate 10%	0.00	0.06	0.13	0.61	0.66	0.78	0.88

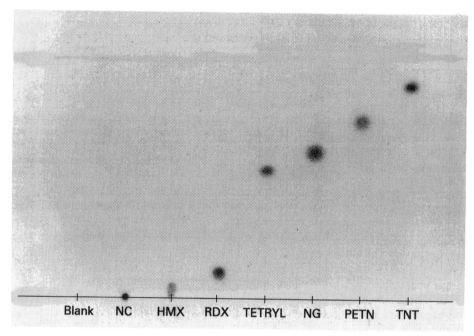

Figure 10.2   TLC plate of standard explosives in toluene/ethyl acetate.

There are no reports of the use of reverse-phase TLC for explosives trace analysis, but there is no reason to believe that it would not give good and useful separations. In 'reverse' phase TLC, the thin layer is non-polar and the eluent is relatively polar and often aqueous. The combination of normal phase and reverse phase TLC would potentially provide a very good confirmatory system.

Visualisation: the Griess reaction

The key to selective and sensitive explosives trace identification by TLC lies in the visualisation reagent. Griess reagent, in a number of versions, has proven to be the most popular means of visualisation. The TLC plate is first sprayed with a strong base such as sodium hydroxide solution (1 M). When the plate is subsequently heated the base reacts with nitrate esters and nitramines to produce nitrite, and subsequent spraying with Griess reagent produces reddish spots. The exact colour is dependent upon the particular version of Griess reagent employed. We use 4% sulphanilamide with 0.4% naphthylethylenediamine hydrochloride in 8% aqueous orthophosphoric acid, which gives magenta spots. Another system is a 50/50 mixture of α-naphthol (0.5 g in 165 ml 30% acetic acid)/sulphanilic acid (1 g in 100 ml 30% acetic acid) which gives orange spots.

This series of reactions is highly specific and there are very few substances other than nitrate esters, nitramines and some nitro compounds which will yield red spots. Of the common high explosives NG, PETN, RDX and HMX all give red spots with the noted Griess reagent. TNT and tetryl give colours upon reaction with the base before Griess reagent is applied. If nitroaromatics only are to be analysed, spraying with bases such as tetramethylammonium hydroxide and 3,3'-iminobispropylamine (Douse, 1982) gives colours much superior to those given by sodium hydroxide. EGDN can be analysed by TLC with Griess reagent visualisation but gives poor and insensitive results because of its comparative volatility.

Quality issues

(1) Contamination   The first quality aspect to be considered is contamination. Contamination of the TLC plate, the eluents and spray reagents would be readily apparent as an overall coloration of the finished plate, but contamination of the apparatus used to apply the samples could produce false positive spots. There are two alternative means to demonstrate the absence of contamination in spotting. If a single micropipette or syringe is to be used for both samples and standards, this should be rinsed and a separate spot of clean solvent applied to the plate before each sample is taken up and spotted. An absence of response from the solvent spots will demonstrate that the micropipette was free of contamination for every sample. An alternative and preferable procedure is to employ disposable micropipettes, one for each sample. In this instance, at least one spot of clean solvent should be applied to every plate, to demonstrate that the supply of micropipettes was free from contamination.

(2) Reference standards   The second quality aspect is the demonstration that the method performed as expected. TLC in conventional glass tanks can give significant variations of R_f from one plate to the next. For this reason, it has always been regarded as essential to include spots of reference explosives corresponding to the explosives being sought on every plate. These standard spots serve two purposes. Firstly, their appearance within the expected range of R_f and with the expected intensity demonstrates that the method has functioned correctly. Secondly, the identity of any suspected explosives spots produced by the samples is determined by direct comparison of R_f values with those of the standards.

(3) Interpretation   The third quality aspect is proper and consistent interpretation of the chromatograms obtained. This is achieved by laying down the basic processes of interpretation in the method documentation. In TLC these include the deductions to be drawn from observation of the spots at appropriate stages during the visualisation process, and of course the comparison of R_f values. When using specific visualisers such as the base/heat/Griess reagent systematic observation of the spots throughout is crucial to a correct interpretation.

A good example of this point cropped up during the author's re-examination of a case which had originally been processed using TLC in the mid-1970s. Study of associated case files revealed a record of interference in the TLC detection of nitroglycerine using base/heat/Griess visualisation. Swab samples taken from black shoes had given spots similar in colour and R_f value to that given by NG. The analyst had, however, been suspicious of the spots and an investigation had been carried out into their cause. It was concluded that the spots were caused by a component present in some black shoe polishes, but that by careful observation they could be distinguished from spots caused by NG. We repeated and carried this investigation further, discovering that there were two potentially interfering substances. These were dyestuffs known as Solvent Yellow 2 (Colour Index 11020, 4-(dimethylamino)azobenzene) and Solvent Yellow 56 (Colour Index 11021, 4-(diethylamino)azobenzene), which are incorporated into some black shoe polishes. Both dyes initially appear on TLC plates as yellow spots. However, on spraying with Griess reagent, which is acidic, a simple pH indicator effect causes the spots to become red. When eluted on plain silica with toluene, the R_f values lie quite close to that of NG.

If the TLC plate is inspected after elution and at each stage of the visualisation process, it is straightforward to recognise that the yellow spots are related to the final red spots, and thus to exclude the latter as NG. The exclusion is even more clear if the plates are fluorescent and are inspected under UV light after elution. The aromatic dyestuffs absorb UV light and quench the plate fluorescence, which NG does not do.

R_f values should be compared according to a preset criterion which allows for the normal slight variations across a plate. We permit a tolerance of ± 0.03 R_f units for conventional

200 mm square plates eluted to a distance 100 mm from the spotting line. If R_f values differ by more than this amount a positive identification cannot be made.

A discussion of R_f values would not be complete without mention of the effects of co-eluting materials. The raw extract from a forensic sample will in most instances contain much larger proportions of materials other than the explosives traces sought. If a strongly concentrated extract of a very 'dirty' sample is applied to a TLC plate, these other materials may interfere in the elution process. The most common effects are depression of R_f values and distortion (streaking) of the spots. Method documentation should draw attention to these possible effects and give procedures to adopt if either is observed. If a reasonably strong spot is found to have been affected the sample solution may be re-analysed after dilution, but for weaker spots a clean-up process may be necessary.

(4) Documentation The final quality factor to consider is clear and adequate documentation of the results. Where TLC is to be used regularly this is most conveniently achieved by the use of a printed form, upon which appearances and R_f values may be recorded. With the base/heat/Griess system the plates themselves rapidly deteriorate after spraying and cannot thus be kept. It is however quite straightforward to make permanent records of plates using a photocopier. Good quality black and white photocopiers record much information, but colour copies are obviously to be preferred. The author has used two varieties of colour photocopier to produce excellent records of TLC plates. The copies should not be regarded as substitutes for the written record, since colours are not copied faithfully and very faint spots may not be recorded at all.

A fast alternative is to record the plates on Polaroid® film.

10.4.2   Gas Chromatography with Chemiluminescence Detection (GC/TEA)

The five most common high explosives have been analysed by gas chromatography since the 1970s. For many years the most popular selective detector was the electron capture detector (ECD), which gives strong responses for the common explosives, but this has been supplanted in recent years by a chemiluminescence detector known as the Thermal Energy Analyser (TEA). For TEA detection, compounds eluting from the GC column are first pyrolysed in a furnace located immediately beside the column oven. Amongst the pyrolysis products yielded by nitrate esters, nitramines and many nitroaromatics, is nitric oxide (NO). In the TEA detector this nitric oxide reacts with ozonised oxygen to form nitrogen dioxide from which near IR light is emitted as a result of a chemiluminescence process. The chemiluminescent emission is detected, after passage through a cut-off filter, by a photomultiplier. The main reason for the switch to chemiluminescence detection (which is considerably more expensive) has been its greater selectivity, which permits trace amounts of explosives to be detected in the presence of much other material.

The advent of fused silica capillary columns greatly improved the analysis of explosives traces. Elution of trace amounts of the more labile explosives such as nitroglycerine from packed columns had sometimes been unreliable and PETN was exceptionally difficult to deal with. The inert surfaces provided by silica columns removed these limitations and in 1981 Douse demonstrated that all of the common explosives could be successfully eluted at sub-nanogram levels.

In common with many other laboratories, we use a combination of silica capillary columns with chemiluminescence detection as the principal tool for explosives trace analysis. Figure 10.3 is a diagram of the system used, which remains essentially as described by Douse (1987) although a number of minor improvements have been made.

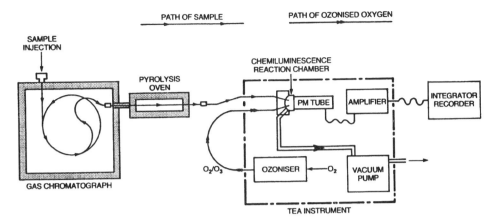

Figure 10.3   Diagram of GC/TEA apparatus.

Conditions for GC/TEA

The conditions used are as follows:

- Common to all columns

Chromatograph: Fisons 'Mega'
  5360 or 8000 Series
Carrier gas: Helium
Syringe: 1 μl, plunger in needle
  type, 70 mm needle
Injection port: Grob Splitless
Injector liner: Glass with silica

Injection temperature: 175°C
Detector: Thermedetec TEA 610
  (Modified)
Pyrolysis temperature: 750°C
Interface temperature: 250°C
Reaction chamber pressure: 0.3–1 mm
  Hg (40–140 Pa)

- Column specific

Column Type	Carrier press kPa	Initial temp °C	Ramp rate °C/min	Final temp °C
BP1	250	80	20	200
BP5	250	80	20	200
CPSIL-19	100	70	20	250

- Details of columns

Column	Details
BP1	SGE type 12Q2/BP1 0.25. 12 metre polyimide-clad silica, 0.22 mm i.d., 0.33 mm o.d., coated with bonded dimethylsiloxane 0.25 μm film thickness.
BP5	SGE type 12QC2/BP5 0.25. 12 metre polyimide-clad silica, 0.22 mm i.d., 0.33 mm o.d., coated with bonded 5% diphenyl-dimethylsiloxane 0.25 μm film thickness.
CP-Sil-19CB	Chrompack catalogue no. 7712, 4 metres cut from 25 metre polyimide-clad silica, 0.25 mm i.d., 0.39 mm o.d., coated with bonded 7% cyanopropyl-7% phenyl-1% vinyl-dimethylsiloxane 0.2 μm film thickness.

Figure 10.4 is a GC/TEA chromatogram of a standard mixture of explosives and Figure 10.5 is an example of a chromatogram obtained from a swab sample.

Quality issues

(1) Contamination Demonstration of the absence of contamination in GC analyses is straightforward, although time-consuming. A blank analysis, consisting of an injection of clean solvent, is carried out before each analysis of a case sample. This blank analysis demonstrates that the entire instrument, the syringe and any other apparatus used in the process of sample injection is free of contamination. Where a sequence of case samples is to be analysed, blank injections may be omitted after sample injections which have given negative results since these act effectively as blanks.

(2) Reference standards To demonstrate that the method has performed correctly, two strategies may be employed. The first is to make injections of a standard explosives solution both before and after the analysis of a case sample or group of case samples. The second is to include with the case sample reference compounds which, by their appearance at the expected retention times in the sample chromatogram, demonstrate that separation and detection have both functioned correctly. The retention times of the reference compounds may also be used as the basis for calculation of relative retention times for explosives standards and sample peaks.

(3) Interpretation Interpretation of the results is based upon direct comparisons between peaks produced by reference explosives and the suspect peaks. Retention time is the most important point of comparison, but the shape of peaks should also be taken into account. A peak which is of comparable height, but is much broader than the corresponding explosive standard peak could have arisen from a large quantity of a substance which gives a relatively low specific response in the TEA detector. The chromatograms obtained from the standard

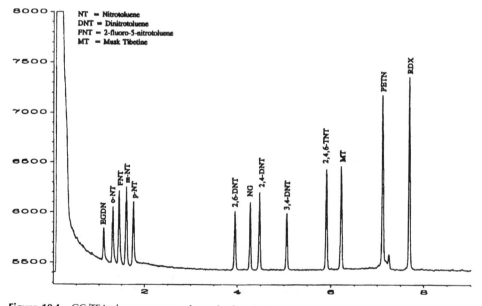

Figure 10.4 GC/TEA chromatogram of standard explosives.

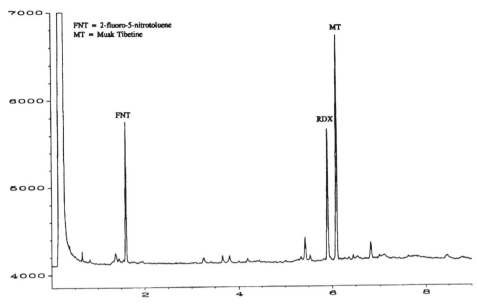

Figure 10.5 GC/TEA chromatogram of sample from swab.

explosives solution used to demonstrate correct functioning of the instrument may also be used to provide comparison peak data.

A systematic scheme for the interpretation of retention times should be devised. At our laboratory, a statistical study for clean solutions of pure explosives indicated that differences in retention time (measured relative to co-injected reference compounds) of up to 0.5% could arise from random instrumental variation. Thus if the relative retention time of a candidate explosive peak in a case sample lies within ±0.5% of the corresponding explosive standard it is regarded as corresponding to that explosive. If the relative retention time differs by more than 1% from the explosive standard it is regarded as a definite non-correspondence. Between these values the candidate is regarded as a possible correspondence, the retention of which may have been shifted by co-eluting materials (as discussed under TLC above). Such shifts have been observed occasionally, and appear to affect the explosive RDX in particular. If other analyses do not eliminate the candidate peak, the co-elution effect may be reduced by diluting the sample solution, or by subjecting it to a further stage of clean-up. (Because the TEA detector is very selective, the co-eluting materials may not necessarily produce any detector response themselves).

(4) Documentation   Documentation of GC analyses is comparatively straightforward and falls into two parts: the raw chromatograms and the interpretation of candidate explosive peaks. The original chromatograms should be plotted either during elution or immediately there-after and of course annotated with the full sample description, the date, time and analyst. Provided that the conditions employed are in accordance with a documented laboratory method, this need only be referred to, but any deviations from standard methods should be recorded in full.

Commercially available integrators and computer data stations measure the retention times, heights and areas of peaks, and print these together with the chromatogram. Software is normally also provided to identify peaks and to make quantitative calculations. In the author's experience, there are not normally facilities provided for the calculation of relative retention times, nor for peak identification based upon such times. These calculations may be carried out manually, and if so are best recorded on a printed form which clearly shows each step. Alternatively one of the many excellent spreadsheet programs now available may be employed. A standardised blank sheet is generated into which the raw peak data are entered, and all calculations are carried out automatically. The calculation spreadsheet used at our laboratory is shown in Figure 10.6. It was critically tested in various ways before being released for use, and all of the cells except those for data entry are locked against modification. The spreadsheet is handled as a controlled document within the laboratory quality system.

10.4.3   Gas Chromatography/Mass Spectrometry (GC/MS)

Gas chromatography/mass spectrometry (GC/MS) was comprehensively reviewed by Yinon in 1987 and is discussed in detail in Chapter 9. Nitrate ester and nitramine explosives yield mass spectra consisting only of low molecular weight fragments when ionised by electron impact, and for this reason GC/MS has not been as widely employed for trace explosives identification as it has for the detection of, for example, polychlorinated biphenyls in the environment. Nevertheless it has much to offer, and the use of chemical ionisation can provide spectra with somewhat less fragmentation.

Quality issues

Quality control issues for GC/MS have much in common with other techniques, and have been specifically addressed by Wu Chen et al. (1990), who gave specific recommendations concerning operation and maintenance of instruments, standards and controls, analytical procedures, etc.

Interpretation   GC/MS does, however, raise the difficult issue of spectral interpretation. Wu Chen and her co-authors (1990) recommend that laboratories should establish criteria for 'what constitutes conclusive identification', but accept that the final interpretation rests with the professional judgement of an experienced mass spectroscopist.

Interpretation of GC/MS data is comparatively straightforward provided that the candidate GC peak is free from co-eluting compounds. In these circumstances, the interpretation can follow an algorithm in which the presence and absence of ions (and their relative intensities) can be compared to the corresponding explosive standard. However, in the presence of co-eluting species, each ion must be ascribed either to a suspected compound or to the co-eluting peak. Examination of the total ion chromatograms assists this process, but the number of potential variables is large and it is hardly practical to write into the standard method an exact procedure which covers every possible case.

It appears that the only practical way to deal with such complex interpretations is for an experienced analyst to document fully the course of each individual interpretation.

10.4.4   Other Analytical Techniques

The principles of forensic quality control referred to above may similarly be applied to other analytical techniques. High pressure liquid chromatography (HPLC), especially when combined with the selective electrochemical detection system developed by Lloyd (1983), or with the TEA, may be used for the identification of explosives traces as discussed in Chapter 8. The combination of supercritical fluid chromatography with TEA detection was reported by Douse (1988), but the unreliability of available hardware has prevented its routine adoption for casework in our laboratory and apparently elsewhere. Improved hardware will surely improve the reliability, and the technique has a number of advantages, for example its ability

330

GC/TEA ANALYSIS - CALCULATIONS
Case No./Job: Example
GC Column Type: BP5
Sample Injected: Control chart standard solution 18/94
Date of injection: 03/02/95
Time of sample injection: 14:05
Time of standard injection: 13:30

	PEAK 1	PEAK 2	PEAK 3	PEAK 4
Suspected identity of peak	NG	TNT	OPETN	RDX
Suspect peak retention (min) RTsusp	2.616	4.587	5.065	5.523
Suspect peak area (int units) Asusp	4355	2694	2785	4265
Reference peak retention (min) RTrefsusp	5.720	5.720	5.720	5.720
Volume injected (µl) Vinj	0.8	0.8	0.8	0.8
Total sample volume (µl) Vtotal	100	100	100	100
Corresponding standard retention (min) RTstd	2.613	4.583	5.060	5.519
Standard peak area (int units) Astd	1646	2085	4478	4476
Ref peak retention for standard (min) RTrefstd	5.716	5.716	5.716	5.716
Volume of standard sol'n injected (µl) Vstd	0.8	0.8	0.8	0.8
Mass of standard explosive injected (ng) Mstd	0.16	0.32	0.6	0.4
Suspect peak relative retention RRTsusp = RTsusp/RTrefsusp	0.45734	0.80192	0.88549	0.96556
Standard relative retention RRTstd = RTstd/RTrefstd	0.45714	0.80178	0.88523	0.96554
Retention time difference % =ABS(RTsusp-RTstd)*100/RTstd	0.11	0.09	0.10	0.07
Relative retention time difference% =ABS(RRTsusp-RRTstd)*100/RRTstd	0.04	0.02	0.03	0.00
Estimate of mass injected (ng) Minj = Mstd*Asusp/Astd	0.42	0.41	0.37	0.38
Estimate of mass in sample (ng) = Minj*Vtotal/Vinj	52.92	51.68	46.64	47.64
Comments				

Figure 10.6 GC/TEA calculation spreadsheet for a routine quality assurance analysis of a mixed explosive standard solution.

to deal with some explosives which are too involatile for gas chromatography. In both cases the quality control procedures closely parallel those for gas chromatography.

10.4.5   Combination of Methods to Provide a Confirmed Identification

As discussed above, a single analysis cannot be regarded as providing sufficient evidence of identity for an explosive trace which is unsupported by other evidence. The principle objective of analyses subsequent to the initial detection of the trace is to confirm or refute its identity as the suspected explosive, but important lesser objectives are to confirm the approximate quantity and to identify the substance if it is found not to be the suspected explosive.

The ideal confirmatory analysis would utilise an analytical system entirely different from that applied in the initial analysis, but this may not be the most practical approach. For reasons of availability, cost and technical feasibility, laboratories often choose to employ a method related to that used in the initial analysis, but differing in a significant respect. To take up the example mentioned in 10.4.1, an initial TLC analysis might be confirmed by a second TLC analysis employing a different eluent. This course of action makes use of existing equipment and skills, and would normally encounter no technical difficulties. Provided that the second eluent is chosen so as to give significantly differing relative R_f values (the use of a normal phase system in the initial analysis and a reverse phase system in the confirmation would ensure this), the probability of a misidentification is considerably reduced. If an alternative specific visualisation reagent were to be employed, this would reduce the probability even further, but unfortunately no reagent for nitrate esters and nitramines has sensitivity and selectivity equivalent to the Griess system.

If gas chromatography is employed for the initial analysis, confirmation can be achieved by subsequent analysis in a different chromatographic column, but again careful attention must be paid to the choice of column so as to ensure that relative retention values change significantly. The polarity of the second column must differ appreciably from that of the first. We use such a system, each analysis being confirmed twice using the three different columns detailed above. Alternative detectors can also be employed, as at the Forensic Science Agency of Northern Ireland, where electrochemical and TEA detectors are used (see Chapter 12, Figure 12.1). A mass spectrometer provides not only a confirmatory detector of high sensitivity and excellent discrimination, but also a means to identify substances which are not confirmed as the expected explosives. We use GC/MS for both purposes.

Confirmatory schemes which mix different types of chromatography can of course be employed, such as that described by Lloyd (1991). Suspect explosive peaks initially separated by HPLC and detected electrochemically are diverted onto a trapping microcolumn. The trapped material is eluted and analysed by GC/TEA. As well as using two completely different analytical methods, this scheme has the advantage that the second analysis is performed only on the sub-fraction of the sample which gave the initial positive response. This further increases the certainty of identification.

Analytical protocol for identification of traces of organic high explosives

Figure 10.7 is a flow chart of the analytical scheme routinely used in our laboratory.

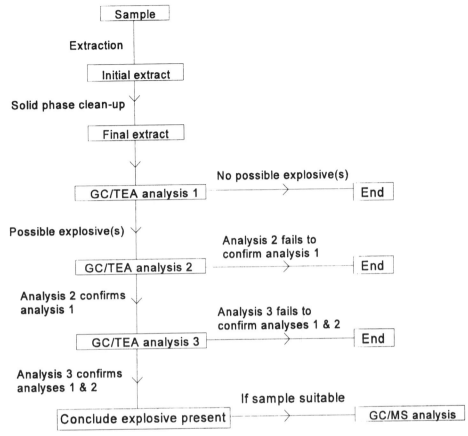

Figure 10.7   Flow chart of a routine analytical scheme for high explosive residues.

10.5   Training of Analysts and Proficiency Testing

10.5.1   *Training*

An important aspect of a quality system is that all those handling and analysing case samples should have been trained and undergone some kind of formal test of their proficiency. The facts of training and testing also need to be documented.

Effective training can take a variety of forms, ranging from attendance at external courses to supervised 'on-the-job' experience. The former is generally used to provide basic knowledge and breadth, whilst the latter is used to impart specific knowledge of local techniques. At our laboratory, the following scheme is used to train analysts in specific methods:

- *familiarisation*:   the trainee reads the written method and watches the method in progress.
- *detailed study*:   the trainer works through the entire method in front of the trainee, giving a detailed explanation and answering any questions which may arise. The roles of trainer and trainee are then reversed and the trainer checks that the trainee has achieved a basic competence, particularly in respect of safety.
- *practice*:   the trainee is given the opportunity to practice until confident; (at this stage forensic case materials may only be processed by the trainee under the close supervision of the trainer).

333

- *proficiency test*: the trainee carries out the method in front of the trainer without assistance. The trainer ensures that all aspects of the trainee's performance are satisfactory, and if this is so, records that the trainee is now qualified to carry out the method.

After a period of regular experience in the method (a minimum of 6 months) the qualified analyst becomes competent to train others and this qualification is also recorded.

10.5.2   Proficiency Testing

In order to ensure continuing competence, some form of proficiency test should periodically be undertaken by each qualified analyst. In the interests of efficiency, these are probably best organised such that all analysts qualified in a particular method analyse one (or more) of a set of test samples prepared together by an independent preparer (in the author's laboratory this is normally the quality officer). The quality system should define courses of action (such as retraining) to be taken if any analyst fails to achieve a result within the expected limits of error for the method. Over a period of time the accumulated evidence of such formally tested competence considerably enhances the credibility both of individual analysts and the laboratory as a whole.

The value of participation in externally prepared proficiency test schemes has been referred to in 10.2.3 above.

10.6   Contamination Control

10.6.1   Literature

Although practising forensic scientists are well aware of the problem, there has been relatively little consideration in the open literature of contamination in forensic work, and in particular very little discussion relating to explosives traces. Lloyd (1992) has briefly discussed the problem of contamination in explosives trace work and concluded that trace analysis work should be conducted in laboratories remote from sites at which intact explosives are handled or used. We consider this an extreme and somewhat misleading view, believing that effective control and monitoring measures are essential in all cases regardless of the perceived proximity of explosives.

Cook (1981) has discussed the general problem of contamination in forensic work, making brief reference to explosives residues. He gave a list of basic precautions (mainly in the context of fibres work) which included cleaning of work benches, the use of disposable paper work surfaces, separation in space of examinations and changing of laboratory coats between examinations. In a model study aimed at the control of carcinogen contamination, Hill et al. (1978) observed the inadvertent release of 0.2–1.3 mg of solid into the surroundings during the preparation of standard solutions from the solid. In a medical context, Kennedy and Stevens (1988) have suggested the use of tracer materials, including a fluorescent dye, to indicate to laboratory workers the spread of contamination.

Moore et al. (1986) carried out a controlled study of fibre movement between working areas in a forensic laboratory. Some transfer, particularly of cotton fibres, was observed between benches within the same room but they concluded that current quality assurance procedures in the laboratory gave a considerable safety margin.

10.6.2   Sources of Explosives Contamination

Forensic laboratories or departments which take explosives cases normally have to deal both with bulk explosives and items for trace examination. In such laboratories the bulk explosives will severely contaminate any surface which they contact. Bulk explosives which contain volatile explosive substances such as nitroglycerine (NG) may also contaminate materials placed within the same closed environment (such as a cupboard or vehicle interior) by vapour transfer. Surfaces contaminated from bulk explosives will include work benches, floors and equipment such as glassware. The skin and clothing of personnel handling the explosives also become contaminated, and their movement can spread the contamination away from its source, although at progressively lower levels.

Solid explosives such as trinitrotoluene (TNT), pentaerythritol tetranitrate (PETN) and cyclotrimethylene trinitramine (RDX) are highly persistent, and can potentially be detected at nanogram levels even after several stages of inadvertent transfer from an initial bulk source. Even if bulk explosives have only entered the laboratory on a single occasion, contamination may remain for a long time.

Any laboratory which accepts items for trace examination will sooner or later encounter one which is very heavily contaminated. For example, the concentrated extract from an item submitted some time ago to our laboratory for trace explosives analysis was found to be virtually neat nitroglycerine! Unless appropriate precautions are in place, such an item can represent a contamination risk for all subsequent items. The problem might not even be recognised if the item is brought into the laboratory and for some good reason not eventually tested for explosives.

If laboratory personnel visit an environment heavily contaminated with explosives, such as a military base, and return to the laboratory they may bring contamination with them. Similarly, military visitors and 'bomb squad' members may bring contamination into the laboratory, and any material associated with their means of transport should not be used to package exhibits! Thus careful precautions and monitoring are necessary in all explosives trace laboratories, whether or not bulk explosives are known to be nearby.

10.6.3 Contamination Control Measures

The various control measures which may be taken are considered in turn below, and the way in which each is applied in our laboratory is explained.

Segregation of activities and control over the movement of personnel

The single most important control measure is physical separation of all activities involving explosives in bulk amounts from the trace activities. A minimum requirement is different rooms divided by a solid and impermeable wall, but the use of separate buildings or separate zones within a larger building is preferable. Complete separation onto different sites appears ideal in principle, but will normally be impractical and carries the disadvantage of splitting explosives expertise.

Such segregation would be largely nullified if free movement of personnel were permitted between the 'bulk' and 'trace' zones, since this would provide a very effective vector for the movement of contamination. If the laboratory is large, with sufficient staff, it is practical to designate the separate zones and allocate different personnel to them on a semi-permanent basis. In smaller organisations such a division is unduly restrictive and impractical, so rules must dictate under what circumstances personnel may move from 'bulk' to 'trace'.

In our laboratory, bulk and trace explosives are examined in separate buildings. Staff are not permitted to enter the trace environment on any working day in which they have visited the bulk explosives examination building (or other bulk areas such as explosives ranges and magazines). They must have changed into clean clothing if they wish to use the trace environment on the following day, and of course thoroughly washed themselves.

Shoes

Shoe contact is perhaps the most effective way to transfer explosives into a trace environment. We take two measures to eliminate this transfer route. Firstly, 'sticky' mats consisting of a series of layers, each of which has a moderately sticky surface, are placed at the entrances to the trace environment. When the exposed layer has become soiled it is stripped away to reveal a fresh layer beneath. Persons entering the trace environment must tread upon a such a mat. Secondly, disposable overshoes which completely cover normal footwear are put on at entry to the trace environment.

An alternative scheme is to maintain entirely separate footwear for use in the trace environment, and to locate a 'swing-over' bench across the entrance. Persons entering remove their footwear whilst sitting on the bench, then swing their legs over the bench to put on the trace environment shoes.

Clothing

Clothing can also transfer explosive traces. Whilst a complete change of clothing might seem desirable this leads to practical complications, and in most laboratories overclothing is worn. We wear disposable oversuits which are close-fitting at wrists and ankles (the ankle fit overlaps the overshoes with no gap). Staff must also remove all watches and other wrist jewellery before entering the trace environment, since these could be exposed at the wrist joint.

Hands

Not only must hands be washed on entry to trace environment, but also it is important to ensure that the materials used for washing are not open to contamination. For this reason it would be unwise to place the washing facilities outside the trace environment. They should be placed either within the trace environment or preferably in entrance lobbies. Disposable gloves are worn for all manipulations of forensic case materials, but as an additional precaution it is probably best to wear gloves at all practicable times.

Movement of materials and equipment

Materials entering the trace environment could also act as a vector for the migration of explosives. It is a reasonable assumption that new materials arriving from suppliers are free of explosives contamination, but there is a possibility that outer packaging could be contaminated. To insure against this possibility, one person opens the packaging in the entrance lobby and a second, clad in protective clothing, takes the materials directly into the trace environment. We have found suppliers generally very obliging in the provision of additional outer packaging (usually at no extra cost) when they understand the reasons for the request.

Equipment for the trace environment is best purchased new, but there will be occasions when used equipment must be taken into the trace environment. Such equipment should be thoroughly cleaned immediately before and after entry and then swab sampled to confirm that no explosive traces are present.

Long-term storage both for casework materials and equipment is best provided elsewhere, since space in the trace environment is expensive to provide and maintain. Such items should be wrapped in at least two impermeable layers, so that on re-entry to the trace environment the outer layer may be removed and discarded. We use heat sealed nylon bags.

Cleaning

The trace environment should be kept reasonably clean and tidy. Particular attention must be paid to the workbenches upon which case materials are examined and extracts processed.

In our laboratory these are cleaned weekly. Other bench surfaces and floors are also cleaned on a regular basis. Immediately before use with a case, a workbench is cleaned by the operator and covered with disposable glazed paper. The bench is cleaned again immediately after use.

Separation of activities in time

The normal forensic practice of case separation in time should be carried over into trace environment operations. Thus, items from different cases should not be exposed simultaneously upon the same workbench.

Disposable minor apparatus

Single-use, disposable, apparatus is used wherever practicable, and is certainly to be recommended for direct contact with case materials. In our laboratory scalpels, forceps and virtually all glassware used for taking samples from case items and working these up for analysis are single-use. The only exceptions are very expensive and rarely-used pieces of larger glassware. These are cleaned and checked before re-use. Separate supplies of items such as notepaper, labels, pens and hand calculators are of course required within a trace environment. Another sundry but essential item is a kit of basic tools such as pliers and screwdrivers.

Major apparatus

Larger apparatus can be divided into two groups based on cost.

Apparatus of moderate cost and frequent use, such as ultrasonic baths, heating blocks and centrifuges, is best purchased new and dedicated to the trace environment. Whilst it is possible to move such items in and out of the environment, the costs of the cleaning and checking necessary on each re-entry will soon outweigh the purchase cost.

Apparatus of relatively high cost, in particular analytical instruments, may be expensive to tie-up in the trace environment and here a balance must be struck. If the anticipated trace workload is high and continuing, then it is practical to purchase and maintain a complete set of trace instrumentation. This has the advantage that the trace analysts can make most effective use of their time without frequently having to enter and leave the trace environment. If, however, the anticipated work is more occasional in nature, the instruments can be located outside the trace environment, and thus available for other use. Provided that the final, relatively concentrated extracts to be analysed are exported from the trace environment within closed septum vials, and that the analytical system is suitably blanked before each case analysis, the risk of undetected contamination at this stage should be very slight indeed.

Standard solutions

Standard solutions are required for the calibration of trace analytical equipment, and represent a potential source of contamination. However, since the equipment is so sensitive, such solutions are normally very dilute and only a recognisable spill will release significant levels of contamination. However, the work of Hill et al. (1978) indicates that the preparation of standard solutions from solids can release very significant quantities of solid into the surrounding environment. Clearly such preparation cannot take place in the trace environment.

The obvious procedure is to prepare concentrated stock solutions and intermediate dilutions outside the trace environment and only to move very dilute solutions into it. For example, in our laboratory a mixed standard solution containing a range of explosives at levels between 1 μg/ml and 8 μg/ml is prepared by successive dilution outside the trace environment. Aliquots of this solution are transferred from the final volumetric flask to smaller trace environment glassware, taking the flask into the trace environment only briefly for this purpose and taking care to isolate it whilst in the trace environment. This solution is used as such for some calibrations and is diluted by a further factor of 10 within the trace environment for others. These dilute standard solutions are stored in a separate refrigerator and only 1–2 ml at a time is taken out for instrument use. The use of a fairly complex mixture as a standard is seen as an advantage since if any contamination from the standard solution ever did occur the simultaneous presence of so many explosives would immediately alert the analyst.

10.6.4 Monitoring

There are two essential monitoring systems: (1) specific tests made at the time case samples are processed; and (2) regular tests made independently of casework.

Specific monitoring

At least one control sample is processed and analysed in parallel with the 'live' samples. Control samples are exposed to the same environment as the live samples, and should acquire approximately the same degree of environmental contamination, if any. This provides the most basic, and extremely important, level of specific monitoring.

Arguments can, however, be made that controls, particularly if their number is small, might by chance acquire less contamination than the live samples. Such arguments can be countered if it is shown that the relevant environment was free of contamination at the time the samples were processed. This may be achieved by the taking of appropriate swab sampling and analysis. We sample the paper-covered workbench surface and the operator's suit front and gloved hands immediately before case materials are worked upon.

Regular monitoring

Specific monitoring will detect any contamination which may have affected case samples, and because of its specific nature it carries the greatest evidential weight, but in the unlikely event that a specific monitoring sample proves positive the value of the case samples will be lost. Therefore, there is a case for regular monitoring of key areas to ensure that this never happens.

There are other good reasons to carry out regular monitoring. Firstly, surfaces which would never come into direct contact with case materials, such as floors, can be tested. Although it is not normally appropriate to sample such surfaces as part of specific monitoring, it might be suggested that the level of contamination upon them was so high as to pose a threat to case samples. If they are never monitored this suggestion cannot be countered with positive data. Secondly, a regular monitoring programme, if suitably planned, can provide a very wide-ranging check not only upon the trace environment itself, but also upon the materials and instrumentation used therein. The regular monitoring tests will thus provide a natural (and indeed essential) part of the overall laboratory quality scheme.

We allocate one person-day per week for regular monitoring. Swab samples are taken from all of the casework bench surfaces, and from representative areas of other surfaces. The samples are taken, processed and analysed using materials from stocks used in casework, and using casework apparatus and methods. One 'spiked' swab containing a range of explosives at low levels is also processed and serves as a regular check upon the efficiency of recovery and analysis procedures.

Since regular monitoring is likely on occasion to reveal some detectable contamination, a policy should be set in advance for actions to be taken in such an event. In our laboratory, contamination below 10 ng is recorded but no further action taken. Contamination higher than this level requires immediate cleaning of the affected area, followed by re-sampling to confirm the effectiveness of the cleaning. Checks are also made to establish whether any casework could have been affected by the contamination. On the very few occasions when significantly higher levels have been detected, an inquiry has been instituted into the source and procedures modified so as to eliminate such a source in future.

Although the 10 ng action level would be significant in casework, it is in practice very conservative. This is the total amount of explosive detected in a monitor sample, which will have been taken by systematic solvent swabbing of an entire bench surface, or perhaps one square metre of floor. Even if all other precautions failed, the proportion of such contamination which could conceivably be transferred to a case sample is very small; certainly less than 10%.

10.7 Future Developments

The technology of trace explosives detection has advanced enormously over the last 20 years, and there is every reason to expect that this advance will continue. The potential of such techniques as supercritical fluid chromatography and capillary electrophoresis is being explored, and methods of widespread applicability will surely arise. Supercritical fluid chromatography offers potential for the analysis of thermally labile explosives, and the pioneering work of Douse (1988) will perhaps soon be brought to fruition. Capillary electrophoresis is already being used for the analysis of inorganic ions of explosives relevance (McCord et al., 1994) and can in principle be applied to organic explosives by the use of micellar techniques (Chapter 8). Its practical implementation to analytical protocols will probably have to await the availability of suitable selective detectors.

Whatever new techniques are developed, the principles of quality control will remain unchanged and will be applicable to these methods. A major change would arise if methods of much greater sensitivity were developed – perhaps with the ability to detect femtogram amounts or less. Two quality responses would be necessary in that case. Firstly, precautions to prevent laboratory contamination would have to be considerably increased, and would become very burdensome. Secondly, the interpretation of such ultra-trace detections would have to be very carefully addressed, so that invalid conclusions would not drawn.

References

AMAS, S. A. H. and YALLOP, H. J., 1966, Identification of industrial explosives of the gelignite type, *Journal of the Forensic Science Society*, **6**, 185–188.

AMAS, S. A. H. and YALLOP, H. J., 1969, A test for cyclotrimethylene trinitramine, *Analyst*, **94**, 828.

ANON, 1992, *Accreditation for forensic analysis and examination*, Document NIS46, Edition 1, NAMAS Executive, Teddington, UK.

BEVERIDGE, A. D., 1992, Development in the detection and identification of explosive residues. *Forensic Science Review*, **4**, 18–49.

BEVERIDGE, A. D., PAYTON, S. F., AUDETTE, R. J., LAMBERTUS, A. J. and SHADDICK, R. C., 1975, Systematic analysis of explosive residues, *Journal of Forensic Sciences*, **20**, 431–454.

BRUNELLE, R. L., GARNER, D. D. and WINEMAN, P. L., 1982, A quality assurance programme for the laboratory examination of arson and explosives cases, *Journal of Forensic Sciences*, **27**, 774–782.

COOK, R., 1981, The problem of contamination, *The Police Surgeon*, **20**, 65–66.

DOUSE, J. M. F., 1981, Trace analysis of explosives at the low picogram level by silica capillary column gas–liquid chromatography with electron capture detection, *Journal of Chromatography*, **208**, 83–88.

DOUSE, J. M. F., 1982, Trace analysis of explosives in handswab extracts using Amberlite XAD-7 porous polymer beads, silica capillary column gas chromatography with electron capture detection and thin-layer chromatography, *Journal of Chromatography*, **234**, 415–425.

DOUSE, J. M. F., 1985, Trace analysis of explosives at the low nanogram level in handswab extracts using columns of Amberlite XAD-7 porous polymer beads and silica capillary column gas chromatography with thermal energy analysis and electron capture detection, *Journal of Chromatography*, **328**, 155–165.

DOUSE, J. M. F., 1987, Improved method for the trace analysis of explosives by silica capillary column gas chromatography with thermal energy analysis detection, *Journal of Chromatography*, **410**, 181–189.

DOUSE, J.M.F., 1988, Trace analysis of explosives by capillary supercritical fluid chromatography with thermal energy analysis detection, *Journal of Chromatography*, **445**, 244–250.

HILL, R. H., GAGNON, Y. T. and TEASS, A. W., 1978, Evaluation and control of contamination in the preparation of analytical standard solutions of hazardous chemicals, *Journal of American Industrial Hygiene Association*, **39**, 157–160.

JENKINS, R. and YALLOP, H. J., 1970, The identification of explosives in trace quantities on objects near an explosion, *Explosivstoffe*, **6**, 139–141.

KENNEDY, D. A. and STEVENS, J. F., 1988, Tracing of laboratory contamination: quality control approach, *Lancet*, **1**, 471–472.

LLOYD, J. B. F., 1967, Detection of microgram amounts of nitroglycerin and related compounds, *Journal of the Forensic Science Society*, **7**, 198.

LLOYD, J. B. F., 1983, High performance liquid chromatography of organic explosives components with electrochemical detection at a pendant mercury drop electrode, *Journal of Chromatography*, **257**, 227–236.

LLOYD, J. B. F., 1991, Forensic explosives and firearms traces: Trapping of HPLC peaks for gas chromatography, *Journal of Energetic Materials*, **9**, 1–17.

LLOYD, J. B. F., 1992, HPLC of explosives materials, *Advances in Chromatography*, **32**, 173–261.

LLOYD, J. B. F. and KING, R. M., 1990, One-pot processing of swabs for organic explosives and firearms residue traces, *Journal of Forensic Sciences*, **35**, 956–959.

MCCORD, B. R., HARGADON, K. A., HALL, K. E. and BURMEISTER, S. G., 1994, Forensic analysis of explosives using ion chromatographic methods, *Analytical Chimica Acta*, **288**, 43–56.

MIDKIFF, C. R. and WASHINGTON, W. D., 1974, Systematic approach to the detection of explosive residues III: Commercial dynamite, *Journal of the Association of Official Analytical Chemists*, **57**, 1092–1097.

MIDKIFF, C. R. and WASHINGTON, W. D., 1976, Systematic approach to the detection of explosive residues IV: Military explosives, *Journal of the Association of Official Analytical Chemists*, **59**, 1357–1373.

MOORE, J. E., JACKSON, G. and FIRTH, M., 1986, Movement of fibres between working areas as a result of routine examination of garments, *Journal of the Forensic Science Society*, **26**, 433–440.

STROBEL, R. A. and TONTARSKI, R. E., 1983, Organic solvent extracts of explosive debris: clean-up procedures using bonded phase sorbents, *Proceedings of the International Symposium on the Analysis and Detection of Explosives*, pp 67–70; Washington, DC: US Government Printing Office.

WALLACE, J. S. and MCKEOWN, W. J., 1993, Sampling procedures for firearms and/or explosives residues, *Journal of the Forensic Science Society*, **33**, 107–116.

WU-CHEN, N. B., CODY, J. T., FOLTZ, R. L., GARRIOT, J. C., PEAT, M. A. and SCHAFFER, M. I., 1990, Report of 1988 ad hoc committee on forensic GC/MS: recommended guidelines for forensic GC/MS procedures in toxicology laboratories associated with offices of medical examiners and/or coroners, *Journal of Forensic Sciences*, **35**, 236–241.

YINON, J. (1987), Mass spectrometry of explosives, in Yinon, J. (Ed.) *Forensic Mass Spectrometry*, pp 105–130, Boca Raton, Florida: CRC Press.

YINON, J. and ZITRIN, S., 1981, *The analysis of explosives*, Oxford: Pergamon Press.

YINON, J. and ZITRIN, S., 1993, *Modern methods and applications in analysis of explosives*, Chichester: John Wiley and Sons.

Appendix: Precautions

The following is an excerpt from a standard method in use at the author's laboratory

1. Scope

This procedure details the precautions to be taken by personnel before entry to, and whilst working in, the S12 forensic explosives trace suite and the H16 car bay. Precautions to be taken upon entry of materials and equipment to the areas are also given. Regular sampling of the trace suite over a period of some 2 years has established that these precautions prevent significant contamination.

2. Precautions before Entry of Personnel

Entry to the trace suite is not permitted to any person who may be significantly contaminated with explosives. In practice this means anyone who has on that day handled bulk explosives, visited a magazine, range or bulk processing area. In addition anyone who has recently undertaken any of these activities and has not subsequently changed their clothes and washed thoroughly may not enter.

Only approved persons previously trained in trace operations may enter the trace areas unescorted. A list of approved persons is displayed at the entry point. Visitors may enter the trace suite with the permission of Head of Chemistry and Research or his deputy, provided that they are escorted by an approved person. Escorts must ensure that visitors observe all precautions given in this procedure.

3. Entry Procedure

3.1 S12 Trace Suite

Outer clothing such as overcoats and jackets should be removed and hung on pegs in the corridor outside the trace suite. (The temperature within the trace suite is maintained at a nominal 20°C.)

Normal entry to the trace suite should be made through doors 53 and 54. Doors 50 and 51 should only be used in exceptional circumstances.

Tread on the 'sticky mat' at the entry point with both feet (not on the floor), and take two overshoes from the supply on the shelf in the lobby between doors 53 and 54. Put on one overshoe and put that foot on the lobby floor. Then put on the second overshoe and stand on the floor. Put on an oversuit from the supply on the shelf. Remove any watches or other wrist jewellery and leave them in the lobby. Record your entry in the day book in the lobby at a point on the page approximately indicating the time of day.

Enter the trace area by standing on the second sticky mat beyond the inner door 54 with both feet. Before touching anything go to the sink and wash your hands thoroughly, drying them with disposable paper towel. Put on disposable gloves.

11

Analysis of Low Explosives

EDWARD C. BENDER

11.1   Introduction

Low explosives include both propellant powders and mixtures of chemicals that normally deflagrate (burn very rapidly with a propagation velocity less than the speed of sound) rather than detonate (explode by detonation with a propagation velocity greater than the speed of sound). From 1990 to 1994, 44% of all bombings in the United States used low explosive fillers. When flammable liquids are removed from the statistics, 64% of all bombings reported involved low explosives (ATF Explosive Incident Report, 1994).

Low explosives most commonly react by oxidation/reduction ('redox'). Formulations typically consist of an oxidizer such as nitrate, chlorate or perchlorate, and a fuel such as charcoal, sugar, dextrine, red gum etc., or metals such as aluminum or magnesium. Some organic compounds, like nitrocellulose, carry the requisite ingredients internally – the nitrate ester group oxidizes carbon and hydrogen atoms in the same molecule. Low explosives also include additives for various purposes such as sulfur and antimony trisulfide (stibnite) which may be added to a formula as 'tinders' to lower the ignition temperature of the composition and make it more reactive. Other materials stabilize the mix, bind the components together, enhance smoke or color output, or prevent caking. Many ingredients serve more than one function. Dextrine for example can be used as a fuel and also to bind the composition together. Detailed discussion of explosive chemistry and solid state reactions may be found in books by McLain (1980) and Conklin (1985).

This chapter describes the low explosive formulations and products most commonly encountered in casework and discusses the analytical methods and protocols best suited to identify them before and after explosion.

11.2   Improvised Explosive Devices (IEDs)

Low explosives burn rapidly when ignited but most low explosives require containment to generate the gas pressure necessary for an explosion. Commercial examples are bullet cartridges cases and cardboard tubes sealed at one end as used in fireworks and model rockets. A home-made bomb is commonly known as an improvised explosive device (IED). Two essential components for a IED containing a low explosive are a container to confine the explosive and a fuse to ignite it. A high explosive does not require confinement to explode.

343

11.2.1   Containers

The 'classical' pipe bomb is a steel pipe with threaded end caps. Polyvinyl chloride (PVC), acrylonitrile-butadiene-styrene (ABS) and copper pipes with flattened ends are also used.

Damage to container fragments can often suggest whether the cause was a high or low explosive. Low explosives typically produce larger pipe fragments than high explosives, but the line is not precise. Figures 11.1 and 11.2 illustrate damage to similar steel pipes caused by low and high explosives respectively. The most obvious difference is fragment size. High explosive damage to metal is discussed in detail by Baker and Winn in Chapter 13, Sections 13.2 and 13.4. Typical effects are rolled edges and gas wash produced by hot gases and micro-craters produced by the high velocity impact of small particles. A blue annealing color caused by the direct heat of the detonation or high speed impact may also be observed. The low explosives tend to split pipes rather than shatter them. Smokeless powder produces damage which is usually considered borderline between high and low explosives. If initiated by flame, the damage is characteristic of low explosives. However, when initiated with a blasting cap, the damage has many high explosive similarities. Explosives tend to take on the velocity of their initiators if they have the energy to do so.

11.2.2   Fuses

Explosives require a fusing system for ignition. Time-delay fuses also enable the bomber to flee the scene.

The simplest and most commonly encountered non-electric fusing system is black powder safety fuse which burns at a fixed rate, typically 4 ft/min. Chemicals which produce flame when they are brought into contact are also used in IEDs. Delay can be achieved by mechanisms such as acid eating through rubber or metal, or by a liquid slowly soaking through

Figure 11.1   Damage to a pipe (10 inches long by 1.5 inches diameter) caused by a low explosive (black powder).

Figure 11.2 Damage to a pipe (10 inches long by 1.5 inches diameter) caused by a high explosive (RDX: military C-4).

paper towels. 'Underground' literature, the Internet and military 'survival' publications contain many such formulations.

Electric fusing systems require a power source (battery), wire, a switch (timer, radio-control, trip-wire etc.), and the ignition element (source of flame or sufficient heat to ignite the explosive, e.g. a resistance wire).

Fuse components and assemblies are limited only by the imagination of the fabricator of the bomb. However, the more complex the system, the more potential forensic evidence may remain.

11.3 Analysis of Unreacted Low Explosives

Suspected low explosives are analyzed systematically to identify the oxidizer and fuel and determine the hazard of the composition. Rigorous safety procedures must be followed, in particular: never grind or attempt to ignite unknown explosive material except in very small amounts (a few milligrams), and always use safety glasses or shields.

Figure 11.3 illustrates a general scheme for analysis of unreacted low explosives. The primary components of the scheme are discussed below.

11.3.1 Visual and Microscopic Examination

A low power stereomicroscope is used to determine the homogeneity and morphology of the material. Commercial low explosives such as Pyrodex®, black powder and smoke-less powders are readily recognizable. The components of homemade mixtures may be

345

PRE-BLAST ANALYSIS

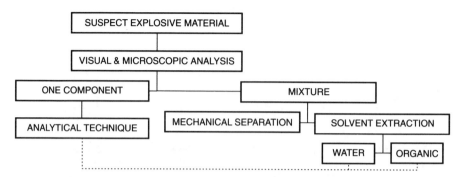

Figure 11.3 General analytical scheme for unreacted low explosives.

recognizable if they have distinctive color or morphology. If only one material is present, it can be directly analyzed (see 11.3.5 below) but if there are several components, then separation may be required.

11.3.2 Separation

Separation is achieved mechanically or by solvent extraction. Mechanical separation is preferred because solvent extraction, particularly with water, can change the chemical composition of extracted components by reactions such as ion exchange. That is, the chemicals recovered when the solvent is evaporated may not be the chemicals which were dissolved. An example is the recovery of the mixed salt sodium carbonate sulfate from aqueous extraction of a mixture of sodium sulfate and sodium carbonate (Beveridge et al., 1975).

Crystalline particles can often be recovered from a mixture by using tweezers or a scalpel. The mixture is spread out on a sheet of paper or in a petri dish and the recovery is carried out while viewing through a stereo microscope.

If mechanical separation is infeasible, solvent extractions can be employed. The typical protocol is to extract the mixture sequentially with:

(1)   a non-polar organic solvent (e.g. ether or pentane);

(2)   a polar organic solvent (e.g. acetone, methanol); and

(3)   water.

The resultant organic and water extracts, and residual insoluble material, are then analyzed.

11.3.3 Analytical Techniques

Many different analytical techniques are available to analyze low explosives. Those which are most widely used are briefly discussed below with respect to sample preparation and applications. The combinations of techniques most frequently used in forensic laboratories are also discussed. Analytical conditions are given in the Appendix.

Infrared spectroscopy (IR)

IR is discussed in detail in Chapter 9. Fourier transform infrared spectrometers (FTIR) have largely replaced dispersive instruments in forensic chemistry laboratories.

IR is most commonly applied to identify unreacted inorganic oxidizers, tinders, and organic fuels in low explosive mixtures. It is common practice to analyze a suspected low explosive mixture by IR before any separation is attempted. The technique is well suited to this application, since most oxidizers have very strong adsorption bands in the infrared region. Several examples are given in the following sections.

Some components of low explosives and their residues, however, cannot be identified by IR. These include metals (e.g. aluminum and magnesium) and carbon which do not transmit IR radiation, and alkali halides (sodium and potassium chloride) which do not significantly absorb IR radiation. Halides are not typical components of low explosives but are primary reaction products of chlorate and perchlorate oxidizers. Hygroscopic residues such as wet sugars or inorganic compounds such as calcium nitrate are very difficult to analyze by IR. The water spectrum can obscure bands of interest and if incorporated as water of hydration, may shift the frequency of the absorptions from what is observed in dry samples.

Thus, it is sound analytical practice to combine IR analysis of low explosives and their residues with elemental analysis in order to ensure that all components are identified. A further advantage of elemental analysis is provision of data which can aid interpretation of the IR spectra.

The usual methods of sample preparation for IR analysis are: (1) potassium bromide (KBr) disc; and (2) diamond cell.

A KBr disc is prepared by grinding a small quantity of the explosive or residue with KBr and pressing with a hand or hydraulic press to form a clear disc. The disc may then be placed in the beam of a dispersive or FTIR instrument and the transmission spectrum recorded. If diffuse reflectance is used, the sample simply is ground and placed in a sample cup with KBr (Miller et al., 1990).

Analysis using a diamond cell requires only that the sample be placed between two diamonds and pressed into a thin layer. The diamond cell may be used with wet samples, but water complicates interpretation; low explosive samples should always be dried prior to IR analysis.

If the sample is an organic low explosive such as nitrocellulose, addition of a drop or two of a solvent like methylene chloride during grinding with KBr may facilitate forming a homogeneous disc. Another possible procedure is to cast a thin film onto a sodium chloride (NaCl) or KBr disc by evaporation of a solvent. Acetone, tetrahydrofuran, or ethyl acetate are suitable for this technique. With a diamond cell, a portion shaved with a scalpel may be placed directly onto a diamond face.

Amorphous compounds such as oils or petrolatums can be extracted by a non-polar organic solvent like pentane or ethyl ether. The evaporated extract is then placed between two KBr or NaCl discs, or between two diamonds, as a thin film.

Elemental analysis by scanning electron microscope/energy dispersive X-ray analyzer (SEM/EDX)

SEM-EDX is a non-destructive technique that can identify a wide range of elements especially when fitted with a light element detector (a beryllium window detector detects elements with atomic number $(Z) > 11$; a windowless light element detector detects elements with $Z > 5$). The instrument measures characteristic X-rays emitted by elements when a sample is bombarded by electrons from the scanning electron microscope. The detection limits are very low. If the sample can be 'seen' with an SEM its qualitative composition can be determined in a few moments.

The primary applications are to identify metallic fuels and to aid interpretation of the results of FTIR and X-ray powder diffraction (see below) analyses by identifying the major elements present. SEM-EDX should be performed before the sample is altered by any extraction technique in order to obtain a general elemental profile of the bulk material. Sample preparation requires no more than placing a sample in carbon glue or paste on a stub. Most instruments can produce elemental maps which show the distribution of specific elements. This is useful if the examiner wishes to determine the homogeneity of the sample.

X-ray powder diffraction (XRPD)

XRPD is one of the most valuable techniques for the identification of crystalline components of inorganic low explosive mixtures. It is non-destructive and can give definitive identification, especially for pure compounds.

Sample preparation begins with grinding the mixture into a fine powder. The sample must be dry. With trace quantities, the sample is placed in a capillary tube which is inserted into a Debye–Scherrer camera. Larger samples may be placed into a welled holder, onto a frosted glass slide or sprinkled on a glass slide coated with grease then exposed to the X-ray beam. The key to sample preparation is random orientation of the crystals to the X-ray beam. Preferred orientation can cause difficulty in the identification of XRPD patterns but it is a common occurrence and with experience can readily be dealt with.

After the XRPD pattern is generated, the identification is made by comparison with an authentic pattern generated by the Joint Council Pattern Diffraction Society (JCPDS). This can be done by a manual search of patterns catalogued by 'Hanawalt' or 'Fink' techniques. (JCPDS Powder Diffraction File search manuals are published by the International Center for Diffraction Data, 1601 Park Lane, Swarthmore, PA 19081-2389, USA.) Computer searching algorithms are more commonly employed and usually are faster. Unknown patterns also can be compared to previously identified samples that are stored in a computer data base, or simply matched from hard copies or films.

Elemental analysis of the sample (SEM/EDX or X-ray fluorescence) is an ideal complementary technique.

There are techniques which enable hygroscopic samples to be analyzed by XRPD. The powders must first be dried with a heat lamp or in a vacuum desiccator. The dried powder should be smeared onto a frosted glass slide and coated with petrolatum or oil or sealed in a Mylar tent. Identifiable diffraction patterns can be generated using this technique although preferred orientation can be severe. Alternatively, trace quantities can be sealed in a capillary tube for analysis with a Debye–Scherrer camera. If an evacuable X-ray diffraction unit is available then these techniques become unnecessary.

High performance liquid chromatography (HPLC)

HPLC is discussed in detail in Chapter 8. Although most of the components found in low explosive mixtures are not amenable to 'classic' HPLC analysis, there are notable exceptions. The most important application to analysis of low explosives, and which is discussed below, is the separation and identification of additives in smokeless powders (Bender, 1983; De Bruyne et al., 1989; Stine, 1991; Robertson and Kansas, 1990). Other applications include separation and identification of dicyandiamide and sodium benzoate in Pyrodex® using weak ion exchange columns (Bender, 1989a), and identification of ions in solution by use of an ion exchange column and a UV detector (Bender, 1989b) (see later).

Another type of HPLC, 'size exclusion chromatography' has been applied by Lloyd (1984) to detect and differentiate nitrocellulose from different sources (Chapter 8, Section 8.5).

Gas chromatography (GC)

GC is discussed in detail in Chapters 8 and 10. It is not widely used for analysis of low explosives because few components are volatilizable organic compounds. The most valuable application is probably analysis of smokeless powder additives (Selavka et al., 1989).

High temperature gas chromatography (Bender et al., 1993) can be used to characterize oils and petrolatums that are sometimes found in improvised mixtures. If it is suspected that the mixture contains a heavy hydrocarbon such as a lubricating oil or Vaseline®, then an ether or pentane extract of the mixture can be directly analyzed by GC (see Chapter 8, Figures 8.4 and 8.5).

Gas chromatography/mass spectrometry (GC/MS)

GC/MS is primarily applied in the identification and comparison of smokeless powder additives (Martz and Lasswell, 1983; Kee et al., 1990). This will be discussed in Section 11.7.

Spot tests

Instrumental techniques have largely displaced chemical spot tests in explosive analysis. However, spot tests do permit rapid and low-cost screening if there is sufficient sample. Because much has been published on this subject, and few improvements have been or will be made in this area, Table 11.1 is limited to a chart of the more useful reagents (Amas and Yallop, 1966; Feigl and Anger, 1972; Parker et al., 1974; Midkiff, 1979; Abramovitch-Bar et al., 1993). More extensive information can be found in the references. Spot tests can be performed both on the low explosive mixture or its residues.

Ion chromatography (IC)

IC is discussed in Chapter 8 (Section. 8.6). It is one of the most useful techniques for analysis of inorganic low explosives and their residues (Reutter et al., 1983).

Sample preparation consists of centrifuging or filtering a water extract (using pre-washed membrane filters not ordinary filter papers: these may add as contaminants the very ions sought). The unevaporated extract is diluted and injected into the ion chromatograph. Quantitative ion chromatograms can be obtained both for cations and anions, though not simultaneously.

The primary advantages of IC are sensitivity and no potential loss during evaporation of components present in low concentration (Reutter et al., 1983). This is especially important post-blast because the major components of the sample typically are reaction products. Ion

Table 11.1 Some chemical spot tests for ions

| | REAGENT | | | | | | |
ION	Silver blue	Barium (modified)	Methylene	Griess sulfate	Nessler's	Aniline	Nitrate chloride
Cl^-	wht.ppt	—	—	—	—	—	
ClO_3^-	—	—	—	—	—	—	yellow
			orange	—			
ClO_4^-	—	—	purple	—		—	
NO_3^-	—	—	—		(w/Zn+)*	—	lt.yellow
NO_2^-	—	—	—		orange†	—	—
SO_4^{2-}	—	wht.ppt		—	—	—	
CO_3^{2-}	wht.ppt‡	wht.ppt‡	—	—		—	—
S	blk.ppt	—	—	—		—	—
NH_4^+	—	—	—	—		orange	—

* See Amas and Yallop (1966); Abramovitch-Bar et al. (1993).
† The observed color depends on the composition of the modified Griess reagent (see Chapter 8, Section 8.4.1.2).
‡ Dissolves with addition of H^+.

chromatograms for commonly encountered low explosives and their residues are illustrated and discussed in following sections.

The primary disadvantage of IC is that it is a separation technique and does not produce definitive identification. An orthogonal technique such as capillary electrophoresis is well suited to confirming the presence of the ions. Other widely used confirmatory approaches are evaporation of the sample and analysis of the residue by IR or XRPD combined with SEM/EDX.

Capillary electrophoresis (CE)

CE is discussed in detail in Chapter 8 (Section 8.7). The mechanism of CE separation is based on differences in charge to mass ratios of the sample ions (McCord and Hargadon, 1993) whereas the separation procedure in IC is based on interactions between the ions and the stationary phase. This difference in separation mechanism makes CE a very suitable complementary technique for IC.

Thin layer chromatography (TLC)

TLC, like spot tests, has largely been superseded by instrumental methods for identification of inorganic species. However, there remains one very useful application to low explosives: distinguishing single base (NC) smokeless powder from double base (NC, NG) powder. The sample is extracted with acetone, spotted onto a TLC plate, eluted with an appropriate solvent system and visualised with modified Griess reagent as described in Chapters 8 and 10.

Another application is identification of smokeless powder additives (Archer, 1975).

Combined techniques

There is no one best method which is applicable to all low explosive samples. The first step in analysis, after reviewing all of the investigative data, is to examine the material with a stereo microscope and then decide on which analytical methods to use. The selection will be determined by experience, the nature of the material in question, and judicial requirements. Commonly used protocols for inorganic mixtures are:

- XRPD and SEM/EDX of solids;
- IR and SEM/EDX of solids;
- ion chromatography and capillary electrophoresis on unevaporated aqueous extracts;
- ion chromatography on unevaporated aqueous extracts and then IR and SEM/EDX on the residue recovered by evaporation of the water.

It is most important to emphasize experience. Books, journals and other publications can provide a list of available methods, but the criteria for method selection are developed from experience on an explosives range and the laboratory. Test explosions which generate 'real world' samples from known mixtures provide the means to objectively determine the usefulness of various techniques alone and in combination.

Thus, any protocol described in the following section on residue analysis may constitute but one of several equally valid approaches.

11.4 Analysis of Residues from Low Explosives

Post-blast identification of low explosives can be quite difficult – residues may contain not only traces of unreacted explosives but also combustion products and unrelated contaminating material from the environment. If only combustion products are isolated, it must be

Table 11.2 Reaction products from components of low explosives

Reaction products	Possible original component
Chloride (Cl^-)	Chlorate (ClO_3^-); perchlorate (ClO_4^-)
Chlorate (ClO_3^-)	Perchlorate (ClO_4^-)
Nitrite (NO_2^{-1})	Nitrate (NO_3^-)
Carbonate (CO_3^{2-}) bicarbonate (HCO_3^-)	Sucrose ($C_{12}H_{22}O_{11}$), charcoal, organic acids, e.g. ascorbic acid, salicylic acid
Metal oxides	Metals, e.g. Al, Mg
Sulfate (SO_4^{2-})	Sulfur

determined what reaction caused their formation (Beveridge et al., 1983; Bender 1989b). Some primary reaction products and possible sources are shown in Table 11.2.

11.4.1 Crime Scene

Thorough processing of the crime scene is essential (see Chapters 4 and 5). The value of laboratory examinations is directly related to the forensic significance of the materials submitted for analysis. Submissions to the laboratory should include all material bearing explosive damage or deposits of foreign substances, debris from areas surrounding the 'seat' of the explosion and all recovered device components. Because of the intense heat generated in explosions, very little unreacted explosive may be found on device fragments; however, combustion products may be found if they are stable at high temperatures.

11.4.2 Analytical Protocols

Figure 11.4 illustrates the examination procedure typically followed to isolate low explosive residues from debris. Once the residue is recovered, the protocol is the same as that for pre-blast analysis (Figure 11.3). Post-blast residues typically include products of reaction and traces of unreacted starting materials. Thus, insoluble material should also be analyzed by elemental analysis combined with XRPD or IR (Beveridge et al., 1975, 1983) since metal oxide products of low explosives are not soluble in water.

The individual techniques are applied as described in 11.3.3 above. Experimental conditions for the individual methods are described in the Appendix.

Visual and microscopic examination

The first step in examination of debris from a bombing is visual observation. From knowledge gained at the scene, and the appearance of the debris, an assessment usually can be made as to whether the damage was caused by a high or low explosive (see Chapters 4 and 5).

Unconsumed particles of black powder or smokeless powder frequently survive an explosion and can be found close to the crater area. They have high evidentiary value. Thus, the protocol at the scene should include sweeping or vacuuming of the general area with clean implements to recover this evidence. Container fragments rarely bear unreacted particles because of the flame and heat effects of the explosion. However, container fragments often bear solid reaction products.

POST BLAST ANALYSIS

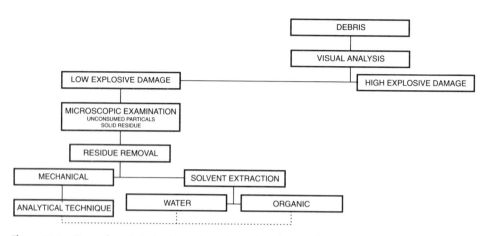

Figure 11.4 General analytical scheme for residues from low explosives.

Debris is examined under a stereo microscope for unconsumed chemicals. These should be removed with tweezers or a scalpel without addition of solvent. The appearance of grains of commercial powders can be altered by the heat of the explosion – black powder particles frequently have rounded edges and smokeless powders often lose their graphite coating and appear a green or yellow color. Particles can also melt or shrink to the point where their original shape is lost.

Solvent extraction

If no unconsumed explosive is observed, it may be sought, along with combustion products, using solvent extraction. Water is the most useful solvent because most combustion products are ionic solids. However, it is good systematic analytical practice to sequentially extract residue with organic solvents and water as described in 11.3.2. Cold rather than hot water should be used since most compounds of interest are relatively soluble in cold water whereas hot water more readily dissolves common contaminants such as calcium sulfates (gypsum board, plaster) and calcium carbonate (concrete).

Water extracts can be analyzed directly by ion chromatography (Reutter et al., 1983; Bender, 1989b, Abramovitch-Bar et al., 1993) and capillary electrophoresis (McCord et al., 1994). Alternatively or additionally, they may be analyzed by evaporating the water and analyzing the resultant solid by XRPD or IR combined with elemental analysis (Beveridge et al., 1975) using SEM/EDX as described in Section 11.3.3.

11.4.3 Protocols Applied to Specific Explosives

The following sections describe the results obtained by applying the protocols shown in Figures 11.3 and 11.4 to the most commonly encountered low explosives and their residues. Both pre- and post-blast situations are discussed since the analytical results often differ.

As noted, the physical observation of damage is an important part of any examination, so photographs of damage caused to similar pipes by several of the explosives discussed below are included. Each pipe, 10 inches long with a $1\frac{1}{2}$ inch diameter and sealed with threaded cast iron end caps, was initiated with military safety fuse inserted through a hole in one end-cap unless otherwise indicated.

The explosives discussed are:

- black powder and substitutes;
- smokeless powders;
- flash powders;
- improvised (home-made) low explosive mixtures.

11.5 Black Powder

As noted in Chapters 1 and 2, black powder is one of the most important discoveries of all time. The present formulation of potassium nitrate/sulphur/charcoal in a ratio by weight of 75/15/10 dates from the 16th century. The burning rate is largely controlled by grain size, configuration and composition.

Commercially, black powder has been superseded as a propellant by nitrocellulose-based smokeless powders, and replaced as a blasting and excavation agent by high explosives and blasting agents such as ammonium nitrate/fuel oil (ANFO). Current usage has dwindled to fireworks, fuses, specialty blasting (slate, marble, granite), igniter charges for large caliber guns and hand-loading for primitive weapons. Although its legitimate uses have decreased, black powder is still used in a significant number of improvised explosive devices. From 1989 to 1993, black powder has been the explosive material in 15% of all destructive devices reported (ATF Explosive Incident Report, 1994). It is widely available in gun stores [in the US] and is relatively easy to 'home-make' – although such manufacture is extremely hazardous since black powder is very sensitive to heat, flame and friction.

11.5.1 Physical Appearance

Commercial black powder is composed of glazed, black, irregularly shaped grains. The mixing of the potassium nitrate, charcoal, and sulfur is so complete that the individual components are not visible through a stereo microscope. Grain sizes are designated in blasting powders as 1FBB–4FBB, in fireworks powders as 1FA–7FA, Meal D, Fine Meal and Extra Fine Meal, and in sporting powders as Fg–FFFFg (GOEX Product Pamphlet, 1995). In each case the higher the number of 'F's the smaller the grain size and the faster the powder burns. The sporting powders (Figure 11.5) are the most frequently encountered black powders because of their availability. The grain size of unburned black powder grains is best assessed by comparing them side-by-side with an authentic sample.

11.5.2 Analytical Identification

To identify black powder, it is necessary to identify the three components: potassium nitrate (occasionally sodium nitrate), sulfur and carbon. Several approaches may be taken.

(a) A *small* sample can be ground carefully and a portion analyzed by XRPD or a combination of FTIR and SEM/EDX. The infrared spectrum (Figure 11.6) shows only adsorption bands which are from potassium nitrate (1769, 1383 and 825 cm^{-1}). SEM/EDX identifies potassium and sulfur, and if equipped with a low element detector, may also identify carbon, nitrogen and oxygen. The X-ray diffraction pattern identifies potassium nitrate and sulfur.

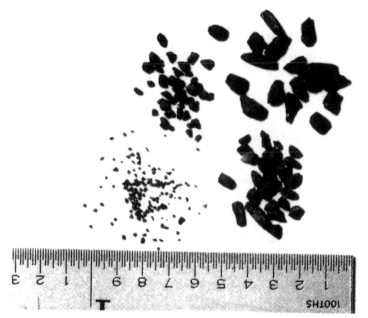

Figure 11.5 Grains of black powder.

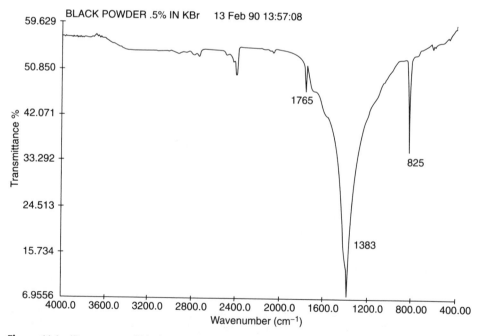

Figure 11.6 IR spectrum of black powder (potassium nitrate).

(b) Sulfur can be identified by extraction in a fume hood with ether or carbon disulfide (both solvents are very volatile and flammable) and subsequent analysis of the resultant pale yellow solid by IR (weak absorption at 470 cm^{-1} due to S_8) SEM/EDX or XRPD. Pyridine/NaOH is a suitable spot test for sulfur (Parker et al., 1974).

(c) Black powder burns in a unique manner, so if there is sufficient sample, particles can be ignited and observed (*ignition susceptibility test*). Sulfur may be determined indirectly in this manner, since a major product is potassium sulfate which is readily identified by its IR spectrum (Figure 11.7; absorptions at 1117 and 620 cm^{-1}). Likewise, carbon forms carbonate which may be present in sufficient concentration to be recognized by absorptions at 1398 and 880 cm^{-1} (Figure 11.7).

(d) Ion chromatography and/or capillary electrophoresis of a water extract of unreacted black powder shows the presence of nitrate and potassium ions.

(e) Quantitative analysis may be achieved by sequentially extracting a weighed sample with carbon disulfide to remove the sulfur, followed by water to remove the potassium nitrate. The residue is charcoal.

Home-made black powder is normally very easy to distinguish from the commercial product using only a stereo microscope, since its individual components are readily discernable as white/colorless nitrate, yellow sulphur and black charcoal. Also, home-made mixtures do not burn as rapidly or evenly as the commercial product and sometimes are difficult to ignite.

11.5.3 Black Powder Residue

Figure 2.1 illustrates typical pipe damage caused by black powder. The pipes usually break at the seam into only a few pieces and the faceplates of the end caps blow out. The breaks are straight with no noticeable metal thinning. There is usually considerable residue left as black/grey powder – commercial black powder burns to yield approximately 43% gases and 56% solids by weight. The other 1% is water.

In the absence of contaminating solids originating from the environment or produced during the explosion (e.g. zinc oxide from galvanized pipes), black powder residues may be recovered mechanically. Otherwise, water extracts the potassium salt reaction products.

The recovered residues are most commonly analyzed by a protocol which includes two or more of: spot tests, IR, XRPD, SEM/EDX, IC and CE. The results obtained by the application of each is described below.

Residue composition can be quite variable. To illustrate this, a sample of commercial black powder and a 'home-made' sample were ignited unconfined in the air. The latter was prepared by mixing the ingredients in the traditional $KNO_3:S:C$ ratio of $75:15:10$, and moistening and grinding them in a mortar and pestle. The residues were analyzed by several techniques. Figures 11.7 through 11.12 illustrate the results of IR, XRPD and IC analyses. The results show that the residue from the commercial sample was primarily potassium sulfate and carbonate whereas the home-made powder produced a high yield of potassium nitrite and unreacted nitrate as well as sulfate. This suggests that a different reaction mechanism may be involved in the burning of the two types of powder.

Spot tests

Residue can quickly be screened for carbonate, sulphate, sulphide, nitrate and nitrite by spot tests.

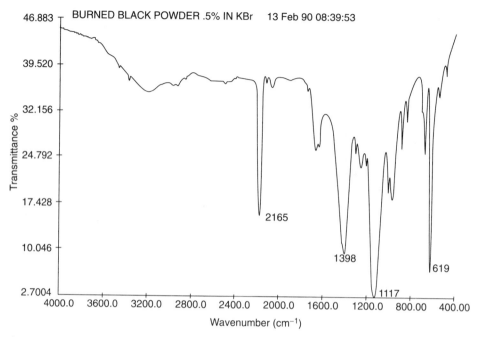

Figure 11.7   IR spectrum of burned commercial black powder residue.

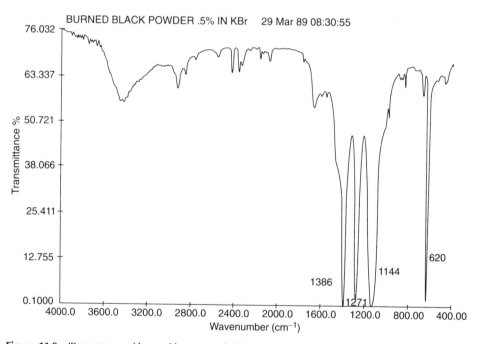

Figure 11.8   IR spectrum of burned home-made black powder.

Addition of an aqueous solution of barium chloride to a water extract of the residue yields a white precipitate of barium sulphate and carbonate. If dilute nitric or hydrochloric acid is added, the carbonate dissolves releasing carbon dioxide which is observed as effervescence. The barium sulphate is unaffected.

356

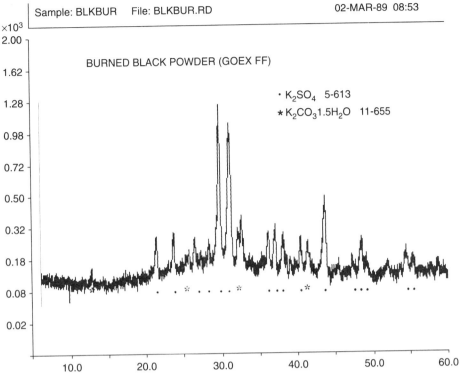

Figure 11.9 X-ray powder diffraction pattern of burned commercial black powder identifying d-spacings from potassium sulphate and carbonate (conditions: Appendix).

Addition of a silver nitrate solution to the extract will form a black precipitate of silver sulfide. A white silver carbonate precipitate will also form but will be masked by the sulfide.

Nitrite can be indicated by modified Griess reagent; when added to the residue the solution will turn pink if nitrite is present; if no colour change occurs, powered Devarda's alloy or zinc may be added, and if nitrate is present, the solution then will turn red (Amas and Yallop, 1966; Abramovitch-Bar et al., 1993).

IR spectroscopy

The major product in the infrared spectrum of black powder residue is potassium sulfate; however, other products frequently are present. Figure 11.7 illustrates the characteristic bands of potassium sulfate (1117 and 619 cm^{-1}) and potassium carbonate (1398 and 880 cm^{-1}) from burned commercial black powder. Figure 11.8 shows the FTIR spectrum of home-made black powder residue in which the primary absorptions besides sulfate, are unreacted nitrate (1386 and 825 cm^{-1}), and its reaction product potassium nitrite (1271 cm^{-1}).

In practice, analysis of an unknown residue by IR should be combined with elemental analysis, XRPD or a technique such as ion chromatography, to determine if alkali halides (e.g. NaCl and KCl) are present. As noted above, these compounds are transparent to IR in much of the normal analytical range of 4000–250 cm^{-1}. A significant concentration of either would indicate the residue probably did not originate from black powder (see Section. 11.6.1 'Post-blast residue').

357

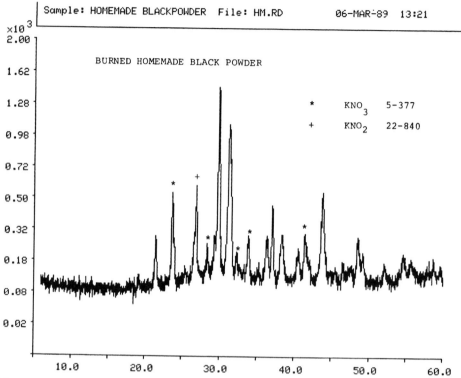

Figure 11.10   X-ray powder diffraction pattern of burned 'home-made' black powder identifying d-spacings from potassium nitrate and nitrite (conditions: Appendix).

X-ray powder diffraction (XRPD)

XRPD will allow the identification of potassium sulfate and occasionally potassium carbonate and potassium sulfide if present in sufficient concentration (approx. 10%). The compounds identified depend on the environmental exposure of the residue; for example, potassium sulfide is rapidly converted to the sulfate by exposure to air.

Figure 11.9 shows the diffraction pattern for residue from burned commercial black powder. The primary d-spacings are assigned to potassium sulphate and carbonate. The numbers in the figure refer to the listing for the compounds shown in the Powder Diffraction File.

Potassium nitrate and its combustion product potassium nitrite are also identifiable by XRPD. Figure 11.10 shows a diffraction pattern for burned home-made black powder in which the d-spacings characteristic of potassium nitrate and nitrite are identified. The characteristic pattern of potassium sulphate (see Figure 11.9) is also present.

The information gleaned from XRPD analyses is similar to that from IR, except that XRPD identifies alkali halides.

Ion chromatography (IC)

IC is more sensitive than IR or XRPD. Figure 11.11 shows the ion chromatogram (suppressed system: Chapter 8, 8.6.1) of the burned commercial black powder. The anion chromatogram contains the expected large amount of sulfate plus cyanate, nitrate, nitrite,

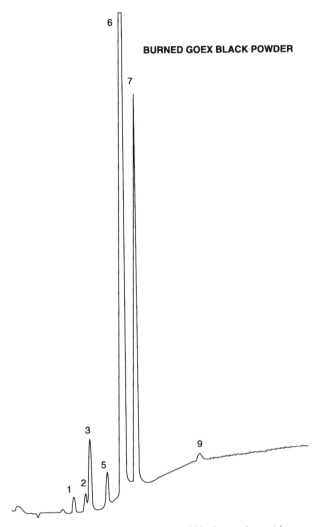

Figure 11.11 Ion chromatogram of burned commercial black powder residue on a suppressed system (conditions: Appendix). Ions are: (1) chloride; (2) nitrite; (3) cyanate; (5) nitrate; (6) sulfate; (7) sulfide; (9) thiocynate.

sulfide and thiocyanate. The small amount of chloride indicated is environmental contamination. If the sample was from a well-contained pipe bomb, the thiocynate ion concentration would be less. Suppressed systems do not identify carbonate or bicarbonate, since these ions are part of the eluent system.

Product formation also depends on reaction pressure. If black powder is not fully confined, thiocyanate will be observed. This is readily observed by ion chromatography. Figure 11.12 shows the suppressed chromatogram for the home-made black powder.

Unsuppressed systems permit identification of carbonate. However, under the experimental conditions used (Appendix) carbonate is eluted in reduced form as the bicarbonate (HCO_3^-) ion and sulfide is eluted in reduced form as the hydrogen sulfide (HS^-) ion. The chromatograms for burned commercial and homemade powders are shown in Figures 11.13 and 11.14.

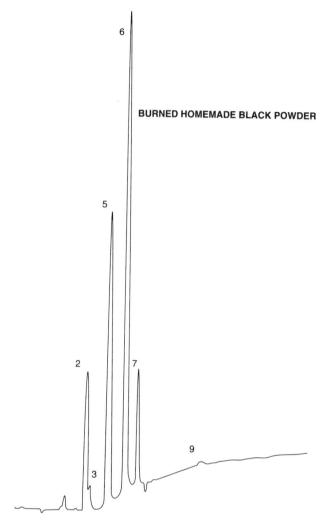

Figure 11.12 Ion chromatogram of burned home-made black powder residue on a suppressed system (conditions: Appendix). Ions are: (2) nitrite; (3) cyanate; (5) nitrate; (6) sulfate; (7) sulfide; (9) thiocynate.

Capillary electrophoresis

CE analysis complements IC by detecting the same ions but not necessarily in the same order of elution (McCord and Hargadon, 1993).

SEM/EDX

Elemental analysis by SEM/EDX identifies potassium and sulfur.

11.6 Black Powder Substitutes

Several products have been developed to replace black powder. Since these products do get used as pipe bomb fillers on occasion, they are discussed below.

360

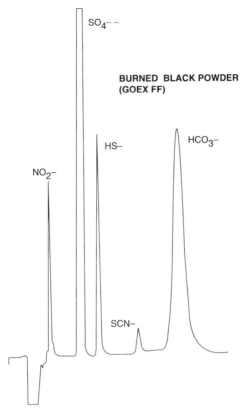

Figure 11.13 Ion chromatogram of burned commercial black powder residue on an unsuppressed system (conditions: Appendix).

11.6.1 Pyrodex®

The most widely used black powder substitute currently available is Pyrodex® which was invented by D. E. Pawlak and M. Levenson in 1978 (US Patent 4,128,443). Pyrodex consists of a mixture of potassium nitrate, sulfur, charcoal, potassium perchlorate, dicyandiamide (DCDA), and sodium benzoate. It is manufactured in Herrington, Kansas and is distributed by the Hodgdon Powder company of Shawnee Mission, Kansas.

Appearance

Pyrodex is grey in color and the grains microscopically appear as nonhomogeneous aggregates of crystalline material.

Analysis

Pyrodex is easily identified by FTIR or XRPD. The infrared spectrum (Figure 11.15) shows adsorption bands for every ingredient except sulfur and charcoal: potassium nitrate (1766, 1384 and 825 cm^{-1}); potassium perchlorate (1086 and 626 cm^{-1}), DCDA (3338, 3189, 2207 and 2163 cm^{-1}), sodium benzoate (1597, 1553 and 709 cm^{-1}).

361

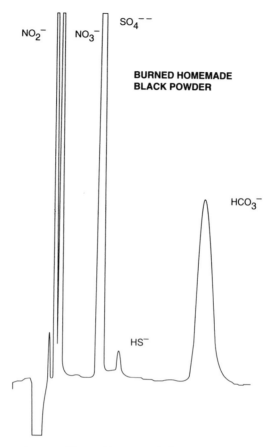

Figure 11.14 Ion chromatogram of burned home-made black powder residue on an unsuppressed system (conditions: Appendix).

X-ray powder diffraction allows positive identification of every component except charcoal. SEM/EDX identifies Na, K, S and Cl, and if equipped with a low element detector may also identify C, N and O.

HPLC can also be used to indicate the presence of DCDA and sodium benzoate (Bender, 1989a). This method is particularly effective with post-blast residue.

Post-blast residue

Pyrodex causes damage similar to black powder and likewise produces a large amount of residual material. The most abundant products are potassium carbonate, potassium sulfate and potassium chloride.

However, the presence of these compounds does not *always* mean that Pyrodex was present, because combustion of black powder in a polyvinyl chloride (PVC) pipe yields potassium chloride as a product through reaction with the pipe. In such instances, further analysis by HPLC for residual dicyandiamide and sodium benzoate is required to establish the source of the residue as Pyrodex. This can be achieved by injecting a water extract into the unsuppressed HPLC system described in the appendix (Bender, 1989a). Both the sodium benzoate and DCDA must be identified to confirm the residue as Pyrodex.

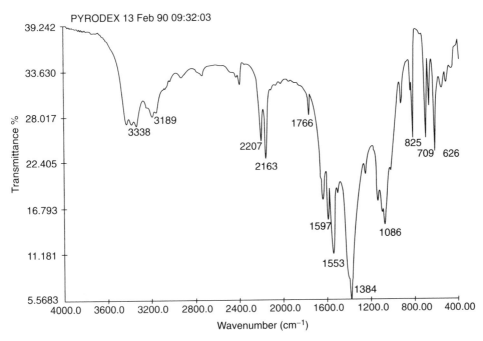

Figure 11.15 IR spectrum of Pyrodex®.

(1) *Spot tests* Unlike black powder, Pyrodex residue contains chloride. Otherwise the primary residue components are the same. The silver nitrate test for chloride can be obscured by a black precipitate of silver sulfide. To minimize this effect, nitric acid is added to the test solution and the solution is heated to convert the insoluble sulfide to soluble sulphate. When silver nitrate is added to a solution treated in this way, a white precipitate of silver chloride will appear. The chance of finding any unreacted perchlorate by a spot test is extremely remote.

(2) *IR* Because KCl is transparent to infrared radiation at the wavelengths commonly used, the IR spectrum of Pyrodex residue may be indistinguishable from that of black powder. However, combination of IR with elemental analysis or ion chromatography (and/or capillary electrophoresis) ensures, by identifying a strong chloride concentration, that the residue is not mistaken for black powder. Figure 11.16 illustrates the IR spectrum of a sample of burned Pyrodex. The primary components are potassium sulfate (1119 and 620 cm^{-1}), potassium carbonate (1401 cm^{-1}) and potassium cyanate (2230 cm^{-1}).

(3) *XRPD* XRPD generally identifies potassium chloride and potassium sulfate as the primary components of the residue. Other components are typically present of concentrations of <10% and are not detected by XRPD.

(4) *Ion chromatography* Ion chromatography is a useful technique to confirm the presence of Pyrodex. Although many ions present are also found in black powder (Figures 11.11–11.14), the additional presence of chloride and especially perchlorate suggest combustion products of Pyrodex (Figures 11.17 and 11.18).

363

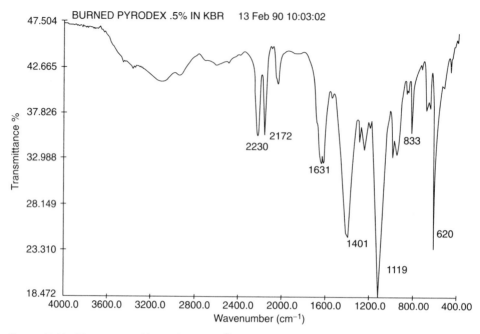

Figure 11.16 IR spectrum of burned Pyrodex.®

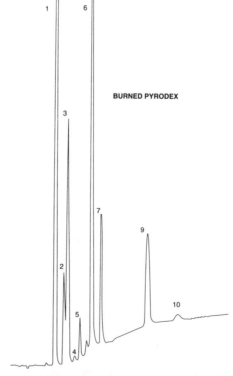

Figure 11.17 IC chromatogram of burned Pyrodex® using a suppressed system (conditions: Appendix). Peaks identified are: (1) chloride; (2) nitrite; (3) cyanate; (4) chlorate; (5) nitrate; (6) sulfate; (7) sulphide; (9) thiocyanate, (10) perchlorate.

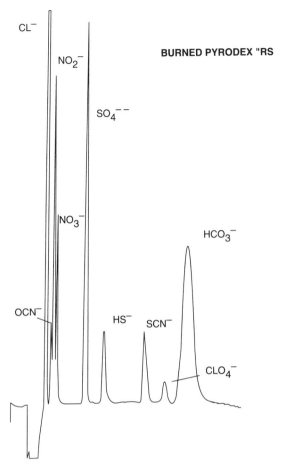

Figure 11.18 IC chromatogram of burned Pyrodex using an unsuppressed system (conditions: Appendix).

(5) *Capillary electrophoresis* CE separates the same components as IC. The electropherogram of Pyrodex residue has been illustrated and discussed by McCord and Hargadon (1993).

(6) *SEM/EDX* Elemental analysis of Pyrodex residue identifies potassium, sodium, sulphur and chlorine as the primary components.

11.6.2 Golden Powder®

Golden Powder® is a black powder substitute which is derived from potassium nitrate and ascorbic acid. No commercial source currently exists. This material was produced as homogeneous grains in various shades of brown, and had a distinct 'vitamin' odor. Since the mixture can be home-made, the analysis is briefly described.

Analysis

Golden Powder is readily identified by FTIR or XRPD. The IR spectrum of this material is dominated by the potassium nitrate adsorption bands at 1384 and 825 cm^{-1}. Ascorbic acid

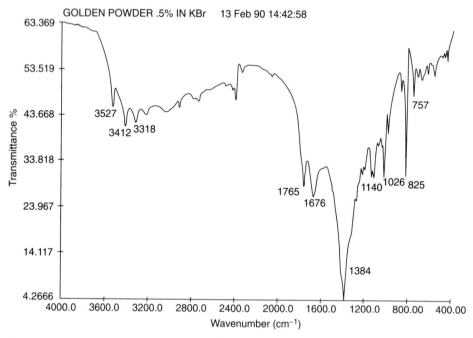

Figure 11.19   IR spectrum of Golden Powder®.

absorption bands at 3527, 3415, 3318, 1765, 1676, 1140, 1026 and 757 cm^{-1} confirm the identification (Figure 11.19). XRPD also identifies these two components.

Post-blast residue

A pipe bomb with Golden Powder as the main charge exhibits the same damage as black powder. Residue consists mostly of potassium carbonate and potassium nitrate as illustrated by the XRPD pattern of the post blast residue (Figure 11.19). An anion analysis by unsuppressed IC indicated the presence of nitrite in addition to carbonate and nitrate. The FTIR spectrum identified potassium carbonate, nitrite and nitrate.

The absence of sulfur-containing ions, i.e. SO_4^{2-}, SCN^-, HS^- and S^{2-} distinguishes Golden Powder residues from black powder or Pyrodex. However, residue of similar composition can also be produced by potassium nitrate and sugar (Beveridge et al., 1983). Consequently, in the absence of unreacted ascorbic acid, identification of the source of the residue may be tentative.

11.6.3 Black Canyon Powder®

Black Canyon® Powder is a potassium nitrate/ascorbic acid propellant which is commercially manufactured by Legend Products Corporation of Las Vegas, Nevada. This powder is grey to brown with grains that have two flat sides. The grains are approximately 8–12 mesh and have a 'vitamin odor'. The results of analysis of this material are the same as for Golden Powder.

11.7   Smokeless Powder

Smokeless powder probably has the highest evidentiary potential of all explosive materials encountered by the forensic chemist. Brand identification can generate investigative leads if the source of the powder can be traced. Powerful associative evidence can be generated by the comparison of powder from a device to powder in a suspect's possession. The development of smokeless powders as propellants is traced in Chapter 1.

All smokeless powders contain nitrocellulose. They are divided into three classes by the chemical composition of their primary energetic ingredient(s):

- single base powder (NC);

- double base powder (NC, NG);

- triple base powder (NC, NG and nitroguanidine [$HN = C(NH_2)NH(NO_2)$], a nitramine).

Smokeless powders are used as gun propellants. Triple base powders are used in large calibre munitions and are rarely if ever encountered in IEDs. Three types of powder which are used in IEDs are discussed in this section. These are bulk sporting powder, and single and double base powders.

11.7.1   *Bulk Sporting Powders*

The first smokeless powders were known as 'bulk sporting powders'. They burned too quickly to be used in rifles, but found applications in shot guns, and as a filler in hand grenades and blank cartridges. Bulk sporting powders are not currently manufactured in the United States, but recently have been encountered in criminal bombings so their discussion is warranted.

Appearance and composition

Bulk sporting powders are irregular grains of congealed fibers (Figure 11.20) which usually are dyed. Common colors include light pink, green, and buff. They consist of approximately 84–89% nitrocellulose and may also contain organic materials like paraffin oil and diphenyl-amine and inorganic coatings like barium and/or potassium nitrate.

Extraction and analysis

Organic additives like diphenylamine and hydrocarbon oil can be removed by extracting with a solvent like methylene chloride or chloroform. The extracts may be analyzed by GC/FID, GC/MS or IR (see below under single and double base powders).

The sample is then extracted with acetone, ethyl acetate or THF to dissolve the nitrocellu-lose which can be analyzed by IR and TLC as described below. The residual inorganic salts will appear as a white insoluble crust which can be filtered or picked out with a scalpel blade. The salts are analyzed by SEM/EDX combined with IR or XRPD.

11.7.2   *Single Base and Double Base Smokeless Powders*

Commercial smokeless powders for rifle, pistol and shotgun propellant use are widely avail-able in bulk form (at least in North America) for use in hand loading. An excellent description of many available powders is the publication *Propellant Profiles* (1988). Because of this avail-ability, these powders frequently are encountered in IEDs, especially pipe bombs. As noted above, single base powder contains only NC as an energetic material whereas double base

Figure 11.20 Bulk sporting powder particles.

powder contains NC and NG. Also, both classes of powder contain chemical additives. Identification of these can greatly assist in individualizing the powder.

Additives

In addition to the primary energetic ingredients, smokeless powders contain an 'additive package' of chemicals which serve various purposes.

- *Stabilizers*: Prevent decomposition of nitrocellulose by scavenging the nitric and nitrous acids which are produced during nitrocellulose decomposition and which catalyse further decomposition if not removed. The most common are diphenylamine (DPA) and methyl and ethyl 'centralite' (*N,N'*-dimethyl diphenylurea and *N,N'*-diethyl diphenylurea respectively).

- *Gelatinizing agents/plasticizers*: Reduce the amount of volatile solvents necessary to colloid the nitrocellulose, or in the case of some double base propellants allow manufacture with no volatile solvents at all. These include nitroglycerin, phthalate plasticizers, dinitrotoluene and ethyl centralite.

- *Surface coatings*: Modify burning rate (vinsol, 2,4-DNT), affect flow or electrical properties (carbon black, graphite) or serve as flash suppressants (zinc powder, potassium sulphate).

The combination of the morphology and chemical additives define the smokeless powder. A list of additives and decomposition products which may be identified by routine analysis are listed in Table 11.3.

Table 11.3 Typical smokeless powder additives

Additives	Decomposition products
Diphenylamine	2-nitrodiphenylamine
Dibutyl phthalate	4-nitrodiphenylamine
Dimethyl phthalate	n-nitrosodiphenylamine
Diethylphthalate	2,4-dinitrodiphenylamine
2,4-Dinitrotoluene	2,4-dinitrodiphenylamine
2,6-Dinitrotoluene	2,2-dinitrodiphenylamine
Trinitrotoluene	4,4-dinitrodiphenylamine
Methyl centralite	
Ethyl centralite	
Nitroglycerin	
Potassium sulphate	
Graphite	

Further decomposition of diphenylamine and the centralites can occur which would expand the list Table in 11.3. A more complete list of decomposition products and the mechanisms of their formation are given elsewhere (Selavka et al., 1989; Curtis and Berry, 1989; Curtis 1990; Bellerby and Sammour, 1991; Bellamy and Sammour, 1993).

Appearance

In general, powder morphology indicates whether the powder is double base or single base. All ball powders are double base. Most tube and cylindrical powders are single base (notable exception – Hercules Reloader® series). Most disc powders are double base (notable exception – some IMR powders, e.g. IMR PB, SR series). Figure 11.21 illustrates some of the sizes and shapes which to a large degree determine the burning rate of the powder.

The morphology of smokeless powder is its most important feature for brand identification (Selavka et al., 1989; Wallace and Midkiff, 1993). First, the shape of the sample is described, i.e. disc, tube, ball, etc., then measurements of the important dimensions, i.e. length, diameter, thickness and web are determined (see Figure 11.21). Some gunpowders also have identifiers or color coded grains which are dispersed throughout to allow easy verification of the powder brand by a reloader.

After the physical characteristics of the powder have been determined, a chemical analysis of the additive package is performed as described in the next section. When all of this information has been collected, it is compared to a data base or reference collection of authentic samples.

11.7.3 Analysis

The analysis of smokeless powder usually has one or more of the following goals:

- defining the material as an explosive substance;
- determining the manufacturer or source;
- comparison of one powder to another.

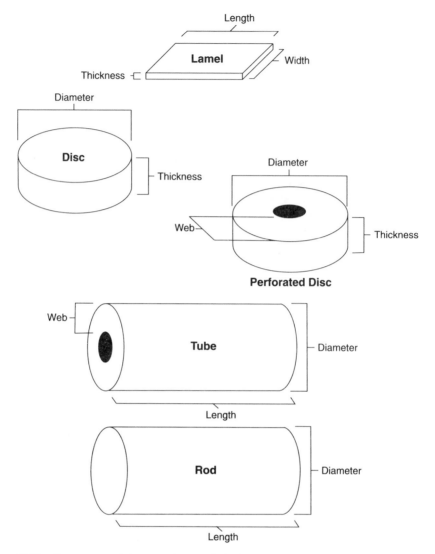

Figure 11.21   Common smokeless powder morphologies.

Identifying the material as an explosive substance involves identifying NC (and NG if double base). The other goals require identification of additives. Effective methods for these purposes are described below.

Nitrocellulose (NC)

All smokeless powders contain NC. The best method for identifying NC is to produce the IR spectrum of a film pressed in a diamond cell or cast from acetone, tetrahydrofuran, methanol or ethyl acetate onto a diamond or onto an NaCl or KBr disc (Kee et al., 1990). To obtain a pure spectrum, the film should be extracted with methylene chloride or chloroform to remove additives prior to IR analysis. The technique is non-destructive and NC can be recovered for further analysis.

TLC is a useful complementary technique (Chapter 8, see 8.4.1 'Solvent systems'). In many commonly used TLC solvent systems NC has an R_f of zero, but Douse (1982) reported that when developed with acetone:methanol 3:2 on a silica gel plate, NC had an R_f of 0.64.

Interpretation of the significance of finding traces of NC by solvent extraction of debris, clothing and skin (no unreacted particles observed) must be undertaken with caution since NC can be a component of paint, nail polish, varnish and collodion. The same caution applies to diphenylamine, a common additive, which occurs widely in the environment, e.g. in tire rubber (Lloyd, 1986, 1987). Distinguishing between the different kinds of nitrocellulose is difficult. Lloyd (1984, 1986) has addressed this issue in some depth and has applied size exclusion chromatography to distinguish propellant and non-propellant sources of NC.

Nitroglycerine

If unreacted smokeless powder is removed from an unexploded device and it is only necessary to determine if it is an explosive material, this can be done simply by analyzing a sample by IR to identify the nitrate ester content and then analysing an acetone extract by TLC to determine if it is single or double base (Beveridge et al., 1975). The cited publication used a solvent system of benzene:hexane (1:1). For health reasons (Peak, 1980) the developing solvent system was changed to toluene:hexane (1:1). In this system, on heat-activated silica gel plates, NC has an R_f of zero and NG an R_f of approximately 0.3 (A. D. Beveridge, personal communication). There are many other equally effective solvent systems (see Chapters 8 and 10).

NG can also be detected by many instrumental techniques including GC/MS, GC/TEA, HPLC/TEA, HPLC/PMDE and others (see references in Chapter 8 and in Beveridge, 1992) but the IR/TLC combination is perfectly adequate.

The same procedure may be applied to unreacted particles recovered after an explosion.

Additives

If the purpose of the examination is to identify the manufacturer and/or to compare two samples of powder, then the morphology should be determined and the additives identified. Identification of the manufacturer is usually possible if the powder is intact.

The components analyzed fall into two categories: (1) those that were originally formulated into the powder; and (2) the decomposition products of these components.

The formulated ingredients may include components from powders which did not meet specifications and are reblended into new products. Nitric and nitrous acids from breakdown of nitrocellulose cause the stabilizers to be nitrated and nitrosated. The presence and relative abundance of the stabilizers and their decomposition products permits sourcing of some powders, whereas other powders have additive formulations and morphologies which are so similar that brand identification is impossible (e.g. Hercules Unique® and Herco®) – but powder to powder comparisons are still valuable.

The additive package is first removed by extraction with a suitable solvent such as methylene chloride or chloroform. These solvents remove the additives but leave the nitrocellulose intact. The solvation process should be closely observed because it can give hints to the composition of the additive package. If the solvent turns instantly brown it is an indication of a Hercules Inc. powder additive. Gradual formation of a yellow color indicates a diphenylamine stabilization system; the absence of color implies that no diphenylamine is present (actually the decomposition products of diphenylamine are colored).

Many techniques have been applied to additive analysis. These are referenced below. The original papers should be referred to for specific analytical conditions.

(1) *TLC* Archer (1975) published a system of six solvent systems designed to identify 21 smokeless powder additives including diphenylamine and derivatives, nitrotoluenes and substituted ureas (centralites). He used heat-activated silica gel plates which were developed with six solvent systems, five of which included benzene. Benzene is now contraindicated for health reasons, and toluene has been proposed as a substitute (Peak, 1980). On this basis, the systems become: (A) toluene; (B) toluene: light petroleum (bp 40–60): ethyl acetate 12:12:1; (C) toluene: light petroleum (bp 40–60); (D) toluene: chloroform 1:1; (E) chloroform; (F) toluene: methanol 4:1. He visualized the plates by use of UV radiation and four spray reagents: vanillin, tetramethylammonium hydroxide, Griess reagent and potassium dichromate.

(2) *GC/FID* GC with a flame ionization detector (GC/FID) was one of the first techniques to be used for additive analysis. Most recently, Selavka et al. (1989) have described its use for analysis of ethyl centralite and dibutyl phthalate using a 0.25 mm × 15 m methyl silicone column.

(3) *GC/MS* GC/MS is an effective method for additive analysis. Martz and Laswell (1983) used capillary column GC/MS analysis of additives to establish a data base based on total ion chromatograms (see Chapter 9). The separation achievable is shown in Figure 11.22 for a chloroform extract of a Winchester double base powder (conditions: Martz and Laswell,

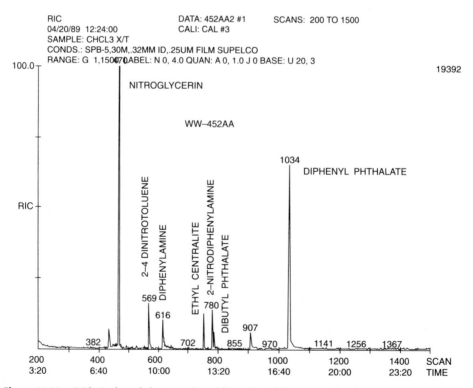

Figure 11.22 GC/MS of smokeless powder additives (conditions: noted on spectrum).

1983). Kee et al. (1990) have used capillary column GC/MS to identify NG, DPA, dibutyl phthalate and DNT additives in individual propellant articles recovered from suspects' clothing as gunshot residue. Like Martz and Laswell (1983), the procedure was based on interpretation of total ion chromatograms.

(4) *HPLC* Several different HPLC systems have been used and have the advantage of quantifying the thermally unstable compounds, notably *N*-nitrosodiphenylamine. Bender has developed a gradient elution system using a UV detector which separates and detects most commonly used additives (Figure 11.23; conditions: Appendix). This system has replaced the GC/FID technique used previously (Selavka et al., 1989) as the analytical procedure underlying the ATF Laboratory smokeless powder data base (Wallace and Midkiff, 1993).

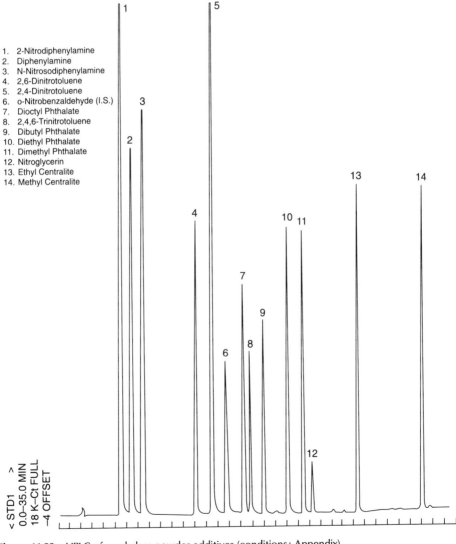

Figure 11.23 HPLC of smokeless powder additives (conditions: Appendix).

De Bruyne et al. (1989) have used HPLC with a C-8 reversed phase packed column, acetonitrile/water as a mobile phase and a UV diode array detector to separate and analyse NG, diphenylamines, alkylphthalates and ureas (centralites). The diode array detector provided the extra dimension of a UV spectrum to complement the retention time.

Lloyd (1986) has used HPLC ODS-Hypersil column and a methanol aqueous phosphate buffered eluent with a pendant mercury drop electrode detector (PMDE) to identify DPA and its products DPA-NO$_2$ and DPA-NO, 2,4-DNT, NG and its breakdown products 1,2-glyceryl dinitrate and 1,3-glyceryl dinitrate.

Stine (1991) and Robertson and Kansas (1990) have respectively used a reversed phase C-18 column with an isocratic acetonitrile/water (65:35) mixture and a gradient acetonitrile/water mixture to analyze analyse decomposition products of diphenylamine.

HPLC is the technique of choice to compare samples. An example of a comparison of smokeless powders of the same brand but from different lots ('ADI'and 'ICI') is shown in Figure 11.24 (conditions: Appendix). The first three significant peaks to be eluted (2–4 min) in both samples are *N*-nitrodiphenylamine, diphenylamine and *N*-nitrosodiphenylamine. The primary differences are due to the presence of significant concentrations of 2,6-dinitrotoluene at about 8 min and 2,4-dinitrotoluene between 9 and 10 min in the chromatogram of the ADI product only.

(5) *Capillary electrophoresis*   CE is the latest technique to be applied to smokeless powder additive detection. Details, references and experimental conditions are described by McCord and Bender in 8.7.2 of Chapter 8, and illustrated in Figure 8.15

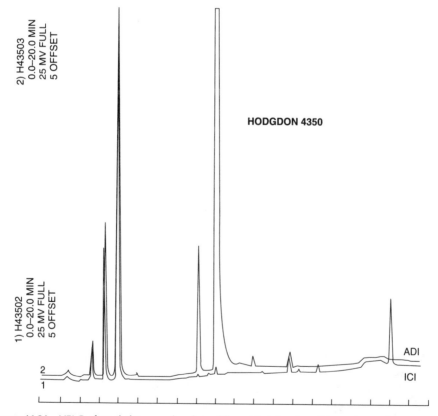

Figure 11.24   HPLC of smokeless powders from different batches (conditions: Appendix).

Figure 11.25 Double base smokeless powder pipe bomb (10 inches long; 1.5 inches od) initiated by flame.

(6) *SEM/EDX* Selavka et al. (1989) have noted the usefulness of SEM/EDX for identification of inorganic additives. An example is identification of potassium and sulfur from potassium sulfate.

11.7.4 Post-blast Examination of Smokeless Powder

Smokeless powder is the most energetic of the low explosives and detonation velocities are possible with appropriate combination of initiator and containment. Generally when smokeless powder is initiated with a spit of flame or hot bridge wire in a pipe bomb the fragments exhibit 90° breaks, bear little or no residue and show no corrosion. Figure 11.25 illustrates fragments from a pipe filled with double base powder and initiated by flame. It is difficult to tell the difference between single and double base powder damage in this example.

When the powder is initiated with a detonator the pipe fragments show considerably more damage and the difference between single and double base powders is more apparent. The double base powder pipe bomb fragments show damage more characteristic of high explosives, i.e. blueing, edge feathering, stretching etc.. The 90° breaks are still seen and the pipe nipple tends to tear into thin strips (Figure 11.26). Single base powder initiated with a blasting cap does not acquire the velocities of double base although the pipe damage is greater than damage due to flame initiation (Figure 11.27). In pipes with a diameter greater than a critical size (about 2 inches) double base powders can attain detonation velocities regardless of the type of initiation. Factors which influence this phenomenon include powder density, nitroglycerin content, and pipe strength.

Figure 11.26   Double base smokeless powder pipe bomb (10 inches long; 1.5 inches od) initiated by blasting cap.

Post-blast analysis of smokeless powder is made easy if intact grains are recovered. The immediate area of the blast should be swept or vacuumed. Only in the most severe conditions will there be no powder remaining. Intact grains are analyzed in the same manner as pre-blast samples.

Care must be given when comparing post-blast samples with unburned powders. The recovered grains should be microscopically inspected for damage. Graphite loss, edge rounding and perforation expansion indicate exposure to heat. Damaged powder should be separated from unaffected grains before extraction of the additive package for analysis.

Pipe or container fragments should be examined for signs of smokeless powder if no particles are recovered from blast seat debris. Undamaged grains sometimes can be found in pipe threads; however, the extreme heat of the metal from the explosion usually produces melted, unglazed fragments resembling yellowish-green plastic (nitrocellulose). These fragments should be removed and extracted with solvent such as methylene chloride or chloroform to remove the remaining additives which are analyzed as described in Section 11.7.3. The insoluble nitrocellulose is then analyzed as described in Section 11.7.3 'Nitrocellulose'.

'Ghosting' can also be found on the container. This is an ash outline of burned smokeless powder grains. The original grain configuration can be determined from these impressions. When smokeless powder burns or explodes, almost all the combustion products are gases. However, even if no visual trace of smokeless powder is found on the container, sequential solvent extraction by methylene chloride or chloroform, and acetone, should be conducted to extract any residual additives and NC respectively.

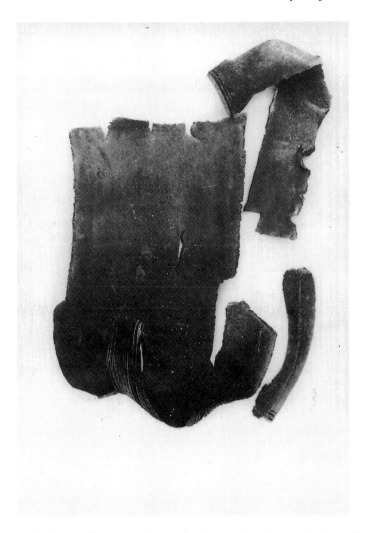

Figure 11.27 Single base smokeless powder pipe bomb (10 inches long; 1.5 inches od) initiated by blasting cap.

11.8 Flash Powder

Photoflash powder, flash and sound mixtures are some of the most dangerous low explosive compositions. They contain powdered metal mixed with a strong oxidizing agent and hence are extremely susceptible to ignition by sparks and can detonate when unconfined. When ignited they produce a bright flash of light accompanied by a loud sound referred to in the fireworks industry as a 'report' or 'salute'.

Photoflash compositions are used in aerial shells or when a loud sound is required, e.g. firecrackers. Military pyrotechnics such as the M115A2 artillery projectile ground burst simulator, the M116A1 grenade simulator, and the M80 explosive simulator (formally called firecracker M80) contain flash powder main charges. These items contain from 3 g (M80) to 56 g (M115A2) of flash powder which makes them quite formidable devices. The M115A2 simulator is considered dangerous to a 50 ft radius. Another military application of flash compositions is night illumination for aerial photography (Conklin, 1985).

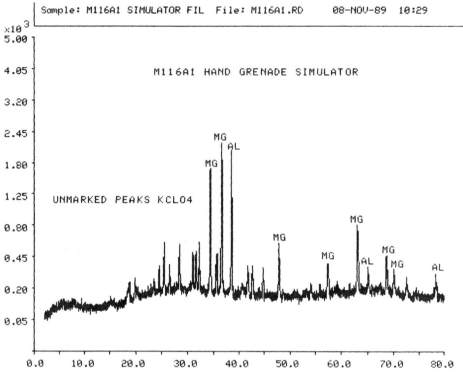

Figure 11.28   XRPD of a flash powder (conditions: Appendix).

11.8.1   Composition

The oxidizers include potassium perchlorate, potassium chlorate and barium nitrate. Metals such as aluminum or magnesium are normally the fuels. A tinder such as sulfur or antimony trisulfide may sometimes be formulated into the composition to reduce the ignition temperature (Meyers, 1978; Midkiff, 1989). The addition of sulfur to a mixture of potassium chlorate and aluminum will change a difficult-to-ignite composition to a very reactive material with an ignition temperature of less than 200°C. Many different formulations of these components are used.

Great care must be exercised when handling these compositions. If the powder is to be ground (not recommended) only small amounts should be present at any one time. Flash powder is best stored in static-proof bags.

11.8.2   Analysis

Analytical schemes for flash powders have been published by Meyers (1978) and Midkiff (1989). A complete analysis of photoflash powder can be done by XRPD. The oxidizers, fuels and tinders are all crystalline and give identifiable diffraction patterns. Figure 11.28 illustrates the XRPD pattern for a mixture containing potassium perchlorate, aluminum and magnesium. FTIR identifies the oxidizer only, but the combination of FTIR and elemental analysis (SEM/EDX) may permit full characterization of the formulation. In addition to the explosive material, extenders like grit-o-cob may be found in some devices. As the name suggests it is

378

simply ground up corn cobs. An appropriately sized sieve will remove the grit-o-cob from the composition which will facilitate the analysis.

Flash and sound mixtures leave solid residues primarily as oxides and halides. The potassium perchlorate mixtures may produce some potassium chlorate; both chlorate and perchlorate mixtures yield potassium chloride and an insoluble metal oxide. The identification by XRPD of aluminum oxide, spinel (magnesium aluminum oxide) and barium aluminum oxide greatly aids determination of the starting components. With pipe bomb fragments, residual material can often be scraped off, but metal oxides form at very high temperatures and when they are deposited can securely imbed in the pipe. As noted earlier, galvanized pipe typically yields white zinc oxide which initially can be mistaken for residue.

Cardboard devices which contain flash powder (firecrackers, M80s, military simulators, etc.) will show traces of aluminum-colored residue on fragments produced from the explosion. This material can be removed and analyzed by the above techniques. If insufficient solid residue is present for FTIR or XRPD then solvent extraction may be performed. An ion chromatography analysis of the aqueous extract combined with elemental analysis of residue from evaporation of the water and insoluble material can be performed. IC will show significant amounts of chloride (chlorate or perchlorate mixtures) and traces of the original oxidizer as shown in Figure 11.29.

11.9   Improvised Low Explosives

There are essentially two types of improvised low explosives: (1) commercially available products which are altered or adapted to act as explosives; and (2) combinations of chemicals.

11.9.1   Commercial Products

The following products were selected based on the frequency with which they are seen in criminal cases.

Road safety flares

One of the most commonly encountered non-explosive pyrotechnic devices is the highway fusee or road flare. They are sold to the public without restriction. Fusees contain a pyrotechnic mixture in a cardboard tube and usually are equipped to be ignited by a scratching mechanism like striking a match. They can be used to simulate dynamite in hoax devices or ignited and used to start fires. The contents can also be removed and used as a source of oxidizers for homemade low explosives.

Composition and analysis   The common red 'fusee' normally contains strontium nitrate, potassium perchlorate, sulfur, sawdust and some oils. The composition appears lite yellow with the sawdust clearly visible and there is a distinct odor of hydrocarbon oil. The pre-ignition analysis can be accomplished by FTIR combined with SEM/EDX or XRPD. The strontium nitrate (which burns with a red flame) can be easily identified by either combination.

Combustion remains of fusees are often recovered at the scene of an arson. Strontium sulfate and strontium carbonate are the primary products. These compounds have little solubility in water and will survive even after the scene has received a copious dousing from the

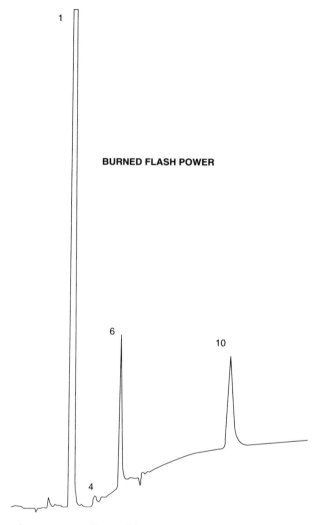

Figure 11.29 Ion chromatogram of burned flash powder. Ions are: (1) chloride; (4) chlorate; (6) sulfate; (10) perchlorate (conditions: Appendix).

fire department. The residue appears as a grey-white mass which can be directly analyzed by FTIR combined with SEM/EDX or XRPD.

The fusee material can also be modified to produce a suitable explosive material. The mixture taken directly from the tube is very hard to ignite and burns slowly. When an active metal such as aluminum is added the composition burns quickly and ignites easily. If a metal is added to the fusee composition, the corresponding strontium metal oxide is formed. The addition of aluminum to the fusee filler forms $Sr_3Al_2O_6$.

Chlorates

Solid oxygen sticks from welding torches provide a good source of sodium chlorate. The sticks are pressed mixtures of sodium chlorate, manganese oxide, quartz, iron and fiberglass which are sold in various configurations such as cylinders or rectangles. The sticks are dark grey and when viewed microscopically show characteristic fiberglass bundles. These mixtures

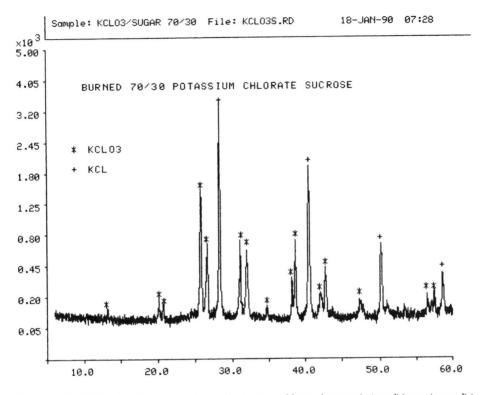

Sample: KCLO3/SUGAR 70/30   File: KCLO3S.RD   18-JAN-90 07:28

BURNED 70/30 POTASSIUM CHLORATE SUCROSE

✳ KCLO3

+ KCL

Figure 11.30   XRPD of residue from a burned potassium chlorate/sugar mix (conditions: Appendix).

require the addition of a fuel before they will function in an IED. The sticks can be ground into a powder then sugar or aluminum, etc. is incorporated as fuel.

The sodium chlorate can also be extracted with water and recrystallized before the fuel is added. The source of such chlorate can be deduced if manganese dioxide powder is present. The manganese dioxide in the solid oxygen sticks is so finely divided that its removal is difficult.

If the solid oxygen sticks are ground and a fuel added, some of the starting material will survive after combustion. For example, the fiberglass bundles can be microscopically observed in the residue and elemental analysis would show the presence of manganese (but manganese can also originate from destroyed batteries). The products formed after combustion depend on which fuel was used. An active metal would produce the corresponding metal oxide and sodium chloride (similarly to flash powder). If sugar is the fuel, the only solid product is sodium chloride.

(1) *Chlorate–sugar mixtures* A widely used low explosive composition is a potassium or sodium chlorate/sugar mix. This explosive is commercially produced as a component of exploding money dye packs (these dye packs normally contain potassium chlorate, sugar, 1-methylaminoanthraquinone, and CS tear gas) and in colored smokes. The advantage of this composition is the low temperature at which it burns. This allows the effective volatilization of the dye/lachrymator or dye package with little decomposition.

Chlorate/sugar explosives are also improvised and used as fillers in improvised explosive devices. Sodium chlorate is available as a weed-killer in the UK.

The chlorate/sugar compositions can be identified by FTIR or XRPD.

(2) Damage and residue   Potassium chlorate/sugar causes much the same damage to a pipe as the equivalent quantity of black powder. It usually yields significant amounts of black residue with a 'caramel' smell which is corrosive and quickly rusts pipes.

The sole solid combustion product is the corresponding chloride (Beveridge et al., 1983) which contraindicates IR spectroscopy as a sole analytical technique. XRPD can readily identify the sodium or potassium chloride as illustrated in Figure 11.30 Perchlorates and sugar will also give chloride (and often chlorate) upon combustion. IC and FTIR are helpful in identifying which anion was initially present.

Residual sugar often forms a syrup which complicates IR interpretation and makes XRPD analysis impossible. HPLC techniques can be applied using refractive index or amperometric detectors rather than a UV detector (Appendix). The pulsed amperometric detector is especially sensitive and selective for this analysis. The residue is simply extracted with water and injected into the HPLC (Zook et al., 1996). An example of this analysis is shown in Figures 11.31 (conditions are given on Figure 11.31). The residual sugars can also be converted into

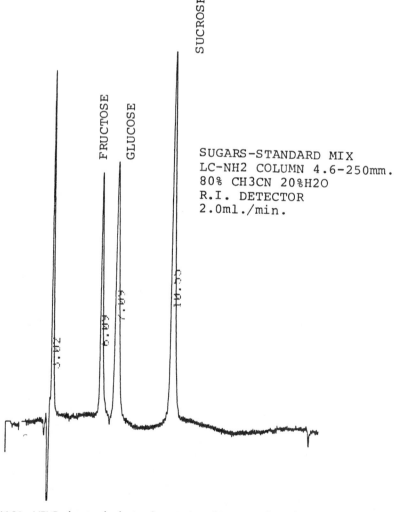

Figure 11.31   HPLC of a standard mix of sugars (conditions noted on chromatogram).

their trimethylsilyl derivatives and analyzed by GC or GC-MS (Beveridge et al., 1983; Nowicki and Pauling, 1988).

The residue can also be hydrolyzed with sulfuric acid and tested with the assorted reducing sugar tests which are available (e.g. Fehling's).

Match heads

Match heads which contain chlorate/sulfur mixtures have been used in pipe bombs. Glatt-stein et al. (1991) have described a analytical protocol involving spot tests for chlorate (aniline sulfate/concentrated sulfuric acid) and sulfur (lead acetate, hydrazine hydrate) supported by SEM/EDX.

Fireworks

Chemicals from fireworks are sometimes used as pipe bomb fillers. Fung (1985) has described analysis by IR of two such formulations from whistling fireworks: potassium perchlorate/ potassium benzoate and potassium perchlorate/sodium salicylate.

11.9.2   *Chemical Combinations*

Many chemicals can be combined to form low explosive mixtures. There is no substitute for forensic scientists continually experimenting with various mixtures which they read about and determining the composition of the residues in order to build up a data base for use with 'unknowns' in casework.

An example is shown in Figure 11.32 which illustrates damage to a series of 6 inch long by 1 inch diameter pipes by home-made mixtures of potassium nitrate with sugar, aluminum and sulfur/charcoal, potassium chlorate with sugar, aluminum and sulfur, and potassium per-chlorate with sugar and aluminum. The residues were isolated and analyzed (Beveridge et al., 1983, and personal communication). These are the most commonly encountered oxidizer/fuel mixtures. But there are many others, one being calcium hypochlorite.

Calcium hypochlorite

Swimming pool chlorinators contain a powerful oxidizer, calcium hypochlorite. If combined with brake fluid or glycerin, the mixture will spontaneously ignite (Kirkbride and Kobus, 1991). The reaction time varies widely according to temperature and formulation, but nor-mally the delay before ignition will be a few minutes. If this mixture is placed with flammable materials a fire will result. The products of this reaction are primarily calcium chloride hexa-hydrate and calcium carbonate. Sodium chloride will also be present because it is in the original chlorinator. Because all of the reaction products are water soluble, the residue is easily washed away when the fire is being arrested. If the residue survives it will appear as a white/grey mass. It can be identified by XRPD.

11.9.3   *Chemical Reaction Bombs*

The term *chemical reaction bomb* is used to describe reactions which produce gas without a source of ignition. The gas pressure can increase to a point where a sealed container will 'explode' producing fragmentation and releasing the contents. Examples are:

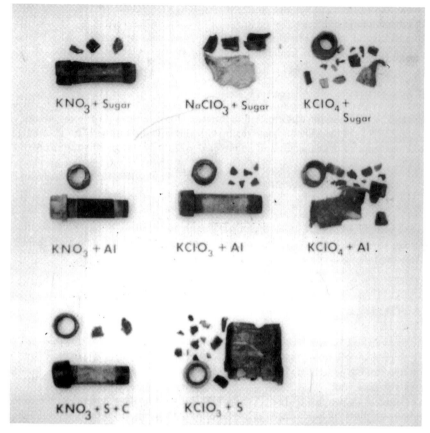

KNO₃ + Sugar   NaClO₃ + Sugar   KClO₄ + Sugar

KNO₃ + Al   KClO₃ + Al   KClO₄ + Al

KNO₃ + S + C   KClO₃ + S

Figure 11.32   Damage to 6 inch by 1 inch pipes produced by various combinations of chemicals (A. D. Beveridge with permission).

- hydrochloric acid and aluminum;
- sodium hydroxide and aluminum;
- calcium hypochlorite and soft drinks or icing sugar;
- dry ice.

Each combination produces gas. The hydrochloric acid or sodium hydroxide can damage the skin or eyes of people in the vicinity.

Hydrochloric acid and aluminum

When aluminum is put in hydrochloric acid, hydrogen is generated according to the reaction:

$$2Al + 6HCl + 6H_2O \rightarrow Al_2Cl_6 . 6H_2O + 3H_2\uparrow$$

The reaction rate is controlled by the concentration of the HCl, the surface area of aluminum exposed to the acid, and the temperature. The strength of container determines the time at which it will rupture. This reaction is in general very unpredictable. After the device has functioned unconsumed aluminum foil is usually present along with an acidic and yellow-tinted liquid. This liquid should be collected and dried. The residual material will be aluminum chloride hexahydrate which is a very hygroscopic salt. It can be analyzed by XRPD if precautions are taken to limit the uptake of water. FTIR will also give an identifiable spectrum. With the presence of unconsumed aluminum, an acidic pH, and an elemental analysis the starting materials can be determined.

Sodium hydroxide and aluminum

Sodium hydroxide and aluminum will also react to generate hydrogen gas. A probable mechanism is:

$$2Al + 2NaOH + 10H_2O \rightarrow 2Na[Al(OH)_4(H_2O)_2] + 3H_2\uparrow$$

The sodium hydroxide (lye) can be added to the aluminum or purchased already mixed, e.g. Drano® contains small pieces of aluminum which give it a bubbling action. After the reaction the liquid will contain black particles which eventually settle. The liquid will have a basic pH. Liquid form Drano® bombs will also contain some blue/green color from a dye in the formulation. Some residual aluminum may also be recovered.

Calcium hypochlorite and soft drinks or icing sugar

These chemicals can react to produce chlorine gas which can build up sufficient pressure to rupture a container.

Dry ice

Solid dry ice left to warm in a container, or combined with warm water, can generate sufficient pressure to rupture some containers. There is no residue.

References

ABRAMOVITCH-BAR, S., BAMBERGER, Y., RAVREBY, M. and LEVY, S., 1993, in Yinon J. (Ed.), *Advances in Analysis and Detection of Explosives*, pp 41–54, Dordrecht: Kluwer Academic Publishers.

AMAS, S. A. H. and YALLOP, H. J., 1966, The identification of industrial blasting explosives of the gelignite type, *Journal of the Forensic Science Society*, **6**, 185–188.

ARCHER, A. W., 1975, Separation and identification of minor components in smokeless powders by thin-layer chromatography, *Journal of Chromatography*, **108**, 401–404.

ATF Explosive Incident Report, 1994, Bureau of Alcohol, Tobacco and Firearms, Washington DC: US Dept of Treasury.

BELLAMY, A. J. and SAMMOUR M. H., 1993, Stabilizer reactions in cast double base rocket propellants. Part IV: A comparison of some Potential secondary stabilizers for use with the primary stabilizer 2-nitrodiphenylamine, *Propellants, Explosives, Pyrotechnics*, **18**, 223–229.

BELLERBY, J. M. and SAMMOUR, M. H., 1991, Stabilizer reactions in cast double base rocket propellants. Part 1: HPLC determination of stabilizers and their derivatives in a propellant containing the stabilizer mixture para-nitro-N-methylaniline and 2-nitrodiphenylamine Aged at 80°C and 90°C, *Propellants, Explosives, Pyrotechnics*, **16**, 235–239.

BENDER, E. C., 1983, Analysis of smokeless powders by HPLC using UV/TEA detection, *Proceedings of the International Symposium on the Analysis and detection of Explosives*, pp 309–320, Washington, DC: US Government Printing Office.

BENDER, E. C., 1989a, The analysis of dicyandiamide and sodium benzoate in Pyrodex by HPLC, *Crime Laboratory Digest*, **16**, 76–77.

BENDER, E. C., 1989b, Indirect photometric detection of anions for the analysis of low explosives, *Crime Laboratory Digest*, **16**, 78–83.

BENDER, E. C., CRUMP, J. and MIDKIFF, C. R., 1993, The instrumental analysis of intact and post blast water gel and emulsion explosives, in Yinon J. (Ed.), *Advances in Analysis and Detection of Explosives*, pp 179–188, Dordrecht: Kluwer Academic Publishers.

BEVERIDGE, A. D., 1992, Development in the detection and identification of explosive residues, *Forensic Science Review*, **4**, 17–49.

BEVERIDGE, A. D., GREENLAY, W. R. A. and SHADDICK, R. C., 1983, Identification of reaction products in residues from explosives, *Proceedings of the International Symposium on the Analysis and detection of Explosives*, pp 53–58, Washington, DC: US Government Printing Office.

BEVERIDGE, A. D., PAYTON, S. F., AUDETTE, R. J., LAMBERTUS, A. J. and SHADDICK, R. C., 1975, Systematic analysis of explosives residues, *Journal of Forensic Sciences*, **20**, 431–454.

CONKLIN, J. A., 1985, *Chemistry of Pyrotechnics*, New York: Marcel Dekker.

CURTIS, N. J., 1990, Isomer distribution of nitro derivatives of diphenylamine in gun propellants: nitrosamine chemistry, *Propellants, Explosives, Pyrotechnics*, **15**, 222–230.

CURTIS, N. J. and BERRY, P., 1989, Derivatives of ethyl centralite in Australian gun propellants, *Propellants, Explosives, Pyrotechnics*, **14**, 260–265.

DE BRUYNE, P. A. M., ARIJS, J., VERGAUWE, D. A. G. and DE BISSCHOP, H. C. J. V., 1989, The HPLC determination of some propellant additives, *Proceedings Third Symposium on Analysis and Detection of Explosives*, pp 27-1-27-15, Berghausen: Fraunhofer Institut für Chemisvche Technologie (ICT).

DOUSE, J. M. F., 1982, Trace analysis of explosives in handswab extracts using Amberlite XAD-7 porous polymer beads, silica capillary column gas chromatography with electron capture detection and thin layer chromatography, *Journal of Chromatography*, **234**, 415–425.

FEIGL, F. and ANGER, V., 1972, *Spot Tests in Inorganic Analysis*, Amsterdam: Elsevier.

FUNG, T., 1985, Identification of two unusual pipe bomb fillers, *Canadian Society of Forensic Science Journal*, **18**, 222–226.

GLATTSTEIN, B., LANDAU, E. and ZEICHNER, A., 1991, Identification of match head residues in post-explosion debris, *Journal of Forensic Sciences*, **36**, 1360–1367.

KEE, T. G., HOLMES, D. M., DOOLAN, K., HAMILL, J. A. and GRIFFIN, R. M. E., 1990, The identification of individual propellant particles, *Journal of the Forensic Science Society*, **30**, 285–292.

KIRKBRIDE, K. P. and KOBUS, H. J., 1991, The explosive reaction between swimming pool chlorine and brake fluid, *Journal of Forensic Sciences*, **36**, 902–907.

LLOYD, J. B. F., 1984, Detection and differentiation of nitrocellulose traces of forensic interest with reductive mode electrochemical detection at a pendant mercury drop electrode coupled with size-exclusion chromatography, *Analytical Chemistry*, **56**, 1907–1912.

LLOYD, J. B. F., 1986, Liquid chromatography of firearms propellant traces, *Journal of Energetic Materials*, **4**, 239–271.

LLOYD, J. B. F., 1987, Liquid chromatography with electrochemical detection of explosives and firearms propellant traces, *Analytical Proceedings*, **24**, 239–240.

MARTZ, R. M. and LASSWELL, L. D., 1983, Identification of smokeless powders and their residues by capillary column gas chromatography/mass spectrometry, *Proceedings of the International Symposium on the Analysis and detection of Explosives*, pp 245–254,

Washington, DC: US Government Printing Office.

McCord, B. R. and Hargadon, K. A., 1993, Explosive analysis by capillary electrophoresis, in Yinon, J. (Ed.), *Advances in Analysis and Detection of Explosives*, pp. 133–144, Dordrecht: Kluwer Academic Publishers.

McCord, B. R., Hargadon, K. A., Hall, K. E., Burmeister, S. G., 1994, Forensic analysis of explosives using ion chromatographic methods, *Analytica Chimica Acta*, **288**, 43–56.

McLain, J. H., 1980, *Pyrotechnics*, Philadelphia: Franklin Institute Press.

Meyers, R. E., 1978, A systematic approach to the forensic examination of flash powders, *Journal of Forensic Sciences*, **23**, 66–77.

Midkiff, C. A., 1989, Identification and characterization of flash powders, *Proceedings Third Symposium on Analysis and Detection of Explosives*, pp 17-1–17-17, Berghausen: Fraunhofer Institut für Chemische Technologie (ICT).

Miller, R. A., Allen, R. A., Bartick, E. G. and Merill, R. A., 1990, 'Infrared analysis of low explosives by diffuse reflectance spectrometry', presentation at the annual meeting of the Mid-Atlantic Association of Forensic Scientists, Fredericksberg, VA, May.

Nowicki, J. and Pauling, S., 1988, Identification of sugars in explosive residues by gas chromatography-mass spectrometry, *Journal of Forensic Sciences*, **33**, 1254–1261.

Parker, R. G., Stephenson, M. O., McOwen, J. M. and Cherolis, J. A., 1974, Analysis of explosives and explosive residues. Part 1: chemical tests, *Journal of Forensic Sciences*, **20**, 133–140.

Pawlak, D. E. and Levenson, M., (1978), US Patent 4,128,443.

Peak, S.A., 1980, A thin-layer chromatographic procedure for confirming the presence and identity of smokeless powder flakes, *Journal of Forensic Sciences*, **25**, 679–681.

Propellant Profiles, (1988), 2nd Edition, Prescott, AZ: Wolfe Publishing.

Reutter, D. J., Buechele, R. C. and Rudolph, T. L., 1983, Ion chromatography in bombing investigations, *Analytical Chemistry*, **55**, 1468A–1472A.

Robertson, D. and Kansas, L., 1990, Surveillance of the Army's Propellant Stockpile: Analysis of Stabilizer Content by High Performance Liquid Chromatography, *Technical Report ARAED-TR-90020*, New Jersey: Picatinny Arsenal.

Selavka, C. M., Strobel, R. A. and Tontarski, R. E., 1989, Systematic identification of smokeless powders: an update, *Proceedings Third Symposium on Analysis and Detection of Explosives*, pp 3-1–3-27, Berghausen: Fraunhofer Institut für Chemische Technologie (ICT).

Stine, G. Y., 1991, An investigation into propellant stability, *Analytical Chemistry*, **63**, 475A–478A.

U.S. Department of Treasury, 1994, *ATF Explosive Incident Report*, Bureau of Alcohol, Tobacco and Firearms: Washington D.C.

Wallace, C. L and Midkiff, C. R., 1993, Smokeless powder characterization; an investigative tool in pipe bombings, in Yinon, J. (Ed.), *Advances in Analysis and Detection of Explosives*, pp. 29–39, Dordrecht: Kluwer Academic Publishers.

Zook, C. M., Patel, P. M., LaCourse, W. R. and Ralapati, S., 1996, Characterization of tobacco products by high performance anion exchange chromatography – pulsed amperometric detection, *Journal of Agricultural and Food Chemistry*, **44**, 1773–1779.

Appendix: Analytical Conditions

Ion Chromatography:

Suppressed

Column	Waters' IC Pak™ Anion HR, 100 mm × 4.6 mm
Mobile phase	20–100% B in 10 min
	100% B for 20 min
	A = 5% acetonitrile/H_2O
	B = 5% acetonitrile/H_2O, 50 mM NaOH
Detector	Conductivity at 30 usfs

Non-suppressed

Column	Vydac 302IC®, 25 cm × 4.6 mm, 10 μm
Mobile phase	1.5 mM Isophthalic acid, pH 4.6 w/KOH
Detector	Indirect photometric UV at 280 nm

High performance liquid chromatography:

Smokeless powders

Column	Supelcosil®, 15 cm × 4.6 mm, 3 μm, silica
Mobile phase	A = 3 ml 1,4-dioxane in 1l isooctane
	B = 0.5 l methylene chloride, 0.5 l isooctane, 50 ml dioxane

Time	%A	%B
0.0	100	0
4.0	100	0
15.0	85	15
40.0	0	100

Detector	UV at 254 nm
Injector	Rheodyne 7125, 5 μl loop
Flow Rate	1.0 ml/min.

DCDA/Sodium benzoate

Column	Polypore H®, 10 cm × 4.6 mm, 10 μm
Mobile phase	0.1 N H_2SO_4 in 10% methanol at 1 ml/min
Detector	UV at 230 nm
Injector	Rheodyne® 7125, 10 μl loop

Sugars

Column	Dionex CarboPac™ PA1, 4 mm × 250 mm
Mobile phase	200 mM sodium hydroxide at 1 ml/min
Detector	Pulsed amperometric, Au electrode

Infrared spectroscopy
Nicolet FTIR
64 scans, 4 cm^{-1} resolution
13 mm KBr pellet ($\approx$ 0.2 mg/200 mg KBr)
DTGS detector

X-ray powder diffraction
Phillips 3100 XRG
Scintillation detector
Long fine focus Cu Anode X-ray tube at 45 kV, 35 mA
Sample mounted on glass slide

12

The Significance of Analytical Results in Explosives Investigation

GERARD T. MURRAY

12.1 Introduction

Explosives, and their use, are a widespread and everyday feature of modern society, essential in a variety of mining, civil engineering and military projects. In some areas, their availability and ease of acquisition have made explosives very common indeed. They have also long been regarded by the criminal fraternity as an extremely potent tool, to effect entry and for the destruction of property and people. As the subversive use of explosives increases, particularly with the establishment of a concerted and sustained terrorist campaign of violence, so the impact and consequent demands on the various investigative agencies, including, of necessity, a forensic science organisation, increase dramatically.

The illegal use of explosives ranges from the inclusion of fireworks, black powder and small arms propellant in relatively amateurish devices to the more sophisticated, organised criminal or politically motivated use of high explosives.

The purpose of the forensic investigation of an explosives-related incident is two-fold; first, the identification of the explosive(s) used, together with any associated hardware such as batteries, timer, container, and wiring, and second, if possible, to assist in identifying the involvement, or not, of a person, or persons, in the perpetration of the crime. It is probably true to say that in any explosion, regardless of the nature of the charge or its size, there is never 100% consumption of the explosive. Some material remains, as residues or breakdown products, for recovery and detection.

Explosives, in common with other forms of forensic evidence, are subject to transfer to contact surfaces, such as hands or clothing, and consequently such residues are also available for recovery and detection. However, it is the interpretation of the results of examinations carried out on such materials, and the significance attached to them, which can lead to difficulties.

The integrity of the various techniques, procedures and protocols employed to achieve these end results must be verifiable at every stage, from recovery of potential evidence to the presentation of such evidence, and available for scrutiny to those with legitimate right of access, such as defence representatives. Furthermore, in presenting this evidence, all relevant information which may have a bearing on the findings must be disclosed.

A number of factors, in particular cross-contamination, can influence the significance of analytical data in relation to explosives residues. An awareness of these, and their attendant potential problem areas, should enable the scientist to make a more meaningful assessment of the results.

12.2   Types of Explosives Encountered in Terrorist Bombings

Down through the years, but particularly since the discovery of nitroglycerine in the mid-19th century, a wide range of explosives has been developed to perform specific tasks. It is not the purpose, in this instance, either to give an exhaustive listing or comment in depth on the characteristics of a very diverse range of compounds and mixtures but rather to focus on those types of explosives which the forensic scientist is likely to encounter. These will, in general, be dictated by local availability but may be subject, on occasions, to outside influence.

There are, essentially, two groups of explosives with which the forensic scientist will have to deal: commercially produced explosives used industrially and by the military, and impro-vised, or home-made, explosives which are often used in improvised explosive devices.

12.2.1   Commercial/Military Explosives

Commercial blasting explosives commonly include:

- nitrate esters, such as nitroglycerine (NG), ethylene glycol dinitrate (EGDN) and penta-erythritol tetranitrate (PETN);
- nitroaromatics, such as trinitrotoluene (TNT);
- nitramines, such as cyclotrimethylene trinitramine (RDX);
- ammonium nitrate-based slurry and emulsion explosives.

The organic explosives are either volatile liquids (NG, EGDN) or solids (PETN, TNT, RDX) at normal temperatures; developments in the application of these explosives are described in Chapter 1.

With the exception of certain specialised applications, the majority of these explosives are, for practical purposes, encountered in combination with other materials. For example, NG, a particularly sensitive and powerful explosive, is modified and moderated by mixing with nitrocellulose and ammonium nitrate in blasting gelatines, EGDN being added as an explo-sive 'anti-freeze' to prevent crystallisation of the NG. PETN and RDX may be encountered either by themselves, or in combination, mixed with plasticisers and oil in a variety of mould-able plastic explosives, mainly for military applications, while PETN, TNT and RDX can be cast as explosive charges. PETN and RDX are used, in powder form, in detonating fuse.

The current trend in commercial explosives for industrial blasting operations is toward ammonium nitrate-based slurry and emulsion explosives with fine aluminium powder as the fuel. Also to be considered in the context of commercially produced explosives are com-mercial black powder, shotgun cartridge and small arms ammunition propellants and fire-work compositions, all of which can be, and have been, employed in improvised explosive or incendiary devices.

12.2.2   Improvised/Home-made Explosives

Improvised, or home-made, explosives represent, in the main, the standard oxidant/fuel com-bination, classic examples being sodium chlorate weedkiller with sugar, ammonium nitrate fertiliser with fuel oil (ANFO) and potassium nitrate, sulphur and charcoal (black powder). Other combinations, which have been used in Northern Ireland, include sodium chlorate/nitrobenzene, ammonium nitrate/nitrobenzene, ammonium nitrate/aluminium and ammon-ium nitrate/icing sugar – all very viable, powerful explosives. Once again, the potential range of such explosives will be determined by local availability of the constituents and, in an attempt to deny suitable ingredients to terrorist groups in Northern Ireland, regulations were introduced (1972–76) controlling a range of oxidising agents (ammonium nitrate, sodium chlorate, sodium chlorite, sodium and potassium nitrite, sodium and potassium nitrate), together with nitrobenzene.

Since they are relatively easy to prepare, having obvious due regard for safety considerations, some single compound high explosives, such as TNT and RDX, may also be encountered in the home-made context.

Because of the inherent hazards associated with primary explosives, e.g. lead azide, lead styphnate, mercury fulminate, etc., these would not normally be encountered although, over a 4 year period (1976–80), home-made mercury fulminate was used in improvised detonators encountered in Northern Ireland. It re-appeared again, in a more sophisticated form of improvised detonator, in October 1990.

12.3 Recovery and Processing of Residues

One very important feature of any explosive, as far as residues are concerned, is the persistence of the explosive, or individual components, either amongst post-blast debris or on a contact surface. The volatility of materials such as NG and EGDN means that, with the passage of time, they will evaporate and be lost unless absorbed into fabric or like material, whereas solids, particularly if bound together in a suitable sticky or plasticised matrix, will remain indefinitely, provided nothing is done to remove them. This persistence is the raison d'etre for a number of analytical procedures designed to identify such species following an explosion or associated with a person(s) suspected of having been involved with explosives.

Following an explosion, two aspects are of prime importance: (1) the time which elapses between the explosion and recovery of residues from the scene; and (2) how the residues are recovered.

It may be possible, at an early stage, to deduce some information about the explosive itself from the nature and extent of the damage, whether high or low order, charge size, experience of similar incidents where the explosives have not functioned, etc., as discussed in the chapters dealing with scenes of explosion. But often it is not until much later that the explosive may be positively identified through laboratory analytical procedures.

It is likely that any residues will be present in very small (typically nanogram) amounts and consequently great care must be exercised in their recovery and subsequent processing. The spectre of cross-contamination always looms large in many areas of forensic examinations, not least in explosives residue analysis, and is certainly one of the main influences on whether the validity of results will stand or fall. The introduction of contamination, or even the possibility of contamination, at any stage could render the exercise pointless.

12.3.1 Recovery of Post-blast Residues

In the post-blast situation, it must be assumed that residues are present, probably in very small amounts and they must therefore be recovered quickly in case: (a) they are volatile, and (b) any cross-contamination occurs. This can be achieved by taking samples of debris from the explosion site and/or swabbing suitable surfaces. But who is going to do this and what will be used?

Irrespective of who does it, that person, or persons, must be demonstrably free from explosives residues. The materials used to sample and package debris, the swabbing kits, even writing implements if used, must be totally (and demonstrably) free from explosives. The packaging materials are also very important since, if residues are present, they must be retained within the packaging for subsequent detection. Volatile species can permeate poly-thene bags and paper is similarly not a suitable packaging medium. The Northern Ireland experience has been to use nylon bags to package submitted material and this also helps prevent contamination of the contents from other sources.

12.3.2   Recovery of Residues From a Person Suspected of Handling Explosives

When a person is suspected of having handled explosives, the following factors are important:

(1) Type of explosive: This will determine the potential availability of residues. Volatile materials will tend to evaporate or be absorbed through the skin relatively quickly while others will remain much longer.

(2) Degree of contact: The more contact with an explosive, the more will be transferred and be available for recovery and detection. This can obviously be minimised by putting a barrier between the explosive and the contact surface, for example, by wrapping the explosive and/or wearing protective gloves.

(3) Time since contact: This is more important in relation to volatile residues which may disappear relatively quickly.

(4) Treatment of surfaces prior to examination: Washing a surface can remove residues. Even the most persistent can be removed by repeated, vigorous washing over a period of time.

(5) Protection of surfaces prior to examination: Once a suspect has been apprehended, it is important that nothing is done to remove any potential explosives residues prior to examination. This can be achieved by covering the suspect's hands with paper bags, secured at the wrists with elastic bands, and putting a protective cape over the upper body.

(6) More than one suspect: If there are a number of suspects, each will require a separate person to carry out the swabbing procedure, to prevent any cross-contamination.

These factors all have a bearing on the interpretation and significance of analytical data which results from such examinations.

Hand swabbing kit

Associated with each suspect is a hand swabbing kit (see Chapter 10, Figure 10.1) together with various articles of clothing and footwear, typically an outer garment (jacket, pullover), inner garment (shirt, T-shirt), trousers/skirt, socks and shoes. Other garments may be sub-mitted and these may be examined at the discretion of the scientist in charge of the case.

On submission, the swab kit is immediately sent directly to the Explosives Microchemistry laboratory for examination. This laboratory deals solely and exclusively with explosives residue analysis.

Clothing and footwear is sent to a 'search laboratory' where all forensic residue work (including explosives residues, firearms discharge residues, fibres, biology, glass) is carried out, in areas which are free from any contamination.

Specific quality control procedures for the swabs are described below.

12.3.3   Precautions to Prevent Cross-contamination in Handling in the Field

The potential for cross-contamination can arise at any stage where there is exposure to a contaminated environment and this can have very serious implications, particularly in cases where suspects are involved. Between recovery and submission to the laboratory, any surface or environment with which the samples may come into contact, for example, on a bench top, in a vehicle, etc., must be free from explosives, to prevent any possibility of cross-contamination.

Any arresting officer, vehicle in which a suspect is transported, the place where he/she is detained prior to examination and any person who swabs the suspect and takes possession of his/her clothing must be demonstrably free from explosives.

12.3.4 Precautions to Prevent Cross-contamination in Handling in the Laboratory

Recovered samples are processed to identify the presence, or absence, of species of explosives significance. To have any validity, explosives residue analyses must be carried out in areas which are free from explosives, and remote from bulk explosives examination, and by personnel not associated with bulk explosives. At the Forensic Science Agency of Northern Ireland (FSANI), formerly the Northern Ireland Forensic Science Laboratory (NIFSL), examination areas are divided into Red and Blue zones. Bulk explosives are examined in Red zones while residue analyses are confined to Blue zones which are free from contamination and regularly monitored to ensure that this status is maintained. Staff movement between the zones is strictly controlled.

Once in the laboratory environment, every person associated with explosives residue samples and the areas where they are processed must be demonstrably free from explosives. The procedures and protocols to examine the samples must have in-built safeguards to ensure the integrity of all operations and the final results. All this can be achieved by the strictest quality assurance and monitoring procedures. If not, or the system fails in any way, the ultimate evidential value of any analysis results could be drastically reduced or of no value at all.

Protocols employed at FSANI have been developed and modified over a number of years to address potential contamination situations and are similar to those described in Chapter 10. They are illustrated by looking at what is involved in processing a suspect.

In the case of examination for explosives residues, the anti-contamination/quality assurance protocols require swabs to be taken of:

- the gloved hands of the person(s) carrying out the examinations;
- the bench top areas where the examinations are carried out;
- the outer surfaces of the packaging containing the garments/footwear, as controls.

An additional, unused, swab is included as a control of the swabs used. The outer and inner surfaces of the garments, and pockets if necessary, are swabbed together with the soles, uppers and inner surfaces of footwear. All swabs are then passed to the Microchemistry laboratory for processing.

The submitted hand swab kit contains:

- an unused control swab;
- a control swab from the samplers (usually a Scenes of Crime Officer) gloved hands;
- a control swab from the sampling area (where the suspect was examined);
- a swab from the suspects right hand;
- a swab from the suspects left hand;
- a nail scraping from the suspects right hand;
- a nail scraping from the suspects left hand.

Analytical protocols

Of the wide range of analytical techniques available, those which have been most commonly applied to the analysis of explosives residues include thin layer chromatography (TLC), gas chromatography (GC) with flame ionisation detection (FID), electron capture detection (ECD), mass spectroscopy (MS) and thermal energy analysis (TEA), and high performance liquid chromatography (HPLC) with electrochemical detection using a pendant mercury drop electrode (PMDE). These involve the separation of species of interest and detection of these species, primarily by comparison with known standards. Confirmation of the identity of a species is achieved by employing a combination of techniques and such confirmation can only be regarded as positive if the results fit within agreed criteria. These techniques are particularly applicable to organic residues. Quality control procedures are as discussed in Chapter 10.

Inorganic species, on the other hand, may be conveniently analysed using ion chromatography (IC) and infrared spectroscopy; additionally, atomic absorption (AA) spectroscopy or SEM/EDX can be used to detect elemental mercury (if, for example, mercury fulminate is suspected) and other elements. Examples of protocols using these techniques for the analysis of low explosives (propellants; improvised/home-made chemical mixtures) are given in Chapter 11. The typical protocol used in the FSANI for analysis of an unknown residue suspected to be a high explosive is shown in Figure 12.1. Details are discussed below.

Swabs processed in the microchemistry laboratory are extracted and the extracts screened initially for organic species such as nitrobenzene (NB), EGDN, NG, nitrotoluenes, TNT, PETN and RDX, using a capillary GC/TEA system. Prior to extraction, the swabs are spiked with 20 μl of an ether solution of 1,3-dinitrobenzene (10 ng/μl), as an internal standard. Extraction is with 2 ml (2 × 1 ml) of ether.

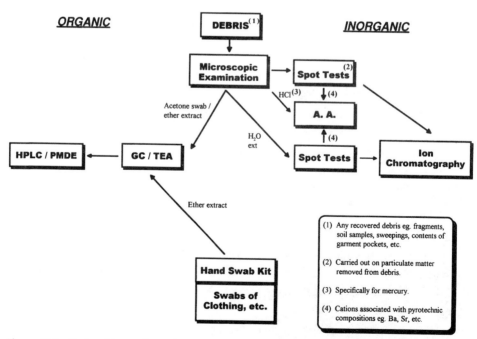

Figure 12.1   Protocol for analysis of an unknown residue suspected to be a high explosive.

The chromatography sequence for a suspect swab kit, for example, would be typically:

- standard
- standard
- standard
- blank
- swab of gloved hands of person extracting swabs
- swab of area where swabs were extracted
- blank
- unused control swab
- blank
- swab of sampler's gloved hands
- blank
- right hand swab
- blank
- left hand swab
- blank
- right hand nail scraping
- blank
- left hand nail scraping
- blank
- control swab of sampling area
- blank
- standard
- standard
- standard.

The standard referred to above is a mixture of EGDN, NB, orthonitrotoluene (ONT), NG, 2,6-dinitrotoluene (DNT), 2,3-DNT, 2,4-DNT, 3,4-DNT (at a concentration of 0.5 ng/5 μl injection), plus TNT, PETN, and RDX (1 ng/5 μl), with 1,3-dinitrobenzene (1 ng/5 μl), as the internal standard. The dinitrobenzene also provides a convenient means of assessing the efficiency of the extraction procedure. Figure 12.2 shows a typical GC/TEA output.

The sequence for clothing/footwear swabs is similar, with standards at the beginning and end, swabs of the examiner's hands and examination area(s), swabs of the bags in which the items have been packaged, general swabs of the items themselves (including pockets where applicable) and blanks between the case samples.

A general purpose swabbing kit is also available and includes a control swab (not used), a swab for the sampler's gloved hands, six general purpose swabs and two spare swabs.

Any sample with a peak which falls within 2 standard deviations (2 SD) of a compound in the standard mix, and is of similar concentration to the standard, is cleaned up using a solid phase extraction procedure and then analysed by HPLC with electrochemical detection. Any sample with a peak which falls within a window of 1% around the retention time of a compound in the standard mix, which, by either concentration or dilution, will see a shift in its retention time, is concentrated or diluted and re-run on the GC/TEA system. If the sample then has a peak which falls within 2 SD of a compound in the standard mix, and is of a similar concentration to the standard, it is cleaned up by a solid phase extraction using a 1:3

INJECTION REPORT

Injection C: <CASEWORK> 3 2971694,1,1

Acquired on 14-Sep-94 at 16:19:54
Modified on 20-Sep-94 at 11:46:38
Reported on 20-Sep-94 at 11:50:34

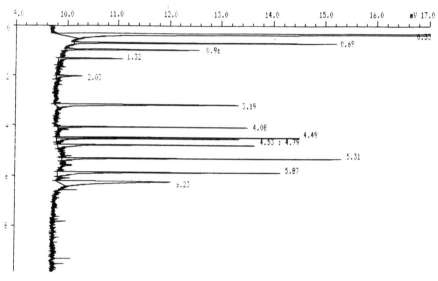

SAMPLE INFORMATION

```
Sample name.............................: STD.
Sample ID...............................:
Sample type.............................: Standard
Sample amount...........................: 1.0000
Number of injections....................: 1
Bottle Number...........................: 1

Retention time update...................: Initialise
Calibration level identifier............: 1
Response factor update..................: Initialise
```

PEAK INFORMATION

Peak	RT mins	Hght uV	Area uVs	ng/5ul	Peak name
2	0.691	5084	10078	0.654	EGDN
3	0.960	2630	5309	0.971	NB
4	1.323	1271	3297	0.754	ONT
6	3.187	3562	10606	0.484	NG
7	4.081	3694	5165	0.540	2,6 DNT
8	4.491	4723	5919	0.590	2,3 DNT
9	4.535	3566	5293	0.582	2,4 DNT
10	4.785	3833	6131	0.646	3,4 DNT
11	5.313	5412	8086	1.225	TNT
12	5.867	4320	10346	1.038	PETN
13	6.228	2207	12396	0.949	RDX
Total		40302	82628	8.432	
Residual		30513	95155	0.000	

Figure 12.2 GC/TEA analysis of a standard mixture.

mixture of Chromosorb 104 and Amberlite XAD4 (40 mg) as support with acetonitrile as the eluent, and then analysed by HPLC.

On the HPLC/electrochemical detection system, any sample with a peak which falls within 2 SD of a compound in the standard mix, and is of a similar concentration to the standard on both the GC and HPLC systems, is regarded as positive for that compound.

All control swabs associated with positive samples are extracted and analysed using the GC/TEA system.

From an operational point of view, the processing of one suspect, with an outer garment, inner garment, trousers, socks and shoes, can entail at least 41 chromatography runs (each of approximately 10 min), initially, to screen the submitted items. Re-runs, for concentration effects, etc., and subsequent confirmation (or not) of potential peaks of interest could add significantly to this figure. Obviously, the more suspects, the greater the resource commitment. One case dealt with by the laboratory in 1989 involved 13 suspects and an estimated 3500 chromatography runs alone.

At every stage, progress must be monitored and verified to ensure the integrity of the results. This is the groundwork which will lead, ultimately, to the identification, or not, of species of interest and the integrity of such findings.

12.4 Interpretation and Significance of Results

The purpose of the foregoing analytical procedures is to establish: (1) whether or not species of interest are present (interpretation); and (2) what this presence, or absence, means (significance), thereby assisting the investigating agencies and, ultimately, helping a court to come to an informed judgement of the scientific evidence presented.

The use of analytical data to indicate contact with explosive is valid only if it can be related back specifically to an explosive as the source of the species detected. Where this cannot be achieved, the influence of other factors must be considered. At all times, the onus is on the forensic scientist to present all findings, positive and negative, together with any other background information available which may have relevance to these findings. For example, to state, in relation to positive NG on a hand swab, that NG is a high explosive and a constituent of a range of commercial explosives (which, of course, it is) and that the presence of this material on the swab is indicative of contact with such an explosive is deliberately misleading since NG is also available, medicinally, in a range of products used to treat vascular ailments. Taken in isolation, the NG detected could just as easily have originated from a medicinal preparation and this must be put forward as a possible option. Such weight as might be put on the latter possibility is for the court to decide by considering all of the evidence. Failure to consider all possible options at the time can lead to serious problems in the future.

The value of any analytical data is only as good as the systems through which casework material is processed. A meaningful assessment of such data must take into account not only the results in relation to the case samples but also adherence to the various procedures and protocols employed, with their in-built control measures; the Quality Assurance factor.

12.4.1 *Pertinent Factors Influencing Interpretation and Significance*

Some key questions which the forensic scientist must ask him or herself follow.

Are explosives' residues present?

This is the first question to be addressed and one which will, without doubt, attract considerable initial attention, particularly in the courtroom situation.

(a) No

Essentially, this means that no species of interest are present on the material examined, be that hand swabs/clothing, swabs of surfaces close to an explosion or debris recovered from the seat of an explosion, or that detected species did not fit the criteria for identification. It does not necessarily mean that there was no involvement with explosives; this is only one of the possible options which are:

- Involvement: There was no involvement with explosives.

- Contact: There was no direct contact between the areas examined (hands/clothing) and commonly encountered explosives; no contact, no transfer, therefore no material available for detection. A physical barrier, such as gloves or wrapping around explosives, can prevent such transfer and, consequently, no material will be available for detection.

- Time and treatment factors: The absence of explosives residues could be due to their having evaporated or having been absorbed through the skin, or having been physically removed from surfaces by washing, prior to examination. The more volatile species will be particularly susceptible to these considerations.

    In the case of swabs or debris from a scene, experience in the Northern Ireland situation has shown that explosives residues, particularly organic residues, are not always detected.

(b) Yes

Explosives residues are present if detected species comply with the criteria set for identification. For example, in the FSANI protocol, species must fall within 2 SD of known reference materials on two discriminating analytical systems.

However, the presence of species of interest does not, of course, address the mechanism by which they may have come to be on the item or surface examined and, again, there are a number of options, namely:

- Involvement: There has been involvement with explosives, either through direct contact or presence in an environment associated with the illegal use of explosives, and the residues have not been removed.

- Cross-contamination:
    (a) There has been inadvertent cross-contamination at some stage during sample recovery, suspect examination or laboratory processing.
    (b) There has been innocent cross-contamination where, for example, a person who has had no involvement with explosives has come into contact with a surface which, at some stage previously, has been contaminated with explosive.

What are these materials and where do they occur?

The range of materials of interest include single explosives (nitrocellulose NC), NG, PETN, TNT, RDX), constituents of compound explosives (NB, nitrotoluenes, various oxidants and fuels) and breakdown products associated with explosives, such as ammonium, potassium, sodium, bicarbonate, carbonate, perchlorate, chlorate, chloride, nitrate, nitrite and sulphate ions.

However, before commenting on the significance of detected species, it is imperative to know what is their (local) availability outside explosives usage. For example, NG and PETN are just two of the many nitro-compounds used in medicine; NC is found as a coating on many surfaces including playing cards as well as a primary ingredient in single- and double-base propellants. NB is used in the manufacture of dyestuffs and has been used in soaps and polishes; nitrotoluenes are also used in the dyestuffs industry. The use of RDX (cyclonite) as a rat poison is referred to in the Merck Index, although it is certainly not used as such in the British Isles and this reference may be no more than just that.

PETN has been used extensively in terrorist devices in Northern Ireland, as detonating fuse and, particularly since 1985, in plastic explosive. A recent survey, to gauge the local availability of medicinal PETN, showed that over the period 1989–92, of the 15–17 million prescriptions issued each year, 400–700 involved PETN preparations. Over the period 1991–93, an average of 421 such prescriptions per year were issued, representing a yearly total of 1.55 kg of PETN. Over the same period, approximately 1200 m of detonating fuse were recovered from terrorist sources, representing 12 kg of PETN. This does not take account of the detonating fuse involved in devices which have exploded nor the PETN content of some explosive charges, such as Semtex H, which has been used extensively in a variety of devices, large and small.

Whilst the detection of a single species, such as NB or NG, would not conclusively point to contact with explosive, the presence of a combination of species associated with explosives, for example EGDN, NG and nitrotoluenes, would strongly support such a contact. Less common species, such as RDX or RDX with PETN, would also indicate contact with explosives.

Inorganic compounds of use in explosives formulation include ammonium nitrate, widely used as a fertiliser, sodium chlorate weedkiller and sodium chlorite and sodium nitrite, both used in the food industry. In fact, most of the common oxidants have a variety of legitimate industrial uses although in Northern Ireland the more important ones, from an improvised explosives point of view, are controlled by statutory regulations and are not available in the unmodified form except under licence and, even then, mostly in quite small amounts insufficient for widespread subversive use.

Also to be considered, particularly when dealing with subversive activity such as a terrorist bombing campaign, is the responsive use of explosives by EOD or bomb disposal personnel whereby small charges are employed to disrupt suspected improvised explosive devices or to effect entry into vehicles which may have contained such devices or been used to transport explosives. In such circumstances, it is essential to know the composition of any explosive charge(s) used in the render safe procedures so that this can be related to analytical data arising from subsequent examination.

What are the background levels?

The terrorist use of explosives, both commercial and improvised, in Northern Ireland has been prolonged and widespread. Therefore, to make any valid assessment of the significance of positive findings, it is necessary to have some base line or reference against which to compare the results. What species are likely to be encountered? These are, primarily, the organic constituents of commercial and improvised explosives for which the case samples are screened, i.e. EGDN, NB, NG, nitrotoluenes, TNT, PETN and RDX. However, a number of these have non-explosive applications, as mentioned above, and it is essential to know not only what their local availability is but also how much, if any, is present in the environment. This can only be achieved by actively examining and monitoring the laboratory environment, holding centres where suspects are processed and other areas of the general environment. Any significant background levels of these species could have a dramatic bearing on the results of laboratory analyses.

In the case of inorganic species, the picture is even less certain. These will invariably be detected as cations or anions, either unreacted material or combustion/decomposition products, and there is no mechanism to marry up pairs of ions to identify the original material. If they are associated with debris which has been intimately involved in an explosion, it may be inferred that a particular oxidant/fuel combination was used but this must be backed up by negative controls from the general background environment. For example, improvised explosive based on ammonium nitrate fertiliser has been used extensively throughout Northern Ireland since 1970. But ammonium nitrate fertiliser is very popular and widely used by the farming community. So, if ammonium and nitrate ions are detected after an explosion, and such ions are present in background controls, the results may not have any great significance for the identification of the explosive. Furthermore, it would be difficult to equate the presence of such species on a person's hands with involvement in explosives manufacture or construction of an explosive device.

As for combustion/decomposition products associated with 'inorganic' explosives, for example black powder,

$$10S + 30C + 20KNO_3 \rightarrow 14CO_2 + 10CO + 10N_2 + 6K_2CO_3 + 3K_2S_3 + K_2SO_4$$

these ions tend to be so common as to be virtually meaningless, in isolation, although again a combination of species associated with explosives damage, whilst perhaps not of the same probative value as would be the case with components of commercial blasting explosives (NG, EGDN, DNTs etc), might none the less point to the use of such explosives.

How much is present?

Whilst it is a relatively straightforward matter to determine how much of a particular species has been detected, quantification isn't necessarily of great significance since it is only an indication of what was present on the surfaces examined. For example, in the case of hand swabs, the swabbing procedure is generally comprehensive and all of the hand area is covered. However, in the examination of a garment, it is not an operationally practical option to divide the surface into a grid and swab each individual section separately in order to 'map' the results. Therefore, only a general swab is taken and, quite conceivably, this could either hit or miss an area of high concentration.

In most cases, quantification only serves to underline the fact that very small amounts (nanograms) of material are involved and can, of course, be used to reinforce the 'innocent contact' argument, particularly in relation to the more persistent explosives species.

How could material be present where it was found?

Having identified species of interest on a surface, be that a suspect's hands or clothing, perhaps the next most potentially contentious issue is how did it get to where it was found? There are two options here, each of which depends on the ability of species to transfer between surfaces in contact with each other. The first option is direct contact; the area examined has been in direct contact with a source of the species of interest, for example, explosive. The second option, which follows from the first, is that the area examined has been in contact with a surface which itself has been in contact with, for example, explosive (indirect contact/cross-contamination). Furthermore, depending on the nature and persistence of the original source, there is the distinct possibility of secondary, tertiary (or further) transfer, which will obviously depend upon the amount involved in the original contact.

Of course, these options must also take into account the possibility of contact with non-explosives sources, such as the medicinal preparations referred to previously. The scientist will not necessarily be in a position to unequivocally identify which option is valid for a particular situation, especially when dealing with persistent materials, although other corroborating factors, as mentioned above (combination of species), may point to the species originating from explosives rather than, for example, medicinal use.

Allied to the second option is the possibility of inadvertent cross-contamination in the handling and processing of a suspect, the forwarding of material to the laboratory and the subsequent laboratory examination of swabbing kit(s) and clothing. Regular monitoring of holding centres, where suspects are processed, and the laboratory environment, particularly areas where residue work is carried out, coupled with strict, verifiable control procedures, should ensure that these areas are contamination-free and not the sources of species which may be detected.

Above all else, within the laboratory environment itself, explosives residue analysis MUST NOT be associated in ANYWAY with bulk explosives analysis.

12.5 Summary

Explosives residue analysis is, without doubt, a very powerful tool in the overall investigation of explosives-related crime. It potentially offers a means of identifying the explosive(s) used and the involvement, or not, of suspects in such criminal acts. To this end, there are two sides to the explosives residue coin, namely identification and significance. In the final analysis, positive findings mean that material of interest has been identified, irrespective of how that material came to be where it was subsequently found. Negative findings mean that no material has been identified either because it was not present or detected species fell outside the criteria for positive identification. Significance, on the other hand, combines the presence, or absence, of such species with other factors, including their general occurrence, background levels, transfer mechanisms, laboratory procedures/protocols and Quality Assurance, to help in coming to a more meaningful assessment of the findings.

The results of analytical procedures cannot be taken in isolation; this is only asking for trouble, both at the time and in the future. The forensic scientist *must* consider *all* possible options and present these as part of his/her overall conclusions.

And if there is one single thread of paramount importance which runs through the whole of explosives residue analysis, it must be Quality Assurance. The value of Quality Assurance can never be over-emphasised; Quality Assurance means confidence in the procedures and protocols and, ultimately, the results. Without a Quality Assurance regime to back up the whole process, from start to finish, explosives residue analysis might as well not be undertaken because, potentially, the results/findings could easily fall at the first hurdle.

To achieve maximum effectiveness, the analytical results must be able to stand on their own, against all arguments, and the significance attached to them must be a true reflection of all possible contributing factors.

The system in operation at FSANI may seem cumbersome at times but the confidence generated shows that it has been a very worthwhile investment of time and effort.

Further Reading

CROWSON, C. A., CULLUM, H. E., HILEY, R. W. and LOWE, A. M., 1996, A survey of high explosives traces in public places, *Journal of Forensic Sciences*, **41**, 980–989.

HILEY, R. W., 1996, Diisopentamethylenetetramine – a potential interference in the detection of explosive traces, *Journal of Forensic Sciences*, **41**, 975–979.

13

Evidence of Explosive Damage to Materials in Air Crash Investigations

MAURICE T. BAKER AND JOHN M. WINN

13.1   Introduction

The Defence Research Agency (DRA), formerly the Royal Armament Research and Development Establishment, (RARDE), has contributed to many investigations over more than 25 years where misuse of explosives in civilian aircraft was suspected. An international reputation has been established since the late 1960s in the art of recognising and characterising the unique signatures of a high order detonation on metallic and non-metallic materials found in aircraft.

The Metal Physics Section of the DRA Structural Materials Centre (SMC) supports programmes throughout the DRA using a range of sophisticated metallographic, optical, electron optical and X-ray equipment. In addition, the section has pioneered the use of microstructural examination and analysis techniques for investigating aircraft losses where sabotage may have been involved. Prior to the formation of the DRA, the UK Ministry of Defence (MOD) laboratories at Woolwich had built up a wealth of unique experience in explosives science/technology and associated forensic activities (safe blowings, etc.) over very many years. This was paralleled by extensive research and development into characterisation techniques such as X-ray diffraction, X-ray fluorescence analysis and electron probe microanalysis. Subsequently, the Metal Physics laboratories at Fort Halstead were among the first in the world to take delivery of a commercial scanning electron microscope (SEM) in the early 1960s.

All the relevant expertise was thus in place when the Materials Division at Fort Halstead was approached by the Air Accident Investigation Branch (AAIB) at Farnborough for assistance into the loss of the Comet aircraft G-ARCO in the Mediterranean Sea in 1967. This incident is well documented (Clancy, 1968). The Comet investigation was the first use of the SEM in explosive sabotage investigations and marked the start of worldwide requests for RARDE/DRA assistance in air accidents.

Tardif and Sterling (1967) and Newton (1985) have published accounts of aircraft sabotage investigative methods, but there have been none in recent years. This chapter expands and updates the subject and is based on the experience of one of the world's leading laboratories in this field.

403

13.1.1 *Number and Scope of Investigations*

In mid-1995, 24 investigations had been completed related to incidents in UK, across Europe, the Middle East, Far East, Africa and the Caribbean. The client base includes the insurance market such as Lloyds of London Aviation Department, associated lawyers, loss adjusters, overseas Governments and police forces, overseas investigating commissions and AAIB. The DRA expertise is specifically limited to possible use of explosives.

Table 13.1 summarizes the incidents investigated by RARDE/DRA since 1967.

Whilst we do not necessarily attend all such incidents, in fact only those where we are requested to do so, it is noticeable that there are variations in the frequency of incidents, even in those reported worldwide.

It is tempting to speculate on the reasons why. The vast improvement in airport security in the late 1970s and early 1980s is obviously relevant and it is now quite difficult to get an explosive device onto an aircraft! Conversely, the greater availability of military explosives like SEMTEX, which need no containment such as metal boxes or tubes, makes it easier to get well disguised devices past X-ray machines. It is now quite feasible to construct a bomb from cardboard and a low vapour pressure explosive such that it will elude sniffer devices and X-ray equipment. Only the timer, batteries and wires will give it away. This calls for better

Table 13.1 Details of incidents investigated by RARDE/DRA since 1967

No.	Airline	Aircraft	Identification	Year	Location
1	BEA	Comet IVB	G-ARCO	1967	Mediterranean Sea nr Rhodes
2	Aer-Lingus	Vickers Viscount	EI-AOM	1968	Irish Sea
3	Cathay-Pacific	Convair 880	VR-HFZ	1972	South Vietnam
4	TWA	Boeing 707	N-8734	1974	Ionian Sea
5	Middle Eastern	Boeing 720B	OD-AFT	1976	Saudi Arabia
6	Cubana	Douglas DC8	CUT-1201	1976	Caribbean Sea nr Barbados
7	Air India	Boeing 747	VT-EBD	1978	Arabian Sea nr Bombay
8	Private	Piper Aztec	5-YACS	1978	Nairobi
9	Philippines	BAC 1-11	RP-C11844	1978	Manila
10	Pakistan Int	Boeing 707	—	1979	Saudi Arabia
11	Itavia	Douglas DC9	1-TIGI	1980	Tyrrhenian Sea
12	Private	Cessna 412A	YV-314	1980	Lisbon
13	Air India	Boeing 747	VT-EFO	1985	Irish Sea nr Cork
14	—	Lockheed Hercules	—	1986	Angola
15	Air Lanka	Lockheed Tristar	—	1986	Sri Lanka
16	TAAG	Boeing 737	D2-TBX	1988	Angola
17	Pan-Am	Boeing 747	—	1988	Lockerbie, Scotland
18	Trans-Afrik	Lockheed Hercules	S9-NA1	1989	Angola
19	Partnair	Convair 580	LN-PAA	1989	North Sea nr Denmark
20	Philippines	Boeing 737	E1-B2G	1990	Manila
21	Trans-Afrik	Lockheed Hercules C130	—	1991	Angola
22	Lauda Air	Boeing 767	—	1991	Thailand
23	Adria Airways	Airbus, DC9 and Dash 7	—	1991	Yugoslavia
24	Private	Jetstream Super 81	—	1994	Aden

X-ray equipment and improvements in training and alertness of the staff. It is safe to say we have not seen the last of such incidents yet!

13.1.2 Rapid Response Team

DRA includes a Forensic Explosives Laboratory (FEL) (see Hiley, Chapter 10) whose main function is the investigation of all terrorist incidents in the UK except Northern Ireland. This group operates a world class facility for the detection and quantitative measurement of traces of all types of explosives using state-of-the-art techniques.

A rapid response team system for air crash work comprises scientists from FEL and SMC, covering residues from explosives and explosive damage to materials respectively. The team has contractual arrangements with major clients (e.g. Lloyds) so that departure to a crash site can be achieved with minimum delay. It is of the utmost importance that a presence at the scene be established as soon as possible as the items most likely to be of interest from a forensic point of view (e.g. baggage, bodies, etc.) tend to be the first to be moved or interfered with, whether by official searchers, police, etc. or by looters, as can sometimes happen.

The DRA team respond only when requested, and all too frequently at a late stage when various 'experts' and others have already been involved. It can be frustrating in the extreme to find that possible sources of evidence may have been (sometimes unwittingly) spoilt. In a recent investigation, having requested sight of victims' baggage items, we were shown to a secure store to find that all cases had been emptied and their contents sent to the dry cleaners. Similarly, in another investigation, baggage which actually showed evidence of damage by high velocity fragments, had only been hand searched and not radiographed for fragments until 8 years later when we discovered the contents in loose heaps on a hangar floor.

The 'Chemistry + Materials' team system has functioned well over the years. One investigator concentrates on identifying any wreckage or other items which may be suitable for examination for explosives traces, while another carries out a similar function covering the materials aspects, seeking out evidence of fragmentation damage etc. It can happen that circumstances are such that both areas of expertise are not required. If, for instance, wreckage has been submerged in a deep sea environment for extended periods, or subjected to extremes of heat from pre- or post-impact fire, it is unlikely that residue detection work would be worthwhile. However, it is surprising what can be found with the modern techniques now in use and every case should be considered on its merits.

Contamination

In recent years, sensitivity limits for the detection of explosives have increased dramatically, such that contamination from the immediate environment, handling, storage, etc. presents real problems. It is now regarded as essential for items that are to be checked for traces to be double- or triple-wrapped, preferably in nylon bags, and sealed on-site by the investigating team. The wrapping is not removed until the items are actually within the confines of the trace laboratory.

Field kits

Experience has shown that a lightweight 'field-kit' should be kept available at all times. Essential items are protective clothing (according to nature of site) including disposable face masks, small focusing torch (flashlight), compact zoom camera, wide selection of sample bags, bottles, labels, etc. and note-books or small dictation recorder. Heavier bulky items such as portable microscope, camcorder or larger cameras with several lenses can be a mixed blessing when faced with long treks in the jungle!

Diplomacy

It has also been found essential to have clear, visual, documentary material on hand showing the experience and expertise of the team, the type of evidence which is being sought and on whose behalf. Bureaucratic problems are common in many countries. Tact and diplomacy are paramount and often it is necessary to establish credibility before being allowed access to a crash site. Where access to recovered bodies is involved, religious aspects, specific local customs and other constraints also apply. The wishes and needs of the bereaved must be respected at all times.

13.1.3 In-house Facilities

The DRA is the largest physics-based research organisation in Western Europe. It exists primarily to serve the needs of the UK MOD by providing the research-based scientific and technical advice necessary to ensure that the MOD can be an expert customer in the procurement of a multi-billion pound equipment programme. The DRA researches all forms of guns, ammunition, missiles and armours and, understandably, is well equipped with a whole range of trials and test facilities. These include full and model scale firing ranges, enclosed test gun facilities covering a range of velocities and 'bomb' chambers.

These facilities have been used extensively to support DRA air crash investigations and to carry out some limited research activities. DRA investigators have indeed been fortunate in this respect and much of our basic data characterizing the effects of explosives on aircraft materials has been obtained using the excellent and almost purpose-built facilities at Fort Halstead.

Experiments have covered such topics as:

- recovery and characterisation of fragments from a variety of 'rigged' explosives and containers;
- test explosions in aircraft baggage containers;
- assessment of explosive damage to a range of metallic and non-metallic aircraft materials;
- characterisation of cratering and penetration damage using projectiles of various shapes, sizes, materials and velocities against an assortment of targets.

Most trials have been in support of individual air crash investigations and have been documented in those reports. Some data have, however, been separately published (Bourne et al., 1984).

The importance of this support cannot be over-emphasised and the authors acknowledge the help and co-operation of their colleagues over many years in carrying out explosions under controlled conditions involving different materials in a variety of configurations which has provided much valuable information. In more recent times we have been familiarising ourselves with the effects and damage patterns associated with ground-to-air and air-to-air missile attack as this aspect seems to be of increasing importance in our work. Further reference to some of the results from in-house trials and experiments will be found in later sections of this chapter.

13.2   Seeking Explosive Evidence

13.2.1   Obvious signs

The basic task of DRA investigators is to determine if an explosion occurred on an aircraft. If no explosive residues are found, then the problem becomes one of determining if there is physical evidence of an explosion.

When an explosive device is detonated, proof positive signatures will always remain. The resultant shockwave causes very severe damage to all materials within the seat of the explosion. Such materials usually consist of the bomb mechanism (timer, batteries, wires, etc.), the container, if one is used, and/or packaging and may extend to part of the fabric of the aircraft itself. Fragments are expelled from the explosion centre with great velocity, driven by hot, expanding gases. The shockwave in the explosive travels at velocities typically from 1000–8000 ms^{-1}, with military explosives producing velocities at the top end of this range. The emergent gases and fragments are somewhat slower, 500–2000 ms^{-1}, depending on the mass and size of the particle and the power of the explosive. To place these figures in perspective, Table 13.2 shows comparative velocities for other events likely to be associated with the subsequent crash and break-up of an aircraft.

The figures in Table 13.2 are approximate but show roughly an order of magnitude difference. Any evidence shown to be unique to the thousands of metres/second range can lead to only one conclusion. The task is to identify this high velocity damage among all the low velocity crash damage. When this is completed successfully it is possible to set about identifying the probable location of the explosion and analysing the materials involved in order to determine, if possible, how and where the explosive was introduced into the aircraft. The material examination task is actually complete, however, when it has been shown that an explosion has occurred. It then becomes a job for forensic scientists and police.

In seeking out the evidence one naturally looks for the most obvious signs. These include straightforward blast damage, heat scorching, general blackening and even smell. Normally, the first reaction by an explosives chemist when handed a suspect item is to smell it! With increasing experience, investigators can rapidly recognise damage associated with break up, impact and pre- or post-impact fire. Impact often produces penetration effects where, for instance, airframe components puncture parts of the fuselage skin. These can easily be confused, by the inexperienced, with explosive or missile damage.

Generally, the message is vigilance for obvious signs before reaching for the hand lens. It is important to remember, when searching a wreckage trail, that the first pieces to be lost from a damaged aircraft are most likely to be the ones carrying the required evidence (see 'trajectory analysis', Chapter 6, Section 6.4.8). The effects of wind, tide and ocean currents must also be taken into account when searching under the sea. Much time, effort and expense may be saved, when seeking evidence where an explosion is suspected, by collecting debris most assiduously from the very beginning of the wreckage trail. It is here that those pieces most affected by an explosion will have been expelled from the aircraft before the subsequent break-up and crash occurs. These fragments may be small but are crucial! They may be a piece of aircraft skin or suitcase but they will carry the important signatures which tell us that an explosion has occurred in the aeroplane.

13.2.2 Effects of Fragmentation

In finding and identifying evidence of an explosion, the best sources of relatively undamaged fragments are the soft yielding materials such as baggage, seating, furnishing materials and

Table 13.2 Comparative velocities

Type of event	Approximate velocity (m/s)
Wreckage impact with ground or ocean	50–200*
Compressor blade failure	up to 300
Fragments generated by high explosives	500–2000 or higher

* At these velocities, impact with the sea is little different from impact with concrete since water is essentially incompressible and massive damage will result at the immediate point of impact.

bodies. Many of our 'successes' have been based on features present on fragments recovered from such sources. A strong stomach is essential when working alongside forensic pathologists. Examination of bodies, victims' clothing and personal effects is a harrowing but essential part of the work.

Radiography is invaluable for locating concealed fragments and may provide some additional information as to the direction or location of the source from damage tracks. As a rule of thumb the more powerful the explosive, the smaller the fragments and the greater their initial velocities. Unique features mainly associated with metallic materials (which have come to be known as signatures) for which we look and which provide *proof-positive* evidence of explosion include the following:

- *rolled edges* produced by the action of hot gases on the thin fracture edges of a spalled fragment;

- *gas-wash* produced by the same action of the hot gases which formed the rolled edges. This was a feature first seen and named for an effect observed on the inner surface of gun barrels after firing even a few rounds;

- *impact pitting* which is a feature identical in appearance to cratering on planetary bodies. The high velocity impact of minute fragments of metal and unburned explosive produces characteristic pits with raised rims.

Additional features may also emerge. For example, a melt-structure may be seen in the bottom of impact pits on clad aluminium alloys commonly used in aircraft. Sometimes it may be seen in the region of a rolled edge or over a larger area of surface. Whilst this feature is not regarded as a *signature* it is taken as an *indication* and may be of use to assist in confirming that a signature is genuine rather than spurious. It is worth mentioning that, although signatures are normally sought on fragments, they may sometimes be found on larger witness materials at the scene of the explosion, such as parts of the aircraft structure. In a recent investigation, an explosive 'cladding' effect was observed in the bottoms of craters when they were examined in section. The 'cladding' was derived from a sheet of aluminium situated between the explosive and the witness material.

Further examples of positive signatures have been identified on man-made textiles and all these features will be covered in detail in Sections 13.4 and 13.5.

13.2.3 Structural Effects in Metals

A further category of evidence which can be useful is a series of structural features which are sometimes observed in certain metals. These are indicative of high strain rate deformation but they are *not* necessarily unique to explosive events. These include:

- mechanical twinning;
- adiabatic shear;
- micro-recrystallisation.

These features are useful supporting evidence and are covered in more detail in Section 13.6.

13.3 Laboratory Techniques

13.3.1 Handling of Recovered Items

The basic procedures for handling forensic samples are followed, but following good forensic practice when working on crash sites spread over large areas in hostile environments can be difficult. Small items can be bagged and labelled on-site, but larger items need to be identified clearly for later removal by freight transport. It is essential that all items are photographed and comprehensive written or recorded notes kept. At sea, it can happen that an item is trawled up and then later dropped onto the sea-bed at another site. If it is later recovered, its presence at the wrong site may be very confusing.

Special precautions apply to items identified for explosives analysis – double wrapping and sealing in nylon bags is preferable. It is important to obtain as much information as possible about the immediate environment of the wreckage, be it on-site or previously recovered and stored, e.g. on a military base. Even details of recent usage of the aircraft, say by military personnel, can be relevant when considering possible contamination in conjunction with currently achievable explosives sensitivity limits. These topics are covered in detail in Chapters 4, 5, and 6 of this book and will not be discussed further here.

The arrangements now in place in DRA are such that items are received into an established forensic explosives laboratory system with proper examination, logging, photographic and other facilities. Covered, secure storage is needed for large items and sometimes unobstructed covered areas are required for examination. We were recently faced with detailed examination of virtually the complete cabin carpet (in many pieces) from a DC9 which were required to be set out and located in their correct positions! Incidentally, the carpet seems an excellent witness material for explosions in or below the cabin. It is far more durable than the expanded material of the floor and does not break up on impact. *May we enter a plea here for airlines to mark all their carpet strips with their positions, port and starboard, front and rear, etc.?* Radiography is an essential tool in these investigations; for items of baggage, seat cushions, clothing, and many others it has become a routine technique. X-ray examination of recovered bodies is desirable, but can present special problems. Many of the recovered fragments examined in our laboratories over the years have been located in this way. It is most useful if two images are obtained from different angles. This makes it possible to determine that the fragment has penetrated the skin rather than sitting on the surface. Such fragments are removed by pathologists as soon as possible and hopefully without damage by forceps, etc. The fragments should be washed in water and dried as quickly as possible to avoid corrosion and degradation of surface features.

13.3.2 *Initial Examination*

Initial laboratory examination normally consists of a thorough visual check, noting features or areas of interest, followed by examination using a standard low power bench stereo optical microscope of the type shown in Figure 13.1.

These examinations should precede more advanced techniques such as scanning electron microscopy. With good illumination, it is often quite remarkable what can be observed on fragments and surfaces when searching for areas of interest such as explosive impact cratering. The overall cleanliness of the sample can also be ascertained. Fused-on fibres or pieces of fabric or foam can be important.

We have only limited experience with the modern CCTV (deep field video microscope) systems now available and have found them valuable, particularly on non-metallic materials (fabrics, etc.) where the ability to produce a high quality image in colour is useful. The lens at the end of a flexible fibre optic is also ideal for looking inside otherwise inaccesible components. A lightweight portable version, preferably providing a stereo image, would be ideal for field use.

A further essential requirement is the ability to prepare samples from bulk materials using normal metallographic techniques, then to examine and photograph them by optical microscopy. We have found the standard preparation techniques for polishing and etching used in most materials technology laboratories to be quite adequate for our needs. A surgical microscope, Figure 13.1, is an ideal way of examining large areas at low magnification, for example, the cabin carpet referred to above.

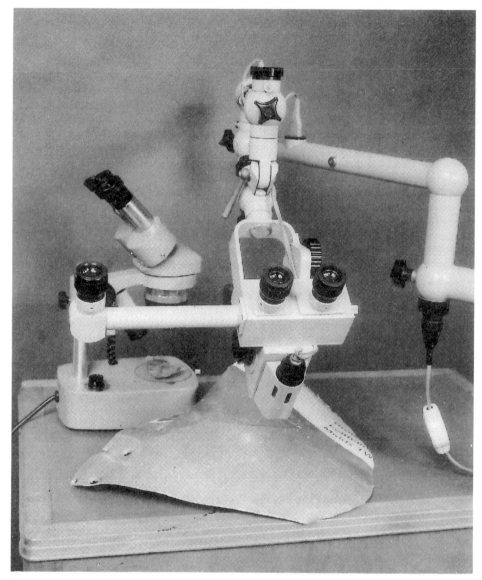

Figure 13.1   Low-power bench binocular and surgical microscopes.

13.3.3   *Scanning Electron Microscopy*

By far the most useful instrument for the identification of positive explosive signatures on metallic and non-metallic aircraft materials is the scanning electron microscope (SEM). We can examine and analyse surface features and topography in detail at high magnification and great depths of field. The availability of this technique from the early 1960s was one of several factors which resulted in Fort Halstead becoming the centre for air crash sabotage investigations in the UK. Instruments have ranged from a Cambridge Stereoscan Mk I (used for the Comet work in 1967) to the Leica-Cambridge S360, one of a suite of instruments in use today. The S360, Figure 13.2, is a state-of-the-art research instrument fitted with a PGT IMIX energy dispersive X-ray analysis system (EDX).

410

Figure 13.2 Leica-Cambridge scanning electron microscope.

The contribution that the SEM technique has made to the understanding and recognition of explosives signatures cannot be overstated and many examples of its use are shown in later sections.

13.3.4 *Electron Probe Microanalysis*

The technique of electron probe microanalysis (EPMA) has also been associated with Fort Halstead since its inception in the late 1950s. One of the first working EPMA instruments was constructed here using an old RCA electron microscope. This followed the pioneering work by Duncumb at Cavendish Laboratories. Again, successive instruments have been installed, culminating in the current Cameca SX50 machine in use today. The SX50 has three wavelength dispersive (WD) spectrometers and an energy dispersive (ED) system. It is equipped with anti-contamination devices and light elements can be detected.

EPMA enables accurate, non-destructive, quantative analytical data to be obtained from a small area on a polished sampled down to 1 or 2 μm in size. Scanning X-ray images can be prepared showing elemental distributions over an area and line scans trace changes of composition with distance across the sample. Many of these operations can be carried out using a modern SEM/EDX system, but we have found the superior analytical sensitivity and accuracy of EPMA to be essential in much of our work. This applies particularly to analysis of fragments and deposits in impact craters and their likely origin with respect to the aircraft. It could, for instance, be vitally important to know whether a fragment of steel contains 0.1 or 0.2% of manganese or silicon and this is not easily achieved using the EDX technique.

As would be expected in air crash investigations, much time and effort has been expended studying aluminium and aluminium alloys by EPMA. Here, care is required in interpretation of analytical data. Most of the alloying elements tend to be concentrated in separate phases so analysis figures obtained from 'clean' metal may not necessarily correspond with published bulk analyses. This is a problem not easily overcome when examining a fragment less than 1 mm in size. We have experimented with methods where quantitative analysis from clean metal and second phases are combined with area fraction measurements with some success (N. D. Harrison, 1992, unpublished RARDE paper).

Some examples of how EPMA data has formed an essential part of our investigations are described later in case studies.

13.3.5   Other Useful Techniques

We would suggest that, almost as essential as SEM and EPMA for air crash investigations, would be the availability of classical powder X-ray diffraction equipment (XRD). We operate a Siemens D500 diffractometer fitted with PC control data system, and still have available Debye–Scherrer powder cameras for identification of very small amounts of material. XRD is a useful back-up technique and the use of ED elemental analysis, combined with the complete JCPDS diffraction pattern data on disc, provides rapid and powerful identification of corrosion products and a variety of other materials.

SMC Metal Physics also has a surface science capability with X-ray photoelectron and Auger electron spectroscopies combined on a VG Escalab Mk 2 instrument. This has been used occasionally for studies of thin surface deposits and other surface specific features. Electron spectroscopies have not seen extensive application in our air crash investigations and their early evaluation for the detection of explosive traces was not encouraging.

Other techniques such as IR spectroscopy, mass spectroscopy and many others are also available to us elsewhere in the DRA and have been invaluable for analysing plastics and polymers when trying to identify small pieces of lining or fibres from seats, etc.

13.4   Positive Explosive Evidence on Metals: The Explosive Signature

Here we introduce and explain the concept of an **explosive signature**. In the early days at the former RARDE, SEM examination of fragments derived from an explosion revealed features not seen before. These features became known as 'explosive signatures'.

An *explosive signature* is a feature showing a positive and unique indication that an explosive detonation has occurred in the immediate vicinity of the witness material.

There is no alternative explanation. Any such signature, no matter how small the actual fragment or piece of evidence, is sufficient to prove that such an explosives event occurred. Only one signature on a single item is sufficient evidence and *cannot* be ignored. However, when only a single item is involved it must be certain that the continuity link with the aircraft is sustained and that, for example, the item has not inadvertently been introduced from another source. When two or more items bearing signatures are involved it becomes very much less likely that such 'contamination' can have happened and, conversely, more likely that the evidence pointing to an explosion is reliable. Over the years we have successfully used 'at least two' as a firm working rule.

The unique signatures which we seek on metals and which are proof-positive that an explosion has occurred are summarised in the following sections.

13.4.1 Fragments

Fragments which derive from an explosion usually have a distinctive 'look' or appearance which is immediately recognisable to the experienced searcher notwithstanding the place where they may be have been found.

As stated in Section 13.2.1, minute high velocity fragmentation is always associated with an explosive detonation. Fragments can originate from the explosive itself in the form of unburned particles, detonators and the immediate containment environment of the explosive. Also, so-called secondary fragments can be generated where two witness surfaces in close proximity 'see' the explosion. As a general rule, fragment size tends to decrease with increasing power of the explosive. Fragment velocities are obviously related to detonation shock wave velocities, but are less easy to determine quantitatively. What is certain, however, is that evidence of fragmentation in the form of small penetrations, cratering and general 'peppering' are a good indication that explosives may be involved. They are certainly easier to interpret than the more obvious large holes which may also be present. Selections of typical fragment shapes and sizes derived from explosive events are shown in Figure 13.3.

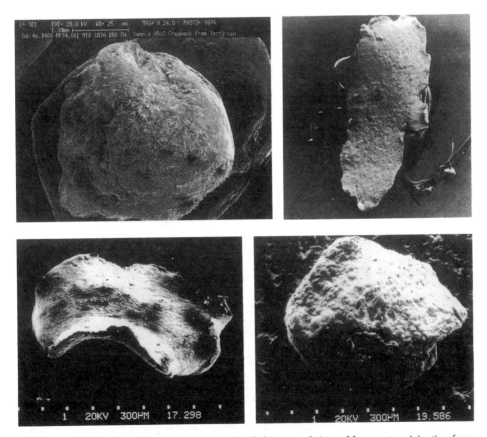

Figure 13.3 Selection of fragments showing typical shapes and sizes of fragments originating from explosions.

413

13.4.2 Rolled Edges

Rolled edges are produced by the action of hot gases at high velocity and pressure from an explosion on tumbling fragments of metal, which results in heating, softening and turning-over of the edges of the metal fragments. A selection of typical rolled edges is shown in Figure 13.4 where their characteristic smooth, curled-over appearance may be seen. Great care and experience are required to distinguish the genuine effect from, say, mechanical damage to the fragment which could even be as a result of careless handling during recovery.

13.4.3 Gas-wash

Gas-wash is produced by the same action of the hot gases which forms rolled edges. The gases have a scouring action and an overall smoothing or eroding effect is observed on all surface features such that any fracture features will be completely or almost completely smoothed away. The effect may be seen not only on fragments (where it will always be present on genuine rolled edges) but also on witness surfaces of the aircraft structure, etc. The

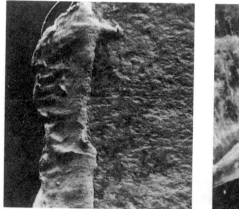
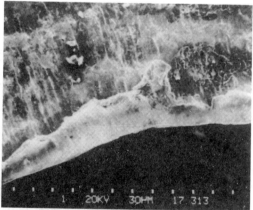
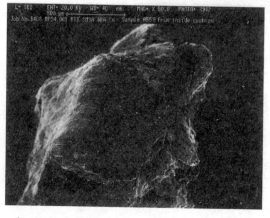

Figure 13.4 Selection of typical rolled edges.

presence of gas-wash on rolled edges is essential when seeking proof that the features are genuine, it is only evident on materials which have been enveloped in the fire-ball of the explosion. Typical examples of gas-wash are shown in Figure 13.5.

13.4.4 *Impact Craters*

These resemble lunar craters when viewed by optical microscopy. They are caused by the high velocity impact of small particulate matter expelled from an explosion with the witness material. They may even be seen on fragments themselves derived from an explosion when they are impacted by other much smaller particles. Particles may be metal, plastic or even unburned explosive. Frequently, traces of the impacting material may be present within the craters. The craters may vary in size from microns up to several millimetres across. All show characteristic raised rims around the crater. The principal features are shown in the sectional sketch, Figure 13.6 and typical craters are shown in Figure 13.7.

The formation of the craters corresponds in many ways to the planetary situation and the subject is covered in an excellent book by Melosh (1989). The basic physics states that if a particle of such minute mass possesses sufficient kinetic energy to form an impact crater its velocity can only have resulted from an explosive detonation.

Where two surfaces in close proximity (e.g. double skinned materials) witness an explosion, it is possible to observe a secondary fragmentation effect. A high velocity fragment projectile impacting the first surface will create a shower of secondary fragments which form a

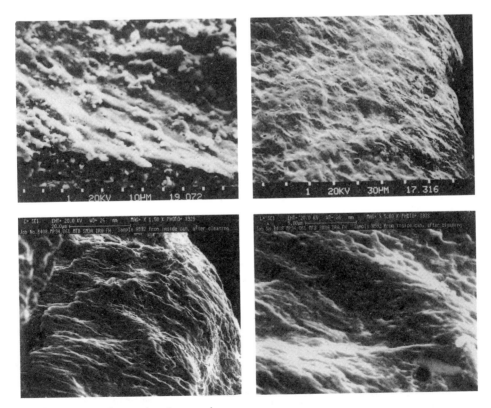

Figure 13.5 Typical examples of gas-wash.

IMPACT CRATER OR PIT   (SECTIONAL VIEW)

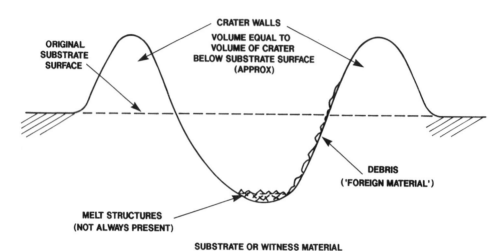

Figure 13.6   Impact crater or pit (sectional view).

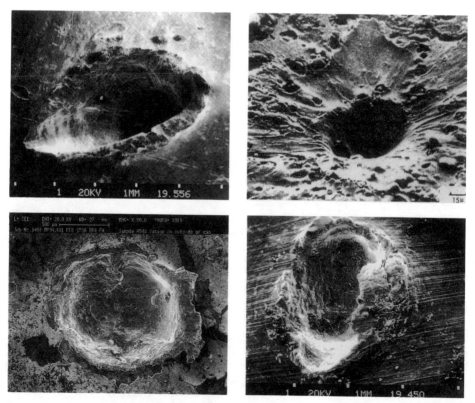

Figure 13.7   Typical impact craters.

halo of craters around the principal impact site on the second surface. The effect is illustrated on a sample of inner skin from an aircraft wing, Figure 13.8.

In the initial examination of craters the accumulated debris may hide important features. Ultrasonic cleaning of the 'as-received' fragments in detergent may reveal concealed signatures like the crater shown in Figure 13.9, which was nearly camouflaged by deposits. Additional features may also emerge. For example, a melt-structure may be seen in the bottom of pits or it may sometimes be seen in the region of a rolled edge, Figure 13.10. On occasions it

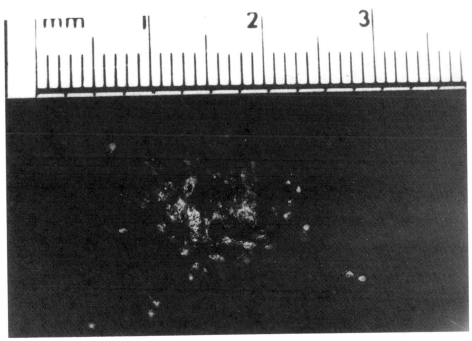

Figure 13.8 Secondary fragment impact sites forming halo around principal fragment impact site.

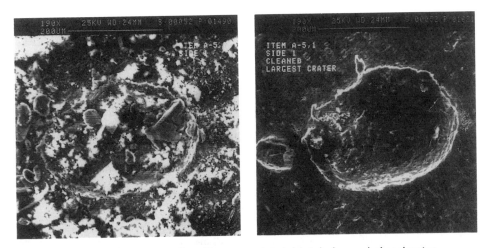

Figure 13.9 Crater, almost camouflaged by accumulated debris before and after cleaning.

417

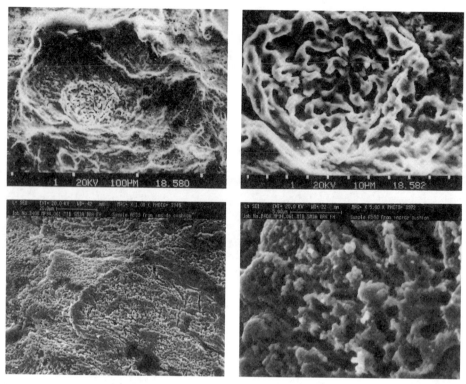

Figure 13.10 Melt structures observed in bottom of craters and on rolled edges.

may be seen over a larger area of surface. Whilst this feature is not in itself regarded as a signature it is taken as an indication and may be of use to assist in confirming that a signature is genuine rather than spurious in origin.

13.4.5 Explosive 'Cladding'

In a recent investigation, an effect was observed which we have called explosive 'cladding'. Deposits were observed in the bottoms of craters when they were examined in section. The electron probe microanalyser was used to quantitatively analyse the deposits and to obtain digital X-ray images, which show the distribution of elements present. We were thus able to demonstrate quite conclusively that the composition of the layer corresponded to an aluminium alloy panel situated close to the explosion and between it and the cratered witness surface. Elements from the panel such as chromium are present in the deposited layer which are not in the substrate and conversely, the zinc content of the substrate is high but there is virtually none in the material of the layer. It is believed that high velocity fragments derived from the panel formed the craters and that deposits of the material from the fragments, which may have melted following impact, then formed the layer.

Further research to determine the extent of material deposition in explosive cratering would be of interest. The results from experiments using different materials/distances/angles could ultimately be of great value in locating seats of explosions.

13.4.6   *Quantitative Microanalysis*

Although not strictly in the category of a signature, accurate non-destructive quantative microanalysis is important. As described above, the electron probe microanalyser has been used extensively in our investigations for accurate and reliable analysis of fragments, layers, small items of shrapnel from warheads, etc. For accurate results, a flat polished surface is essential but this need only be a few square microns in area. It is remarkable what can be achieved without destroying evidence. The ability to produce rapid, high quality scanning X-ray images to show the distribution of elements detected and quantitative X-ray line scans, tracing changes of composition with distance on a micron scale, are also of great value.

Many of these features now appear on modern EDX systems, but, possibly because we have 'grown up' with the technique, the confidence level for quantitative analysis seems that much higher with EPMA. An example is shown in Section 13.9 where the use of EPMA enabled small fragments of steel to be positively identified as originating from a specific type of missile warhead.

13.5   Positive Explosive Evidence on Fabrics

Three primary effects of proximity to an explosion are sought on fibres:

- 'explosive flash melting' or 'transient heating';
- globularising or melting of the ends of fibres;
- inter-penetration of fabrics.

13.5.1   *Explosive Flash Melting*

Our introduction to signatures on synthetic, woven materials arose from an investigation into the loss of a Cubana Airlines DC8 in 1976. The aircraft crashed into deep water off the coast of Barbados. Some bodies and a quantity of flotsam were all that was recovered. Only one piece of metallic evidence was identified.

One of two new effects observed came to be known as 'explosive flash melting' or 'transient heating'. This occurred in a fire retardant blanket made from Nylon 6.6 material and was visible at the high points of the weave, Figure 13.11. This is a definite signature as we have been unable to reproduce the effect by other means. In-house trials were able to show further examples where the overlaying fibres of the woven cloth were melted and swollen, but the bulk of the material was unaffected. This effect is only observed in a narrow band of distances from the seat of an explosion. Our findings in this instance were supported by the discovery of significant quantities of explosive traces in the plastic material of a suitcase. There was also a rolled edge on the suitcase hinge.

13.5.2   *Globularising of Fibres*

The second effect is seen as globularising or melting of the ends of fibres (Figure 13.12). It is an adiabatic heating effect and no scorching or internal gas bubbles are seen – unlike normal heating. The effect is again only seen on 'man-made' fibres. Natural fibres such as wool tend to char, although they may also melt (Figure 13.13). The globularising effect is seen when intermediate and high velocity fragments pass through synthetic woven materials (Figure 13.14). It is sometimes familiarly known as the 'toffee apple' (apple on a stick) effect

Figure 13.11   Explosive flash-melting on nylon 6.6.

and is a very useful signature. It was seen in a DC9 investigation where high velocity fragments had passed through items of luggage.

Up until now all our experience has been with nylon fabrics but we are confident other synthetic materials which exhibit a melting point will behave in the same way.

A related effect, we believe, was also observed in the latter investigation where a piece of black nylon cloth was found to exhibit fibres with swollen or globularised ends. The fibres occurred in patches and only those which stood up vertically from the surface were affected. The weave of the cloth was not damaged. It was concluded that the effect was caused by flash heating from an explosion and that it occurred at an oblique angle to the cloth which was hanging in folds such that only those upstanding fibres on the tops of the folds witnessed the flash. It is thus a combination of the explosive flash melting and globularising effects.

In a recent investigation into the loss of the Lauda Air Boeing 767 in Thailand, a torn black nylon holdall bag (Figure 13.15), showing features of interest was among items recovered from the jungle crash site. It was observed that the torn edges had been affected by heat but it was clearly not the simple adiabatic effect described above. A degree of swelling along a significant length of the fibre was observed (Figure 13.16). A series of trials was conducted in which cut fibre ends were exposed to various combinations of mechanical damage and heat, and it would appear that fibre ends subjected to transient heat tend to show a form of globularising plus swelling or thickening of the fibre 'stems'. It may be concluded that the extent of thickening of the stems is consequent on the duration of] the heating effect. It appeared that the nylon holdall may have been subjected to whiplash in the aircraft slip stream but that hot gas or flame had played over the material at the same time. This is an important result which may assist in distinguishing this type of damage from explosive flash

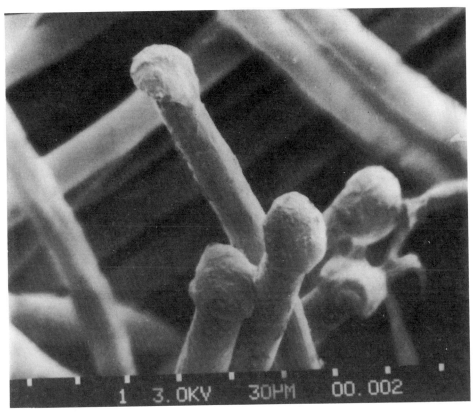

Figure 13.12   Globularising of fibre ends in nylon 6.6.

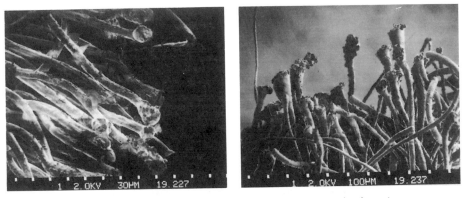

Figure 13.13   Melting of wool-fibre ends showing charring effect in a wool-nylon mixture.

heating or the more familiar 'toffee apple' effect associated with high velocity fragment penetrations observed in many previous investigations where explosives have been involved.

Firing trials also confirmed that globular ends can be produced at penetration velocities below the explosives range but still above crash or engine break-up velocities and their presence alone cannot be taken as proof-positive. They are however less well formed than the

421

Figure 13.14 Globularising of fibre ends resulting from high velocity fragment penetration.

Figure 13.15 Torn black holdall bag recovered from Lauda-Air 767, Thailand, 1991.

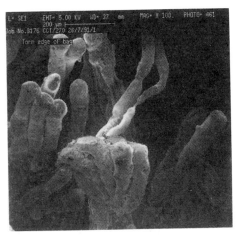 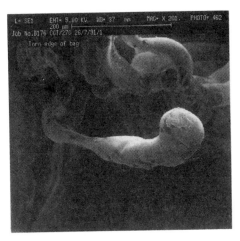

Figure 13.16 Extended swelling from fibre ends, found on holdall bag.

truly globular ends and it may be possible to relate the shape to the fragment velocity. However, the presence of 'classical' globular fibre ends around a hole can be taken as strong evidence of high velocity fragment penetration, especially in conjunction with other unique signatures. The present investigations, however, have shown that the effect must be treated with some caution, particularly where there is evidence of exposure to heat. With care and experience it is possible to differentiate the various effects.

13.5.3 Inter-penetration of Fabrics

The intrusion of fragments of shredded textile material into woven fabric is positive evidence of an explosion. The shredded material is expelled from the seat of the explosion and the small pieces become embedded in other textile materials close by. It is believed that such fragments of shredded material do not travel far before losing their velocity due to their low mass and density – hence they quickly lose their ability to penetrate. An example of this is shown in the optical view (Figure 13.17), in which a woven material is seen penetrating the weave of a doll's dress which was found among some recovered wreckage. If we are correct in assuming that this effect does not 'carry' very far it clearly points to the explosion being inside the aircraft. In our view, further study is required into the signatures created by explosive events on samples of woven fabrics. Examination of clothing and baggage items has often shown 'fusing' together of fabric fibres and it would, for example, be valuable to obtain some idea of the maximum distance from the seat of an explosion at which these effects can be induced.

13.6 Structural Features in Metals

There are also structural features, sometimes observed by metallographic examination, which are useful as supporting evidence or 'indicators'. Contrary to earlier published work these features are not unique to explosive events. They are however indicative of high strain-rate shock at normal temperatures and can be of great value when combined with other signatures.

423

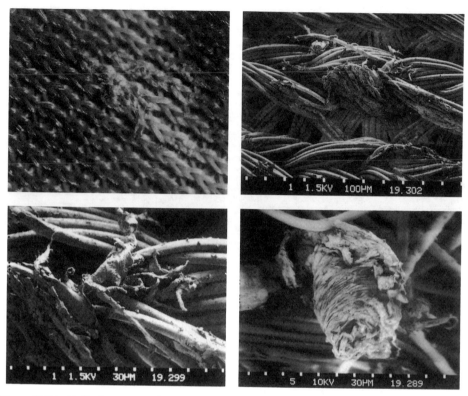

Figure 13.17 Doll's dress material, interpenetration and fusing of pink fabric.

13.6.1 Mechanical Twinning

Mechanical twinning is frequently found in body-centred cubic metals. Figure 13.18 shows an example in mild steel but it can also be found in copper, titanium, etc. The twins are often difficult to find. Unfortunately, there is no record of its occurrence in aluminium and its alloys.

Twinning occurs when deformation rates are too high for normal slip processes to operate. However, in aluminium, the slip processes are so easy and can occur on so many easy-glide planes that twinning is apparently not an option. Different materials clearly have differing strain rate sensitivities for the formation of mechanical twins and the only safe procedure is to carry out strain-rate trials on material of the same composition and in the same condition as the recovered exhibit.

13.6.2 Adiabatic Shear

Adiabatic shear occurs when the heat generated by plastic deformation has insufficient time to dissipate. It results in a microstructural modification. The effect can be made to occur at

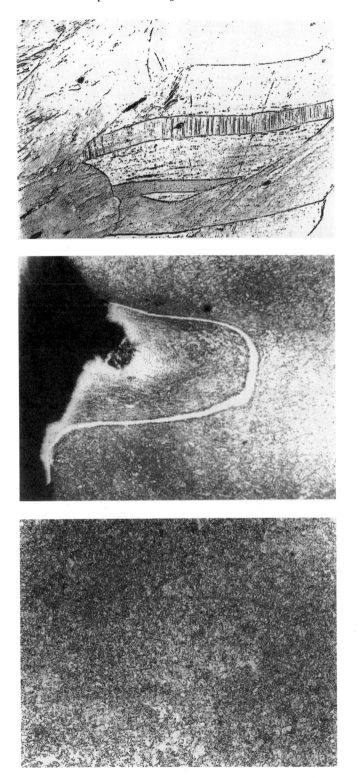

Figure 13.18 Examples of crystal twinning, adiabatic shear and microrecrystallising.

relatively low strain rates in the laboratory, nevertheless it is still regarded as a useful indicator when found in aircraft debris.

13.6.3 Micro-recrystallisation

A related effect is that of micro-recrystallisation, which occurs when the recrystallisation temperature is exceeded in the region where adiabatic shear has occurred. It results in an ultra-fine grain size. Care is needed here as some alloys are designed to have very fine grain sizes, but they are usually uniform in structure.

Examples of micro-recrystallisation, abiabatic shear and mechanical twinning are shown together in Figure 13.18.

13.6.4 Grain Deformation

An effect sometimes seen in thin sheet which has a large grain size has become known as the 'orange peel' effect (Figure 13.19). The individual grains exhibit convex deformation on the

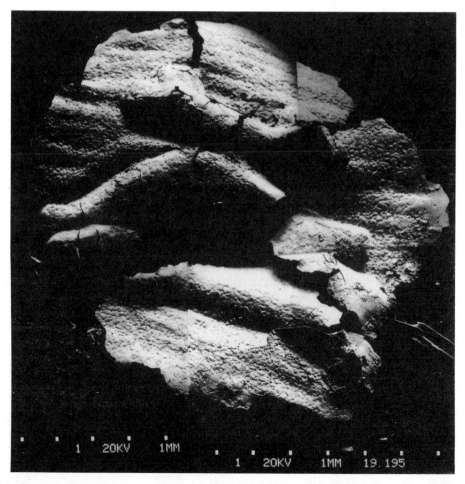

Figure 13.19 Deformed grains in thin sheet metal – the 'orange peel' effect.

side facing away from the explosion. The metal in our example is gold plated brass found in an item of luggage. We have no further experience of this effect, but it has been documented previously (Newton, 1985).

13.7 Environmental Considerations

The metallic components of an aircraft are normally well protected by anti-corrosion protocols when the plane is in-service. Following a crash, however, the combination of structural damage and the nature of the environment may stimulate corrosion. Whilst the 'best' situation for a crash site is probably a remote desert, the wreckage is more frequently located in a tropical rain forest or in the sea.

Technology exists to recover wreckage from the sea, but the rate of damage to structures is high due to currents, erosion by sand and gravel and attack by marine life. Perhaps more importantly, the effects of tidal flow and wave action ensure a steady supply of dissolved oxygen in the water. This depolarises the cathodic part of most corrosion reactions enabling the anodic reaction (the dissolution of metal), to occur at a more rapid rate.

For metals such as stainless steel and titanium the supply of oxygen is advantageous as it tends to promote protective oxide formation by maintaining the metal in its passive state. For metals such as magnesium and aluminium, which also rely on oxide films for protection, the presence of chloride in the sea-water interferes with oxide formation and a severe localised form of attack known as *pitting corrosion* tends to occur (Figure 13.20).

13.7.1 *Pitting Corrosion*

The chloride ions pass through the oxide films at thin places – for example, at sites where intermetallic compounds outcrop the surface. The attack may be either on the intermetallic

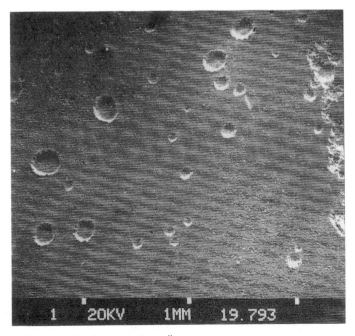

Figure 13.20 Pitting corrosion in magnesium alloy.

particle itself or, if it is electrochemically cathodic with respect to the substrate, the substrate is attacked around it until the particle is undermined and becomes isolated or falls out. The corrosion products produced within this corrosion pit are frequently a solution of metal chlorides. If there is little movement of the water, these salts may approach saturation and achieve high densities. There will be a tendency for these metal chlorides to diffuse out either by convection, or sometimes by gravity, if they are very dense. Once outside the environs of the pit they will encounter OH^- ions produced by the reduction of dissolved oxygen on the cathodic area surrounding the pit (like a battery, the two reactions of anodic dissolution of metal and cathodic reduction of oxygen must balance).

The interaction of the emerging anodic products with the OH^- ions results in the precipitation of metal hydroxides around the periphery of the pit, just within the cathodic area. They may be oxidised to higher valence states if sufficient oxygen is available. They are frequently non-stoichiometric and hence difficult or impossible to identify by X-ray diffraction. Elements such as manganese may diffuse further afield before precipitating as dark brown deposits of sooty appearance which can be mistaken for evidence of fire when found on white paint.

Where corrosion attack is rapid the pit becomes deep and the deposited material occurs in the mouth of the pit to form a roof or mound over the top of the pit. The more normal type forms a wall of deposit around the outside of the sheltered central pit which is often hemispherical and may even be chemically polished and shiny inside. This type of pit closely resembles the form of an impact crater to the unwary and a degree of experience is needed to determine the true origin of the pit.

If in any doubt, it is of great value to examine the feature by means of a stereo microscope or, in the SEM, to produce 3D stereo-pairs by obtaining two photographs of the feature with a tilt angle difference of 8° (Figure 13.21). At the periphery of the corrosion pit the surface oxide films may be thickened and curl upwards at the edges like crater walls, due to release of stresses where the film is broken. When seen in 3D, the resemblance to a crater wall of an impact crater is only superficial. Where overlapping craters are seen, corrosion pits tend to

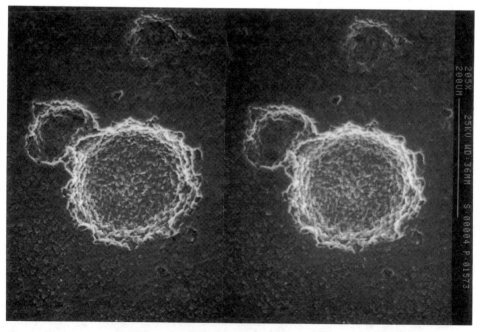

Figure 13.21 Stereo view of corrosion pitting in magnesium alloy from overhead trunking.

break through and fuse together with little or no intervening wall, whereas the true impact craters will be formed in succession with the last crater having a complete wall around it. However, if the overlapping corrosion pits are of differing depths the deepest pit will appear to be complete when seen in 2D and the image then appears to be much the same as for the impact craters (Figure 13.22). Seeing the same image in 3D can then be quite a shock.

Gentle cleaning with ultrasonics and detergents may reveal melt structures in the base of a true impact crater. However, the corrosion pit may contain dendritic structures deriving from an original cast structure and only a careful search for etch pits and corrosion attack will enable the researcher to tell them apart from these features. Research was carried out at DRA some years ago to explore the effects of seawater exposure on explosive signatures (G. Weston and S. King, 1988, unpublished RARDE paper). The results showed that impact craters remain essentially free of corrosion and corrosion products for a considerable period of exposure and that features such as flow ridges and raised edges around the craters remain identifiable. Whilst the onset of general corrosion is largely controlled by the availability of dissolved oxygen, the crater floor is usually covered by a relatively thick and protective air-formed oxide film and remains intact until the surrounding areas show quite severe degrees of corrosion. The use of chemical cleaning agents such as sulphamic acid and ethyl-enediaminotetracetic acid (EDTA) should only be considered as a last resort.

In deep sea, the same corrosion patterns still hold but, owing to the lower temperatures, near stagnant conditions and reduced rates of oxygen supply, the corrosion rates are much reduced. However, they are not negligible and attack will occur at fracture edges and damage sites. Under-film attack also occurs because of oxygen differential effects when wreckage is partially buried in silt. Intergranular attack is also common in skin alloys, especially when intermetallic precipitates are present in the grain boundaries. Although such attack may not completely destroy the integrity of the metal, it may spread throughout large areas via the grain boundaries, weakening the structure.

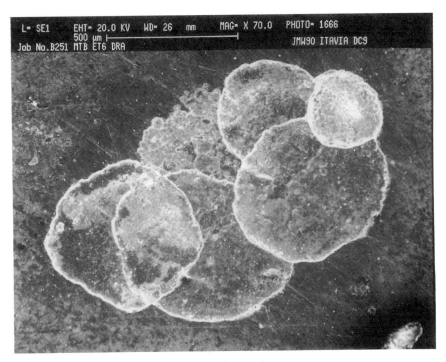

Figure 13.22 Overlapping corrosion pits in stainless steel.

13.7.2 *Wreckage Recovered From Sea Water*

It is important to wash all wreckage recovered from the sea with fresh water as soon as possible and then to keep it dry to prevent corrosion continuing whilst awaiting examination. Washing reduces the concentration of chloride ions and increases the resistivity of the electrolyte, whilst storing in humidity controlled conditions of less than 60% relative humidity will cause most salt solutions to dry-out, and even intergranular corrosion will slow down or stop if the humidity is kept sufficiently low. If de-humidifiers are not available and long term storage in dry conditions cannot be guaranteed, then the use of a de-watering agent or temporary protective coating would appear to be essential. However, care must be taken when selecting them as they must be cleaned off before samples can be examined in the SEM. Some coating materials are designed for use on ferrous alloys and contain inhibitors in the form of amines. These can cause severe corrosion on non-ferrous alloys, especially copper, and should be avoided. The simplest process is to follow the fresh water washing and air-drying by the application of a layer of hydrocarbon or paraffin wax to protect from the air. 'WD-40' is also suitable for this purpose. Untreated aircraft skin which dries out tends to effloresce due to formation of aluminum salts including aluminum sulphate.

13.8 The Need for an 'Expert Eye'

13.8.1 *Misinterpretation of Features*

The investigator must have an educated or 'expert eye' in order to be able to recognise those features which are the true signatures and indicators of an explosion having occurred. There are many traps for the unwary such as drill turnings in the form of small fragments. These seem to spring from every nook and cranny when an aircraft crashes. Few SEM operators or metallurgists have much cause to examine drill turnings in the normal course of their duties and their appearance is therefore 'strange'. Such artifacts present great difficulty as it is 'strange objects' for which the investigator is searching!

When under pressure from a panel of 'experts' or a commission who are convinced, by whatever reasoning, that they know what happened, deciding on the authenticity of such features becomes doubly difficult. Other features such as mechanically curled edges and corrosion pits, as discussed above, present equal difficulties in rapidly distinguishing them from explosion-produced rolled edge and impact craters when under such people-pressures.

Whilst it is possible to understand the concepts involved in the production of explosive signatures and other indicating features which have been discussed, by reading published papers and examining photographs, there is no substitute for experience.

We recommend that anyone who seeks to carry out this work should become thoroughly familiar, by test explosions and subsequent examinations, with the forms of damage known to have been derived from explosions. It is equally useful to be conversant with other types of deformation which can be expected to be found in the wreckage of an aircraft. The need for stereo microscopy and stereo-imaging techniques is re-emphasized since 2D images can be surprisingly misleading.

13.8.2 *The Consequences!*

When all the evidence has been examined, the decision has to be made which will state whether or not there was evidence of an explosion. The final decision will always be made under some pressure, including media, diplomatic and legal interests as well as the matter of payment by the relevant underwriters – so it is important to be certain of the evidence.

Our rule is that there must be 'at least two' explosive signatures before we will issue a positive opinion that an explosion occurred.

Only once have we been faced with the situation of having a single piece of positive evidence. Our report drew attention to the lack of corroborative evidence whilst stating that the single fragment had witnessed an explosion. It was later determined that an earlier investigation had been carried out and a 'control' specimen was inadvertently left amongst the samples. This example is referred to in more detail in the case studies. Whilst this may seem trivial it should be remembered that a positive confirmation of an explosion combined with loss of life will result in a murder investigation and the inevitable intense political, legal and media activity.

13.9 Case Studies

The choice of material from the investigations carried out at DRA, which we are able to present here, is limited to some extent by legal constraints – e.g. where the evidence is *sub judice* (as in the case of Lockerbie), or where it seems likely that there may be an appeal against a court decision or where it may affect compensation payments to victims' relatives. Nevertheless it is hoped that the following short summaries of typical results from actual investigations will be of interest.

13.9.1 *Air Lanka Tristar, Sri Lanka, 1986*

This incident occurred on the runway of Colombos' Katunayake airport in May 1986 when the aircraft was destroyed by an explosion and some 20 people were killed (Figure 13.23). Selected items from the wreckage were recovered by investigators. All the items were bagged

Figure 13.23   Air-Lanka Tristar, Sri Lanka, 1986. Photograph: Eric Newton.

separately and checked with a commercial explosives detector when received at Fort Halstead. Those which gave indications of a positive result were then subjected to a detailed chemical examination for explosive traces. None was detected.

The selected items which displayed evidence of damage which could possibly have derived from an explosion were examined initially by optical microscopy. The observed features were not, however, positively explosive-related. Subsequently, radiography of pieces of ceiling panel revealed the presence of some 17 trapped fragments. When extracted and examined by scanning electron microscopy, many of them showed the typical appearance of fragments from an explosion with explosive signatures, such as rolled edges, gas-wash and explosive pitting. It was subsequently shown that a device had been placed amongst spares in a storage case in the aircraft luggage hold. This is a good example of a straight-forward investigation where positive evidence was quickly and easily found, using the techniques described above. It contrasts starkly with the next example where a long and arduous search was carried out and a number of strange features were investigated before a satisfactory conclusion could be reached.

13.9.2 Convair 580, Denmark, 1989

On 8 September 1989, a Convair 580 LN-PAA, a twin propeller aircraft which had recently undergone a complete refurbishment and belonged to the Norwegian company Partnair, was carrying 55 passengers and crew when it crashed into the Skaggerrak just north of Denmark. The plane descended from its cruising altitude of 22 000 ft, breaking up as it fell. No message was received from the pilot. The wreckage was found on the sea-bed at a depth of 40–90 m and scattered over approximately 6 km². Most of the wreckage and all of the bodies were recovered.

Figure 13.24 Partly re-assembled fuselage of Convair 580, Skaggerak, 1989.

The NATO exercise 'SHARP SPEAR' was in progress at the time, but no evidence was found to substantiate rumours that the plane has been shot down. The rumours were also repudiated by NATO Defence Command. A Norwegian F16 pilot who passed the airliner 9 min before the incident was not part of the exercise. He pointed out that it was against regulations to carry weapons when flying over other countries. The radar record showed he did not deviate from his course.

RARDE was approached by the National Bureau of Crime Investigation based in Oslo for assistance and a team visited Oslo to examine the wreckage. The fuselage had been partly re-assembled in a hangar on a military base (Figure 13.24). Other pieces had been assembled in an area surrounding the hangar. Several items of baggage and fragments of metal recovered from two of the bodies were also examined. Nineteen items selected from wreckage, personal effects and fragments from the bodies, plus X-ray radiographs were received initially. Additional samples followed.

After a preliminary examination, most of the samples were found not to show any evidence of involvement in an explosion. Several samples were screened for explosive residues by solvent swabbing, followed by a gas chromatography with a thermal energy analysis detector (GC/TEA). One item gave a positive result and carried traces of RDX at the 6.0 ng level. Seven samples were selected for further study; four had been found in or on the bodies of two of the passengers – two pieces of tinned copper wire, a small flake of aluminium and a fragment of glass. Following detailed examination by stereo microscope, the two pieces of wire and the glass fragment were discounted as not showing evidence of explosive effects. The flake of aluminium, however, bore impact craters, rolled edges and gas-wash (Figure 13.25).

Of the remaining three samples, the first was the sample on which RDX had been detected: a piece of overhead trunking from the ventilation system which also carried lighting and the flight-attendant call button. After cleaning and SEM examination, craters were found which were judged to be corrosion pits (see Figures 13.20 and 13.21). The material was mag-

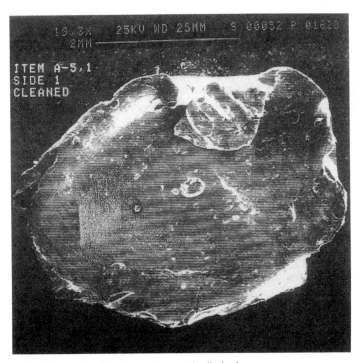

Figure 13.25   Aluminium flake showing craters, and rolled edges.

nesium alloy and the corrosion was the result of 3 weeks immersion in the sea.

The second sample, when thoroughly examined was found not to carry evidence of explosive damage. However, the depth of penetration of a piece of plastic sheet material through the hard plastic skin and into the aluminium honeycomb structure was curious. Trials carried out by firing slivers of plastic card from a Browning gun could not reproduce the effect. It was concluded that the card material may have been caught up in the slipstream as the plane broke up and that it was probably spinning rapidly when it struck the composite honeycomb partition.

The third sample consisted of a section of aluminium/balsa wood sandwich (this was an old aircraft) which was possibly part of a shelf or similar fitting. A single penetration was present which had a multi-petalled exit hole which could be regarded as significant. The event causing this damage was clearly dynamic. The failure mode on the entry side of the sheet was by shear which indicates the projectile was travelling at a medium to high velocity. Firing trials, again utilising the 0.5 inch Browning gun firing metal cubes of steel, aluminium and magnesium from sabot rounds were carried out against undamaged areas of the board. These trials were only partially successful, in fact the best match was obtained when the metal cube did not separate from its sabot. It was concluded that the actual projectile which produced the penetration was probably a composite item of steel and plastic with a mass of approximately 5 g and a velocity of between 400 and 700 ms^{-1}. Velocities in this region are greater than those associated with aircraft break-up or impact but lower than those from explosive events. These results were not readily explained. The value of on-site facilities for undertaking trials is certainly appreciated in such circumstances.

The work was reported in two parts. It is a good and thankfully, rare example of problems associated with multiple investigators. The Norwegian police, after receiving our first report, discovered that the fragment of aluminium which we had examined and which carried the signatures was actually a 'control' or comparison sample of a piece of detonator cap. It had been inadvertently left behind on a microscope stub by a previous investigator. This was the 'single positive result' case referred to above. The true sample was very much smaller and had been fixed in silver-filled adhesive. It had gone unnoticed and had been discarded when changing to a Cambridge stub for examination in the RARDE SEM. Photography of the true sample taken by the previous investigator however showed no evidence of it having witnessed an explosion.

A further series of small fragments was received along with other pieces of ventilation ducting and two pieces of bright aluminium foil with insulating foam backing. One sheet had a large dark area in its centre. All these samples proved totally 'innocent' and were reported as such.

It was suggested that the hole in the aluminium/balsa composite shelf material was possibly caused by a whiplash effect involving a bolt or fitting on a wire and it was therefore concluded that no explosion had occurred and that the exceedingly low level of explosive traces found on one item were due to contamination possibly at the military site where the wreckage had been stored.

This is a useful reminder that, along with fatigue cracks and loose tools, etc., the presence of explosive traces may not necessarily have anything to do with the cause of the accident. This also served to highlight potential problems associated with the very high detection sensitivity now being achieved in the DRA laboratories. It is now a working rule that all items selected for these tests are double wrapped in nylon bags at the crash site. It is of interest to refer to the conclusions of the accident inspectors in their report, where the causal factors are listed as follows:

(1) Maintenance instructions in use for this particular aircraft did not reflect the current aircraft configuration.

(2) Vital parts of the tail structure had failed and caused loss of control of the aircraft (the flight crew did not identify the problems in time to take corrective action).

(3) The vertical stabiliser was attached to the fuselage with pins and sleeves which did not comply with the specified values for hardness and tensile strength, and were therefore not airworthy.

(4) Abnormal wear had developed in the vertical fin attachments.

(5) Undampened oscillations in the elevator contributed to the destruction of the empennage.

(6) The left-hand main A/C generator was inoperative and could not be repaired in time for the flight. It had been decided that, as a substitute the auxiliary power unit (APU) should be used in flight. This had been agreed by the Pilot-in-Command. The APU had been installed with a front support of inferior quality and unknown origin which was believed to have fractured, the resulting vibrations from the APU contributed to the failure of the tail structure).

(7) Faulty, out-of-date maintenance instructions and inadequate maintenance procedures had left the problems in the APU's front support undetected.

13.9.3 Lockheed Hercules, Angola, 1989, 1991

On 8 April 1989, a Lockheed L382 Hercules Transport Aircraft, S9-NAI was descending to the runway at Luena, Moxico Province, Angola, when, at about 700 ft a loud noise, resembling an explosion, was heard and the overheating warning for No. 2 Engine was activated followed by fire warnings for No. 1 and No. 2 Engines. Both engines were immediately closed down causing the aircraft to veer off the course from that needed to successfully land on the airstrip. The decision was therefore made to crash land. The crew were able to escape through a window but the aircraft and its cargo of 20 000 litres of aviation fuel were destroyed by fire.

Lloyds Aviation Department requested assistance in determining whether the aircraft was lost due to accident involving fire alone, or if damage may have been inflicted by small arms fire or missile attack. Due to the prevailing conditions in this area, the DRA team did not attend but a representative from Lloyds brought many colour photographs of the wreckage showing various damage features including perforations and slots in the metal of the wings and engine nacelles. Although such photographs may have suggested the likelihood of high velocity impact by fragments, it was impossible to dismiss the possibility that the damage features were inflicted by undergrowth, etc. during the crash landing. It was therefore necessary for a representative of the insurance underwriters to make a subsequent visit to the crash site, at some personal risk, to bring back a number of items of interest selected by DRA from the photographs.

A total of 10 items were brought back for examination. All were double wrapped to prevent contamination. Three were selected for explosives residues tests using gas chromatography with thermal energy analysis detection. These comprised:

(1) an area from the underside of No. 2 Engine turbine casing;

(2) part of No. 1 Engine nacelle;

(3) part of main wheel-gear door (port side).

No traces of explosives residues were detected on the first two samples (1 and 2). However, low levels of cyclonite (RDX) and nitroglycerine were detected on the outer surface of sample 3. The actual levels were 8.0 ng and 2.7 ng respectively.

All the samples were examined by eye and hand lens. Seven of the 10 items appeared to bear damage characteristics which could be regarded as significant. The main damage features consisted of multiple petalled holes and long, narrow slots up to 25 mm in length (Figure 13.26). Some of the items additionally appeared to contain trapped fragments. Three of the seven items were selected as possible sources of micro-signatures required for proof

435

Figure 13.26 Damage to part of engine nacelle.

that an explosion had occurred in their proximity. A certain amount of experimental work was necessary to extract the required information.

Item 3, on which explosives residues had been detected, appeared to be heavily peppered with pits and impact craters (Figure 13.27). After a careful survey with a stereo microscope, one crater was discovered which contained an embedded fragment.

A delicate drilling and sawing operation allowed removal of a small area from the item, complete with crater and fragment. It was then possible to examine the whole sample in the SEM. The presence of a paint film on all the metal surrounding the crater induced charging problems in the electron beam of the SEM making it difficult to obtain photographs and ED analysis of the impact area. Optical microscopy did not provide sufficient depth of field to enable a fully in-focus view of the crater and fragment to be obtained. It was therefore decided to metallurgically section the fragment, whilst still in situ within the crater, which would then show the configuration of the feature and the structure of the fragment whilst at the same time facilitate both EDX and EPMA (Figure 13.28).

Figure 13.27 Peppering on landing gear wheel door.

Item 9, the de-icer can and bellows from the port wing, showed a number of oblique damage sites. Because of the very shallow angles of attack to the bellows, the damage tracks traversed several bellows rings before coming to a stop. In one case, because the bellows are double-skinned and sinusoidal in contour, a fragment was found to have passed through 13 thicknesses of metal before being stopped. This is shown in (Figure 13.29), complete with trapped fragment which was subsequently removed and analysed.

In addition, at least one of the possible impact penetrations in the main body of the can showed rust staining although the can itself was of stainless steel. The area of can containing the hole was carefully drilled and cut around the periphery with a fine saw. The appearance of the hole with rust nodule is shown in (Figure 13.30). After an attempt at cleaning away the rust had failed, the rust nodule was micro sectioned and examined in the SEM. A ferrous fragment with low alloy content was revealed lodged in the stainless steel material of the can.

Item 10 was a portion of metal skin from the under-wing surface of the port wing between the engines. Careful examination did not reveal any embedded fragments present but there were numerous penetrations which appeared to align with holes in a second piece of skin material of similar size and shape which, it was understood, was the inner layer of skin of the wing. The holes in the inner layer were much larger and were generally multi-petalled. With the two pieces of metal skin held in their attitude of probable alignment, it was possible to insert wooden rods through the holes in both sheets, such that they were all roughly parallel, (Figure 13.31), suggesting a common source for the penetrating fragments.

A small area was selected from the outer surface of the outer skin layer and examined in the SEM. Typical examples of impact craters were found together with an area of surface melting (Figure 13.32).

The three fragments from the wheel gear door, the de-icer can and the bellows were analysed in the SX50 microanalyser together with the materials of the door and the de-icer can. Material from the warheads of two Soviet missiles (SA7 and OG9) and from an Amer-

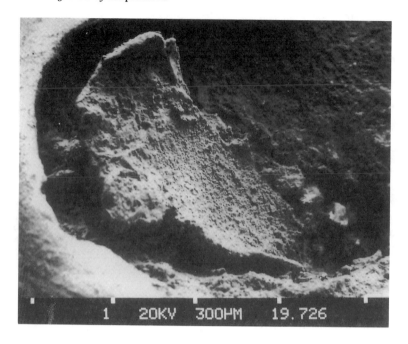

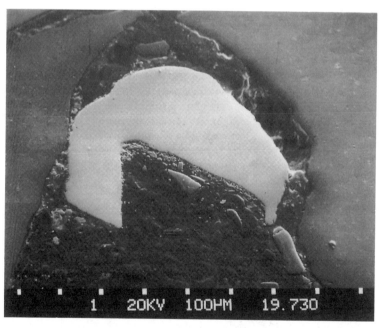

Figure 13.28 Embedded fragment as found, and after sectioning.

ican 'Stinger' missile warhead were also prepared and analysed. The results are shown in Table 13.3.

The results are certainly interesting. The Stinger was discounted as it was found to be a titanium aluminium alloy. The Cr, V and W concentrations in the fragments and the Soviet warheads were all similar.

Table 13.3 Quantitative EPMA analysis results from recovered fragments and control samples

Sample	Element Concentration Wt%											
	Ni	Mn	Cr	V	W	Cu	Si	Al	Mo	Co	Ti	Fe
Fragment from wheel gear door	1.04	0.74	1.23	0.10	0.86	0.13	0.99	0.02	0.02	0.14	—	Bal
Fragment from de-icer can	0.99	0.69	1.08	0.08	0.74	0.11	0.90	0.08	0.01	0.13	—	Bal
Fragment from bellows	1.02	0.70	1.13	0.09	0.79	0.08	0.88	0.04	0.02	0.14	—	Bal
SA7 warhead	0.96	0.68	1.11	0.09	0.92	0.07	0.86	0.04	0.02	0.14	—	Bal
OG9 warhead	1.05	0.74	1.20	0.10	0.87	0.12	0.99	0.02	0.01	0.14	—	Bal
Stinger warhead	0.02	0.00	0.08	4.53	—	0.02	0.04	6.29	—	0.01	Bal	0.19

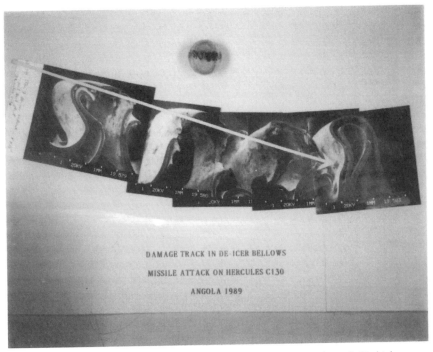

DAMAGE TRACK IN DE-ICER BELLOWS

MISSILE ATTACK ON HERCULES C130

ANGOLA 1989

Figure 13.29 Fragment embedded in de-icer can bellows after passing through 13 thicknesses of metal.

A comparison of the microstructures of the three fragments with those of the warheads from the SA7 and the GO9 showed them all to be tempered martensite. They all had hardness values greater than 400 VPN except the GO9 which was somewhat softer. The GO9 is a more unlikely choice as the weapon is vehicle launched, whereas the SA7 is man portable and more suited to the terrain.

In addition, the recovery and identification of traces of RDX explosive and nitroglycerine on the outside surface of Item 5 were conclusive proof that an explosion occurred outside the aircraft, in close proximity to the port wing. This result is particularly interesting as nitroglycerine is not a normal component of military explosives. It is more likely that the presence

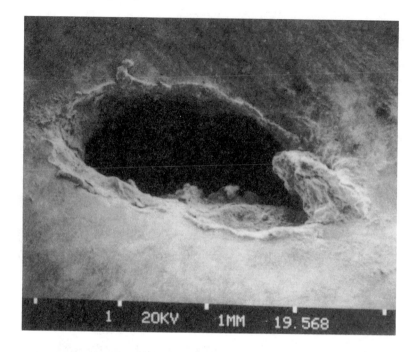

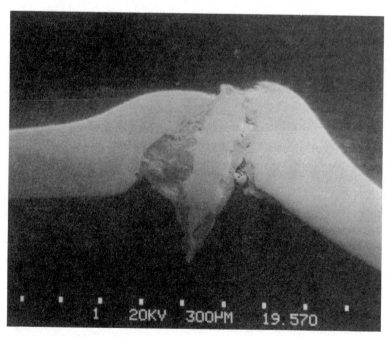

Figure 13.30 Hole with rusted nodule in de-icer can, as found, and after sectioning to show steel fragment.

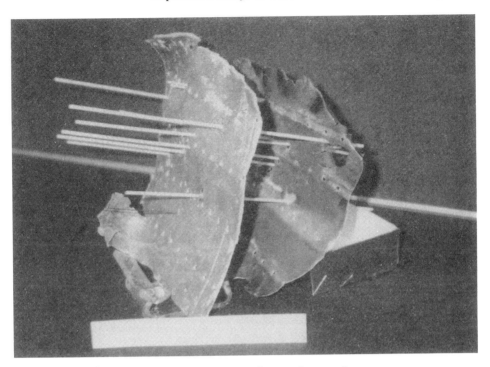

Figure 13.31 Alignment of perforations in inner and outer wing coverings.

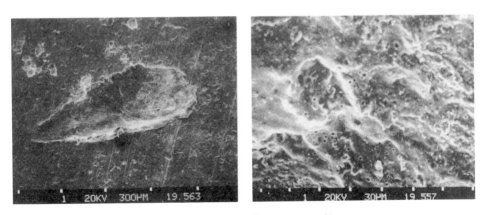

Figure 13.32 Typical impact crater and surface melting on outer skin.

of this material may have derived from unburned rocket motor fuel, especially as the incident occurred at low altitude (700 ft) when only a small proportion of the propellant would have been burned. The explosive filling in the SA7 warhead is RDX whereas the OG9 is filled with TNT. No traces of TNT were found.

It was therefore concluded that the aircraft was brought down by an external explosion and that the fires in engines 1 and 2 were the result of damage by fragments from the warhead of a missile. The composition and microstructure of the fragments together with the explosive traces suggest that a Soviet manufactured weapon, probably an SA7 man-portable missile was responsible.

As a corollary to the above investigation, Lloyds again approached DRA regarding the loss of another Hercules aircraft in Angola in March 1991. Amongst the pieces of wreckage brought back for examination was a large, crumpled steel sheet which had been identified as part of No. 1 engine exhaust duct. Tests for explosive residues proved negative on all samples. Also there was no evidence of high explosive damage, except on the steel sheet from the engine duct. This duct had several large holes and numerous perforations, some with multiple petals. An area was found with micro-perforations and craters. A piece was cut to a suitable size for examination in the SEM and ultrasonically cleaned to remove loose soil and debris.

Many examples of high velocity penetrations were seen in the SEM together with typical impact micro-craters. Although it had been cleaned the metal surface was unusually rough and ED analysis of the steel showed it to be a variety of stainless steel. The rough surface was, however, found to be a layer of deposited material largely composed of titanium and aluminium. Thus the analytical studies of warhead materials for the previous Hercules investigation were vindicated!

References

BOURNE, B., JONES, P. N. and MARKHAM, J. A., 1984, Microstructural Features of Shock-Loaded Metals, Inst. Phys. Conf. Ser No.70, *3rd Conf. Mech, Prop. High Rates of Strain*, Oxford. Published by Inst. of Physics, Bristol & London.

CLANCY, V. J., 1968, Comet G-ARCO: Solving the Riddle, *New Scientist*, **39**, 614, 533.

MELOSH, H. J., 1989, *Impact Cratering, A Geological Process*, Oxford: University Press.

NEWTON, E., 1985, Investigating Explosive Sabotage in Aircraft, *The Institute of Aviation Safety*, **3**, 43–48.

TARDIF, H. P. and STERLING, T. S., 1967, Explosively produced fractures and fragments in forensic investigations, *Journal of Forensic Sciences*, **12**, 247–272.

The Use of Vibration Spectrograms in Aircraft Accident Investigation

FRANK SLINGERLAND

14.1 Introduction

Was it a 'bomb' or a structural failure? Whenever an aircraft is destroyed at altitude, hundreds of millions of dollars in claims hang on the answer to that question. The answer is difficult to obtain because of the wide dispersion of the wreckage trail, especially over deep oceans. However, the aircraft's flight recorders are equipped with radio and sonar transponders, and have been relatively easily recovered, even at ocean depths below 4000 m. The present chapter describes a new method, developed by the writer since 1985, for using one of the flight recordings for accident diagnosis. The cockpit area microphone (CAM) signal, recorded on the cockpit voice recorder (CVR), contains fuselage vibration data from which the damaging event can be identified and located, and eventually, its strength and effects evaluated.

14.2 Waves

Most children are fascinated by the wave patterns which run to and fro when a stone is dropped in a quiet pool or ditch. Mechanical systems having mass and elasticity also generate waves of vibration under sudden impact – e.g. a piano string struck by its hammer. An aircraft fuselage, like the hollow tube of a doorbell chime, shows similar waves of vibration when subjected to positive or negative pressure impulses.

14.2.1 Fuselage Vibration Modes

An aircraft fuselage is a much more complex structure than a hollow tube, having skin thickness tapers, ring frames and stringers and tapered ends. However, for the kinds of vibrations which can be excited by large-area pressure loads and which lie within the 100–3000 Hz frequency range of the CAM, it can be visualized as a classical thin shell with negligible bending stiffness (Leissa, 1973). The radial vibration modes of such a shell are formed by integral numbers (0, 1, 2, 3, etc. – call them n) of sine waves around the circumference, combined with sine waves of any axial wavelength along its length. The speed of travel (*group*

443

velocity) of these waves depends on their value of n and on their axial wavelength; for any mode, their frequency also depends on axial wavelength. Such waves are called dispersive.

For each mode (value of n) there is one frequency or wavelength component which travels fastest; other components at higher or lower frequency travel more slowly. For typical aircraft dimensions these modes travel at maximum speeds of 800 ms^{-1} to 1500 ms^{-1}. The $n = 0$ mode is an exception – it is the ring or breathing mode which 'piggybacks' (couples) on the axial vibration so as to propagate at *only one* frequency and wavelength, and at very high speed – typically 3000 ms^{-1}. Sketches of these modes are shown in Figure 14.1.

The $n = 0$ and $n \geqslant 2$ modes lie within the frequency range of the CAM. The $n = 1$ mode represents bending vibration of the whole fuselage at low frequencies, and is not normally recorded. The speed of travel of these modes as a function of frequency is shown qualitatively in the graph of Figure 14.2.

The horizontal axis of the graph could also represent transit time in milliseconds from a more distant source. The floor or floors of the aircraft complicate the simple mode patterns of Figure 14.1 by restricting radial motion at the floor ends. Each n-mode separates into two versions having different frequency at any axial wavelength. For example the $n = 0$ ring mode comes in the two types shown in Figure 14.3.

Judging by their observed frequency, it is always the asymmetric version which is excited by the damaging force. In the Boeing 747 it appears that the three modes, $n = 0, 2, 3$ are observed in CAM records; in smaller aircraft, only $n = 0$ and 2.

14.2.2   Exciting Forces

The most common forces acting to destroy an aircraft in flight are of two types: (1) those caused by a sudden breaking open of the pressure hull due to a structural fault (e.g. loss of a cargo door) and (2) those caused by detonation of an explosive charge ('bomb') planted within the aircraft by a criminal. Both are very large. The loss of a cargo door or rear bulkhead represents the sudden release of a force of about 3 MN (30 tonnes). The impact of an explosive blast wave on a cargo bay pressure bulkhead can reach 40 MN. Note that in both cases the region of application of the force is roughly the same size as the fuselage radius; thus low values of n are more efficiently excited than large ones.

14.3   Vibrations Recorded by the CAM/CVR

The CAM was designed to hear the airborne sound of speech by the aircrew, and is usually mounted on the 'eyebrow' panel above their heads. How then can it detect vibration? When

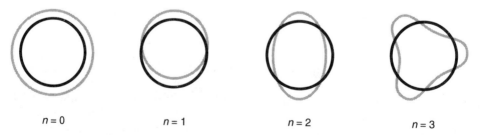

$n = 0$ $\qquad\qquad$ $n = 1$ $\qquad\qquad$ $n = 2$ $\qquad\qquad$ $n = 3$

Figure 14.1   Vibration modes in a cylindrical shell.

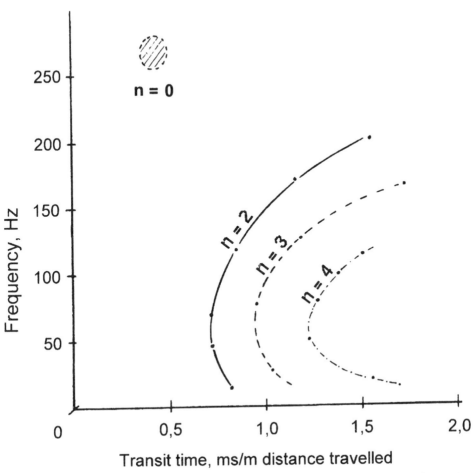

Figure 14.2 Group velocities of vibration waves in a cylindrical shell (aluminium, 6.4 m diameter).

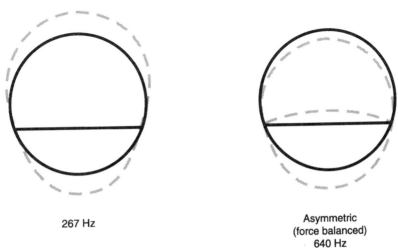

267 Hz

Asymmetric
(force balanced)
640 Hz

Figure 14.3 Qualitative $n = 0$ modes of floored shell (frequencies calculated for Boeing 747 radius).

the fuselage surface close to the microphone vibrates in the radial direction, it acts as a loudspeaker, feeding sound to the microphone. Also the microphone itself is shaken by the vibration of that part of the fuselage on which it is mounted, which creates alternating pressures on its diaphragm recorded as sound. The CAM is in fact a very sensitive vibration detector, responding to otherwise inaudible hand-slaps on the fuselage. By contrast there have been 'bomb' explosions and tire bursts which were not heard by the aircrew above the ambient noise.

Because of their higher propagation speed – three to eight times the speed of airborne sound – vibrations are the first damage messengers to reach the CAM (these speeds have been experimentally confirmed by test explosions on board a mothballed aircraft). In fact in the five disasters analysed to date, no sign of an airborne sound signal can be discerned in the CAM recording or in the spectrogram, even when one knows exactly where in the recording to look for it. It is dissipated by floors and partitions; it is absorbed and diffused by multiple reflections from seats. Only in a bare test aircraft, and only for small explosions in the passenger space, can the acoustic signal be observed.

14.3.1 Typical CAM Record

Figure 14.4 shows the voltage:time trace of the CAM record from the Pan Am Lockerbie disaster of 21 December 1988 – the only one for which written permission to publish has been received. Like all such records analysed to date it begins with a few cycles of clean sine wave (the $n = 0$ mode!) and then breaks into more complicated waveforms. Its signal-to-noise voltage ratio is about 9:1, other such records range from 7.5:1 to 12:1. Unlike other records, it terminates after about one-tenth of a second because of disruption of power to the CVR. When played back through a speaker it sounds like 'tch!' Most other recordings have not involved CVR power loss, and continue recording up to 0.4 s before the wave reflections are attenuated by damping in the structure. The recorded sound is more like 'boom', and it includes up to 11 reflected wave passes past the CAM. The voltage–time amplitude history contains no useful information, since it is subject to automatic gain control (AGC) in the CVR with a time constant of about 35 ms.

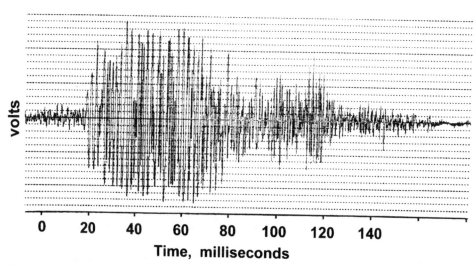

Figure 14.4 Voltage–time trace of Lockerbie CAM recording (final 0·13 s).

14.3.2   Spectrogram

Referring back to Figure 14.2, it is clear that even on first pass the CAM signal at any moment will usually contain components of more than one mode. The real situation is even more complex, since several wave reflections of each mode are all present at the CAM at any instant, as they bounce to and fro off the ends of the fuselage. The only way to sort out the data is to form a 'spectrogram', which is a '3-D' graph of signal strength versus *both* time and frequency. Such a graph is like a contour map of terrain, where the North axis represents frequency, the East axis represents time, and the elevation contours represent signal strength. For ease of reading, the different strength levels are coloured in using a hot-metal scale: white-hot for high levels to blue-cold for low levels. The Pan Am Lockerbie spectrogram is shown in Figure 14.5.

The spectrogram can be formed in various ways, but the most successful to date has been to use a spectrum analyser to perform a Fourier transform on successive 20-ms Hanning-windowed time samples, overlapped 10 ms. The spectrogram level contours in Figure 14.5 were then drawn in manually, but they are now produced by computer. At first glance Figure 14.5 looks like a meaningless scramble of contours, but when examined more closely, a series of mountains and hills, and bumps on the edge of sidehills, can be discerned in the signal terrain. These can be connected, five or six at a time, to define successive passages past the CAM of the waves of two modes. They resemble the theoretical wave contours of Figure 14.2, and are shown by the heavier lines traced onto the spectrogram, looking like parabolas lying on their sides. The parabolas close in with each successive wave as the slower components fall increasingly behind the fastest components at the nose of the parabolas.

The next step is to analyse the spectrogram to determine what happened when the vibrations were recorded.

14.4   Analysis

14.4.1   Was it a 'Bomb'?

The first attempt at an answer made use of the direction, positive or negative, of the very first excursion of the ring mode as it arrives at the CAM ahead of the crowd. Since this is a pure-tone non-dispersive wave, its initial motion, radially inward or outward, at the source will carry forward to the CAM with the same phase. For example if the source is an explosion in the passenger space, the initial movement of the fuselage at that point will be outward, driven by explosive overpressure. Then the initial movement at the CAM, carried by the ring mode, will also be outward, and will induce a negative-going pressure on its diaphragm. The overpressure which caused this signal will not yet have arrived at the CAM since it travels much more slowly.

The next problem is: 'which way is up', i.e. which direction is positive on the CAM recording? The analyst is usually working with a great-grandchild copy of the original CAM tape, and there is no guarantee that upward voltage still represents positive pressure. This can be resolved by examining any vowel in the pre-event speech of the aircrew, since the larynx always generates positive pressure pulses. An example is shown in Figure 14.6.

Each cycle of the 'O' sound shows an initial upward movement, hence up is positive. (In fact none of the five records analysed has shown a phase inversion.)

Another problem is the asymmetric nature of the ring mode (Figure 14.3) in the presence of a floor (the presence of a second floor in the Boeing 747 does not cause a second asymmetry). The negative-going initial excursion we observed in the example cited above could be caused <u>either</u> by an explosion in the passenger space *or* a structural-failure-type opening in the baggage space causing decompression. No way has been found to resolve the

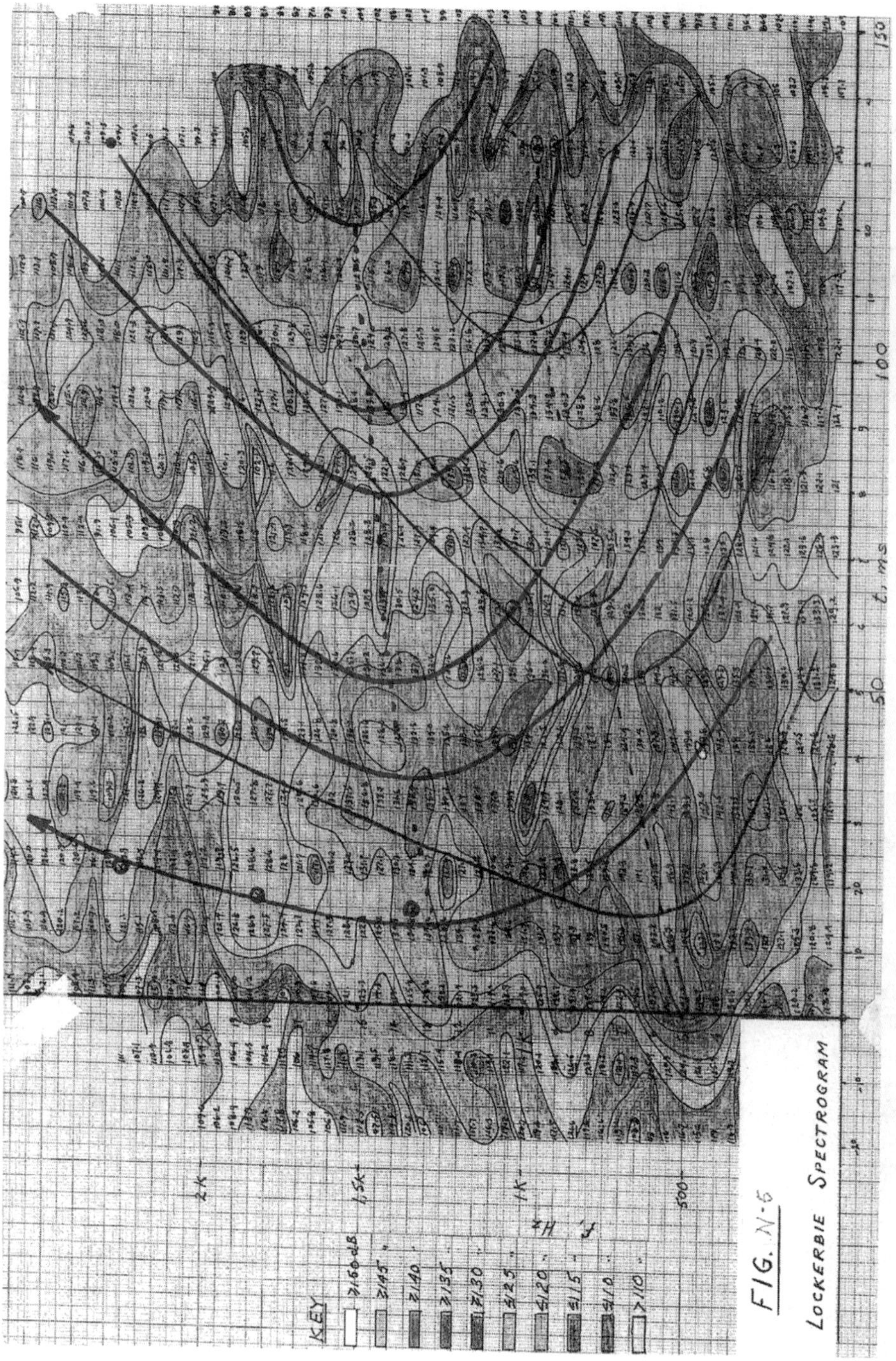

Figure 14.5 Lockerbie spectrogram.

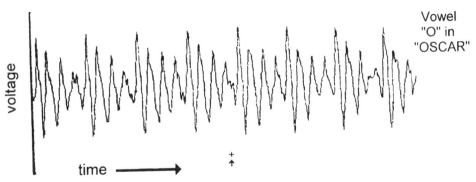

voltage

time ⟶

+
↑

Vowel
"O" in
"OSCAR"

Figure 14.6 Signal polarity from speech breath impulses.

ambiguity of initial ring-mode phase. However it has another use as we shall see in the next section.

A better method of identifying the cause is to examine the frequency trends of the successive waves in the spectrogram shown in Figure 14.5. Dotted lines have been traced through the noses of the waves, showing frequency increases of 20% and 60% between first and last wave. In other longer records one observes a rise, then a fall in frequency to less than the initial values. Why? Shell theory provides the reason – the frequencies of radial vibration modes depend directly on the pressure difference across the shell. A typical effect in aircraft is 50% increase per atmosphere. Just as a guitar string rises in pitch when tightened, a fuselage rises in pitch when tightened by internal pressure. A localized zone of pressure increase or decrease also translates into a frequency increase or decrease. Thus the only possible cause of the frequency increases showing in Figure 14.5 is the sudden overpressure caused by the detonation of a 'bomb'. The subsequent drop in frequency, hence pressure, in other records indicates that the explosion perforated the pressure hull, causing decompression. These trends hold whether the event occurred in the cargo or passenger spaces.

Accidents involving structural failure causing pressure loss show a steady drop in frequency right from the start, with no initial rise. These two trends are shown in Figure 14.7 for five accidents comprising three 'bombs' and two decompressions. The two types of event are clearly distinguishable.

Returning to the ring mode initial phase approach, we now know by frequency trends whether the event was an explosion or a structural opening. Thus (in the hypothetical case assumed) we can rule out a decompression in the cargo bays; thus the event is located in the passenger cabin. It remains to locate it in the fore and aft direction.

14.4.2 Where was the Bomb?

The first wave contours of Figure 14.5 started out from the source as vertical lines – all frequency components were excited at the same instant. Thus their observed curvature should indicate how far they have travelled to reach the CAM. However we do not have enough measured data on the variation of wave speed with frequency for each mode in each aircraft and altitude. Another approach would be to measure the time lag between arrival of vibration signals and arrival of the associated airborne sound. But again measured wave speeds are unavailable, and the acoustic signal is usually squelched by AGC action and its propagation path is unknown.

The axial location of the source can still be determined from the time spacing of successive reflections of the vibration waves. The first wave to arrive is the one which travelled directly

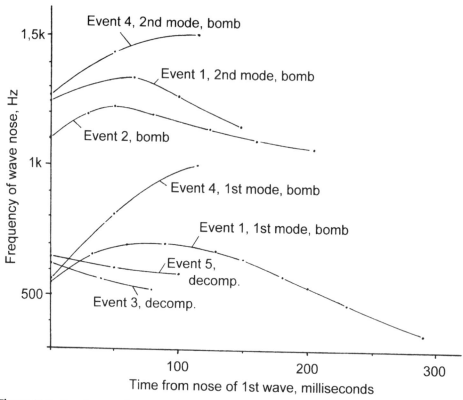

Figure 14.7 Spectrogram frequency shifts for various events (event 4 is the Pan Am Lockerbie incident).

from source to CAM. The second arrival is the wave which propagated aft, reflected off the aft end and then returned to the CAM. The third wave is the first one after it has made a full round trip. The sequence then repeats. The time delay between first and second waves is thus the time to travel from source point to the rear and back to the source point. The ratio of this time to the round trip time represents the distance of the source forward of the rear, as a fraction of total fuselage length. Corrections must be applied since the fuselage is not a uniform tube. Nose and tail tapers are easily handled since shell theory shows propagation speeds to be inversely proportional to fuselage radius. When this is applied it is found that the equivalent echoing length of a fuselage can with negligible error be taken to be the physical distance between fore and aft pressure bulkheads. The waves also speed up when passing through the stiff wing box area. An empirical correction for this effect was obtained from a survived explosion accident. The calculated source position could be made to coincide with the known position by assuming a 39% speedup over the whole length of the wing root, and this correction has served well on three different sizes and models of aircraft. When wave time ratios are measured repeatedly over the whole spectrogram and averaged, the fore and aft position of the source is determined within ±1 m in nearly all cases.

14.5 Special Cases

Not all accidents are 'classic' from a spectrogram viewpoint. For example in one case, many lines of evidence converged to suggest that an explosive charge detonated into an almost empty cargo hold. Once free of its container the blast wave propagated forward to the forward bulkhead of the hold and reflected with approximately four times incident over-pressure. In this case the spectrogram analysis located the 'source' within 33 cm of this bulk-head (the site of the largest forces) with only minor waves emanating from the actual charge location. A second accident involved the loss of the rear cargo door of a DC-10. Wreckage analysis showed that the floor collapsed downward under the pressure difference between passenger and cargo spaces. The spectrogram showed two trains of waves from two suc-cessive events, with an upward jump in frequency between them, presumably due to the jamming of the collapsed floor into the cargo hold. The analysis also suggested that vibration waves are slowed down near the aft bulkhead in this aircraft, presumably by the mass of the rear-mounted engines.

14.6 Performance of the Spectrogram Method

The techniques described above have now been applied by the writer to five aircraft accidents – three 'bombs' and two decompressions. They have also been applied by other accident investigators to the loss of an Italian aircraft.

In every case the analyses have agreed with the known cause and location of the source, or with the convergent results of other lines of investigation. In two analyses conducted by the writer the location of the source was withheld as a test by authorities, but correctly estab-lished by the new method.

14.7 Further Developments

The rates of frequency rise and fall in the spectrogram contain information on the explosive yield of the device and the size of hole it created. The methods described by Avery (1981) could be used to determine explosive weight and perforation area. Already fuselage breakup is readable from the sudden increase in the rate of the downward frequency trend.

It is also possible to relate the frequency range covered by the first dispersive wave to the duration of the force pulse which excited it: $\Delta t = (\Delta f)^{-1}$. This in turn can be related to explosive mass by the simple formula for the duration of the positive phase of a blast wave:

$$t^+ = m^{1/3}$$

for time t (ms), and mass m (kg).

However, not all of t^+ is effective, and further study is needed of the dynamic interaction of blast waves with vibration waves, and of pulse-stretching by baggage. A new algorithm for spectrogram generation – the Wigner distribution (Wahl and Bolton, 1990) may improve spectrogram readability.

The method should be applicable to damaging forces (e.g. from missiles) applied outside the fuselage, or on a wing. There are at least 30 explosion events for which CVR/CAM records are held by various national authorities. The application of spectrogram analysis to them would greatly increase our knowledge of the technique.

If/when fully mechanized on computer, vibration spectrograms offer rapid diagnosis of an aircraft accident within a few hours after CVR recovery.

Acknowledgements

The method described herein was developed and applied while the writer was in the employ of the National Research Council of Canada. The support of the Council and its officers is gratefully acknowledged. Special thanks are due for the encouragement of Mr. John Garstang of Transportation Safety Board of Canada, and of Mr. Bernard Caiger of the NRC Flight Recorder Laboratory.

Figures 14.2, 14.3, 14.4 and 14.7 are reprinted with permission from SAE paper No. 912041 © 1991 Society of Automotive Engineers Inc.

References

AVERY, J. G.,]1981, Design manual for impact-damage-tolerant aircraft structures, *AGARD report AG-238*, p. 33.

LEISSA, A. W., 1973, Vibration of shells, *NASA Special publication No. 288*.

WAHL, T. J and BOLTON, J. S., 1990, Identification of structure-borne noise components by the use of frequency-time filtering, *AIAA 13th Aeroacoustics Conference, Talahassee Fl.* Paper AIAA-90-3937.

15

Forensic Pathology of Victims of an Explosion

JAMES A. J. FERRIS

15.1 Introduction

Death and injuries from the effects of both explosive devices and the explosive substances can occur in a variety of circumstances including military, terrorist, and civil. Although the forensic pathologist, medical examiner, and coroner are concerned primarily with the investigation of death and the collection and documentation of injuries and evidence relating to these deaths, trauma surgeons and emergency physicians have an equally important role in the examination of non-fatal victims. The principles involved in the investigation of these injuries are the same, whether or not the victim has died, although identification of the victim is an additional problem for the death investigator. These topics are well illustrated in an epidemiological study by Mallonee et al. (1996) of injuries and deaths in the 1995 Oklahoma City bombing.

Because of the commercial nature of military explosions, the injuries sustained are to some extent predictable, however, in the case of terrorist devices, explosives may be home-made or combined with commercial explosive materials, and the interpretation and documentation of injuries under these circumstances may help with determining the nature of the explosive device. Civilian incidents are often industrial, as may occur in underground hard rock mining, chemical storage facilities, grain storage buildings, and explosive and fire-work manufacturing plants (Isa and Moe, 1991; Benmeir et al., 1991; Hubbard et al., 1992; Origitano et al., 1992; Zane et al., 1992). Accidental explosions in the home may be associated with gas leaks and the storage of explosive materials, such as propane. The cause and incidence of such explosions is dealt with elsewhere in the book (Chapter 7).

The primary role of the medical investigation of explosive trauma is to document the injuries and collect evidence that will assist with the investigation of the nature and source of the explosion. The examination and documentation of injuries sustained by victims of explosions is essentially similar, whether on not the victim dies. For the purposes of this chapter, it is the role of the pathologist and the procedures followed at postmortem examination which will be considered in detail.

15.2 Postmortem Examination

The postmortem examination in cases of explosive injury can be divided into six separate phases.

- identification;
- radiological examination;
- collection of surface evidence;
- documentation of injuries;
- identification of natural disease;
- collection of internal samples.

15.2.1 Identification

Although the majority of explosion victims' bodies remain relatively intact, extensive disruption can occur, and in the case of individuals who have been in immediate proximity of the explosion, there may be total fragmentation of the body. The collection, assembly and separation of these remains becomes an important initial step in the autopsy process.

The first priority is to establish how many bodies are represented and to attempt to allocate the correct fragments of remains to the right individuals. Where the bodies of several victims have been disintegrated by the explosion, the pathologist may be confronted with a large pile of human tissue, and it may be very difficult, if not impossible, to allocate all of this tissue to specific individuals.

As in all identification cases, the first priority is to determine sex, age, height and general stature. Secondary identification features can then be sought. It is important to recognize that clothing and other personal possessions which may be attached or embedded within the remains may prove valuable aids to identification.

Unlike many forensic pathology identification cases, because of the disruption of the entire body, structures such as prostate, uterus, breasts, etc., can be used for determining sex (Marshall, 1976). Hair, colour of skin fragments and eyes, may also provide additional valuable identification information, and may assist with racial typing. Fingerprinting of bodies can often be performed even in instances of severe fragmentation. Small portions of friction skin surface, if recognized, can often allow for positive identification.

The role of the forensic odontologist in examining recovered teeth and jaw fragments will often provide additional information of identification, in cases where jaws and teeth have been fragmented. The skill of the forensic odontologist will often allow for matching of fragments and identification of the total number of victims involved.

In some cases, basic knowledge of forensic anthropology and osteology will help with the identification and classification of bones. Significant age characteristics identifiable on radiologic examination of bone, and other forensic anthropological information can be obtained from standard forensic pathology and forensic anthropology texts.

Blood and tissue typing, and genetic typing ('DNA fingerprinting') are valuable but time consuming identification techniques. Tissue samples can be compared with known blood samples and in the case of genetic typing, the samples can be compared with DNA material from parents or siblings in a form of 'reverse' paternity testing.

15.2.2 Radiological Examination

All bodies should be X-rayed prior to postmortem examination, to identify any radiopaque materials. Not only will this assist in detecting lethal missiles not visible from the surface, but, in victims who have been in close proximity to the explosive device, small fragments of trace metal that had formed part of the bomb mechanism can be identified and recovered (Figure 15.1).

X-ray examination may also reveal evidence of other injuries such as gun-shot wounds sustained prior to the explosion, since in many cases of terrorist explosions, the victims may

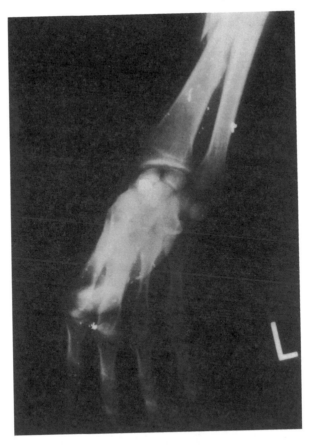

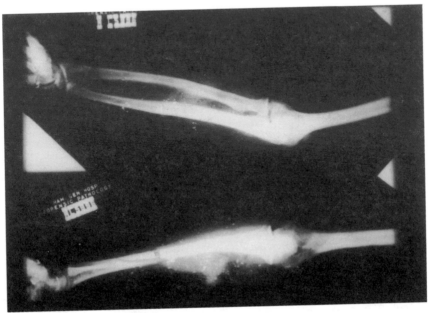

Figure 15.1 Fragments of a device embedded in an arm and hand.

have been murdered or tortured prior to death. Radiological examination allows for identification by comparison with prior X-rays, and also the identification of any previous skeletal injuries or bony abnormalities, which may provide evidence for identification. Matching of bone fragments by radiological examination may also be possible, although in most cases, physical matching of bone fragments is best done after removal of soft tissue.

15.2.3   Collection of Surface Evidence

Traces of powder, bomb chemicals, and bomb fragments may be on the surface of the body or the body fragments. All of this material should be photographed, collected and preserved in accordance with the normal procedure for the handling of such evidence.

15.2.4   Documentation of Injuries

Since the nature and extent of injuries is an integral part of the incident reconstruction, it is important for the investigator to know not only the definition of the terms used, but to have a basic understanding of the manner and mechanism by which injuries are produced. All wounds are important, however apparently insignificant. They may demonstrate the direction of the impact forces, the distance from the point of explosion and in some instances the magnitude. The features of injuries that the pathologist will wish to note include:

- the size, shape, number and location;
- the presence of foreign material such as oil or paint;
- the nature and extent of internal damage;
- any vital reaction to the injury or other evidence of healing or medical treatment.

15.2.5   General Classification of Injuries

All injuries can be classified into five general groups:

- blunt impact injuries;
- incised wounds;
- penetrating injuries;
- burns;
- fractures.

Blunt impact injuries

Blunt impact injuries are caused by blunt non-penetrating force and there are three types which may occur singly or in combination: bruises, abrasions, and lacerations.

(1) *Bruises*   Bruises (contusions) are produced when blood leaks into the subcutaneous or deeper soft tissues as a result of some blunt impact of sufficient force to rupture blood vessels, usually capillaries. A rapid accumulation of blood is often referred to as a haematoma. Interpretation of bruises requires great caution since the size of a bruise cannot usually be related to either the force or location of the impact.

Since skin bruising is only the external evidence of internal bleeding that has extended to just beneath the skin, a bruise may take several hours to develop and may not appear immediately beneath the actual site of impact. Bruises change colour with ageing but such changes may differ from one individual to another and may be subject to many variables. In some instances, the bruise may involve only the most superficial layers of the skin and may be made up of tiny ruptured capillaries. This type of bruising is sometimes referred to as petechial bruising and may be significant since these injuries may appear to reproduce specific textured surfaces such as fabric marks.

(2) Abrasions Abrasions are superficial blunt impact injuries in which the outer layers of the skin have been compressed and scraped off by movement of the injuring surface across the skin. Unlike bruises, abrasions occur at the site of impact. The degree of pressure required to produce an abrasion is determined by the roughness of the impacting surface. The presence of clothing may prevent or modify an abrasion.

A 'friction' or 'sliding' abrasion is the common graze or scratch. The cuticle layer of skin becomes heaped up by the rough surface and the direction of movement of the skin over the impacting surface may be determined. These type of abrasions are typically seen when the blast victim is projected against stationary objects. 'Pressure' abrasions are associated with much less movement and may reproduce some of the features of the contact surface.

During life, abrasions exude serum and then form a scab. If the victim dies shortly after injury then the area of injury dries out and the skin becomes parchmented causing the abrasion to become more clearly defined and appear much more severe.

(3) Lacerations Lacerations are tears or splits of the skin, usually occurring over bony prominences where the skin is less mobile. For example, lacerations of the scalp and face are common, whereas lacerations of the abdomen are relatively rare. Lacerations are caused by crushing or shearing forces and the margins are irregular and invariably bruised or abraded. The softer tissues beneath the skin are torn apart and foreign material is frequently found within the wound depths (see Figure 15.2).

Incised wounds

Incised wounds may be found anywhere on the body and have clean-cut margins. They are not usually associated with abraded irregular margins and as their name implies they occur when the skin and tissues are cut by some sharp edge such as broken glass or torn metal. The deep structures are cleanly divided and fatty tissue often bulges into the wound.

Penetrating injuries

Penetrating injuries are defined as wounds produced by the penetration of some object into the tissues where the depth of penetration is greater than the skin length of the wound. Depending on the sharpness of the penetrating object, the margins will be either clean-cut or abraded and in explosions there may be foreign material, such as pieces of metal or glass, in the depths of the wound. Because of the extreme nature of the impact forces in explosions, penetrating wounds may be very deep and it is possible for the body to be completely transfixed.

Burns

Fire associated with explosions can be very destructive and the interpretation of the injuries can be masked to a greater or lesser degree by the extent of burns sustained. In severe cases,

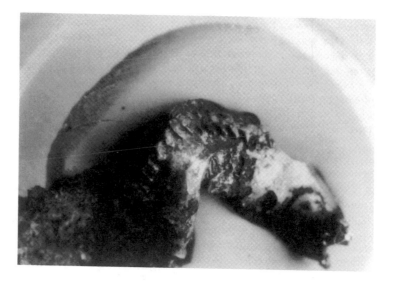

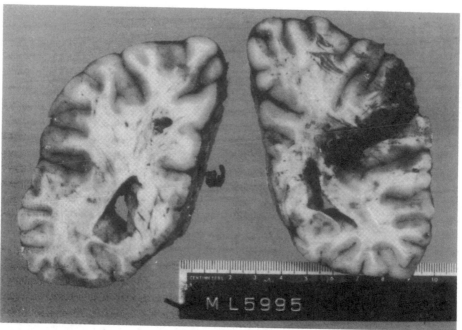

Figure 15.2 Laceration of the brain produced by a bomb fragment (inset).

the intensity of the heat may produce long splits of the skin which must be distinguished from blunt force lacerations or more rarely incised wounds.

Burns can be produced by contact with flame or hot surfaces, by contact with burning fluids and by scorching associated with a flash fire from exploding substances. The severity of burn injury depends not only upon the degree of heat applied to the body but also the length of exposure, so that burns may vary from slight reddening of the skin to complete charring with destruction of the tissues. Figure 15.3 illustrates extensive burning to the face of a bomb victim.

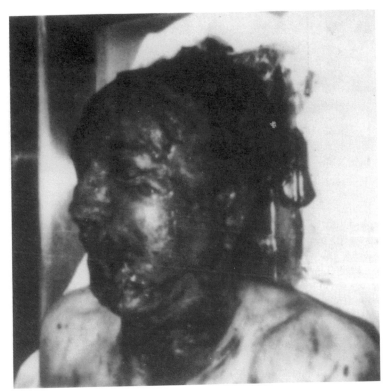

Figure 15.3 Burn injuries to the face of a bomb victim.

There are several classifications used to describe burn severity but in medical-legal practice, burn injuries are usually classified as first, second and third degree. First degree burns are superficial and will heal with no scar. Second degree burns are more severe and will show evidence of blistering and the tissues immediately beneath the skin are usually permanently damaged. Third degree burns are most severe and may be associated with charring and splitting of the skin. Extensive third degree burns are invariably fatal.

Fractures

The primary and secondary impact forces in explosions can and often are very great. When these forces are applied to the human skeleton, fractures and joint dislocations can occur. Such injuries are important because they may be taken as an indication of the severity of the forces applied to the body. Fractures however, do not necessarily occur at the point of impact. Although most bones appear hard and rigid, they are in fact capable of bending and distorting before they fracture. Impacts to the knees can result in the impact force being transmitted along the length of the femur with resulting fractures or dislocation of the hip joint.

Fractures can be the result of both direct and indirect force. Because the body as a whole may be projected some distance and since the body is not a rigid structure but capable of flailing motion, sufficient secondary force may be applied to the joints to cause fracture-dislocation of the neck, arms and legs. The nature of a fracture is also affected by the relative elasticity of the bones. For example, a twisting force can cause a spiral fracture with longitudinal splitting of the bone. A direct impact will usually result in a local fracture. In some cases

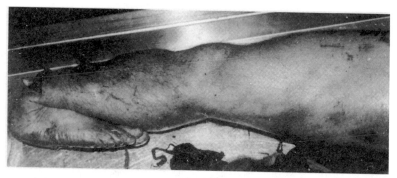

Figure 15.4 Flail injury to a foot.

the force applied may be so severe that the natural elasticity of the bone will be overcome and the bone will shatter in the way a toffee bar will break and not bend when it is struck hard.

Flail injuries: Flail injuries are almost always the direct result of extreme centrifugal forces which cause severe rotational or unilateral flailing of the limbs and spine and will present as fractures, dislocations and fracture-dislocations at joints such as shoulder, wrist, neck etc. These injuries are most frequently seen in aviation accidents where the victims are ejected from an aircraft following in-air breakup or explosive decompression. They may also be seen as a result of ejection following impact with the ground. In some circumstances the flailing may be so severe that limbs may be lacerated near joints and partial amputations may occur. Flailing limbs may strike intervening objects and secondary injuries may be produced. It may be difficult if not impossible to distinguish such flail injuries from injuries produced by impact with the ground. Figure 15.4 illustrates flail injury to a foot.

15.2.6   Identification of Natural Disease

The health of a victim before and after the explosion may have a very significant influence on the results of investigation and reconstruction. Natural disease or drugs, either prescription or non-prescription, may significantly affect the interpretation of the findings and the reconstruction of the sequence of events.

In the pre-explosion phase, any drugs or pre-existing disease which are capable of altering the skills and/or the perceptions of the bomber or victims, may be directly causative. For example, coronary artery disease is not only capable of causing sudden death but can cause sudden collapse or fainting prior to an incident. Other forms of heart disease such as heart valve disease may also be significant. This means that any history of chest pain, indigestion, pain in the arms or fainting may be significant indicators of potential accident causing heart disease. The following is a list of some of the diseases which can cause sudden incapacitation and may be of causative significance:

- diseases of the heart and blood vessels;
- abnormalities of heart rhythm including pacemakers;
- hypertension;
- heart valve disease;
- epilepsy;
- diabetes and/or insulin therapy;
- mental disorders and severe emotional distress.

Not only are alcohol and the common drugs of addiction important to identify as potential causes of accidents, but many prescription drugs also may alter an individual's reflexes, response time and even level of consciousness. Many common 'over-the-counter' drugs such as cold remedies and cough medicines contain antihistamines, and any history which might indicate the use of such substances should be sought. Toxicological analysis is time consuming and very expensive and it is helpful for the toxicologist to be able to target the laboratory analysis at specific drugs or groups of drugs.

In the post-explosion phase, pre-existing medical conditions may result in a change in the nature and effect of injuries and may have serious if not fatal consequences to an individual with heart disease or almost any other significant medical condition. Orthopaedic conditions such as cervical disc lesions and previous injuries may be aggravated by the impact forces sustained in the blast. Examples of previous apparently minor spinal injuries which have been aggravated to the point of paralysis and even death have been reported, and the aggravation of old injuries is a major medical-legal issue.

15.2.7   Collection of Internal Samples

The following samples should be taken and preserved for appropriate forensic examination:

- **for toxicology**: blood; urine; liver; bile and stomach and their contents if required; and vitreous fluid;
- **for tissue matching**: hair; and blood for serology and DNA.
- **other trace evidence**: powder traces; paint fragments; oil or grease stains; and glass and other foreign material.
- **clothing and footwear**.

Toxicology should be a routine part of the postmortem examination. Prescription and non-prescription drugs may be significant causative factors and can be tested for if required. In some jurisdictions, street drugs like heroin, cocaine and marijuana are routinely tested. Some prescription drugs such as amphetamines may have been taken inappropriately or illegally. In explosive fires, carbon monoxide and cyanide should be tested but it should be remembered that carbon monoxide may be inhaled as a result of smoking tobacco and cyanide is a normal constituent of blood.

15.3   Categories of Blast Injury

The pattern, distribution, consequences and medical management of blast injuries varies greatly with the nature of the explosive device (Bajec et al., 1993; Boffard and MacFarlane, 1993; Hull et al., 1994). Explosive blast injuries can be divided into four main categories, as shown in Table 15.1 with the nature and extent of injuries dependent upon the blast wave energy of the explosion (Cooper et al., 1991, Mellor, 1992; Tintinalli et al., 1992).

15.3.1   Primary Blast Injuries

Blast injuries result directly from the sudden changes in environmental pressure caused by the blast wave. The nature and extent of the blast injury will vary depending on the forces involved and the tissues exposed. The blast wave begins as a single pulse of increased pressure that rises to peak levels within milliseconds, and then rapidly falls to a minimum pressure which is lower than the original atmospheric pressure (Cooper et al., 1991; Mellor, 1992;

Forensic Investigation of Explosions

Table 15.1   Categories of blast injury

Category	Mechanism of injury	Type of injury	Primary target organs
1. Primary blast injury	Blast wave	Complete tissue disruption; Partial tissue disruption; Internal organ damage	Ears; lungs gastrointestinal tract; central nervous system
2. Secondary blast injury	Victim struck by flying debris	Injuries and stippling from explosive device; impact from flying debris	Skin; central nervous system; eye; musculoskeletal system
3. Tertiary blast injury	Victim impacted against stationary object; injuries by collapsing buildings	Crush injuries; acceleration and deceleration impact injuries; vagal inhibition	Head injuries; skin; musculoskeletal system
4. Thermal and chemical burns	Inhalation of dust or toxic gases, radiation, etc.	Burns, respiratory failure, etc.	Skin; eye; respiratory system

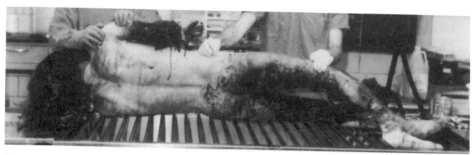

Figure 15.5   Injuries to the victim from a bomb which exploded under his seat.

Tintinalli et al., 1992). The nature and extent of the injuries produced will depend on the duration and the level of the high pressure peak attained in the explosion, the density of the organs involved and the distance from the point of detonation. Solid organs, such as bone, muscle and liver, because they are less compressible, may be less effected by this blast wave. Conversely, air and fluid filled organs, such as lungs, eyes and ears, are compressible, and the blast wave is capable of inducing extensive tissue distortion and tearing. Thus, although primary blast injury may cause localized disruption and disintegration of directly exposed tissue, the most severe damage is to tissues which are loosely supported and where there is an interface between the tissue and environmental pressure.

Disintegration or disruption of tissues tends to occur when the tissues are in close proximity to the explosive device (Marshall, 1976; Knight, 1991; Hull, 1992). The identification and mapping of the parts of the body that are disrupted may not only give an indication of the position of the victim, relative to the direction of the blast waves, and therefore, the explosive device, but will also give an indication as to the proximity of the victim to the explosion. For example, injury to the backs of the thighs and legs with localized disruption of the tissues may suggest that the victim was sitting on the device at the time that it exploded (Figure 15.5). Disintegration of a hand or forearm may indicate that the victim was handling the device at the time of the explosion.

Total disintegration and disruption of the entire body not only indicates extreme explosive forces but that the victim was in close proximity to the device at the time of the detonation.

462

Ear and throat injuries

Hearing loss is the most frequent effect of explosive blasts in non-fatal cases (Garth, 1994; Wolf et al., 1991; Kronenberg et al., 1993). Hearing can be damaged in one of three ways:

- the tympanic membranes may rupture, and this usually occurs in adults at a pressure differential between the middle and external ears of approximately 360 mm of mercury (7 psi) and can be seen as a linear laceration of the tympanic membrane;
- dislocation of the ossicles of the inner ear may be associated with laceration of the tympanic membrane, but can also occur as the sole injury;
- deafness can result from the blast effects on the structures of the inner ear where the pressure changes may directly traumatize the nerve endings. These blast effects may also significantly damage eustachian tubes.

In addition to permanent hearing loss, there may be short term or long term tinnitus and vertigo.

Pulmonary manifestations

While there is controversy as to whether pulmonary damage occurs from the direct transmission of the shock wave through the thoracic wall or through the oral/nasal orifices and air passages, it seems likely that both occur in most cases. Lung tissue can be severely affected by blast injuries, and these injuries are likely to be life-threatening (Cooper et al., 1991; Mellor, 1992; Tintinalli et al., 1992). The blast wave causes wide spread alveolar damage because of its effect on the tissue gas interfaces, and the associated disruption of the alveolar membrane and capillary bed produces interstitial and intra alveolar haemorrhage and edema. There may be internal parenchymal lacerations and in some cases, tearing of the pleura and surgical emphysema of the lung tissue may occur leading to air in the chest cavities and more wide spread subcutaneous surgical emphysema. Although difficult to diagnose, systemic air embolism may also occur.

Individuals who survive major blast lung injury will exhibit shortness of breath and other signs of pulmonary distress, chest pain, coughing blood and evidence of pulmonary edema and haemorrhage. The paranasal and maxillary sinuses are also susceptible to blast injury with intra-sinus haemorrhage.

Gastrointestinal manifestations

Blast injuries to the abdomen can result in rapid expansion of the hollow organs, such as stomach and intestines, with associated haemorrhage into the bowel wall and multi-focal perforations. Since the large intestine usually contains more gas than the small bowel, it tends to be more severely effected. The clinical features include abdominal distention, pain, and on X-ray, free air in the abdomen. Injuries to the esophagus may also occur, although they appear to be less frequent (Guth et al., 1991).

Neurological manifestations

Blast injuries to the central nervous system fall into three main groups:

- injuries to the scalp, skull and brain associated with impacts and penetration with explosive missiles;
- direct shock wave effects, which produce a concussion syndrome and various forms of focal and petechial intracerebral haemorrhage and subarachnoid haemorrhage;

- secondary effects which may be as a result of systemic tissue embolism including fat and air embolism secondary to the disruption of other parts of the body.

15.3.2 Secondary Blast Injuries

Most bodies, unless in extreme proximity to the centre of the explosion, remain relatively intact, and even in those cases where limbs have disintegrated, the remainder of the body may be sufficiently intact to allow for a detailed postmortem examination, not only of the internal organs, but the surface of the body.

One of the characteristic features of bomb blasts is 'body stippling' with the injury triad of bruising, abrasions, and lacerations (Figure 15.6). These injuries are produced by the impacting of the body by tiny explosive fragments and the pattern is often produced by fragments of the bomb itself (Marshall, 1976; Tintinalli et al., 1992).

Postmortem examination in these cases may be primarily directed at the recovery of such debris. Stippling injuries are virtually diagnostic of the victim being in the immediate vicinity of the bomb. Bomb fragments and other fragments created by the bomb blast may be projected great distances, and while large fragments of material may be observed protruding from a body, metal, glass, and other hard fragments may penetrate deeply into the body and give the external appearance of stab wounds. The examination of victims with such injuries requires caution, not only with regard to the detection of the foreign material involved, but the protection of the examiner.

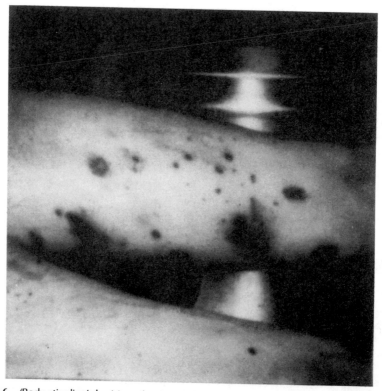

Figure 15.6 'Body stippling': bruising, abrasions, and lacerations.

Pigmentation and diffuse abrasion can be caused by fine debris such as dust, dirt and concrete, impacting against a body in the immediate vicinity of an explosion. The patterned distribution of these areas of pigmentation and tattooing will greatly facilitate the determination of the location and the relative position of the victims. Such areas of pigmentation and tattooing are subject to shadow effects where another part of the body, clothing on the body, or intermediate objects may protect the skin from this fine particulate material.

The body is also subject to non-penetrating impacts from large solid objects. Patterned injuries may be identified corresponding with the shape or texture of the surface of the impacting object and it may be important at the time of recovery of the victims at the scene of the explosion to photograph and identify objects in the immediate area of the victim in order to relate these objects later to the injuries on the body.

The flame and heat generated by the initial explosion may produce flash burns on the exposed surface of the body with singeing of exposed hair. Melting and welding of clothing can also be identified. As with other injuries sustained in the immediate vicinity of the explosion, there will be a directional element to these burns allowing for the positioning of the victim relative to the explosion.

15.3.3 Tertiary Blast Injuries

These injuries fall into two groups:

- acceleration and deceleration injuries caused by the victim impacting against stationary objects;
- injuries caused by the collapse of surrounding structures such as walls and ceilings.

The pattern and distribution of acceleration and deceleration injuries are similar to those sustained by other victims of such forces and include head injury with contrecoup bruising of the brain, laceration and tearing of internal organs including the parenchyma of the lungs and liver, and vascular injuries, including laceration of the aorta with associated traumatic aneurysms, mediastinal haemorrhage or haemorrhage into the body cavities.

The pattern and distribution of fractures may help identify the direction of impact, for example, compression fractures of the spine, fractured dislocations of the posterior ribs near the costal vertebral joints, and spiral or transverse fractures of the long bones.

Crush injuries, particularly involving compression of the chest and abdomen, may produce the characteristic pathological findings of crush asphyxiation, secondary to mechanical respiratory paralysis, multiple cutaneous petechial haemorrhages, and clinical evidence of hypoxia may be evident.

Vagal inhibition

In some cases following explosions, individuals are found dead with little evidence of the direct effects of the explosion or blast on their bodies. There may be no primary blast injuries, no evidence of stippling or penetrating injuries, no evidence of chemical or thermal burns. It is believed that in many of these cases, death is as a result of vagal inhibition. Sudden stimulation of the vagus nervous system can induce reflex slowing or stopping of the heart, leading to sudden syncope collapse and death. Such deaths may in the past have been attributed to shock or cardiac arrest.

The mechanism of death in many of these cases may be as a result of a direct impact to the body, either by some blunt object or by the blast wave itself producing effectively *comotio cordis* – cardiac arrest induced by direct mechanical impact to the heart.

In all such cases, it is important to exclude the existence of any natural disease, cardiac enlargement, alcohol, or drugs, and in view of the possible difficulty in attributing such deaths to natural or other causes, detailed pathological examination of the heart, coronary vessels and cardiac conducting system may be necessary.

15.4 Thermal and Chemical Burns

Many explosions cause fires. Victims, both fatal and surviving, will frequently show evidence of thermal burns as a result of fires secondary to the explosion. The nature and extent of burns will depend entirely on the circumstances of the fire and so the distribution of the burn will not necessarily indicate the direction of the explosion, but will simply reflect the nature of the secondary fire.

It may be important in any investigation to determine whether or not the victim was partially or completely immobilized by the initial explosion and whether or not a secondary fire prevented survival. Similarly, it would be a requirement of the postmortem examination in fatal cases to determine whether or not the individual died as a direct result of the explosion or the secondary effects of burns and inhalation of the products of combustion.

Inhalation burns with associated adult respiratory distress syndrome and respiratory failure pose significant clinical problems and may be superimposed upon the primary pulmonary damage associated with primary blast injury. These conditions can lead to a significant prolongation of the dying process and major medical management problems.

Chemicals may, as a direct effect of their dispersal by an explosion, cause injury to the skin and respiratory passages. Dispersement of corrosive chemicals may not only lead to local damage from the direct effects of the chemicals, but also to systemic toxic complications. Some combustible materials such as gasoline are capable not only of inducing direct chemical burns in the absence of ignition but severe thermal burns following ignition.

In most fires, following explosions, the products of combustion may include carbon monoxide, hydrogen cyanide, and other poisonous gases. Pathological and clinical examination of victims of such explosions must include the retention and proper preservation of samples for toxicological analysis.

15.5 Summary

An injury-producing explosion can be a medical crisis and the medical management and forensic pathological investigation of the victims is complex and time consuming. Reference to the experience of others is an essential component of the preparation for such incidents (Hodgetts, 1993; Karmy-Jones et al., 1994).

References

BAJEC, J., GANG, R. K. and LARI, A. R., 1993, Post Gulf war explosive injuries in liberated Kuwait, *Injury: International Journal of the Care of the Injured*, **24**, 517–20.

BENMEIR, P., LEVINE, I., SHOSTAK, A., OZ, V., SHEMER, J. and SOKOLOVA, T., 1991, The Ural train–gas pipeline catastrophe: The report of the IDF Medical Corps assistance, *Burns*, **17**, 320–322.

BOFFARD, K. D. and MACFARLANE, C., 1993, Urban bomb blast injuries: patterns of injury and treatment, *Surgery Annual*, **25**, 29–47.

COOPER, G. J., TOWNEND, D. J., CATER, S. R. and PEARCE, B. P., 1991, The role of stress waves in thoracic visceral injury from blast loading: modification of stress transmission by foams and high-density materials, *Journal of Biomechanics*, **24**, 273–85.

GARTH, R. J. N., 1994, Blast injury of the auditory system: a review of the mechanisms and pathology, *Journal of Laryngology and Otology*, **108**, 925–29.

GUTH, A. A., GOUGE, T. H. and DEPAN, H. J., 1991, Blast injury to the thoracic esophagus. *Annals of Thoracic Surgery*, **51**, 837–9.

HODGETTS, T. J., 1993, Lessons from the Musgrave Park Hospital bombing. *Injury: International Journal of the Care of the Injured*, **24**, 219–21.

HUBBARD, T. J., DADO, D. V. and IZQUIERDO, R., 1992, Massive craniofacial injuries from recreational fireworks: a report of three cases, *Journal of Trauma*, **33**, 767–72.

HULL, J. B., 1992, Traumatic amputation by explosive blast: pattern of injury in survivors, *British Journal of Surgery*, **72**, 1303–6.

HULL, J. B., BOWYER, G. W., COOPER, G. J. and CRANE, J., 1994, Pattern of injury in those dying from traumatic amputation caused by bomb blast, *British Journal of Surgery*, **81**, 1132–5.

ISA, A. R. and MOE, H., 1991, Fireworks related injuries during Hari Raya festival in hospital, *Medical Journal of Malaysia*, **46**, 333–7.

KARMY-JONES, R., KISSINGER, D., GOLOCOVSKY, M., JORDAN, M. and CHAMPION, H. R., 1994, Bomb-related injuries, *Military Medicine*, **159**, 536–9.

KNIGHT, B., 1991, *Forensic Pathology*, pp 248–251, London: Edward Arnold Publishing.

KONENBERG, J., BEN-SHOSHAN, J. and WOLF, M., 1993, Perforated tympanic membrane after blast injury, *American Journal of Otology*, **14**, 92–4.

MALLONEE, S., SHARIAT, S., STENNIES, G., WAXWELLER, R., HOGAN, D. and JORDAN, F., 1996, Physical injuries and fatalities resulting from the Oklahoma City bombing, *Journal of the American Medical Association*, **276**, 382–387.

MARSHALL, T. K., 1976, Deaths from explosive devices, *Medicine, Science, and the Law*, **16**, 235–9.

MELLOR, S. G., 1992, The relationship of blast loading to death and injury from explosion, *World Journal of Surgery*, **16**, 893–8.

ORIGITANO, T. C., MILLER, C. J., IZQUIERDO, R., HUBBARD, T. and MORRIS, R., 1992, Complex cranial base trauma resulting from recreational fireworks injury: case reports and review of the literature, *Neurosurgery*, **30**, 570–6.

TINTINALLI, J. E., KROME, R. L. and RUIZ, R., 1992, *Emergency Medicine: A Comprehensive Study Guide*, 2nd Edition, pp. 788–91, New York: McGraw-Hill Book Company.

WOLF, M., KRONENBERG, J., BEN-SHOSHAN, J. and ROTH, Y., 1991, Blast injury of the ear, *Military Medicine*, **156**, 651–3.

ZANE, D. F. and PREECE, M. J., 1992, Study of Phillips tragedy gives insights into etiologies of plant blast injuries, *Occupational Health and Safety*, **61**, 34–40.

16

Presentation of Explosive Casework Evidence

JAMES W. JARDINE

16.1 Introduction

The prospect of leading evidence in criminal litigation following the explosion of an explosive device requires extensive organization of subject matter and detailed analyses of scientific evidence. The complexity of the process, and in many instances, the magnitude of the task, make such prosecutions difficult to manage and to make comprehensible.

The testimony and the evidence must be marshalled so that it may be understood by the average juror or judge. The investigative process, from the monumental task of combing the explosion site, the finding, securing and preserving of relevant evidence, and the interviews of relevant witnesses, is by its very nature so complicated and filled with details that is difficult to follow – let alone master. Good advocacy requires diligence from counsel.

To properly prepare, one must take the time to gather all the information available and then marshall the facts and conduct research to obtain a thorough and clear understanding of the applicable legal principles. The key is *preparation*. From a proper base of preparation comes the development of a trial plan.

The trial plan allows counsel to determine an effective strategy for the trial and to anticipate the case for the opposition. The organization and analysis of the factual material is made much easier if counsel is involved early in the investigation.

As described in Chapters 4, 5 and 6, the search teams must be briefed thoroughly, supervised closely, and committed to the painstaking task of documenting, by photographic and written records, the location of all items seized from the explosion debris. There should be liaison, consultation and coordination between the respective scientists and search teams so the investigators know what they should look for in the debris and how the exhibits should be secured to preserve the integrity of the exhibits. The initial consultation must be done by a team of forensic specialists, including at least one multi-dimensional expert who can draw on a wide background and provide informed advice in a number of areas.

By its very nature the investigation will likely call on the expertise of a number of forensic experts with different qualifications and levels of expertise. In addition to experts on explosives and demolition, there will likely be a need for a broadly-based team encompassing areas such as physical matching, tool marks, document examination, chemistry, computer enhancement, pathology, electronics, or expertise with a particular model or make of electronic device, etc. The list may seem endless.

For counsel, it is important to be flexible and adaptable to the circumstances of the investigation, the available personnel and the logistics of the case. Diplomacy, tact and effective communication are necessary. Counsel must be clear, reasonable and fair in explaining the legal requirements with which the investigators, the forensic scientists, and ultimately the prosecutor will have to cope. It is important to remember that the preparation of a forensic expert to give evidence commences at the first consultation or briefing. All meetings, information supplied and advice sought and given, should be fully documented so that the expert can give evidence about everything which was said and done. A fair, balanced and objective approach from the beginning will go a long way towards enhancing the credibility of any expert.

The evidence of forensic scientists who have received and examined the relevant material from the debris, and the broad spectrum of tests and experiments employed must be recorded and capable of being reproduced and/or readily explained to the fact-finder(s) at trial. Explanations of this nature require 'expert' evidence which can simplify and clarify what was done, how it was done, when it was done, and why it was done. The testimony and the real evidence must be simplified to enable counsel, both through the testimony and argument to explain the course of the investigation to enable the fact-finder to understand the process.

In such a prosecution it is necessary for the prosecutor to call at trial a battery of experts to explain the examination of the real evidence at the scene. The opposing counsel may wish to call a number of experts as well. The facts will determine the nature of the defence available and the resultant defence tactics. It may well be that counsel will simply retain experts to assist in the preparation of cross-examination.

The legal requirement for preservation of the exhibit in the same state from the point it was found to all subsequent scientific examination is an essential feature to the establishment of its value as a piece of evidence. This 'continuity' or 'chain of custody' evidence is required for every step of the path the exhibit material follows, including visual examination and any subsequent specific scientific test conducted. Any change in the exhibit or its physical characteristics due to examination or scientific analysis, must be fully documented. Each piece of relevant evidence from the explosion must be interpreted for the trier of fact to enable the tribunal to determine the explosive device utilized, the identity and estimated quantity of the explosive, the nature of the explosion, and the extent of damage caused by the explosion.

In Canada there are limitations on the number of expert witnesses who may be called in a criminal trial. (Canada Evidence Act, R.S., c. E-10, s. 7). In essence, expert evidence is allowed where it is relevant and where it assists the Court in determining the issues and the facts to be decided. The principles of law are well established. For the purposes of this chapter, it is to be taken as given that expert testimony is essential to enable the trier of fact to understand explosive casework evidence and that the Court will grant leave to counsel at the outset of the case to call more than five experts for the prosecution and/or the defence.

I have focussed this chapter on outlining a workable framework to deal with the practical issues and problems with which counsel will have to contend in the presentation of explosive casework evidence at trial. I have used a factual model as an example.

16.1.1 Factual context

For the purposes of providing a context to presentation of explosive casework evidence, I have related the issues to the factual underpinning of the investigation of a fatal bombing at the New Tokyo International Airport, Narita, Japan, and the prosecution *in R. v. Reyat* in which I headed the prosecution team. The explosion at Narita took place in June 1985. The investigation continued right to the trial which commenced in September 1990. The trial concluded in May 1991 with a conviction for manslaughter and several other counts relating to explosives. The preparation of the testimony, which took over six days of trial, of the main forensic witness called by the prosecution is described.

Under international law, an act carried out in Canada (making a bomb) which caused death in Japan placed the duty of conducting the criminal investigation and prosecution on

Canada. This led to five year investigation involving many police and forensic scientists in Canada and Japan, many of whom testified at the subsequent 9 month trial.

At approximately 11:19 p.m. Pacific Daylight Time on 22 June 1985 (which was 23 June 1985, 3:19 p.m. Japan Standard Time) there was an explosion in the luggage handling area of the South Terminal of the New Tokyo International Airport, Narita, Japan.

The nature of the offence, the motive and explanation for why this explosion happened, where it happened, and how it happened, was to be found in the history and timing of the events themselves. The time frame provided the framework of the web of circumstances which pointed at the accused, and through which the evidential foundation could be established for the scientific foundation of the case.

The time frame in point form was as follows:

- 1983: *A* showed an interest in dynamite and discussed the blasting of a rock wall with an acquaintance.

- 1984: *A* was upset at political events in India and he expressed his frustration and upset to a number of people. He expressed an interest in dynamite and was provided with a 'Blaster's Handbook' by the acquaintance.

- 1985, spring: *A* took steps to acquire dynamite and became very interested in the operation of timing devices, particularly those devices connected to an electronic component such as a 'ghetto blaster' or 24 h alarm clock. The documentation of his purchases, his returns, his attempts to connect relays, batteries, and 24 h alarm clocks, demonstrated his interest. He explained himself in different ways to the various people he consulted.

- 8 May 1985: *A* purchased a Micronta clock.

- 1985, mid-May: *A* asked questions about how to connect a relay to the Micronta clock. Around this time he visited an acquaintance taking with him a Micronta clock, a buzzer and a lantern-style battery wired together in sequence. *A* sought help in making the buzzer stay on, rather than sounding intermittently.

- 26 May 1985: *A* obtained from a blaster, six to eight sticks of dynamite and a few days later, two to three blasting caps.

- 4 June 1985: *A* purchased a second Micronta 24 h DC alarm clock, as well as two relays of different types: models 275-003 and 275-8218. *A*, with two others, conducted an explosives experiment.

- 5 June 1985: *A* purchased a Sanyo model 'FMT 611K' tuner and a can of single base smokeless powder. He also exchanged one of the relays purchased on 4 June for a model 275-004 relay.

- 19 June 1985: *A* purchased another 275-004 relay.

- 22 June 1985: *A* purchased two (2) model 732 12 volt lantern batteries before 11 a.m. at an 'Automarine' store in Burnaby, BC (a municipality adjacent to Vancouver).

  The explosive device was checked in at Vancouver International Airport in luggage to be interlined in Tokyo from Canadian Pacific Airlines flight CP 003 to Air India flight AI 301 to Bangkok. The Canadian Pacific flight left Vancouver on 22 June 1985 and arrived in Narita, Japan on 23 June 1985, at approximately 14:47 hours Japan Standard Time. The bomb exploded at about 15:19 hours local time and killed two baggage handlers.

In the three weeks to a month prior to the explosive device being placed on the CP 003 flight, *A* purchased or acquired seven items which were within the very explosive device which exploded at Narita. Moreover, he had access to two further items which were found in the debris. Those nine items found in the explosive debris were consistent with the items purchased, acquired, possessed, and to which *A* had access. The evidential links of the very items to *A*, formed the evidential foundation for the prosecution.

The scientific evidence of the correspondence between the exploded device and the items purchased, acquired, possessed and to which the accused has access, was the part of the Crown's case which provided the *link* or *nexus* between the items which *A* had and the very device which exploded at Narita. The identification by Japanese and Canadian forensic scientists of the nine items: stereo tuner, timer, relay, battery, 'Liquid Fire' ether can, single base smokeless powder, dynamite, blasting cap and adhesive tape (green), provided the nexus between the proven motive of the accused and the items linked to him. The noted timing of the purchases, acquisitions, and possessions was also important.

16.1.2 The Component Parts of the Bomb

The component parts of the bomb which exploded in Narita were as follows:

- container: a 'Sanyo' stereo tuner model 'FMT 611K' packed in the original carton complete with styrofoam packing blocks and the owner's manual;
- power source: an 'Eveready' 12 V lantern battery;
- timer: a 'Micronta' car clock;
- relay: a 'Radio Shack' electrical switch;
- detonator (blasting cap): aluminum cap;
- main explosive charge: dynamite;
- secondary explosive: smokeless powder;
- ether can: 'Liquid Fire' brand starting fluid (diethyl ether).

16.1.3 Explosion

The Narita airport bomb scene was something of a 'worst case scenario' as far as potential residue recovery was concerned, in that water pipes broke, ceiling insulation material fell and fire extinguisher powders were used (see Chapter 4, Figure 4.11). The resultant problems for residue analysis were that: (1) sodium compounds, ammonium compounds, and nitrates, which are frequently found in explosive residues, are soluble in water and (2) even if sodium, ammonium, or sulphate were to have been found, this would have been of no significance since sodium, ammonium, and sulphate (and phosphate and bicarbonate) were all added to the scene in the form of dry chemical fire extinguisher powders.

16.1.4 Search and Seizure

One of the first tasks for counsel dealing with foreign police agencies is mastering the documentation.

Subsequent to the explosion, the area was secured and roped. It was taped off in a grid system which followed the lines on the floor produced by the concrete pour-pattern in the baggage handling area (Chapter 4, Figure 4.12). The area was then photographed by Japanese police officials (Chapter 4, Figure 4.13). All of those actions took place in the afternoon and the evening of 23 June 1985 and into the early morning of the 24th.

The area nearest the explosion was covered with debris 10–15 cm deep (approximately 6 inches). It was wet and contaminated with the ceiling rock-wool insulation which had fallen due to the explosion. It was also wet due to the water sprinklers from the fire system having burst, or through their activation from the force or the heat of the explosion.

Jurisdiction for the search fell to Chiba Prefecture Police. The search master decided to use the floor grid pattern for the purposes of the search and seizure and he divided that area into two main grid blocks, A and B. It was then further subdivided into smaller subsections small (a) (⌐) to (f) (∧) in block A which included the seat of the explosion (Chapter 4, Figure 4.12).

The search was conducted over a number of days. All aspects of the search were fully documented, photographed, sketched, and charted. Each member of a search team was assigned as either a collector, recorder, or retainer. The collector collected the items in numerical order, bagged them, and marked the bag with the identification code. The recorder listed the collected items in a 'Table of Collected Materials'. The retainer then placed the bags into a box and prepared a 'Certificate of Safekeeping' document required by the laws of Japan which listed the items seized and retained. This process was followed for each exhibit seized from the grid floor pattern. In addition, there was a 'Confiscation Report' prepared for identifiable items seized from the luggage and cargo in the luggage sorting area after each of those items was X-rayed and passed through a metal detector to ensure safety and security.

Over 1347 items were collected. By the time the sifting and sorting of the debris was completed, some 1609 items were described and set aside. Over 700 photographs were taken and 24 tables or charts were created setting out and describing the collection of exhibits, and marking them, or their container, with a particular code to describe the place of their collection. There were three sweeps through a particular area on some occasions. For instance, in the A–a first sweep there were 54 items that were identified, photographed and seized. For the second sweep, items 55 to 133 were seized, and in the third sweep, items 134 to 189. In each instance the area was photographed with a civilian witness present, in accordance with the laws of Japan, prior to the judicial police officers seizing the exhibits.

Subsequently the areas were swept and the sweepings were placed in bags. The debris from those bags were sifted and sorted for the smaller, less identifiable items, over some three weeks from 30 June to 23 July. The collection of the larger identifiable items took from 23 to 27 June.

From 27 to 30 June, the debris and dust remaining in each of the grid blocks was collected by sweeping and shovelling it into 44 vinyl and 108 burlap bags. The bags were taken to a large room at a conference centre where sifting and sorting of the collected, scattered materials and the dust and debris which had been swept into the bags, was carried out. For each of the collected materials, there was a unique identification code established.

Items were seized from approximately 23 pieces of luggage in the cargo area. The luggage was searched by members of the team from 25 to 26 June.

At the conference centre the contents of bags of 'Collected Scattered Material' were sifted with flour sifters. They were sorted through screens, examined, and the identification codes expanded as the items were located. For instance, one bag taken from the grid block section A–a–1 would be so designated; however, if nine bags were taken from A–a–4, then each item would be assigned an additional number: A–a–4 Scattered Material (Hisanbutsu) 1, 2, 3, 4, etc. Thus the items were encoded and listed in that chronological fashion. In some instances, the larger items which had been collected as identifiable items and taken to the conference centre were found on closer examination to contain one or two items of note and those items were called 'Collected Materials with Additional Numbers'. So the identification codes for those items would be 'A–a 54', and then 'a 1', or 'a 2', or 'a 3'. Each of the identification codes described not only what had been seized, it also was referable to a particular table in terms of documentation used to record the process.

It was interesting to note that it subsequently took a Canadian forensic scientist/investigator team two 12 hour days to review all of the exhibits.

16.1.5 Post-mortem Examinations

The issues of continuity and preservation of exhibits for examinations at the respective laboratories, including their movement to Canada and to the relevant Canadian laboratories, were problematic. The differences in the legal systems between Canada and Japan were very apparent. The Japanese were content to prepare and provide detailed written materials in report form. The forensic experts were rarely, if ever, questioned or called on to justify the findings or conclusions drawn by them. The logistics of tracking exhibit movement were detailed, time-consuming and administratively difficult. Thus, the written materials and the photographs were invaluable. This was particularly the case with post-mortems, where we discovered that for cultural reasons, materials from autopsy were not retained.

16.1.6 Exhibit Selection and Seizure for Court in Canada

All the exhibits recovered from the scene were kept in one exhibit room created for the purpose at the New Tokyo International Airport Police Station in Chiba Prefecture. An exhibit log was maintained to control exhibit movement in and out of the exhibit room.

The Japanese system of judicial police officers collecting the evidence at the direction of the court, included a legal requirement that a civilian be photographed with the exhibits during the collection process. There were many photographs taken. They were invaluable as evidence to explain the extent of damage and the labour intensive and time-consuming process of search and seizure at an explosive site. They were also very significant as part of the factual foundation for the opinions of the forensic experts.

In order to fully document the process, the investigators in Japan had photographs taken at each stage of the seizure process as well as at almost all of the scientific examinations conducted. The photographic trail of continuity enabled many of the witnesses to refresh their memories during interview. It also provided a simple means of showing the fact-finder what was done, when it was done, and how it was done.

It is my view there can never be too many pictures taken at an explosion site or during the exhibit examination continuity trail.

As the case developed and results from police investigations on three continents and the forensic examinations on two came together, the significance of some of the exhibits began to emerge. The Japanese scientists had concentrated on identifying fragments from the device. Once the components of the device had been identified, it fell to the police investigators to associate the individual components to the suspect. The Canadian police investigators commenced tracking down the identified components. Potential sources for the purchase or acquisition of the items were canvassed. Potential sources for the obtaining of explosives, detonators and timers were searched to see if any items could be associated with the suspect. Canadian scientists were primarily working on relating material from the bomb scene directly to material associated with the suspect. Pertinent exhibits were identified by a forensic scientist/investigator team which examined all seized materials. Exhibits were transferred from Japan to Canada using diplomatic channels. *Notes Verbale* requesting specific items were sent between Ministries of Foreign Affairs and with the able assistance of Police liaison officers the protocol worked well. The actual exhibits were also viewed by counsel at this early stage of the investigation.

16.2 Forensic Expert Evidence

Forensic scientists performed keys roles in this case. In general terms, I see the role of a forensic chemist in a major investigation of this type to be a seven-step process:

(1) pre-lab: scene assessment; consultant on exhibit collection;

(2) forensic examination: evaluate exhibits; analyse exhibits;

(3) deliver initial forensic opinions to investigators;

(4) issue forensic reports to investigators;

(5) consult with the prosecution on the weight/significance of the forensic evidence;

(6) organise testimony;

(7) testify as an expert witness.

When all examinations were complete – and indeed the dynamite was only identified a few weeks before the trial started as a result of new procedures and technology in the Vancouver laboratory (Beveridge, 1991) – the prosecution team was faced with a massive task which included selection and preparation of expert witnesses in forensic chemistry (explosives, fibres, plastics, paint, adhesive tapes, metal, electronic parts, wires), questioned documents (printing, handwriting, stencils, decipherment), explosives (industrial use; formulation; improvised explosive devices), electronics and so on.

Rather than deal with each individually, I shall deal with the subject of expert witnesses first in general terms and then specifically in the preparation of the witness who had to present the most complex testimony.

16.2.1 Selection of Experts

One of the most important, if not the most significant, events in the investigative and prosecution process is the selection of the forensic experts. The initial decisions of exhibit triage determine the flow of exhibits through the investigative process. These may have a significant impact on the weight to be attached to a vital piece of evidence when it is ultimately tendered in evidence at trial.

Any explosive casework investigation will utilize highly sensitive chemical analyses as part of the investigative process. The potential for disaster through inadvertent adulteration of a piece of evidence is very high. Any evidence of improper handling of an exhibit opens the investigation to attack by opposing counsel. Therefore, it is extremely important that the investigation immediately acquire an experienced, careful, and highly qualified team of forensic experts to assist the investigators in securing the scene, deciding the order of seizure of the exhibits, if possible, and directing the preservation of the seized exhibit materials in the exhibit continuity process.

Qualifications

The most important factor in the selection process is to ensure the expert chosen has sufficient academic credentials, related scientific experience, and related forensic investigation experience to demonstrate to the court that the expert opinions offered should be accepted as fact. The reputation of the expert in these areas should influence heavily the selection decision.

Another important consideration in the selection of the expert is the ability of the expert to communicate the complex explanations required in language understandable to a lay person. The potential for a jury trial makes this a very important factor in the selection process. This factor may require the additional cost or time problem of using two experts: one who has the expertise, qualifications and experience to conduct the examinations and arrive at the conclusions; and a second one, who, although of not as strong an academic or scientific background, has the special demonstrated capacity and reputation of being an excellent witness. (For an interesting discussion and analysis of the issues of expertise and the difference between the admissibility of the evidence of an expert and the weight to be attached thereto see, *R. v. Marquard* (1993), 85 C.C.C. (3d) 193 (S.C.C.).)

Moreover, it is important to ensure that the proposed expert has no bias, prejudice or interest, pecuniary or career-path, which could taint the weight to be attached to the examinations conducted, findings made, and conclusions drawn.

16.2.2 Exhibit Selection

The selection of exhibits for trial in an explosive case is a difficult process. The debris from an explosion site is, due to physical results of explosive forces, voluminous. The bits and pieces will range from large items with structural damage, to pieces which can only be microscopically examined. The smallest of items may be extremely significant. The process requires close consultation between the investigators, the forensic scientists and counsel. The exhibits to be selected must be capable of proof. A conservative view of the admissibility of evidence is essential. The opinion of the expert will be formed from his or her scientific examination of the exhibits. If one cannot prove the exhibit came from the explosive site or establish the continuity of the exhibit, it may be inadmissible. The removal of a key piece of evidence may destroy the opinion or conclusion being sought. All the exhibits necessary to elicit the opinion or conclusions favourable to the theory of the prosecution must be selected. One should be certain they will be admissible before they are selected.

The process is circular. The expert should only use as a foundation for his or her opinion exhibits which are capable of proof. Counsel must ensure, through interview, that the opinion of the expert is founded only on material which is capable of being proved at trial.

Early communication identifying potential evidential problems may avoid any potential problems and perhaps save time, effort and/or reputations.

16.2.3 Exhibit Continuity, Encoding Identification Marks and Labelling

The establishment of a system for the marking of exhibits is essentially one of organisation. It does not matter what system is used to keep track of individual exhibits. What is required is that there be a system in place for orderly listing. A combination of letters, numbers or characters is commonly used. Once the system has been determined, the preparation of lists of exhibits and the preparation for the presentation of exhibits is a time-consuming but surmountable task.

16.2.4 Written Forensic Reports

The written report format should include a description of the occupation of the scientist and the individual scientist's academic and professional qualifications. It is important that the initial part of the report or an attached *Curriculum Vitae* (CV) set out any relevant case experience and that it outline relevant forensic experience. The expert's background, experience in preparing reports and giving evidence is an important factor in assessing how effective a witness that person will be during testimony before a court or tribunal. Practical court experience should be outlined as it is one of the most important factors in the interview and preparation process.

The format of the report should clearly establish and demonstrate the objectivity of the examinations conducted and the opinions offered. Such a report affords an excellent opportunity for the forensic scientist to be independent from the police investigation.

The facts of the case should not be intertwined into the scientific report as truth. If the format of the report includes the acceptance of 'information' from the investigator which is not accurate, the whole report becomes biased and almost useless. It is essential to establish the objectivity and independence of the forensic scientist at the outset. Therefore all written forensic material must be scrupulously objective and credible.

The nature of the request from the investigator, and the purpose of the report, should be stated in question form to enable the scientist to write, under a separate and distinct heading, a number of factual hypotheticals for the purposes of giving conclusions and opinions during testimony in court.

The preparation of a forensic scientist for court is a separate and distinct stage of the process from the initial scientific examination and report of findings. The preparation of a 'witness' to give evidence comes much later in the prosecution process and is entirely different from the examinations for scientific purposes conducted during the investigation.

Working from that basic premise, the written report format should be:

(1)   A complete resume of the qualifications and experience of the forensic scientist.

(2)   Continuity of exhibits and description of items received.

(3)   Factual information to set the objectives of the purpose of the examination.

(4)   An outline of the purpose of the scientific examination.

(5)   A description of the technical issues.

(6)   The nature of the examinations including what scientific techniques have been used and the results of the scientific tests.

(7)   Interpretation of the findings.

(8)   The conclusion of the expert.

Scientific reports cannot be used to 'stand alone' and provide the information which counsel requires for trial without:

- communication with the scientist;

- elaboration by the scientist;

- preparation of factual hypotheticals based on the anticipated factual foundation which can be proven at trial.

One must be conservative in this exercise. Any report format incorporating the truth of information provided by investigators, blends together scientific findings with the factual foundation in the trial proper. I am concerned that the scientist, therefore, becomes less an independent observer and too much an investigator with a bias or perceived prejudice in the case. The inclusion of contested facts may make the scientist appear to be an advocate for a position. That approach will defeat the aim of the expert to establish credibility in the case, and repeated reports of this nature will undermine the long-term credibility of the witness. Such an expert is open to being destroyed on cross-examination.

It is important to consider that in a hypothetical outline it is not for the scientist to argue, it is for the scientist to state or opine what characteristics were consistent, and what characteristics were inconsistent with the originating source.

During the testimony of the expert witness, case information evidential issues and result interpretation are the factors which give weight to the testimony. This is not a matter for report writing, unless the hypotheticals are clearly set out both from the investigator requesting the examination, and by the scientist attempting to answer the hypothetical in a written report.

Counsel and the expert witness must, at the outset of the testimony in the courtroom, ensure a clear introduction of the foundation for the opinion and the conclusions drawn.

It is appropriate for the protocol of the laboratory to outline for the scientist how they wish the scientist to address the issues of results of examinations and conclusions based on the case specific factual foundation. It is not appropriate for a laboratory to tell the scientist what to say.

A case-specific example may be of assistance. In a circumstance where a scientist is attempting to match objects found at an explosive site with standards provided by the investigators and linked directly to the accused, the results of the examination are factual statements derived from the scientific examination conducted. If the items are consistent with having come from the same source, the scientist can explain the characteristics of consistency and indicate how many unique and similar characteristics have been found. If the items are not completely consistent, in other words if they do not match exactly, then he or she might wish to explain that fact. An example used in the Narita case was:

> Based on the similarities of characteristics and the chemical composition of the exhibit and the standard, I am of the view that they are completely consistent with having come from a source of the same manufactured item. There are no distinguishing characteristics which cannot be explained by the exposure of the exhibit to explosive forces.

By the use of this wording the forensic expert (scientist) has explained the weight to be given to the conclusion arrived at, based on the result of the examinations and expert opinion.

Each technologist or scientist who has conducted part of an examination by instrument analysis is a necessary part of the report. It is not sufficient, for the purposes of disclosure to the defence, nor is it sufficient as a factual foundation for evidence in court, to have the report summarize the work of another and use it as a foundation for the testimony of an expert. Some reports use terms such as 'examination was conducted using scientific support staff'. The work conducted by others which forms a necessary part of the factual foundation of the expert opinion must be independently admitted into evidence. Those witnesses should be called to provide the proven factual foundation for the expert opinion advanced. (*R. v. Abbey*, (1982) 29 C.R. (3d) 193, 68 C.C.C. (2d) 394 (S.C.C.).)

The report should contain explanations for the reader with appropriate appendices outlining the hypothetical factual foundations and the conclusions which follow. It is my view that case information, evidential issues, and result interpretation are the most significant aspects of the scientific report. It is therefore my view that a separate section in each report should set out, in appendices, a number of hypotheticals based on alternate factual foundations provided by the case information. This approach will enable the scientist to remain independent from the investigation and yet provide helpful information both to the investigators and/or prosecuting counsel to allow them to determine the most effective means of presenting the scientific evidence.

It is important to note that it is the report of the forensic expert that provides the foundation for his or her testimony. The report-writing stage of the investigation is therefore key to the process of witness preparation. The retained expert and the counsel should consult about the format of the report and how the contents are to be expressed, to ensure that problems which could be avoided are not created.

16.3 Witness Preparation

16.3.1 Qualifications

The first step to leading evidence from an expert witness is to demonstrate to the court that the witness has a particular expertise or special knowledge in the area about which he or she is to testify. In some cases, one can expect to tender a *resumé* or *Curriculum Vitae* which will be marked as an exhibit. It is my view you have more effective advocacy if counsel prepares questions outlining the witness' academic training and work experience so that the expert can respond and clarify. This approach enables the expert's qualifications to be brought before the court in such a manner that the expert does not look like he or she is boasting. (For a clear review of the Canadian law in this area one should refer to Sopinka (1995); for Australian law, see Freckelton and Selby (1993).)

The evidence must be directed towards the witness' expertise in the relevant area and clearly outline the fields in which the witness is tendered as an expert. Counsel, as a matter of practice, should tender the witness and outline first the specified field of expertise and the areas on which expert opinion is sought. It is a matter of law for the court to consider the admissibility of the opinion evidence. (*R. v. Morin* (1988) 2 S. C. R. 345, 66 C.R. (3d) 1 (S.C.C.); *R. v. Mohan* (1994) 2 S.C.R. 89 C.C.C. (3d) 402 (S.C.C.).)

16.3.2 Expertise (Fields of Expertise)

A clear statement is required of the fields of expertise for the witness. The weight to be attached to the opinion of the witness is directly related to whether the opinion given is within the expert field of knowledge of the witness. One should not ask the witness to stretch to cover a field in which the witness is not fully qualified.

One can anticipate cross-examination on the field of expertise if the expert is being asked to provide an opinion in a new area of science or expertise. It is also likely there will be full cross-examination if the expert is being asked to stretch either the extent of his or her knowledge, field of expertise, or the opinion being tendered. It is my view neither an expert nor counsel should attempt to stretch; such an approach ruins reputations. One's career is not made by one case. A conservative long-range approach should be followed. In addition, the use of such an opinion will only serve to weaken the presentation of your case by damaging the perspective the fact-finder has of you.

16.3.3 Consolidation

Each of the members of the team of forensic specialists who have conducted a battery of tests or examinations should be asked to prepare consolidated reports. It is frequently the practice of laboratories for the forensic scientist to prepare individual reports for each separate scientific examination. In order to prepare for interview and testimony, each of the scientists should be asked to prepare a consolidated report outlining all exhibits which they have examined and all scientific examinations conducted on the selected exhibits. The preparation of the consolidated report will provide a framework for further preparation of the witness for testimony and court. This is particularly important if there has been a multi-dimensional expert who has provided informed advice in a number of areas as was the situation in the Narita case. One expert alone submitted over 20 reports on various examinations conducted throughout the 5 years between the crime and the trial.

16.3.4 Informing the Witness

As part of witness preparation, the witness should be made aware of where his or her evidence fits in the trial. The witness is then able to understand why their evidence is important and why certain questions will be asked. They can also more easily anticipate cross-examination through understanding what the opposing counsel will be trying to counter.

As part of their preparation the witness should be told what to anticipate from counsel for the other side. They must understand what the cross-examination is likely to entail. At this stage of the preparation of the witness there should be close consultation between the witness, the investigator, and counsel to expose any potential weaknesses in the evidence being advanced, the expert opinion, or the theory of their case. In the Narita case, Canadian scientists worked closely with the prosecution as technical advisors concerning the Japanese forensic reports.

Any potential areas of weakness should be addressed with a view to undertaking further scientific testing, examination, or experimentation. It may also be that they will have to change their approach. They may have to remove exhibits earlier thought relevant and capable of being proven in evidence. It is better to catch the problems early before the trial has begun. It is much more difficult to reassess the case when you are already at trial.

16.3.5 Cross-examination of the Witness

Upon receipt of the consolidated report from the forensic expert, and following interview in which the witness evidence is compared to the whole of the case, counsel should then engage in a cross-examination of the expert by way of interview. Counsel should perform the role of opposing counsel to determine how well the witness will stand up to cross-examination. Figure 16.1 illustrates this process being followed with Japanese scientists in the Narita case. This was initially a difficult area for the Japanese scientists who were not used to having their conclusions challenged. A senior Canadian scientist worked closely with them and with counsel.

The witness should *not* be told what their evidence ought to be. It is proper to alert the witness to the type of language which counsel might employ to modify their evidence. The witness should be warned and alerted to potential techniques which counsel may employ. Expert witnesses should be counselled to refrain from any exaggeration, enhancement or embellishment of their evidence.

16.3.6 Demeanour

The credibility of the expert witness and the effectiveness of the testimony is dependent upon their demeanour, their language and their use of demonstrative evidence to support their conclusions. The witness should be advised to use clear and simple language. The jargon or nomenclature of their area should be avoided. They should be advised to modulate their voice during their testimony.

Any complex ideas should be related to simple examples if possible. Demonstrative devices including photographs, charts, graphs, videotapes or other suitable visual aids should be used whenever possible and marked as exhibits. Considerable attention should be given to the witness' command of the substance of his or her testimony and the vocabulary with which to

Figure 16.1 Witness interviews.

Figure 16.2   Files taken daily to court by an expert witness during 6 days of testimony involving 100 exhibits, 68 photographs and 6 videotapes of test explosions.

express it. The danger of exaggeration or shading details should be clearly explained to the witness.

Concessions in small areas may highlight the witness's objectivity. More may be gained in credibility by a small concession than will be achieved by a defensive posture which exposes the witness to concerns of partiality or bias. Objectivity by the expert witness is essential to the fact-finder accepting the opinion or conclusion. The conclusions expressed by the expert witness should be stated clearly and concisely. Any qualification or caution should be outlined clearly with the accompanying reason for the qualification.

16.3.7   *Preparation by Witness*

Once the consolidated report has been finalized, and pretrial interviews have been conducted, it is time for the witness to organize for the trial testimony. The witness should catalogue all working notes in the order of the outline of testimony discussed and settled on with counsel. Each exhibit, photograph, video, demonstrative aid, chart, and report should be checked and re-checked so the evidence can flow in an orderly manner without undue interruption.

The witness should be able to go to the necessary material without becoming flustered or diverted. There is nothing worse than being deflected from one's true purpose by becoming embarrassed about a failure to organize the material being presented. During many days of testimony a forensic expert may have to handle hundreds of exhibits, photographs, charts, etc. It is key to maintaining one's poise to have the materials organized and set out in an orderly fashion near at hand. It is also of the utmost importance that the forensic expert review all the foundation experiments conducted, opinions sought and expressed, and the conclusions drawn therefrom.

One useful technique used by the main forensic witness in the Narita case was to have a separate file for each exhibit which contained all original material, and a summary file which contained, as photocopies, the charts, etc. which formed the foundation of opinions relating one exhibit to another. This enabled the witness to refer quickly to the summary files for information on any technical issues raised and to access the individual files only if necessary. In this way, the summary files and consolidated report were retained in the witness box and the individual files were nearby in separate file boxes for ready reference (see Figure 16.2).

16.3.8 Preparation by Counsel

Counsel has the same task as the forensic expert with an additional feature. Counsel should, at all times in the preparation of the outline for the questions, maintain an awareness of the whole of the case. Particularly in the preparation of the evidence of a multi-dimensional expert, should counsel ensure the evidence fits the case being presented. Any areas of conflict or potential conflicts should be addressed. They must be set out clearly and faced in the testimony of the expert. If they cannot be discounted then they must be explained. It is at this point that early attention to detail and diligence in the preparation of each and every witness pays huge dividends. This task cannot be undertaken during the trial. It will be simply impossible time-wise to put it all together.

The outline of the testimony is the key to effective presentation of the evidence. The preparation, planning and strategy of the whole of the case should be reflected in the outline of the evidence expected from the forensic expert.

16.4 Testimony

16.4.1 Evidence-In-Chief

Outline

The examination-in-chief or direct-examination of an expert should be carefully planned. In a trial involving a judge alone, counsel should prepare from the consolidated report, an outline of the evidence of the witness. The outline should include the following:

- a list of all exhibits for continuity evidence;
- all information provided to the expert requesting forensic examination and opinions;
- all forensic examinations conducted by the expert;
- all findings or results of analyses or examination;
- all opinions or conclusions arrived at by the expert.

An example of the form of the outline is attached as an Appendix to this chapter. (The actual outline of the evidence of the main forensic witness in *R. v. Reyat* was in excess of 50 pages, legal sized, single spaced. The outline followed essentially the same course as the consolidated report with the relevant exhibits interwoven. Due to space limitations it is not fully reproduced. The format can be seen in the outline.)

The full outline should be provided to opposing counsel and to the judge to enable each to more easily follow the evidence. In a jury trial, one may wish to follow the same practice for the judge, opposing counsel and the jury. A skeletal outline may, in the discretion of the court, be provided to the jury to enable them to follow the evidence. However, it is unlikely that a judge would exercise his or her discretion to allow the full outline to be left with the jury.

The framework of the outline should follow a narrative chronology. In explosion cases this approach is often difficult. Where you have multiple areas of examination, the witness's testimony should be broken into separate areas of examination. In each area counsel should follow a chronology including exhibit continuity, forensic examination, findings, and conclusions for each area. This will enable the fact-finder to more easily follow the evidence. The format for the witness should be flexible and enable counsel to ask follow-up questions to bring out detail and to relate each area of examination for the relevant exhibits.

The successful presentation of a case requires an ordered, thorough and persuasive presentation of the supporting evidence. The following principles should be borne in mind in planning and presenting an explosive case:

(1) Prepare an outline of a coherent, comprehensive plan as to your case in general; ensure that the outline of the particular expert's evidence is consistent with this plan; and prepare point form questions within the context of the comprehensive plan for the expert. Carefully prepare and write out the areas of expertise and the foundation facts for any hypothetical to be advanced for the expected opinion or conclusion.

(2) Ensure that the witness is fully prepared for the testimony to be elicited. Preparation, of course, includes ensuring a clear understanding on the part of the witness of where the witness fits in the case, the testimony expected, and the importance of maintaining objectivity as an expert witness. Essentially this involves, in the first instance, a thorough understanding of the evidence that the expert witness is expected to provide. This should have been fully canvassed during the interviewing process.

(3) Ask only one question at a time.

(4) Make the questions as brief as possible.

(5) Frame the questions in simple, readily understandable language.

(6) Ask leading questions in appropriate areas.

Witness notes

The forensic expert should bring to court all relevant notes and working papers. All working papers, print-outs of computer analyses, handwritten notes made during the course of examination of exhibits, photographs, and background information used to form the basis of their opinion, should be available for review by opposing counsel and the court. In many instances, the demonstrative evidence in the form of visual aids such as photographs, videos, and charts should be marked as exhibits to enable the witness to explain what tests or examinations were conducted, and how they contributed to their relevant conclusions. During the witness's testimony, counsel should extensively use the exhibits already marked, and the demonstrative aids brought by the expert. (Counsel may have the witness bring a cart or dolly with boxes of material set out in order. These materials and the exhibits already marked and present in court should be referred to extensively to keep everyone awake! (see Figure 16.2).)

16.4.2 Cross-examination

There are two principal objectives of cross-examination. The first is to weaken the opposing party's case by discrediting the witness, or by pointing out errors, inconsistencies or confusion in the testimony advanced. This can be done by eliciting information on the background of the witness which may undermine or destroy his credibility, and in particular, eliciting facts which tend to demonstrate bias or prejudice and generally cast doubt on the witness's objectivity. In the case of expert witnesses this may be done very effectively by attacking the background facts or foundation for the expert testimony. If the expert has relied upon facts which are open to question, this will detract from the weight to be given to the expert's opinion.

The second objective is to obtain evidence that supports your case. The expert must be advised to anticipate opposing counsel shading the facts or modifying the foundation for the conclusion. The witness should listen very carefully to the question being asked and to ensure that the answer is responsive to the question and fully explains the witness's underlying rationale for the opinion expressed.

Cross-examination will frequently consist of a series of leading questions which suggest the desired answer. The witness may accept the suggestion in a question, or reject it. He or she may be asked to concede, admit, or agree with something suggested in the examiner's question. The expert should be advised to expect a series of short questions which lead to an inevitable conclusion. Each question should be considered carefully and the expert witness should answer each question bearing in mind the scientific analyses conducted and the opinion earlier expressed. If an explanation is required, the witness should provide it. The expert should not allow counsel to require the witness to answer simply with 'yes' or 'no' if an explanation is warranted.

The witness should have carefully reviewed the contents of all written material which he or she has prepared and which may have been disclosed to the opposing party. If questioned about any of the written material, the witness should go to the original and carefully read the note or the report. Having read the note, the witness should clearly and concisely answer the question asked.

16.4.3   Re-examination

The most effective way to prepare for re-examination is to fully prepare the witness to handle the cross-examination in a manner so that no re-examination is necessary. If the witness has fully and clearly explained any potential confusion or problem raised during cross-examination, then counsel should not have to embark on any re-examination.

However, if new areas of evidence were raised during cross-examination or if further explanation is required in some area of evidence, counsel should be prepared to re-examine in that area. The expert witness should be briefed during the pre-trial interviews on the process. The witness should be alerted to the potential that there will be areas requiring re-examination.

Counsel should ensure the questions are clear and as concise as possible. One should set out the areas requiring further explanation or clarification. A comparative form of question may be most suitable. For instance, you have been asked about X, and Y, but not about Z; would Z affect your examination or conclusion? Please explain.

16.5   Conclusion

The use of expert witnesses is intended to simplify matters for the fact-finder in the trial process. Explosive casework involves complex scientific evidence. There is no short-cut which one can follow in leading evidence from an explosion site. An analogy can be drawn between the presentation of the factual background for the expert in explosive casework and the completion of a giant jigsaw puzzle in the order of magnitude of thousands of pieces. The real evidence should be listed, catalogued, photographed and charted so that the evidence may be more easily followed. Demonstrative evidence in the form of photographs, video tapes, draw-ings and scale replicas or models should be utilized whenever feasible and where permitted by the Court.

It is essential to the process of preparation that the lines of communication between counsel and the expert be established right at the outset of their relationship. The consultive process is two-sided – the expert must explain to counsel what the expert requires, and counsel must communicate to the expert what is legally required.

A mutually convenient compromise which arises from their interchange will set the tone for the success, or lack of success, which they will enjoy in the preparation for testimony and the presentation of the evidence. Based on my experience in the Narita case, I would recom-mend a team approach be established at the outset with the team being comprised of the investigators, the forensic scientists and counsel. This process may be utilized by both the prosecution and the defence. It is the most effective means of understanding the presentation of explosive casework evidence.

References

BEVERIDGE, A. D., 1991, Analysis of explosives, *Proceedings of the International Symposium on the Forensic Aspects of Trace Evidence*, pp 177–190, Washington, DC: US Govern-ment Printing Office.

FRECKELTON, I. and SELBY, H., 1993, in *Expert Evidence*, Freckelton, I. and Selby H. (Ed.), Part 1C, The Rules of Expert Evidence; part 1D, Legal Issues Associated with Technical Areas of Expert Evidence. Sydney: The Law Book Company Ltd.

SOPINKA, J., 1995, The use of experts, in *The Expert, A Practitioner's Guide*, Matthews, K. M., Pink, J. E., Tupper, A. D., Wells, A. W. (Eds.), Chapter 1, Toronto: Carswell.

Appendix. Beveridge Evidence Outline

List of Exhibits

Each of the exhibits with which the witness had involvement was listed. The list of exhibits was divided according to the relevant component device. The breakdown into separate areas allowed the court to more easily follow the exhibit continuity trail.

Tuners	Single Base Smokeless Powder
Timer	Tape
Relay	Cardboard
Batteries	Manual
'Liquid Fire' can	

Explosive Residues/Barium Sulphur/Barium Sulphate/EGDN/NG

The relevant exhibits were listed to provide a check list and to ensure that the witness placed in evidence the relevant material.

Dynamite
EGDN/NG
Jetstream suitcase

Test Explosions

There were eight test explosions from which exhibits were seized to enable the expert to place before the court examples of the nature of the damage caused by different explosive devices and different explosives.

Photographs and Video Tapes

Each of the explosions was photographed and videotaped to afford the court an opportunity to view the differences.

Involvement with Japanese Police and Scientists in the Narita Airport Explosion Investigation

Each of the trips was described to demonstrate the foundation for exhibit continuity, examination of the explosive site as a foundation for opinion, as well as a foundation for the Canadian requests of the Japanese to conduct chemical analyses of selected exhibit materials. This also enabled the Canadian investigators to determine what exhibits would be requested from the Japanese for investigative and trial purposes. Each of the briefings, visits and requests was summarized to set out the nature of the forensic requests. The consultations and briefings were set out with the purpose of explaining what was done, why it was done, and what was found. The initial diplomacy and goodwill continued throughout the case and is an example of international mutual legal assistance.

Briefings
First Visit to Japan
Second Visit to Japan
Request for Exhibits From Japan
Briefing in Vancouver, January 1986

Third Visit to Japan
Fourth Visit to Japan
Fifth Visit to Japan
Sixth Visit to Japan

Analyses

The third phase of the outline of evidence was directed towards the steps taken to examine each of the exhibits. A description of the nature of the examination was set out. If the examination was visual and microscopic it was described. If the examination included scientific testing, chemical analyses or specialized equipment, the method was described. The findings of each of the examinations was summarized. At the end, a concluding paragraph setting out the opinion of the forensic scientist was outlined.

Tuners
   General
      Examination; Findings; Conclusion
   Wire Comparison
      Examination; Findings; Conclusion
Timer
   Crystal Oscillator
      Examination; Findings; Conclusion
Wire Comparison – Timer
      Examination; Findings; Conclusion
Relay
   Copper Wire
      Examination; Findings; Conclusion
   Adhesive Tapes and Wire
      Examination; Findings; Conclusion

Batteries
  Cardboard
    Examination; Findings; Conclusion
  Painted Metal
    Examination; Findings; Conclusion
  Battery Cell Core
    Examination; Findings; Conclusion
'Liquid Fire' Can
  Examination; Findings; Conclusion
Single Base Smokeless Powder
  Examination; Findings; Conclusion
Tapes
  Green Tape
    Examination; Findings; Conclusion
  Clear Tape
    Examination; Findings; Conclusion
  Masking Tape
    Examination; Findings; Conclusion
Cardboard
  Examination; Findings; Conclusion
Residues From Explosives
  Examination; Findings; Conclusion
Test Explosions
  Test Explosions Involving Smokeless Powders, Ether Cans, and High Explosive and Simulated Tuners.
    Examination; Findings; Conclusion
  Tests Involving Single Base Smokeless Powder and 'Liquid Fire' ether cans.
    Examination; Findings; Conclusion
  Test Explosions Involving Baggage Containers.
    Examination; Findings; Conclusion
  Experiment to Determine How Much Dynamite Can be Packed Into a Sanyo Stereo Tuner Model FMT 611K.
    Examination; Findings; Conclusion
  Tests Involving Single Base Smokeless Powder.
    Examination; Findings; Conclusion
  Explosion Tests Involving Single Base Smokeless Powder, Dynamite, Plastic Explosives and Baggage Containers RCMP Explosives Range, Quebec Province, June 4/7, 1990.
    Examination; Findings; Conclusion
  Test Explosions Involving Single Base Smokeless Powder and Dynamite.
    Examination; Findings; Conclusion.

Overall Conclusions/Source of Explosion

Based on all of the examinations conducted (which should be listed) and proven in evidence, what is the conclusion as to the source of the explosion?

The form of this question will be very long. Each of the listed examinations, and each of the factual foundations utilized for the opinion, should be set out so the fact-finder will have the submission in argument clearly underscored.

Subject index